McGraw-Hill Series in Mechanical Engineering

Consulting Editors

Jack P. Holman, *Southern Methodist University*
John R. Lloyd, *Michigan State University*

Anderson: *Modern Compressible Flow: With Historical Perspective*
Arora: *Introduction to Optimum Design*
Bray and Stanley: *Nondestructive Evaluation: A Tool for Design, Manufacturing and Service*
Dally: *Packaging of Electronic Systems: A Mechanical Engineering Approach*
Dieter: *Engineering Design: A Materials and Processing Approach*
Eckert and Drake: *Analysis of Heat and Mass Transfer*
Edwards and McKee: *Fundamentals of Mechanical Component Design*
Heywood: *Internal Combustion Engine Fundamentals*
Hinze: *Turbulence*
Hutton: *Applied Mechanical Vibrations*
Juvinall: *Engineering Considerations of Stress, Strain, and Strength*
Kays and Crawford: *Convective Heat and Mass Transfer*
Kane and Levinson: *Dynamics: Theory and Applications*
Martin: *Kinematics and Dynamics of Machines*
Phelan: *Fundamentals of Mechanical Design*
Raven: *Automatic Control Engineering*
Rosenberg and Karnopp: *Introduction to Physics*
Schlichting: *Boundary-Layer Theory*
Shames: *Mechanics of Fluids*
Sherman: *Viscous Flow*
Shigley: *Kinematic Analysis of Mechanisms*
Shigley and Uicker: *Theory of Machines and Mechanisms*
Shigley and Mischke: *Mechanical Engineering Design*
Stoecker and Jones: *Refrigeration and Air Conditioning*
Vanderplaats: *Numerical Optimization: Techniques for Engineering Design, with Applications*

Also Available from McGraw-Hill

Schaum's Outline Series in Mechanical Engineering

Most outlines include basic theory, definitions, and hundreds of solved problems and supplementary problems with answers.

Titles on the Current List Include:

Acoustics
Basic Equations of Engineering
Continuum Mechanics
Engineering Economics
Engineering Mechanics, 4th edition
Fluid Dynamics
Fluid Mechanics & Hydraulics, 2d edition
Heat Transfer
Introduction to Engineering Calculations
Lagrangian Dynamics
Machine Design
Mathematical Handbook of Formulas & Tables
Mechanical Vibrations
Operations Research
Strength of Materials, 2d edition
Theoretical Mechanics
Thermodynamics, 2d edition

Schaum's Solved Problems Books

Each title in this series is a complete and expert source of solved problems containing thousands of problems with worked out solutions.

Related Titles on the Current List Include:

3000 Solved Problems in Calculus
2500 Solved Problems in Differential Equations
2500 Solved Problems in Fluid Mechanics and Hydraulics
1000 Solved Problems in Heat Transfer
3000 Solved Problems in Linear Algebra
2000 Solved Problems in Mechanical Engineering Thermodynamics
2000 Solved Problems in Numerical Analysis
700 Solved Problems in Vector Mechanics for Engineers: Dynamics
800 Solved Problems in Vector Mechanics for Engineers: Statics

Available at your college bookstore. A complete list of Schaum titles may be obtained by writing to: Schaum Division
McGraw-Hill, Inc.
Princeton Road, S-1
Hightstown, NJ 08520

FUNDAMENTALS OF MECHANICAL COMPONENT DESIGN

Kenneth S. Edwards, Jr.

Professor of Mechanical Engineering, Emeritus
University of Texas at El Paso

Robert B. McKee

Professor of Mechanical Engineering
University of Nevada, Reno

McGraw-Hill, Inc.

New York St. Louis San Francisco Auckland Bogotá Caracas
Hamburg Lisbon London Madrid Mexico Milan Montreal New Delhi
Paris San Juan São Paulo Singapore Sydney Tokyo Toronto

This book was set in Times Roman.
The editors were John J. Corrigan and Scott Amerman;
the production supervisor was Denise L. Puryear.
The cover was designed by Rafael Hernandez.
R. R. Donnelley & Sons Company was printer and binder.

FUNDAMENTALS OF MECHANICAL COMPONENT DESIGN

1 2 3 4 5 6 7 8 9 0 DOC DOC 9 5 4 3 2 1 0

ISBN 0-07-019102-6

Library of Congress Cataloging-in-Publication Data

Edwards, Kenneth S. (Kenneth Scott)
  Fundamentals of mechanical component design / Kenneth S. Edwards,
  Jr., Robert B. McKee.
      p.   cm.
  ISBN 0-07-019102-6
      1. Machine parts—Design and construction.   I. McKee, Robert B.
II. Title.
TJ243.E38   1991
621.8'15—dc20   90-5771

CONTENTS

This book is intended primarily for component design and selection courses in the design sequence of mechanical engineering curricula. While the principal objective has been a text that is easy to read for students, the depth of the topic coverage makes the book a useful professional reference.

Much of the challenge and opportunity in mechanical engineering consists of devising load-bearing elements which combine superior function with minimum cost. Such design requires mastery of both engineering fundamentals and practical optimization techniques. Thus the text aims at giving the reader useful training and insight in optimal design within the context of the design of basic mechanical components.

The book begins with a review of fundamentals. Students are usually astonished to find that they will be expected, as engineers, to define the problem which they will later solve. Thus the first chapter deals with the process of creating designs to answer particular needs. Uncertainty being a fact of engineering life, probability theory is presented in the next chapter. Tolerancing and computer-aided design and manufacture are discussed. A review of the principles of static equilibrium is followed by a detailed presentation of the Mohr's-circle method of analyzing commonly encountered stress situations. A discussion of factors of safety includes factors based on statistics. Materials testing is discussed to give the student an understanding of the meaning of published material properties. Next is an examination of current theories of failure for metals under static loads, including linear elastic fracture mechanics. Methods for designing against fatigue failure include the state-of-the-art local-strain model. The background material concludes with a chapter on practical optimization techniques.

The remainder of the text examines the design of various types of mechanical elements, with necessary theory reviewed in the context of the problem. Optimal proportions are chosen for tension members with the aid of the computer. Column theory, including the powerful energy method, is used to design members loaded in compression. The chapter on torsion includes calculation of the stiffness of members of nonsymmetric cross sections. A beam acting as a vibration isolator is examined, and proportions are chosen for best performance. Several practical methods of obtaining beam deflections and their resonant vibration frequency are presented. Practical ways to handle triaxial stress situations are discussed. An up-to-date review of bolted joints accompanies the principles of design of such elements. Similarly, review of theory is combined with design technique and component selection for composite materials, helical springs, weldments, rolling bearings, and drive belts.

The authors will be grateful for comments and suggestions.

ACKNOWLEDGMENTS

The authors wish to express their gratitude to several individuals for review and suggestions in their areas of expertise: John H. Bickford, Raymond Engineering; Dr. Michael J. Manjoine, Westinghouse Research and Development Center; Dr. Paul W. Wallace, SPS Technologies; and George F. Leon, General Dynamics, Electric Boat Division—all experts in bolted-joint design. John C. Ekvall, Lockheed-California; Drs. Harold S. Reemsnyder, Bethlehem Steel Co.; and Ronald G. Lambert, General Electric Co.—authorities in fatigue theory. Dr. Po-Wen Hu, University of Texas at El Paso, an expert in probability theory. Dr. David K. Felbeck, the University of Michigan, an expert in fracture mechanics. Stephen Landsman, Associated Spring, the Barnes Group, an expert in spring design. And Ted Howe of the Fafnir Bearings Division, the Torrington Co., an expert in rolling-component bearings.

Thanks also are due to numerous persons for contributions to examples and figures and for permission to use them, acknowledged where they occur.

McGraw-Hill and the authors would like to thank the following reviewers for their valuable comments and suggestions: Charles Beadle, University of California, Davis; Gary Gabrielle, Rensselaer Polytechnic Institute and State University; Ed Haug, University of Iowa; Jerald Henderson, University of California, Davis; Jack Holman, Southern Methodist University; Harold Johnson, Georgia Institute of Technology; Robert Lucas, Lehigh University; Robert Pangborn, Pennsylvania State University; Gerhard Reethof, Pennsylvania State University; Charles Reinholtz, Virginia Polytechnic Institute and State University; Joseph Shigley, University of Michigan; Karel Silovsky, South Dakota School of Mines and Technology; Darrell Socie, University of Illinois, Urbana-Champaign; and Ralph Stephens, University of Iowa.

Teaching assistants William Morton and Jenq-Tzong Chern at the University of Texas at El Paso and Frank Stanko at the University of Nevada, Reno, gave invaluable help in editing, trying out problems, and programming solutions. Our thanks also go to Kim H. Pries, systems manager of the engineering computer at UTEP, for the individual attention projects like this require. We are particularly grateful to the mechanical engineering department secretaries at UTEP and UNR for their patience with manuscript revisions: Mary Jean Acosta, Sandra Tipton, and Peggy Hart. To the several others whom we have no doubt overlooked, our apologies and thanks.

Kenneth S. Edwards, Jr.
Robert B. McKee

THE INTERNATIONAL SYSTEM OF UNITS, SI (SYSTÈME INTERNATIONAL d'UNITÉS)

Humans understandably turned first to parts of the body and their natural surroundings for measuring instruments. Early Babylonian and Egyptian records and the Bible indicate that length was first measured with the forearm, hand, or finger and that time was measured by the periods of the sun, moon, and other heavenly bodies. When it was necessary to compare the capacities of containers such as gourds or clay or metal vessels, they were filled with plant seeds which were then counted to measure the volumes. When means for weighing were invented, seeds and stones served as standards. For instance, the "carat," still used as a unit for gems, was derived from the carob seed.

As societies evolved, weights and measures became more complex. The invention of numbering systems and the science of mathematics made it possible to create whole systems of weights and measures suited to trade and commerce, land division, taxation, or scientific research. For these more sophisticated uses, it was necessary not only to weigh and measure more complex things, but also to do it accurately time after time and in different places. However, with limited international exchange of goods and communication of ideas, it is not surprising that different systems for the same purpose developed and became established in different parts of the world—even in different parts of a single continent.

The English System (U.S. Conventional System, USCS)

The measurement system commonly used in the United States today is nearly the same as that brought by the colonists from England. These measures had their origins in a variety of cultures—Babylonian, Egyptian, Roman, Anglo-Saxon, and Norman French. The ancient "digit," "palm," "span," and "cubit" units evolved into the "inch," "foot," and "yard" through a complicated transformation not yet fully understood.

Much of this material is excerpted from publications of the Office of Metric Programs, U.S. Department of Commerce.

Roman contributions include the use of the number 12 as a base (12 inches to the foot) and words from which we derive many of our present weights and measures names. For example, the 12 divisions of the Roman *pes*, or foot, were called *unciae*. Our words *inch* and *ounce* are both derived from that Latin word.

The "yard" as a measure of length can be traced back to the early Saxon kings. They wore a sash or girdle around the waist—which could be removed and used as a convenient measuring device. Thus the word *yard* comes from the Saxon word *gird* meaning the circumference of a person's waist.

Standardization of the various units and their combinations into a loosely related system of weights and measures sometimes occurred in fascinating ways. Tradition holds that King Henry I decreed that the yard should be the distance from the tip of his nose to the end of his thumb. The length of a furlong (or furrow-long) was established by early Tudor rulers as 220 yd. This led Queen Elizabeth I to declare, in the sixteenth century, that henceforth the traditional Roman mile of 5000 ft would be replaced by one of 5280 ft, making the mile exactly 8 furlongs and providing a convenient relationship between two previously ill-related measures.

Thus, through royal edicts, England by the eighteenth century had achieved a greater degree of standardization than the continental countries. The English units were well suited to commerce and trade because they had been developed and refined to meet commercial needs. Through colonization and dominance of world commerce during the seventeenth, eighteenth, and nineteenth centuries, the English system of weights and measures was spread to and established in many places, including the American colonies.

However, standards still differed to an extent undesirable for commerce among the 13 colonies. The need for greater uniformity led to clauses in the Articles of Confederation (ratified by the original colonies in 1781) and the Constitution of the United States (ratified in 1790) giving power to the Congress to fix uniform standards for weights and measures. Today, standards supplied to all the states by the National Bureau of Standards ensure uniformity throughout the country.

The Metric System

The need for a single worldwide coordinated measurement system was recognized over 300 years ago. Gabriel Mouton, Vicar of St. Paul in Lyons, proposed in 1670 a comprehensive decimal measurement system based on the length of 1 minute of arc of a great circle of the earth. In 1671 Jean Picard, a French astronomer, proposed the length of a pendulum beating seconds as the unit of length. Other proposals were made, but over a century elapsed before any action was taken.

In 1790, in the midst of the French Revolution, the National Assembly of France requested the French Academy of Sciences to "deduce an invariable standard for all the measures and all the weights." The Commission appointed by the Academy created a system that was, at once, simple and scientific. The unit of length was to be a portion of the earth's circumference. Measures for volume and mass were to be derived from the unit of length, thus relating the basic units of the system to each other and to nature. Furthermore, the larger and smaller versions of each unit were to be created by multiplying or dividing the basic units by 10 and its powers. This feature provided a great convenience to users of the system, by eliminating the need for such calculations as dividing by 16 (to convert ounces to pounds) or by 12 (to convert inches to feet). Similar calculations in the metric system could be performed simply by shifting the decimal point. Thus the metric system is a *base-10* or *decimal system*.

The Commission assigned the name *metre* (which we spell *meter*) to the unit of length. This name was derived from the Greek word *metron*, meaning "a measure." The physical standard representing the meter was to be constructed so that it would equal one ten-millionth of the distance from the north pole to the equator along the meridian of the earth running near Dunkirk in France and Barcelona in Spain.

The metric unit of mass, called the *gram*, was defined as the mass of one cubic centimeter of water at its temperature of maximum density. The cubic decimeter (a cube one-tenth of a meter on each side) was chosen as the unit of fluid capacity. This measure was given the name *liter*.

Although the metric system was not accepted with enthusiasm at first, adoption by other nations occurred steadily after France made its use compulsory in 1840. The standardized character and decimal features of the metric system made it well suited to scientific and engineering work. Consequently, it is not surprising that the rapid spread of the system coincided with an age of rapid technological development. In the United States, by an act of Congress in 1866, it was made "lawful throughout the United States of America to employ the weights and measures of the metric system in all contracts, dealings or court proceedings." Since 1893, the internationally agreed-to metric standards have served as the fundamental weights and measures of the United States.

SI: The International System of Units

International cooperation aimed at standardization of length and mass units was the purpose of the Metric Convention, a treaty signed in 1875 by 17 countries, including the United States. Established with this agreement were the General Conference of Weights and Measures, to meet every six years; the International Bureau of Weights and Measures (located near Paris); and other machinery to implement the decisions of the General Conference.

In 1960, the General Conference adopted an extensive revision and simplification of the system. The name *Système International d'Unités* (International System of Units), with the abbreviation SI, was adopted for this modernized metric system.

The U.S. Congress passed the Metric Conversion Act in 1975, which declared it to be national policy to coordinate and plan the increasing use of the metric system within the United States. The Department of Commerce's Office of Metric Programs has the role of aiding in this conversion. That name implies a certain confusion of terms, for the scientific and engineering communities seek the adoption not of the metric system, but of SI. Of course, the name was chosen for its familiarity.

Current Status

As of this writing, only Burma and the United States are not "metric," or "SI." But even some traditionally metric nations have not totally adopted SI.

In the United States, the automobile industry has been a leader; more than 90 percent of car components are now specified in SI. Heavy equipment and computer manufacturers substantially use SI. Much of the packaging industry, and film, tires, cigarettes, liquor, pharmaceuticals, wine, and soft drink, use SI. In some states gasoline is dispensed by the liter. International sports use SI units exclusively (track and field events, swimming, etc.).

SI in Machine Design

Those units of SI commonly used in machine design are described below. There are numerous others, naturally, applicable to other fields.

The SI base units are the following:

1. *Length:* the meter (m), which was originally (1793) defined as a certain fraction of the earth's circumference, then later as the distance between two marks on a bar kept at the French Bureau of Standards. It was redefined recently by the Geneva Conference in terms of the distance traveled in vacuum by a certain wavelength of light.
2. *Mass:* the kilogram (kg), equal to the mass of the standard kilogram kept at the International Bureau of Weights and Measures.
3. *Time:* the second (s), defined in terms of the period of a certain radiation of cesium 133.

Other base units not necessary for this text are those for electric current, temperature, luminous intensity, and substance.

Units for developed quantities are derived from the base units. Four of these bear special names and are of importance in this text.

1. *Force:* the newton (N), the force which will impart to a 1-kg mass an acceleration of 1 m/s^2. (The pull of gravity, i.e., the weight, of a large apple is roughly a newton.)
2. *Energy:* the joule (J), or a newton-meter (N·m).
3. *Power:* the watt (W), or a joule per second (J/s).
4. *Pressure or stress:* the pascal (Pa), or a newton per square meter (N/m^2). Engineering stresses usually run in millions of pascals, hence the megapascal (MPa) is most commonly seen.

Note that *weight* has become an obsolete term in this system. One can speak of a kilogram of butter, but the reference is to a mass, not the force exerted on it by gravity, which would be expressed in newtons.

Numerous other units without a name are also derived, e.g., those for velocity, acceleration, torque, density.

It has been recognized that some units in very common use will persist. For example,

Units of time: minute, hour, etc.

Angular measure in degrees

Liquid volume in liters (1000 cm^3 or 10^{-3} m^3)

Tonne (1000 kg = 2200 English pounds, commonly called a *metric ton* or a *long ton*)

Large and small quantities are designated by prefixes already in common use:

Size	Prefix	Symbol
10^9	giga	G
10^6	mega	M
10^3	kilo	k
10^{-2}	centi	c
10^{-3}	milli	m
10^{-6}	micro	μ

(There are others inapplicable in this text.)

The English use of the decimal period is to continue, for example, 1.93, 0.35. Where a number of digits precede or follow the decimal point, they are to be written in groups of three, without commas, such as 2 013 567.2 or 0.005 6. The computer practice of showing multiples of 10 by E (1.03E $-$ 02 = 0.0103) will be common (used in this text). Useful conversion factors are printed on the inside cover of this book. Here are some numbers which you will find useful to commit to memory:

25.4 $\times$ in = mm
2.2 $\times$ kg = lbm
4.45 $\times$ lbf = N
6900 $\times$ psi = Pa

Various journals now require submission of articles in SI. The American Society of Mechanical Engineers foresees a period of some years with the USCS and SI systems both in common use, and it is in line with that view that the examples and problems of this text are divided between the two. There is also the question of getting people used to what sounds right in the new scheme. If a stress of several million pounds per square inch turns up in a problem solution, one suspects an error. What about several hundred megapascals?

FUNDAMENTALS OF
MECHANICAL COMPONENT DESIGN

CHAPTER
1

ENGINEERING DESIGN

This is a book about design, the design of mechanical elements in particular. We should start by clarifying what we mean by *design*. The word *design* is not a euphemism—it is not another way to say *draw*. Design is, to quote Webster, "deliberate, purposeful planning." As so defined, design is so typical of the engineer's work that it could be used as a definition of engineering. It is true that engineers must know mathematics and physics, just as physicians must know biological science and laboratory techniques. But the skills that distinguish the physician are diagnosis and prescription. The skill that distinguishes the engineer is design.

1.1 HEURISTIC METHOD

People's interest in the process of design is at least as old as written history. The Greeks studied design, and Pappus (A.D. 320) formally described a discipline of problem solving that he called *heuristics*. The essence of the heuristic method is to imagine that the problem has been solved, the objective reached. That the problem has been solved implies that certain conditions have been met, and these conditions imply the satisfaction of other precursors. Precursor is added to precursor until the beginning is reached.

The original Buck Rogers of cartoon fiction relied on an engineer named, appropriately enough, Doctor Huer. The good doctor's method of designing a spaceship for a journey to some other planet was to imagine that the ship had just completed the journey. Then Doctor Huer worked his way back to the starting point, keeping a record of what had to be done and what supplies were needed during each stage of the trip.

The first trip to the moon was planned in exactly this way. The end of that journey, of course, was the return to earth. A sea landing was planned, and so the spacecraft had to be watertight. Fuel was needed throughout the trip to keep the air conditioning system going. Before landing, parachutes were necessary, and for reentry the craft needed a heat shield. The craft had to be aerodynamically stable so that the reentry shield would be facing the airflow. At this point, it was only necessary to add the fuel needed for midcourse corrections, and for operation of the cabin air conditioning system, to get a good idea of what the craft had to contain when it left the moon on its return home. Thus the Apollo designers traced the entire mission in reverse, and as they did, the vehicle they planned became larger and more complex. The success of the mission is testimony that their design, their deliberate, purposeful plan, was a good one.

The heuristic method is not limited to engineering. It is used in every field of human endeavor where problems must be solved. The most recent use is in *artificial intelligence*, computer routines which combine the heuristic method and a base of loosely organized and imprecise knowledge to provide expert diagnoses.

Many thinkers have studied the process of design. For example, the French philosopher and mathematician Descartes (1596–1650) wrote that, as a child, he had wondered how scientists had arrived at the discoveries of which he had learned. So he studied the lives of the scientists and relived, in imagination, their discoveries. Gradually he found that he was, in every case, following the same general procedure. The pattern that Descartes observed has received continued study. It consists of several different stages, which we describe below.

1.2   DEFINING THE NEED

The first step in the design process, as with heuristics, is to get a clear idea of what it means to have solved the problem.

> **Example 1.1.** New York (Associated Press), December 26, 1987. Peter Chaconas is the kind of guy who likes to work on cars, hunt deer, dig holes with a backhoe, and think up new designs for tools . . . [including] one of the best selling tool accessories in Black and Decker's history, the Piranha circular saw blade. . . . Now he has invented the Bullet, a bit for home power drills . . . "the first major innovation in the industry-standard product in almost 100 years." . . . His job he describes laughingly as "euphoria." And his approach: "Once you identify a problem, it's easy to fix it."

The original statement of need is usually vague or stated in terms that do not translate directly to technical language. First the statement must be focused into an elemental need: a statement as short and clear as possible. Once a satisfactory elemental need is agreed upon, it can be expanded and technical needs derived.

For the trip to the moon, the elemental need can be simply stated: Get there, land, and get back. When the further requirement of taking people along and returning them safely is added, a host of resulting needs emerges. These were all defined with great care and precision, and redefined at every opportunity. The process of refining the statement of needs still goes on. It will end only if we give up the exploration of space.

It is difficult to get used to the idea that needs must be carefully defined. This is particularly true for engineering students, who are at the end of 12 or more years of solving problems for which all the needs have been defined with great precision by their instructors. That is a shame, because there is no more important step in the solution of any problem than the definition of need.

> **Example 1.2a.** We are going to market a new jack stand to be used by amateur mechanics to support a car on which they are working. The overriding need is that the stand not collapse under the weight of whatever car it supports or tip over as a result of some ordinary disturbance. The jack stand must be adjustable in order to support the range of spring and frame heights that will be encountered. Finally, these requirements must not be conditional on the stand's being expertly placed and adjusted. That is, the jack stand must be as "idiot-proof" as possible.

With the needs stated in a descriptive or qualitative manner, we start to refine them, using more precise descriptions. The final objective is to replace all adjectives with numbers.

> **Example 1.2b.** We stated as an elemental need that the new jack stand should support "a car." Given the popularity of pickup trucks, that need should be expanded to include at least a standard pickup truck. At most, such a vehicle weighs about 5000 pounds (lb) empty, and we can expect that 60 percent of that weight will be supported by the front wheels. A stand placed under one of the most heavily loaded pair of wheels would thus support a load of $5000(0.6)/2 = 1500$ lb. Keeping in mind the consequence of a collapse and the uncertainties of usage, we choose the relatively large factor of safety of 5. Factors of safety are discussed in detail in Chapter 5. Applying this factor to the anticipated load, we get a design load of $1500(5) = 7500$ lb (for strength calculations).
>
> To satisfy the non-tipover requirement, a reasonable upper limit for the lateral load is derived by calculating the impulse delivered by a 200-lb person walking full tilt [at 6.5 feet per second (ft/s)] into the side of the vehicle:
>
> $$\text{Momentum} = \left(\frac{200}{32}\right)(6) = 37.5 \text{ lb·s}$$
>
> Assume that the collision lasts 0.2 s:
>
> $$\text{Force} = \frac{37.5}{0.2} = 187.5 \text{ lb}$$
>
> For a conservative estimate, assume that one jack stand, at its maximum height, must withstand the entire lateral load. Vehicle jacking points range from 8 to 16 inches (in) above the ground. It may not be feasible to build our stand for this much adjustment, but we will consider compromises later. The free-body diagram of Figure 1.1 results.
>
> An equilibrium summation of moments about the corner opposite the lateral load gives, for the width of the base,
>
> $$w = \frac{187.5(16)(2)}{1500} = 4.0 \text{ in}$$

It is always a good idea to check that a calculation does, in fact, represent an extreme case. We had calculated the necessary width of a jack stand on the basis of

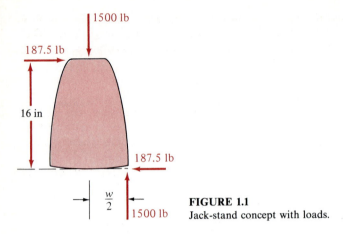

FIGURE 1.1
Jack-stand concept with loads.

its carrying the heaviest weight. Will a stand of that width also be stable under a smaller weight?

> **Example 1.2c.** When the calculation of Example 1.2*b* is repeated with the less heavily loaded wheel of the pickup truck, a larger width is required. In fact, every decrease in the load being supported causes an increase in the necessary width. Rather than carry this to an absurd extreme, we limit our consideration to the most lightly loaded wheel of the smallest automobile that is likely to be encountered: 2200 lb, with 40 percent on the rear wheels. The corresponding load per wheel is thus 2200(0.4)(0.5) = 440 lb. The required width has increased to 14 in.

1.3 GENERATING IDEAS

Completion of the definition of needs will yield a spectrum of design loads, safety requirements, and desirable features. To proceed further, we need ideas. This is the second stage in the design process—the generation of possible solutions. To do this right, it is necessary to resort to some tricks, because our minds are such prisoners of the past. Engineers, especially, have a difficult time generating new ideas because the habits of mental discipline and orderly thought that characterize engineering are naturally opposed to any unproven scheme.

We are indebted to the wild world of advertising for a practical way to break the bonds of the past. It is called *brainstorming*. The object of brainstorming is to create new ideas. To brainstorm, a group of people is assembled so that each person can gain stimulation and ideas from those expressed by others. The brainstorming group should not be limited to those directly involved in the design. The janitor, the receptionist, and the accountant can bring fresh, unprejudiced minds, and good ideas, to the mixing pot.

There are just two rules for brainstorming, and they must be strictly observed: (1) The ideas should satisfy the defined need, and (2) criticism of any sort is forbidden. While Descartes did not apply the word *brainstorm* to this activity, he noticed that successful innovators practiced it. Successful designers, that is, problem solvers, have the ability to suppress criticism and old patterns of thought. They can form new associations and explore, at least mentally, new ways of doing things. They have learned to do their own brainstorming. Anyone can brainstorm, as long as rule 2 is strictly observed: No criticism!

> **Example 1.2d.** Some ideas are required for the configuration of the jack stand of Example 1.2(*c*). First, we repeat the needs: The jack stand must not collapse under a load of 7500 lb, its erection and adjustment should be foolproof, and its base should be 14 in wide.
>
> It is all right to start with conventional ideas:
>
> A tripod with an adjustable top
> A tripod with adjustable legs
> A round or square "box" that adjusts
> An air bag that inflates
> A locking jack with square base
> A hydraulic jack

1.4   CHOOSING THE BEST IDEAS

If brainstorming is successful, designers will have many more ideas than they can use. The best of these ideas must be selected for further development. That bears repeating—the best ideas, *not* the worst, must be selected from the list. It is normal to search for the rotten fruit in the barrel; otherwise, infection will spread from the bad fruit to the good. But in design there is no such thing as a bad idea. There are only ideas and better ideas.

The validity of this philosophy for choosing ideas was proved some years ago by a large-scale experiment with engineers at the General Motors Corporation. The number of *usable* ideas generated by a group that was denied criticism was about double that from a group that was allowed to criticize.

This bears repeating: The pile of ideas must be reduced by picking the best ones, *not* by eliminating the worst.

The rest of the design process is more familiar. We refine the best ideas and add details. As the concept takes shape, we can be more detailed and more quantitative in checking that needs are going to be satisfied.

> **Example 1.2e.** After the list of ideas for a jack stand is scanned, the tripod with an adjustable top is selected first.
>
> Another needs analysis and idea session should be used to develop ideas for the adjustable top. Because the jack must be safe even when it is inexpertly placed and adjusted, it should be designed so that it will support no load at all unless it is in a configuration that is safe.
>
> To continue with this example, we assume that has been done and that the result is a round tube sliding up and down in the tripod. Figure 1.2 shows that the design has taken a form.
>
> Among the materials available for the stand, steel is a good choice because it combines low price and easy formability with high strength and ductility. First we will

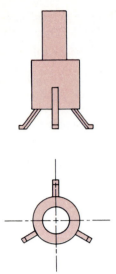

FIGURE 1.2
Jack stand showing base layout.

check that the stress in the tube is not too high. We do this by comparing the normal stress in the tube with a conservative estimate for the strength of the steel to be used. This is the simplest of several models for failure of metals which are discussed in Chapter 7.

The cross-sectional area A of a thin tube can be closely approximated by

$$A = \pi dt \tag{1.1}$$

where d and t are the diameter and thickness, respectively. Using a lower limit of 30 kilopounds per square inch (ksi) for the yield stress of the steel, $\frac{1}{16}$ in for the thickness of the tube, and the design load of 7500 lb, we get an estimated diameter of

$$d = \frac{7500(16)}{30\,000\pi} = 1.27 \text{ in}$$

The next largest standard tube size is $1\frac{3}{8}$, or 1.375, in.

The tube, supporting a load acting along its axis, is acting as a column, and we need to know whether it may buckle under the load. We will determine whether this column is long enough to be in danger of buckling by calculating the slenderness ratio. This is the ratio of length L to the radius of gyration ρ. The latter is the square root of the least moment of inertia of the cross section divided by the area:

$$\rho = \sqrt{\frac{\pi d^3 t}{8}\left(\frac{1}{\pi dt}\right)} = 0.486 \text{ in}$$

With a length of 8 in, the L/ρ ratio is 16.5, well below the slenderness ratio at which buckling is a concern. (Buckling is discussed in detail in Chapter 12.)

Referring to Figure 1.2, we see that if the minimum effective width of the tripod w_{min} is to be at least 14 in, then the radius of the circle that circumscribes the legs must be at least $\frac{2}{3}w_{min} = 10$ in. If the minimum height of this jack stand is to be 8 in, then the maximum height must be less than 16 in. This problem will be reported to the marketing department while the design proceeds.

One more detail will be given form here, with the rest of the design left as a student exercise. We have decided to use a tube whose length is adjustable. We did not decide how this variable length is going to be achieved. This is a design problem in its own right.

For a tube sliding in the base, the length can be adjusted in at least three ways:

1. Tubes of different lengths can be inserted into a socket.
2. The tube can be held by a mechanical catch.
3. The tube can be held by a frictional catch.

Of these, the second will be chosen now because it does not require extra pieces (which might be lost) and does provide reliability against slipping.

Those ideas which cannot be construed as feasible may be replaced by others from the top of the pile. The survivors are again ranked.

1.5   FROM IDEA TO REALITY

The best idea of all the feasible ones is now developed in a sort of dress rehearsal called the *preliminary design*. The idea is to develop the solution, with a minimum of expenditure, to the point where a decision can be made whether to produce the item. One of three possible decisions is made:

1. Go ahead to final design and implementation.
2. Return to, and reevaluate, the brainstorming ideas.
3. Give up the project; it would only lose money.

If the decision is to go ahead, a test model is made. For a simple device like the jack stand, the model would also be a prototype. For other devices, a full-scale operating model might be far too expensive. A good example is an airplane. The first model of an airplane is almost invariably a mathematical one. Following the revisions in design called for by operation of the first model, a scale model might be built and tested—to scale—in a wind tunnel. It is no accident, by the way, that the most powerful computers are housed in the same facility with the largest wind tunnels. Computers are much less expensive to run than wind tunnels.

The results of model tests may now be discussed with the prospective customer— or the surrogate, the sales and marketing force. Changes in specifications may be made and the design stages we have discussed gone through again. This may seem a wasteful process: Why not do it right the first time? The answer is that this is exactly the way to make sure that the design *is* done right "the first time."

PROBLEMS

The first nine problems are primarily conceptual, requiring a minimum of skill in force and stress analysis. Problems 1.10 through 1.13 require good free-body analysis of forces, and the remaining problems call for calculation of stresses.

1.1. Design a book rest for use on your study table or classroom desk. The book rest should fold so that it could be carried with your books.

1.2. Design an educational toy for elementary school children that will teach, without mentioning them by name, the associative and commutative properties of arithmetic.

1.3. Design a book bag which will hold the normal collection of books that must be carried to a day's classes in a position so that any one book can be picked out and which will then hold that book in a position to be read.

1.4. Define the needs and provide several concepts for a device which will retrieve tennis balls from the other side of the court while you practice your serve.

1.5. Design a folding seat which can be carried to places like balloon races and rock concerts where there may be a shortage of chairs. It should be something that could be used on the floor of a gymnasium—that means no spikes on the bottom.

1.6. We have formed the Euteseit Portable Parker company to manufacture and market a device which will hold a bicycle upright between classes and prevent its theft. The device must be easily carried on the bicycle. Assume that the parking place is paved and relatively level. Do not assume that there will be posts or other anchoring points. Restate the elemental needs, develop them, and suggest at least six possible solutions.

1.7. *Class project:* Design a student's desk and chair. Use each other as experimental models to determine the best chair seat and back height and tilt. That is, devise an experimental table and chair whose heights and inclinations, and locations of book holders and book and pencil holders, can be changed. Each member of the class then uses the setup and determines the best combination for her or his use. Using this information, individually design a desk and chair. Keep in mind that this one design must be used by every member of the class. In addition to overall suitability, your design will be graded on its ruggedness, appearance, and economy of construction.

1.8. Toward the end of 1986, the aircraft *Voyager* flew around the world without refueling, the first such voyage ever accomplished. State the elemental need for the design of this craft, and refine that need toward specification.

1.9. Devise a jack stand made of just two pieces: a stand and an extension. The total height must be adjustable from 8 to 14 in, and the stand must be mechanically held at each position. *Hint:* The two pieces may be any shape you wish.

1.10. The Rangatangue Manufacturing Company occupies a large building with a flat concrete floor. Many of its supplies come as liquids in 45-gallon (gal) barrels. These must occasionally be moved by hand, by using a hand truck (Figure P1.10).

FIGURE P1.10
A hand truck.

The operation of the truck is as follows: The operator places the body of the truck against the barrel and fastens the strap (an automotive seat belt) tightly. Then she pulls back on the handles, rotating the hand truck and raising the barrel off the floor. Once the barrel has been lifted, it will be moved from place to place in the position shown in Figure P1.10*b*.

Your assignment is to proportion the truck so that the operator can accomplish the tasks described. Assume that the operator weighs at least 140 lb and can exert a maximum of 60 lb of force on the handles of the truck.

1.11. The lever-type jack sketched in Figure P1.11 is a type used in auto racing. After the end of the J has been placed under the wheel suspension strut, the long end is pulled down. Proportion the jack so that the operator of Problem 1.10 can use it to raise one wheel of a car that weighs a total of 2800 lb, with 60/40 weight distribution. The suspension strut or jacking point is normally 7 in from the floor. The jack must raise it $1\frac{1}{2}$ in, and when the jack is elevated, a line from the jack's axle to the jacking point should be inclined at least 5° from the vertical, as shown in the figure.

Make a scale layout of the jack, calculating the proportions necessary to keep operator effort within specifications.

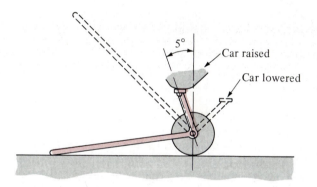

FIGURE P1.11
A racing-type automobile jack.

1.12. Light airplanes weigh about the same as a subcompact car (Problem 1.11) and take about as much force to move. As shown in Figure P1.12 (see page 10), the nose-wheel axle usually projects beyond its yoke so that a towing bar can be fitted to it. Then the airplane can be moved by pulling on the tow bar in the desired direction. Design a human-powered tow motor so that the operator can move the plane by exerting a smaller force, say 25 lb instead of 100 lb, over a greater distance. This might be done in several ways. The plane might be moved by pulling a handle 4 ft to move the plane 1 ft. A long handle might simply be rocked back and forth. The operator might move the plane by stepping on a foot pedal, thus using his or her weight. *Caution:* It is not possible to simply convert a force of 25 lb to a force of 100 lb. Your device must provide a way to obtain the missing 75 lb by a reaction from the floor.

1.13. After the class has reached a consensus on the proportions of the jack in Problem 1.11, determine the sizes of the members by stress analysis. Use hot-rolled low-carbon steel with a design stress of 11E + 03 pound per square inch (psi) for tension and compression, one-half that for shear, and a modulus of elasticity of 30E + 06 psi.

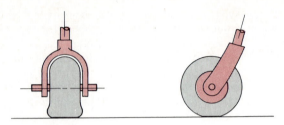

FIGURE P1.12
Nose wheel of a light airplane.

1.14. Refer to the discussion of a jack stand design in the text. Design a mechanical catch to hold the tube wherever desired within its range of adjustment. Start by restating and refining the need to be satisfied—which becomes the specification of the catch.

1.15. Assume that the catch of Problem 1.14 consists of a round pin which is inserted through matching holes in the movable and stationary tubes of the jack stand. Review the needs and select a suitable size for the pin. Assume that the maximum shear stress in the pin is $1\frac{1}{2}$ times the average shear stress. When calculating the compressive bearing stress in the tube wall, for the area being stressed use the thickness of the tube times the diameter of the pin. Use AISI 1040 cold-rolled steel for all parts, and assume that the yield stress in compression is the same as that in tension.

1.16. Some small automobiles use a jack which fits under a rail that runs along the underside of the body. Figure P1.16 shows a clever scissors jack used for this purpose.

When the jack is put into place, its height is 9 in, and since the car is being supported by its springs, the weight to be supported is zero. That load will increase to 1400 lb as the side of the car is lifted off its springs. The height of the jack at that time is 14 in. You may assume that the change of load with height is linear. The relation between force on a power screw and torque can be found in any statics textbook. Proportion the jack so that the force on the crank handle never exceeds 20 lb.

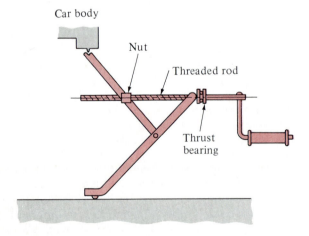

FIGURE P1.16
Scissors jack for a small automobile.

1.17. Determine the initial design specifications for a portable electric log splitter. The basic needs are to split a log up to 12 in in diameter and 24 in long. The splitting is done by forcing a wedge 6 in wide into one side of the log or by forcing the log onto the wedge. The splitter will be powered with an extension cord; that is, it will be plugged into an ordinary 110-volt (V) outlet. To determine the necessary force to split a log, you will need to look up the strength and stiffness of various types of fuel wood and to diagram a plausible mechanism for the failure, that is, the splitting, of the wood.

CHAPTER
2

COPING
WITH
UNCERTAINTY

Steel, aluminum, and other mills turn out tons of product intended to have certain properties. To ensure the quality of the product, samples are frequently taken and tested. For any given metal type, a sizable amount of data will accumulate, and inevitably the results will vary somewhat. What meaning do these variations have? If the supplier advertises a certain strength, we want to know what strength it is—minimum, maximum, or average. We also want to know the extent of the variations from the advertised value, for example, what the chance is that we will get a strength within a certain required range. Similar data can be gathered and similar questions asked about the loads that machine parts are subjected to—there is a difference in the way people use and abuse automobiles, skis, and washing machines. All this puts us into the world of probability, the mathematics of chance.

2.1 BERNOULLI TRIALS AND THE BINOMIAL PROBABILITY DISTRIBUTION

Let us start our study with that simplest of random trials—the flipping of a coin. The chance that a fair coin, fairly flipped, will come up heads is 1 in 2, or 1/2, since there are two possible outcomes: heads (H) or tails (T). This chance, which we call *probability*, is the same for every toss of the coin. Folklore to the contrary, there is no truth to the idea that the chance of a head is higher after a number of tails have been thrown. It has been shown many times that the numbers of heads and tails do not necessarily equalize quickly. Tails can stay ahead of heads, or vice versa, for hundreds of tosses. These facts are well known, of course, to those who make a living at gambling.

Let us calculate the chance of getting at least one head in two tosses of a coin. We might be tempted to add the probabilities; $1/2 + 1/2 = 1$, or certainty. We *know* that is not right! The probability of at least one head in two tosses can be evaluated by counting the possible outcomes, of which there are four: HH, HT, TT, and TH. Three of these four will produce at least one head, so the probability is 3/4. This probability does not mean a lot unless you work with large numbers. If you were to repeat this experiment of two tosses 100 times, in about 75 of the sets you would get at least one head. The larger the number of experiments, the closer the numbers will be. In fact, theoretically you have to make an infinite number of tosses to end up with a least one head in exactly 75 percent of the sets.

This probability can be developed mathematically by looking at the possible outcomes of each throw. If the chance (probability) of a head in one throw is 1/2, then the probability of also getting a head in the second throw is $(1/2)(1/2) = 1/4$. Thus the probability of throwing one or more ("or more" means 2 here) heads is the probability of throwing one head (1/2) plus the probability of throwing two heads (1/4), for a total of 3/4. We can also come to this result through the back door, so to speak, by identifying two outcomes which between them exhaust all possibilities. They are (1) no heads and (2) one or more heads. The sum of their probabilities must be 1. The probability of no heads is designated p(no heads), and it equals p(two tails) $= (1/2)(1/2) = 1/4$. This is the same as the computation for p(two heads) above. Thus, if p(no heads) $= 1/4$, the p(one or more heads) must be 3/4.

The toss of a coin, the bet on a roulette wheel, and the chance that a part made in an automatic machine will be defective are all single trials with only two outcomes and a probability of success which does not change from trial to trial. They are called *Bernoulli trials*, after Jacob Bernoulli,[1] who formulated some of the first rules for gambling games. The probable results of a series of Bernoulli trials are expressed in the binomial probability. We can develop it by going a little further with coin tosses. Let us consider three tosses. The possible outcomes are

HHH   HHT   HTH   HTT

TTT   TTH   THT   THH

---

[1] Jacob Bernoulli was born in Basel, Switzerland, in 1654. At the urging of his father, he first read theology, but his interest in mathematics changed his career. At age 28 he established a seminary for experimental physics at Basel, and five years later he became professor of mathematics at the university there. He was active in the development of calculus (determined the catenary), geometry (first user of polar coordinates), and the theory of probability. The famous Bernoulli probability theorem, which is basic in statistics, was published after his death in 1705.

Other members of the Bernoulli family also achieved prominence in mathematics and science. Daniel, the founder of mathematical physics, is known among engineers for his work in fluids. Johann, brother of Jacob, became professor of mathematics at Groningen, Holland, and, on the death of his brother, succeeded him in the chair at Basel. He was the mentor of Leonard Euler.

In the table below these are grouped by number of heads, and the probability of each number of heads is shown.

Number of heads	Outcomes producing this number of heads	Probability of outcome
0	TTT	$1p(T)^3 p(H)^0 = (1/2)^3(1/2)^0 = 1/8$
1	TTH THT HTT	$3p(T)^2 p(H)^1 = 3(1/2)^2(1/2)^1 = 3/8$
2	HHT HTH THH	$3p(T)^1 p(H)^2 = 3(1/2)^1(1/2)^2 = 3/8$
3	HHH	$1p(T)^0 p(H)^3 = (1/2)^0(1/2)^3 = 1/8$

The formulas of the last column can be written as

$$A p(T)^M p(H)^N$$

where M is the number of tails, N the number of heads, and A the number of combinations yielding the result. Thus, for example, in the second row we want the probability of getting one head. One possibility is the first shown: TTH. The probability of getting a tail on the first toss is 1/2. The probability of getting a tail on the second toss is also 1/2, and the probability of getting a head on the third toss is 1/2. The probability that outcomes 1 and 2 and 3 will occur is the product of the individual probabilities: $(1/2)^M(1/2)^N = 1/8$ ($M = 2$, $N = 1$). But there are two other possibilities for getting one head: THT and HTT, for a total of three. The probabilities for these combinations are also 1/8. The probability of getting the first combination or the second or the third (note we said *or*, not *and*) is the sum of the individual probabilities: $1/8 + 1/8 + 1/8 = A(1/8) = 3/8$ ($A = 3$).

You should check that the formula works for the cases of zero and three heads.

In the case of one head, the head can come from the first, second, or third toss. If we designate the three tosses as the events A, B and C, then we see that one head can result from event A, B, or C. This represents the combinations of 3 things taken 1 at a time. Two heads result from events AB, BC, or AC, the combinations of 3 things taken 2 at a time.

We can now write the binomial probability formula. The probability of x successes in n trials, if the probability of success in any given trial is the constant p, is

$$b(x: n, p) = {}_nC_x p^x(1 - p)^{n - x} \tag{2.1}$$

The first term on the right is the number of combinations of n things taken x at a time, which is given by

$${}_nC_x = \frac{n!}{x!(n - x)!} \tag{2.2}$$

Binomial probabilities are tabulated in handbooks.

Example 2.1. Let us check expression (2.2) by computing the combinations of 3 things ($n = 3$) taken 2 at a time ($x = 2$). We already know it is 3:

$${}_3C_2 = \frac{3 \times 2 \times 1}{(2 \times 1)(1)} = 3$$

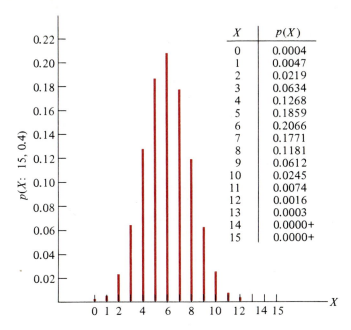

X	p(X)
0	0.0004
1	0.0047
2	0.0219
3	0.0634
4	0.1268
5	0.1859
6	0.2066
7	0.1771
8	0.1181
9	0.0612
10	0.0245
11	0.0074
12	0.0016
13	0.0003
14	0.0000+
15	0.0000+

FIGURE 2.1
Probability of acceptance of x parts in batch of 15 with probability of acceptance = 0.4 in any one test.

Example 2.2. As an example of applying the mathematics of the Bernoulli trial to engineering, consider a stream of parts coming from a production line. Suppose that 40 percent of the parts, on average, pass the acceptance test. We write

$$p(\text{a given part passes acceptance test}) = 0.40$$

We would like to know the probability that, out of a sample of 15 parts, exactly 5 will pass. We did not say *at least 5*; we will get to that later. Equation (2.1) then gives

$$b(5: 15, 0.4) = \frac{15!}{(5!)(10!)} (0.4)^5(0.6)^{10} = 0.186$$

The probability of getting x surviving parts in the acceptance test of the above example is shown in Figure 2.1. The probability of getting, say, at least 5 acceptable parts, which we alluded to earlier, will be the sum of the bars from 1 through 5: $p(\text{at least } 5) = 0.3937$. The total of all the bars is exactly 1 (except for roundoff errors), since that encompasses all possible outcomes.

The chance of getting more than 11 acceptable parts, or none, is extremely small. The highest probability (0.2066) is associated with 6 acceptable parts. This is as it should be; if you multiply the number in the sample (15) by the probability of acceptance (0.4), you get 6. Over a long period, the average number of acceptable parts per sample will be 6. Of course.

The chart of Figure 2.1 is called a *probability distribution*. Since the data are discrete (that is, they have only definite values: 0, 1, 2, etc.), the ends of the bars should not be connected to make a graph, though you will occasionally see it done.

Recall that this probability distribution, given by expression (2.1), is the Bernoulli or binomial. It represents tests (trials) with only two possible outcomes (heads or tails, pass or fail) and with a constant probability of one of the outcomes [in the example, $p(\text{success}) = 0.4$]. That probability is obtained from the past history of testing. You can see that this distribution would have importance in product inspection, that is, quality control.

2.2   FREQUENCY DISTRIBUTIONS, OR HISTOGRAMS

The data we deal with in design, such as values of strength, load, stiffness, etc., will usually be not discrete, as in the preceding paragraph, but *continuous*. For example, the strength of a certain steel may vary from 195 to 240 megapascals (MPa), with any value in this range being possible, such as 220.26. We need to examine how data like these can best be shown.

For the purpose of redesigning the support for a tractor seat, the following weights of drivers have been recorded.

180	178	182	158	193	176	187	206	193
164	207	186	200	171	169	173	158	187
181	178	205	184	161	177	185	184	162
174	198	195	178	167	186	183	181	168

These numbers range from 158 to 207. Although the weights were evidently recorded to the nearest pound, any number is possible, depending on the precision of the measurement. The data are said to be continuous, as opposed to the discrete data of the binomial distribution, where, for example, we might have 5 or 6 successes in 15 trials, but we would never have 5.2 or 6.3.

We could draw a bar graph showing the number of occurrences of each of the numbers listed, but because most of the values are different, the frequencies would be very low. The highest, in fact, is 3 for 178 lb. The graph would not give much of an idea of the distribution. We can improve the display by grouping the data into intervals. Although the intervals do not have to be of equal width, the presentation is usually better if they are. The number of intervals is arbitrary. One author in the field suggests using the square root of the number of data. In this example, we have 36, so that would result in 6 intervals. Another suggestion is 10 to 20 intervals for 50 or more data values, 6 to 10 for less than 50. Six intervals satisfy this approach as well.

If the intervals are to have equal width w, then

$$w = \frac{\text{data range}}{\text{no. intervals}} = \frac{207 - 158}{6} = 8.167$$

This will result in interval boundaries to several decimals. If that is not objectionable, it is then a simple matter to count the occurrences of the data within each interval. It can, in fact, be done by computer. If nicer-looking boundaries are desired, we round the interval width down to 8. To make it impossible for any value to fall exactly on a boundary, we start at the lowest observed value less one-half the measurement precision. Here the measurement precision is 1 lb, and the lowest value is 158 lb. We start, therefore, at 157.5, and the intervals become

$$157.5–165.5$$
$$165.5–173.5$$
$$173.5–181.5$$
$$181.5–189.5$$
$$189.5–197.5$$
$$197.5–205.5$$
$$205.5–213.5$$

Since we rounded the interval width down, we have ended up with 7 intervals instead of 6, which does not matter, and the last interval goes somewhat beyond our highest value, 207. We could center the range better by starting 3 units lower, which produces the intervals of the table shown in Figure 2.2a.

The number of data falling in each interval is called the *frequency*, and the fraction of the total is the *fractional frequency*. Figure 2.2b is a bar graph of these fractional

Interval boundaries	Frequency	Fractional frequency
154.5–162.5	5	0.139
162.5–170.5	3	0.083
170.5–178.5	8	0.222
178.5–186.5	10	0.278
186.5–194.5	4	0.111
194.5–202.5	3	0.083
202.5–210.5	3	0.083
	$\Sigma = 36$ (total data)	$\Sigma = 0.999 \simeq 1$

(a)

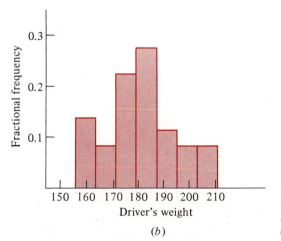

(b)

FIGURE 2.2
Construction of histogram for tractor driver's weights. (a) Frequencies of weights. (b) Histogram.

Driver's weight	Probability density
158.5	.0174
166.5	.0104
174.5	.0278
182.5	.0347
190.5	.0139
198.5	.0104
206.5	.0104

(a)

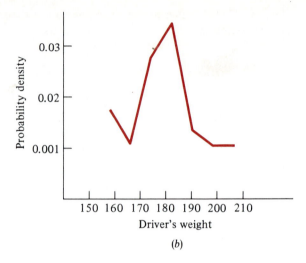

(b)

FIGURE 2.3
Probability density for tractor driver's weights. (a) Table of probability densities of weights. (b) Probability density.

frequencies. It is called a *histogram.* The ordinate may be the frequency or the fractional frequency, which is just a question of scale. The fractional frequency of any interval represents the probability of finding values within the interval, and it follows that all the frequencies must total 1. This is verified in the table in Figure 2.2a. The histogram gives an easily read graphic idea of the distribution of a set of data.

2.3 PROBABILITY DENSITY FUNCTION

If we divide the fractional frequencies (probabilities) shown in the table of Figure 2.2a by the interval width, the result is the probability density. This is the probability per unit value of the variable, hence the term *density.* The results are shown against the center value of the intervals in the table of Figure 2.3a and are plotted in Figure 2.3b. The variation of the probability density with the variable is known as the *probability density function* (pdf) or *distribution.* Of course, with only seven points we do not get a very smooth curve. If we had enough data that smaller intervals were possible, the results would be better. If we could amass an infinitude of data, then we could consider an infinitesimally small interval width of the variable X and take the limit as the width approaches zero, to get the definition

$$\text{Probability density} = \lim_{\Delta X \to 0} \frac{p[(a - \Delta X/2) < X < (a + \Delta X/2)]}{\Delta X}$$
$$(\text{at } X = a)$$

It follows from this definition that the probability of observing values of the variable between any two limits will equal the area under the curve between those limits, as shown in Figure 2.4. Also, the area under the entire curve will equal unity,

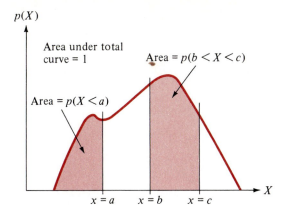

FIGURE 2.4
The probability of finding values of the variable between limits equals the area under the pdf between those limits (any pdf).

since it includes all possible values. We will discuss probability density functions a good bit more, but first we need a few definitions.

2:4 MEAN, MEDIAN, AND STANDARD DEVIATION OF SAMPLE DATA

In the example of the previous section in which the weights of a number of tractor drivers were recorded, there was a good bit of scatter. Of course, that is something we are used to; people differ. There will also be scatter among measurements of a quantity not expected to vary, the maximum torque output of a line of new automobile engines, for instance. One characteristic is present in these and other measurements: You cannot predict the next one. They are thus said to be *random variables*. This does not mean that there is no order present and that we cannot get a handle on it. We can predict the most likely value and the chance of getting a value within a given range of values. We commonly encounter the results of such studies, in life, fire, and automobile insurance rates, for example.

In many, if not most, production situations, it is not feasible to test every part made or every item purchased. It would cost too much, or the test might destroy the part. Thus we test reasonably sized groups, called *samples*, like the batch of 15 of the acceptance test in Example 2.2. Several properties of sample data are important in probability analysis. We will need three of them.

Mean or Average Value $\bar{x}$

The *mean* or *average value* of the data is obtained by adding all the values and dividing by the number of data n:

$$\bar{x} = \sum_{j=1}^{n} \frac{x_j}{n}$$

The mean is the value about which the data are centered, but it does not follow that most of the data must be concentrated around it. For example, in an elementary

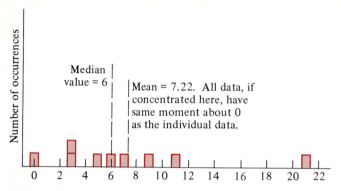

FIGURE 2.5
Mean as center of mass of data. Median value is shown.

school, the children may all be aged 3 through 10 years, while the teachers are 30 years and above. The average age of all the individuals in the building might work out to be 19 years, but there is not one person within 9 years of that age there!

If you think of each data point as a unit mass at distance x from the origin of x, then the mean corresponds to the center of mass of the data. For example, if we have values of x equal to 7, 9, 0, 11, 21, 6, 3, 3, and 5 and we draw the histogram, as sketched in Figure 2.5, then the mean, which is $65/9 = 7.22$, is the point where we could concentrate all the bars and produce the same moment about the origin (or any other point, for that matter) as the individual bars. This is the definition of the *center of mass* and of the mean. Note that the mean does not divide the data equally above and below its value. In this simple example, there are six values below the mean and three above it. Of course, if the values are symmetrically placed about the mean, i.e., if the histogram is symmetric, then there will be an equal number above and below.

Median Value *m*

The *median* is another measure of central tendency. It is the middle value, after the numbers have been put in order, and hence divides the data equally above and below its value. In the example above, the numbers in sequential order are 0, 3, 3, 5, 6, 7, 9, 11, 21, and the median $m = 6$, shown also in Figure 2.5. If the number of data is even, then there is no single middle number; the average of the two middle values is used. You can see that in a large population $p(X < \text{median}) = p(X > \text{median}) = 0.5$.

Standard Deviation *s*

The *standard deviation s* is a measure of the dispersion of the data from the mean value. You might imagine calculating the dispersion by measuring the difference between each particular value and the mean, summing these values, and dividing by the number of values. The fatal flaw in this method is that the pluses and minuses cancel. After all, that is the definition of the mean. However, the minuses can be eliminated by squaring the difference between each observation and the mean. An associated term also emerges, which is the *sample variance*:

$$\text{Variance } s_x^2 = \frac{\sum\limits_{j=1}^{n} (x_j - \bar{x})^2}{n - 1}$$

The positive square root of the variance is the *standard deviation*. Note that the sum of the squared differences is divided by $n - 1$, rather than n, which makes the sample variance a bottom estimator of the population variance.

You will see now that a small standard deviation means that most of the sample measurements do not deviate far from the mean. In a factory that means good quality control; in a classroom test it means that most of the grades are close to the class average.

The mean, median, variance, and standard deviation of a sample (even though the sample may be quite large) are estimates of what would result if an infinite number of observations could be made. This distinction is necessary, because probability theory is based on an infinitude of data. For example, suppose we have 100 values of the tensile strength of a certain steel. The mean value of the data in this sample is an estimate of the mean which would result if we could get an infinite number of measurements of this property. The mean is also the *expected value* if we continue testing; that is, if we had to bet on the outcome of the next test, we would pick the mean of our sample.

Example 2.3. We want the mean and the standard deviation of the 15 grades received in a school test. They are listed in the table below, which is used to make the calculation.

(1) Grade	(2) Grade less mean (grade − 67)	(3) (Column 2)²
56	−11	121
65	−2	4
45	−22	484
89	22	484
98	31	961
34	−33	1089
67	0	0
78	11	121
82	15	225
76	9	81
93	26	676
56	−11	121
45	−22	484
31	−36	1296
90	23	529
$\sum = 1005$		$\sum = 6676$

$$\bar{x} = \frac{1005}{15} = 67 \quad \text{mean} \qquad s^2 = \frac{6676}{14} = 476.9 \quad \text{variance}$$

$$s = 21.8 \quad \text{standard deviation}$$

Example 2.4. It can be shown that the standard deviation of a binomial probability density distribution is $s = \sqrt{np(1-p)}$. In Example 2.2, the production test with 15 samples and a probability of success of 0.4, this works out to be $s = \sqrt{(15)(0.4)(0.6)} = 1.897$. Let us check that by tabulating the probabilities used to construct Figure 2.1 and calculating the standard deviation from its definition.

(1) Number of good parts	(2) p (of getting that many)	(3) Number less the mean	(4) Column 2 × (column 3)²
0	0.0004	−6	0.0144
1	0.0047	−5	0.1175
2	0.0219	−4	0.3504
3	0.0634	−3	0.5706
4	0.1268	−2	0.5072
5	0.1859	−1	0.1859
6	0.2066	0	0
7	0.1771	1	0.1771
8	0.1181	2	0.4724
9	0.0612	3	0.5508
10	0.0245	4	0.3920
11	0.0074	5	0.1850
12	0.0016	6	0.0576
13	0.0003	7	0.0147
14	0.0000+	8	0.0000
15	0.0000+	9	0.0000

Check: $\sum = 0.9999 \simeq 1$

$\sum = s^2 = 3.6000$

$s = 1.897$ Q.E.D.

2.5   UNDERLYING POPULATIONS AND SAMPLES

A bottling company performs a bursting test on 1 bottle in every 10 000 which it receives from a supplier. The company does 10 such tests per week. The 10 tests are a sample, and the results are used to estimate the mean strength and standard deviation of the larger group, the 100 000 bottles received during the week. Naturally, the larger the sample, the better the estimate. The limitation is one of practical economics— bottles are destroyed in the test, and the test takes time and money. We can also say that the 10 weekly bursting strengths are a sample of the population of bursting strengths of all the bottles of the type which the supplier has ever produced, termed the *underlying population*. Most probability theory is based on an infinite underlying population. In any practical situation, an infinitude of data do not exist, hence an "infinite underlying population" is purely an abstract concept. However, the population in a practical setting may be sufficiently large that we can consider it to be infinite. Samples of data are used to estimate the nature of the underlying population, i.e., its mean, standard deviation, and median. We can never know those parameters directly.

   Random variables are normally designated with a *capital letter*, such as X, which we have used earlier. A *specific value* of a random variable is indicated with a *lower-*

case letter: $x = 27$. The mean, standard deviation, and median of sample data are designated by lowercase letters $\bar{x}$, s, and m. The population parameters are generally designated by μ for mean, σ for standard deviation, and M for median.

The distinction between an underlying population and samples drawn from it which are used to estimate the parameters of the underlying population is important in the discussion of population distributions.

2.6 NORMAL PROBABILITY DISTRIBUTION

Earlier we calculated a few binomial probabilities and showed them graphically in Figure 2.1. The binomial can be approximated by a function called the *normal* or *gaussian*[1] *distribution*, which has the form

$$ p(x: \mu, \sigma) = \frac{1}{\sigma\sqrt{2\pi}} \exp\left[-\frac{1}{2}\left(\frac{x - \mu}{\sigma}\right)^2 \right] \qquad (2.3) $$

The left side reads, "the probability density of the random variable X at $X = x$, given its mean value μ and standard deviation σ." Note that the normal distribution is completely defined when the mean and standard deviations are known.

The normal distribution, Equation (2.3), is shown superposed on the binomial probability density function from Example 2.2 in Figure 2.6. As you can see in expression (2.3), the normal distribution is symmetric about the mean; the binomial was not quite so. Also the normal distribution runs off to infinity in both directions, whereas this particular binomial has a spread from 0 to 15 only. The approximation is exact when the number in the group n approaches infinity. On a practical basis, with p the probability of success in any one trial, it is quite good when $np > 5$ if $p \leq 0.5$, or when $n(1 - p) > 5$ if $p > 0.5$. For our 15-specimen example $p = 0.4$, then $np = 6$, and the above criteria indicate a good fit, as we see in the figure.

---

[1] The credit for the normal probability density function is attributed variously to Laplace, De Moivre, and Gauss. The term *gaussian distribution* derives from the extensive reference Gauss made to it in discussing errors of measurement.

Karl Friedrich Gauss (1777–1855) is held by some to rank with Archimedes and Newton as one of the world's three greatest mathematicians. He is often referred to as the founder of modern mathematics. His contributions to astronomy and physics were of equal importance.

Nearly all Gauss' discoveries in mathematics were made when he was between the ages of 14 and 17. At 17 he formulated the method of least squares. From 1831 on, Gauss pursued research in magnetism and electricity with his collaborator, Wilhelm Weber. The theories of magnetism and of the potential were the result, as well as the bifilar magnetometer and the mirror galvanometer. The unit of magnetic flux density has been named the gauss (G); in SI units it has become the tesla (T). Magnetic flux is still measured in a unit called the *weber*.

In 1833 Gauss and Weber connected two buildings by a wire and by means of electromagnets rang a bell at one end when the circuit was completed at the other. The idea of transmitting messages by such means led to publications of their work in 1834. They preceded Morse by 10 years.

Gauss was married twice and had six children, two of whom emigrated to Missouri and none of whom showed their father's talents.

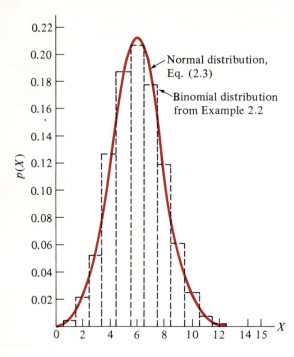

Figure 2.6
Approximation of binomial results from Example 2.2 by normal distribution [see Equation (2.3)].

When we introduced the probability density function (pdf) in Section 2.3, we stated that the probability of the variable falling between two different values equals the area under the pdf curve between those values. This is, of course, also true for normal distributions (Figure 2.7). Since the curve extends to infinity in either direction, the problem of determining the probability of having values less, say, than some value a will involve an integration from $x = -\infty$ to $x = a$. It follows that the area under the entire normal pdf is 1, since that represents all possible outcomes. To prove it, you would integrate Equation (2.3) between $-\infty$ and $+\infty$.

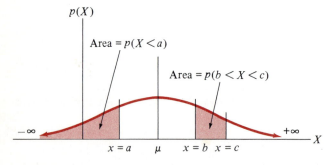

FIGURE 2.7
The probability of finding values of a normally distributed variable between limits.

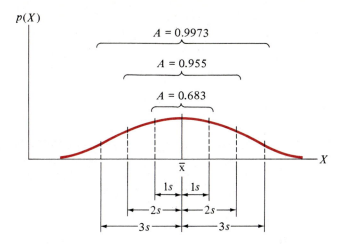

FIGURE 2.8
The probability of finding values within 1, 2, and 3 standard deviations of the mean, for normal distributions.

If you calculate the probability of finding values of X within 1 standard deviation ($\pm 1\sigma$) of either side of the mean, you will get 0.683. Another way of saying this is that we can expect 68.3 percent of the values of X to be within 1 standard deviation of the mean. Going further yields

$$p[(\mu - 2\sigma) < X < (\mu + 2\sigma)] = 0.955$$

$$p[(\mu - 3\sigma) < X < (\mu + 3\sigma)] = 0.9973$$

Thus, if a variable is normally distributed, 99.73 percent of all values will occur within 3 standard deviations of the mean. This is sketched in Figure 2.8. On a practical level, if the quality-control department says it can hold the strength of a part to a given average value plus or minus 9 percent, say, then the standard deviation can be estimated to be 3 percent of the mean value. The assumption is made, of course, that the strengths are normally distributed.

Probability with Continuous PDF

Suppose we want to find the probability, in the acceptance test we have used so much, of getting 8 successes. If we try to find the area under the normal pdf at 8, it is an area with zero width—we get zero probability. The problem is that the normal distribution curve is continuous, whereas the acceptance test had discrete outcomes: 0, 1, 2, etc. To get the probability of 8 successes, we must find the area under the normal curve from $7\frac{1}{2}$ to $8\frac{1}{2}$, that is, we include a half-interval on each side. In doing this we are approximating a rectangle of base 1 and height from the bar chart for 8 successes in Figure 2.1.

We must handle continuous distributions in the same way when calculating probabilities. Take the height of males between 18 and 19 years of age. If the figures are recorded to the nearest millimeter and the heights fall between 1500 and 2500 mm, then there are 1000 discrete outcomes. If we measure to the nearest $\frac{1}{10}$ mm, there are 10 000 outcomes. These numbers depend on the precision of the measurement. The normal distribution curve, being continuous, represents infinite precision, you could say. If we are asked what the probability is of finding a person 1995 mm in height, we must attach a tolerance to the measurement; otherwise, we cannot calculate an answer from the normal curve. We can calculate the probability of finding a height between 1994 and 1996 mm. If we narrow.the range to $\frac{1}{10}$ mm, then the probability is correspondingly smaller. A range of the variable must always be specified when the normal curve (or any other continuous distribution function) is used to calculate probabilities.

2.7   STANDARDIZED VARIABLE

In the last paragraph, in which we introduced the normal probability function, we performed no calculations, offered no examples. If you take another look at the relation in question (2.3), you will see why—it is messy. Fortunately, through a change in variable, the situation simplifies a great deal. The variable which does the trick is

$$Z = \frac{X - \mu}{\sigma} \tag{2.4}$$

This variable measures the distance from the mean in standard deviations. It is normally distributed, and it is unitless, which means that a computer algorithm, or one set of calculations, i.e., a table, can be used for all problems. It is therefore called the *standardized variable*. Sometimes it is termed the Z *score*.

Given the definition of Z, we see that its mean is $\bar{Z} = (\mu_X - \mu_X)/\sigma_X = 0$, and its standard deviation is $\sigma_Z = \sigma_X/\sigma_X = 1$. The normal probability function in terms of the standardized variable then becomes

$$p(z) = \frac{1}{\sqrt{2\pi}} \exp\left(\frac{-z^2}{2}\right) \tag{2.5}$$

which is a great deal simpler than the earlier version. The plot of expression (2.5) is shown in Figure 2.9.

We are also going to encounter the *cumulative (probability) distribution function*, abbreviated cdf. This is simply the integral of the normal distribution (i.e., the area) from $-\infty$ up to the value of the variable. It also appears in the figure. Its value goes from 0 at $Z = -\infty$ to 0.5 at $Z = 0$ (because of the symmetry of the pdf) to 1 at $Z = +\infty$. A cdf exists for any probability distribution. It is very useful because it gives directly the probability of finding values of the variable less than the abscissa.

Now we can do an example or two requiring use of the normal probability function. Actually, we are saved having to perform any integrations or to perform any calculations, since they have been done and can be found in many handbooks.

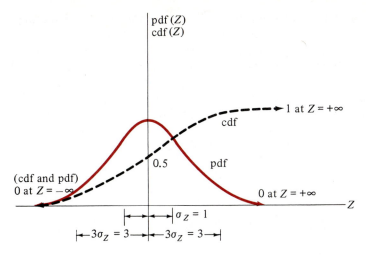

FIGURE 2.9
Normal probability density and cumulative distribution curves for the standardized variable.

Table 2.1 is a typical presentation of the cumulative normal distribution. Algorithms to compute this function are also built into some hand calculators and most computer libraries. We may have to juggle the numbers a little for a particular problem, but what we need is readily obtained, as the following examples show.

Example 2.5. Let us check the statement we made earlier that 99.73 percent of a normally distributed population falls within 3 standard deviations of the mean. We need therefore to find the area under the curve in Figure 2.9 between $Z = -3$ and $Z = +3$.

We get the area A_1 in the sketch of Figure 2.10a directly from the table: $A_1 = 0.99865$. If we subtract 0.5 from A_1, we have A_2 in the sketch of Figure 2.10b: $A_2 = 0.49865$. Doubling A_2 then gives us our probability:

$$p(-3 < Z < +3) = 2 \times 0.49865 = 0.9973 \qquad \text{Q.E.D.}$$

Example 2.6. Returning to Example 2.2, which dealt with the number of survivors in a group of 15 in an acceptance test, we check the fit of the normal distribution by computing the probability of getting 4 to 10 successes. The standard deviation was computed in Example 2.4: $\sigma_X = 1.90$. The mean is $\mu = 6$.

To find the probability of 4 to 10 successes, we need the area under the normal curve from $X = 3.5$ to $X = 10.5$. The corresponding values of the standardized variable are

$$z_{X=3.5} = \frac{3.5 - 6}{1.90} = -1.3158 \qquad z_{X=10.5} = \frac{10.5 - 6}{1.90} = 2.3684$$

Thus we require the area shown shaded in Figure 2.11a. First, we determine the area A_1 of the sketch in Figure 2.11b. From the table, $A_1 = 0.99092$ (interpolated between $z = 2.36$ and $z = 2.37$). The area A_2 in Figure 2.11c must be subtracted from A_1. This area is not given directly in the table. We get it by finding the complementary area A_c shown in Figure 2.11d and subtracting that from unity. From the table, $A_c = 0.90587$,

TABLE 2.1

Cumulative normal distribution $A = \int_{-\infty}^{z} \frac{1}{\sqrt{2\pi}} e^{-u^2/2}\, du$

z	0.00	0.01	0.02	0.03	0.04
0.0	0.500 00	0.503 99	0.507 98	0.511 97	0.515 95
0.1	0.539 83	0.543 79	0.547 76	0.551 72	0.555 67
0.2	0.579 26	0.583 17	0.587 06	0.590 95	0.594 83
0.3	0.617 91	0.621 72	0.625 51	0.629 30	0.633 07
0.4	0.655 42	0.659 10	0.662 76	0.666 40	0.670 03
0.5	0.691 46	0.694 97	0.698 47	0.701 94	0.705 40
0.6	0.725 75	0.729 07	0.732 37	0.735 65	0.738 91
0.7	0.758 03	0.761 15	0.764 24	0.767 30	0.770 35
0.8	0.788 14	0.791 03	0.793 89	0.796 73	0.799 54
0.9	0.815 94	0.818 59	0.821 21	0.823 81	0.826 39
1.0	0.841 34	0.843 75	0.846 13	0.848 49	0.850 83
1.1	0.864 33	0.866 50	0.864 64	0.870 76	0.872 85
1.2	0.884 93	0.886 86	0.888 77	0.990 65	0.892 51
1.3	0.903 20	0.904 90	0.906 58	0.908 24	0.909 88
1.4	0.919 24	0.920 73	0.922 19	0.923 64	0.925 06
1.5	0.933 19	0.934 48	0.935 74	0.936 99	0.938 22
1.6	0.945 20	0.946 30	0.947 38	0.948 45	0.949 50
1.7	0.955 43	0.956 37	0.957 28	0.958 18	0.959 07
1.8	0.965 07	0.964 85	0.965 62	0.966 37	0.967 11
1.9	0.971 28	0.971 93	0.972 57	0.973 20	0.973 81
2.0	0.977 25	0.977 78	0.978 31	0.978 82	0.979 32
2.1	0.982 14	0.982 57	0.983 00	0.983 41	0.983 82
2.2	0.986 10	0.986 45	0.986 79	0.987 13	0.987 45
2.3	0.989 28	0.989 56	0.989 83	0.990 10	0.990 36
2.4	0.991 80	0.992 02	0.992 24	0.992 45	0.992 66
2.5	0.993 79	0.993 96	0.994 13	0.994 30	0.994 46
2.6	0.995 34	0.995 47	0.995 60	0.995 73	0.995 85
2.7	0.996 53	0.996 64	0.996 74	0.996 83	0.996 93
2.8	0.997 44	0.997 52	0.997 60	0.997 67	0.997 74
2.9	0.998 13	0.998 19	0.998 25	0.998 31	0.998 36
3.0	0.998 65	0.998 69	0.998 74	0.998 78	0.998 82
3.1	0.999 03	0.999 06	0.999 10	0.999 13	0.999 16
3.2	0.999 31	0.999 34	0.999 36	0.999 38	0.999 40
3.3	0.999 52	0.999 53	0.999 55	0.999 57	0.999 58
3.4	0.999 66	0.999 68	0.999 69	0.999 70	0.999 71
3.5	0.999 77	0.999 78	0.999 78	0.999 79	0.999 80
3.6	0.999 84	0.999 85	0.999 85	0.999 86	0.999 86
3.7	0.999 89	0.999 90	0.999 90	0.999 90	0.999 91
3.8	0.999 93	0.999 93	0.999 93	0.999 94	0.999 94
3.9	0.999 95	0.999 95	0.999 96	0.999 96	0.999 96

(*continued*)

TABLE 2.1 (*Continued*)

z	0.05	0.06	0.07	0.08	0.09
0.0	0.519 94	0.523 92	0.527 90	0.531 88	0.535 86
0.1	0.559 62	0.563 56	0.567 49	0.571 42	0.575 34
0.2	0.598 71	0.602 57	0.606 42	0.610 26	0.614 09
0.3	0.636 83	0.640 58	0.644 31	0.648 03	0.651 73
0.4	0.673 64	0.677 24	0.680 82	0.684 38	0.687 93
0.5	0.708 84	0.712 26	0.715 66	0.719 04	0.722 40
0.6	0.742 15	0.745 37	0.748 57	0.751 75	0.754 90
0.7	0.773 37	0.776 37	0.779 35	0.782 30	0.785 23
0.8	0.802 34	0.805 10	0.807 85	0.810 57	0.813 27
0.9	0.828 94	0.831 47	0.833 97	0.836 46	0.838 91
1.0	0.853 14	0.855 43	0.857 69	0.859 93	0.862 14
1.1	0.874 93	0.876 97	0.879 00	0.881 00	0.882 97
1.2	0.894 35	0.896 16	0.897 96	0.899 73	0.901 47
1.3	0.911 49	0.913 08	0.914 65	0.916 21	0.917 73
1.4	0.926 47	0.927 85	0.929 22	0.930 56	0.931 89
1.5	0.939 43	0.940 62	0.941 79	0.942 95	0.944 08
1.6	0.950 53	0.951 54	0.952 54	0.953 52	0.954 48
1.7	0.959 94	0.960 80	0.961 64	0.962 46	0.963 27
1.8	0.967 84	0.968 56	0.969 26	0.969 95	0.970 62
1.9	0.974 41	0.975 00	0.975 58	0.976 15	0.976 70
2.0	0.979 82	0.980 30	0.980 77	0.981 24	0.981 69
2.1	0.984 22	0.984 61	0.985 00	0.985 37	0.985 74
2.2	0.987 78	0.988 09	0.988 40	0.988 70	0.988 99
2.3	0.990 61	0.990 86	0.991 11	0.991 34	0.991 58
2.4	0.992 86	0.993 05	0.993 24	0.993 43	0.993 61
2.5	0.994 61	0.994 77	0.994 92	0.995 06	0.995 20
2.6	0.995 98	0.996 09	0.996 21	0.996 32	0.996 43
2.7	0.997 02	0.997 11	0.997 20	0.997 28	0.997 36
2.8	0.997 81	0.997 88	0.997 95	0.998 01	0.998 07
2.9	0.998 41	0.998 46	0.998 51	0.998 56	0.998 61
3.0	0.998 86	0.998 89	0.998 93	0.998 97	0.999 00
3.1	0.999 18	0.999 21	0.999 24	0.999 26	0.999 29
3.2	0.999 42	0.999 44	0.999 46	0.999 48	0.999 50
3.3	0.999 60	0.999 61	0.999 62	0.999 64	0.999 65
3.4	0.999 72	0.999 73	0.999 74	0.999 75	0.999 76
3.5	0.999 81	0.999 81	0.999 82	0.999 83	0.999 83
3.6	0.999 87	0.999 87	0.999 88	0.999 88	0.999 89
3.7	0.999 91	0.999 92	0.999 92	0.999 92	0.999 92
3.8	0.999 94	0.999 94	0.999 95	0.999 95	0.999 95
3.9	0.999 96	0.999 96	0.999 96	0.999 97	0.999 97

Source: From W. M. Hines and D. C. Montgomery, *Probability and Statistics in Engineering Science*, 2d ed., Wiley, New York, 1980, pp. 474–475. Reprinted by permission.

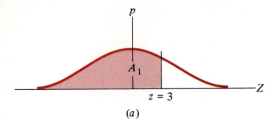

(a)

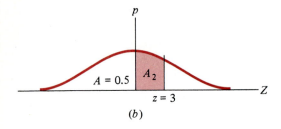

(b)

FIGURE 2.10
Areas involved in calculation for Example 2.5.

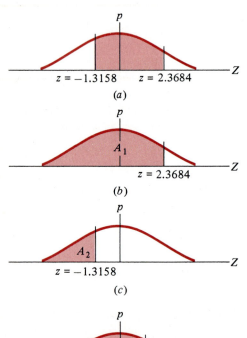

(a)

(b)

(c)

(d)

FIGURE 2.11
Areas involved in probability calculation of Example 2.6. (a) Required area. (b) Area A_1 is found from the table. (c) Area A_2 must be subtracted from A_1. (d) $1 - A_c = A_2$.

then $A_2 = 1 - 0.90587 = 0.09413$. So the probability we want is

$$p(4 \text{ to } 10 \text{ successes}) = A_1 - A_2 = 0.99092 - 0.09413$$

$$= 0.8968 \quad \text{(probability table)}$$

From the table in Example 2.4, the probability of 4 to 10 successes is

$$
\begin{array}{c}
0.1268 \\
0.1859 \\
0.2066 \\
0.1771 \\
0.1181 \\
0.0612 \\
0.0245 \\
\hline
p(4 \text{ to } 10 \text{ successes}) = 0.9002
\end{array} \quad \text{Bernoulli formula}
$$

The difference is 0.4 percent.

2.8   LOGNORMAL DISTRIBUTIONS

Many distributions are obviously not "normal," since they are unsymmetric. The lognormal distribution accommodates some of these cases. It finds application in lifetime studies and is thus useful in the physical and social sciences and in engineering, where it is applied in time-to-failure analysis. In the context of this book, that means fatigue life.

Table 2.2 displays 150 data from a lifetime test. A 10-interval histogram for these data is shown in Figure 2.12. Note that it is heavy on the lower side (called *right-skewed*). This fact suggests the idea of plotting, rather than the variable itself, a function of the variable which compresses the upper end along the abscissa axis

TABLE 2.2
Lifetime test data

499	577	596	625	655	658	661	671	671	684
684	687	700	708	709	709	710	711	718	723
735	740	741	745	751	753	754	767	776	776
777	780	780	781	794	797	809	810	818	822
827	835	835	844	851	851	856	858	871	871
874	875	883	883	897	907	908	928	935	941
962	974	977	977	980	984	986	987	996	996
999	999	1004	1006	1019	1021	1022	1024	1026	1027
1033	1034	1035	1037	1041	1047	1049	1052	1062	1069
1073	1078	1081	1083	1090	1095	1097	1098	1099	1104
1104	1108	1131	1139	1152	1164	1164	1167	1182	1191
1212	1213	1237	1249	1257	1257	1261	1310	1311	1319
1324	1369	1371	1374	1377	1397	1401	1425	1427	1428
1432	1444	1461	1465	1468	1469	1487	1496	1527	1538
1566	1567	1650	1720	1836	1926	1997	2013	2102	2167

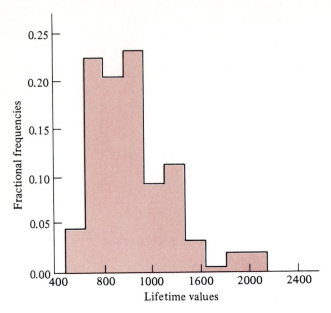

FIGURE 2.12
Ten-interval histogram for data of Table 2.2.

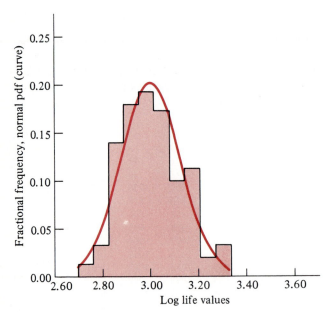

FIGURE 2.13
Ten-interval histogram for logarithms of data of Table 2.2. Normal pdf is superposed.

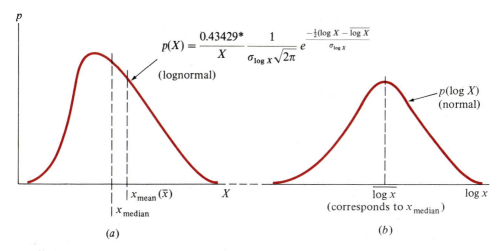

$$p(X) = \frac{0.43429*}{X} \frac{1}{\sigma_{\log X}\sqrt{2\pi}} e^{\frac{-\frac{1}{2}(\log X - \overline{\log X})}{\sigma_{\log X}}}$$

(lognormal)

$p(\log X)$
(normal)

$|x_{\text{mean}}(\overline{x})$ X

$|x_{\text{median}}$

(a)

$\log x$
(corresponds to x_{median})

$\log x$

(b)

FIGURE 2.14

Lognormal distribution.

* $0.43429/x = 1/x \log_{10} e = d/dx(\log_{10} X)$. In making the change of variable from $\log X$ to X, this factor is necessary to assure that $\int_{-\infty}^{+\infty} p(X)\,dx = 1$.

so as to create a more symmetric display. Taking the logarithm of the variable and treating it as a normal distribution is one way of doing this. The result is called the *lognormal distribution*; some statisticians hold that it is as fundamental as the normal distribution.

Figure 2.13 is a histogram of the logarithms of the data of Table 2.2. The shape is now reasonably symmetric. The mean and standard deviation for the logarithms of the data are computed in the usual fashion; they are called the *log mean* and the *log standard deviation*. A pdf based on those values (scaled up by the interval width—recall that we divided the fractional frequency by the interval width to get the probability density) is superposed on the histogram. The fit appears good; how good "good" is we will discuss later.

The procedure, then, for fitting data to a lognormal distribution is to take the logarithm of each value of the variable and then to treat the logarithm as normally distributed. The base of the logarithm does not matter; statisticians use the natural base, engineers generally prefer 10. The pdf can be displayed versus the logarithm of the variable or versus the variable itself. In the latter case, of course, it is not symmetric, as shown in Figure 2.12.

In normal distributions, the pdf was seen to be symmetric about the mean value of the variable, and the mean occurred at the peak of the curve. The mean was thus also the median, the value above and below which 50 percent of the observations would occur. However, that is not the definition of the mean; it is really the median (the middle value) which divides the events 50/50. We tend to forget that fact when the distribution is normal, since the mean and median are the same then, owing to the symmetry. In Figure 2.14 you see the distinction. Also note that the mean no longer falls at the peak.

Example 2.7. A group of specimens fail at the following number of cycles in a fatigue test (same stress): 11 000, 12 200, 13 000, 16 000, 20 000, 28 500, 39 000. Assuming a log-normal distribution, what is the probability of survival to 23 000 cycles?

We construct a table as previously to find the mean of $\log n$ and the standard deviation of $\log n$.

(1) n	(2) $\log n$	(3) $\log n - \overline{\log n}$	(4) (Column 3)^2
11 000	4.0414	−0.2147	0.0461
12 200	4.0864	−0.1697	0.0288
13 000	4.1139	−0.1422	0.0202
16 000	4.2041	−0.0520	0.0027
20 000	4.3010	0.0449	0.0020
28 500	4.4548	0.1987	0.0395
39 000	4.5911	0.3350	0.1122
	$\sum = 29.7928$		$\sum = 0.2515$

$$\overline{\log n} = \frac{29.7928}{7} = 4.2561 \qquad s_{\log n} = \sqrt{\frac{0.2515}{6}} = 0.2047$$

The sample mean and standard deviation are used as estimates of the underlying population.

Converting to the standardized variable, we get

$$Z = \frac{\log n - \mu_{\log n}}{\sigma_{\log n}}$$

The value of z corresponding to a life of 23 000 cycles is

$$z = \frac{\log 23\,000 - 4.2561}{0.2047} = 0.5160$$

Now we have to be a little careful of the semantics. The distribution is for failure times, and our task is to find a probability of survival. The probability of survival to 23 000 means the probability that failure occurs *above* 23 000. Thus

$$p(\text{survival to 23 000}) = \int_{z=0.5160}^{\infty} e^{-z^2/2}\, dz$$

which we find, by using Table 2.1, to be $0.6971 - 0.5 = 0.1971$, or about 20 percent.

Recall that normal probability distribution curves extend from $-\infty$ to $+\infty$. There are many, many distributions where the variable definitely will be on one side of zero and never the other: the life of anything, including people, strengths of materials, diameters of shafts, weights of packages, ocean soundings, etc. The lognormal distribution has the advantage of being applicable only to positive values of a variable; that is, it has a low-end cutoff. This is so because the logarithm of a negative number does not exist.

2.9 DETERMINING THE DISTRIBUTION TYPE

Before calculations are performed which presume a normal or a lognormal distribution of data, the question arises whether the data are, in fact, so distributed. We describe here several ways of answering this question. Note, however, that there are many distributions, and the broader question, when one is faced with a set of data, is: What distribution fits best? Fortunately, in design work most random variables we deal with are either normally or lognormally distributed, so we will look at those distributions first. In the next section we discuss another commonly used distribution.

Histogram

If we have a histogram of a group of data, we can superpose the normal or lognormal distribution curve on it and see how we like the result. A glance at the histogram might even cause us to reject the idea that a presumed distribution fits the data. The normal or lognormal curve is drawn by using the mean and standard deviation calculated from the data.

In the previous section, the histogram for 150 lifetime data was drawn in Figure 2.12. The shape did not suggest a normal distribution. A lognormal pdf was superposed on the (log) histogram, and the result seemed a reasonable fit.

In today's world, histograms are usually the product of computers, which was the case for Figures 2.12 and 2.13 and several others to follow. The programs are frequently part of statistics software packages, but it is not hard to write your own.

Example 2.8. We are grateful to the Aluminum Company of America for supplying us with strength data from production runs of one of their alloys. The following, in numerical sequence, are yield strengths in kilopounds per square inch (ksi).

64.8	65.6	65.8	65.9	66.2	66.2	66.5	66.9
67.4	67.4	67.5	67.6	67.9	67.9	67.9	68.0
68.1	68.1	68.1	68.1	68.2	68.2	68.2	68.3
68.3	68.5	68.5	68.5	68.5	68.5	68.5	68.5
68.6	68.7	68.7	68.8	68.8	68.9	69.0	69.0
69.1	69.1	69.1	69.2	69.2	69.3	69.3	69.3
69.3	69.3	69.4	69.4	69.5	69.5	69.5	69.5
69.6	69.7	69.7	69.8	69.8	69.8	69.9	69.9
69.9	69.9	69.9	70.0	70.1	70.1	70.2	70.3
70.4	70.5	70.6	70.7	70.8	70.9	70.9	71.1
71.4	71.5	71.7	71.9	71.9	72.0	72.3	72.5

The mean value is 69.11, and the standard deviation is 1.536 (2.2 percent of the mean). The histograms for 6 and 9 intervals are shown in Figure 2.15a and b. You will notice that 6 intervals for these 88 data produce a histogram which is barely adequate. (The low-side limit is 1 interval only, which would provide no information at all.) Nine intervals, which correspond to the rule of thumb we mentioned earlier of the square root of the number of data, seem to work out well. If we used more intervals, some would contain but few data. In the limit there would be so many intervals that each would contain a maximum of one point. There would be a number of intervals with no data. Again the shape of the distribution would not be discernible.

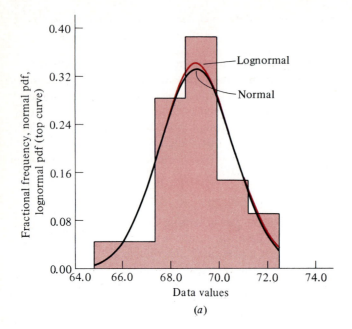

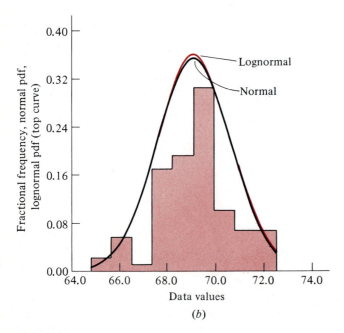

FIGURE 2.15
Histograms with superposed normal and lognormal pdf's. (a) Six intervals. (b) Nine intervals.

The normal and lognormal distribution curves are superposed on the histograms. We make three observations:

1. The data seem roughly normally or lognormally distributed.
2. We get the impression that more data would be desirable.
3. The normal and lognormal curves have surprisingly fallen nearly on top of each other. This point merits some discussion.

Normal vs. Lognormal Distribution

If the data of a sample are closely grouped about their mean (small standard deviation), the distribution curve, plotted as a function of the variable, will be high and narrow and its skewedness will be far less evident than if the data were widely scattered about the mean. Since the principal feature of a lognormal distribution is its skewedness, the data farthest from the mean, i.e., the tails of the distribution, are the main differentiator between normal and lognormal distributions. The size of the tails is determined by the extent to which the data group about the mean, which is in turn measured by the standard deviation.

Figure 2.16a, b, and c shows normal and lognormal pdf's for several values of (normal) standard deviation, expressed as a percentage of the mean value. The difference between the two pdf's becomes discernible at a standard deviation of about 30 percent of the mean. In the previous example, the standard deviation was only 2.2 percent of the mean, and the curves were nearly identical.

Prepared Probability Graph Paper

This paper is available commercially and is so constructed that gaussian (or other) distribution curves plot as straight lines. The data must first be ordered, and the values of the variable are plotted on the abscissa. The ordinate is the cumulative distribution of the variable, i.e., the percentage of the readings falling below any particular point. The ordinate scale is on this page and can be copied to construct normal or lognormal paper. For the former, the abscissa scale is linear; for the latter, it is logarithmic, which you can obtain from a sheet of log paper.

If the data fall in a straight line, the distribution is normal. The mean and standard deviation may be inferred from the plot, as in the following example.

Example 2.9. The first 25 of the yield strength data supplied by Alcoa for Example 2.8, before we put them in numerical order, were these:

69.9	67.9	69.9	69.5	69.5
69.3	69.3	69.9	69.7	68.1
68.2	67.6	69.4	70.2	68.8
69.9	70.5	71.1	71.7	69.8
70.6	69.3	69.3	70.7	70.9

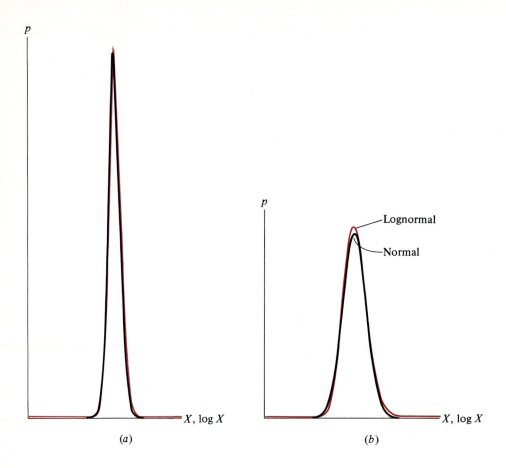

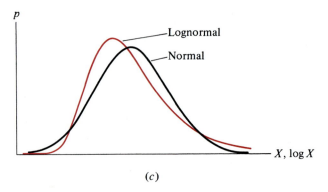

FIGURE 2.16
Normal and lognormal pdf's for several standard deviations. (a) SD = 14%. (b) SD = 27%. (c) SD = 64%.

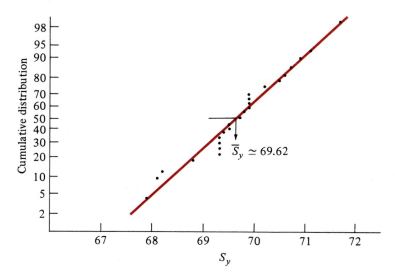

FIGURE 2.17
Probability plot for Example 2.9.

The plot points are prepared by putting the data in order and calculating the cumulative distribution $(j - 0.5)/n$ for each,[1] as shown in the following listing.

j	S_y	Cumulative distribution, %	j	S_y	Cumulative distribution, %
1	67.6	2	14	69.8	54
2	67.9	6	15	69.9	58
3	68.1	10	16	69.9	62
4	68.2	14	17	69.9	66
5	68.8	18	18	69.9	70
6	69.3	22	19	70.2	74
7	69.3	26	20	70.5	78
8	69.3	30	21	70.6	82
9	69.3	34	22	70.7	86
10	69.4	38	23	70.9	90
11	69.5	42	24	71.1	94
12	69.5	46	25	71.7	98
13	69.7	50			

Figure 2.17 is the probability plot, and the data do fall reasonably along a straight line. The mean value can be estimated at the 50 percent mark: $\overline{S_y} = 69.62$. The calculated value (sum/n) is 69.64.

---

[1] For example, the cumulative distribution for the third reading is $(3 - 0.5)/25 = 0.10$; that is, 10 percent of the readings fall at $S_y = 68.1$ or less. You will also find j/n and $j/(n + 1)$ used for the cumulative distribution function (cdf).

Since 68.3 percent of all values fall within 1 standard deviation of the mean, the standard deviation can be estimated as the difference between the 50th and the $50 + 68.3/2 = 84$th percentiles: $s = 70.69 - 69.62 = 1.07$. The calculated value is 1.005. The 7 percent difference owes to taking the difference of two close numbers.

(We know, from working with these data before, that because the standard deviation is small, the distinction between a normal distribution and a lognormal distribution is not discernible in a subjective test.)

Analytical Methods

There are several analytical methods, known as *goodness-of-fit tests*. We describe two: the *chi-square test* and the *Kolmogorov-Smirnov (K-S) test*. Neither analytical nor graphical tests give definitive answers to the fit question. Graphical tests are subjective; i.e., you may reach a conclusion that the data fall in a straight line, and someone else may say the opposite. In the case of analytical tests, numbers are compared, and one number is certainly larger than the other, leading to a conclusion. But there is a probability that the conclusion may be wrong. This works as follows.

An assertion, called a *hypothesis*, is made concerning a sample of data. The assertion might have to do with the distribution of a population of strengths. As an example, suppose a number of tests have been made on a certain steel and the yield strengths recorded. We think the distribution is normal. The hypothesis is that the sample of values of S_y is from a normally distributed population with the mean and standard deviation equal to those of the sample. Such assertions are termed *null hypotheses* and are generally given the symbol H_0.

In deciding whether a hypothesis is true, which will be based on the test chosen (the evidence), it is possible to make two types of errors, known as types I and II.

Type I error involves the probability of rejecting the hypothesis when it is actually true. To be conservative, you want the chance of rejecting a true hypothesis to be very small. Thus this probability is termed the *significance level* of the test, and it usually is given the symbol α. A value of $\alpha = 0.05$ means that 5 percent of the times you reject the hypothesis, it will have been a mistake to do so.

Type II error is the probability of accepting the hypothesis when, in fact, it is untrue. This probability is commonly called β, and $1 - \beta$ is called the *power* of the test, that is, $1 - \beta$ is the probability of correctly rejecting a false hypothesis. Note that $\alpha + \beta \neq 1$.

In this synopsis of goodness-of-fit tests, we limit our concern to type I errors.

CHI-SQUARE TEST. This test verifies a hypothesis that the sample data are from a stated distribution. The data are divided into k intervals. The observed frequency O_i (fraction of the total) of the measured variable in each interval is determined. Also, the expected frequency E_i in each interval, based on the assumed distribution, is calculated. For continuous distributions, E_i is the area under the distribution curve for the interval. The mean and standard deviation may be estimated from the data sample, or they may be known from previous testing. We then compute the quantity

$$\chi^2 = \sum_{i=1}^{k} \frac{(O_i - E_i)^2}{E_i} \tag{2.6}$$

TABLE 2.3
Chi square

F \ α	0.995	0.990	0.975	0.950	0.900	0.750	0.500	0.250	0.100	0.050	0.025	0.010	0.005
1	0.0^4393	0.0^3157	0.0^3982	0.0^2393	0.0158	0.102	0.455	1.32	2.71	3.84	5.02	6.63	7.88
2	0.0100	0.0201	0.0506	0.103	0.211	0.575	1.39	2.77	4.61	5.99	7.38	9.21	10.6
3	0.0717	0.115	0.216	0.352	0.584	1.21	2.37	4.11	6.25	7.81	9.35	11.3	12.8
4	0.207	0.297	0.484	0.711	1.06	1.92	3.36	5.39	7.78	9.49	11.1	13.3	14.9
5	0.412	0.554	0.831	1.15	1.61	2.67	4.35	6.63	9.24	11.1	12.8	15.1	16.7
6	0.676	0.872	1.24	1.64	2.20	3.45	5.35	7.84	10.6	12.6	14.4	16.8	18.5
7	0.989	1.24	1.69	2.17	2.83	4.25	6.35	9.04	12.0	14.1	16.0	18.5	20.3
8	1.34	1.65	2.18	2.73	3.49	5.07	7.34	10.2	13.4	15.5	17.5	20.1	22.0
9	1.73	2.09	2.70	3.33	4.17	5.90	8.34	11.4	14.7	16.9	19.0	21.7	23.6
10	2.16	2.56	3.25	3.94	4.87	6.74	9.34	12.5	16.0	18.3	20.5	23.2	25.2
11	2.60	3.05	3.82	4.57	5.58	7.58	10.3	13.7	17.3	19.7	21.9	24.7	26.8
12	3.07	3.57	4.40	5.23	6.30	8.44	11.3	14.8	18.5	21.0	23.3	26.2	28.3
13	3.57	4.11	5.01	5.89	7.04	9.30	12.3	16.0	19.8	22.4	24.7	27.7	29.8
14	4.07	4.66	5.63	6.57	7.79	10.2	13.3	17.1	21.1	23.7	26.1	29.1	31.3
15	4.60	5.23	6.26	7.26	8.55	11.0	14.3	18.2	22.3	25.0	27.5	30.6	32.8
16	5.14	5.81	6.91	7.96	9.31	11.9	15.3	19.4	23.5	26.3	28.8	32.0	34.3
17	5.70	6.41	7.56	8.67	10.1	12.8	16.3	20.5	24.8	27.6	30.2	33.4	35.7
18	6.26	7.01	8.23	9.39	10.9	13.7	17.3	21.6	26.0	28.9	31.5	34.8	37.2
19	6.84	7.63	8.91	10.1	11.7	14.6	18.3	22.7	27.2	30.1	32.9	36.2	38.6
20	7.43	8.26	9.59	10.9	12.4	15.5	19.3	23.8	28.4	31.4	34.2	37.6	40.0
21	8.03	8.90	10.3	11.6	13.2	16.3	20.3	24.9	29.6	32.7	35.5	38.9	41.4
22	8.64	9.54	11.0	12.3	14.0	17.2	21.3	26.0	30.8	33.9	36.8	40.3	42.8
23	9.26	10.2	11.7	13.1	14.8	18.1	22.3	27.1	32.0	35.2	38.1	41.6	44.2
24	9.89	10.9	12.4	13.8	15.7	19.0	23.3	28.2	33.2	36.4	39.4	43.0	45.6
25	10.5	11.5	13.1	14.6	16.5	19.9	24.3	29.3	34.4	37.7	40.6	44.3	46.9
26	11.2	12.2	13.8	15.4	17.3	20.8	25.3	30.4	35.6	38.9	41.9	45.6	48.3
27	11.8	12.9	14.6	16.2	18.1	21.7	26.3	31.5	36.7	40.1	43.2	47.0	49.6
28	12.5	13.6	15.3	16.9	18.9	22.7	27.3	32.6	37.9	41.3	44.5	48.3	51.0
29	13.1	14.3	16.0	17.7	19.8	23.6	28.3	33.7	39.1	42.6	45.7	49.6	52.3
30	13.8	15.0	16.8	18.5	20.6	24.5	29.3	34.8	40.3	43.8	47.0	50.9	53.7

[†] α is the probability of committing a type I error.

From: C. M. Thompson, *Biometrika*, vol. 32, 1941, as abridged by A. M. Mood and F. A. Graybill, *Introduction to the Theory of Statistics*, 2d ed., McGraw-Hill, New York, 1963.

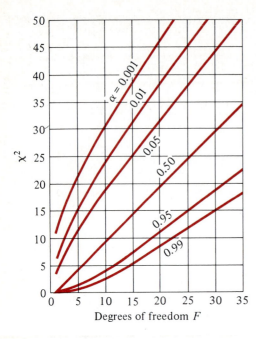

FIGURE 2.18
Chi-square function.

Such parameters are called *statistics*. This one is known as *chi square*. Notice that it measures the relative deviation of the observed values from the expected values.

The result of Equation (2.6) must be compared with a calculated value of χ^2 for the significance level α chosen. If Equation (2.6) yields less than the calculated value, the hypothesis is accepted.

The calculated values of χ^2 for a given value of α depend on the *degrees of freedom F* in the measurements:

$$F = k - p - 1$$

with k being the number of terms in Equation (2.6) (the number of intervals) and p the number of parameters estimated from the sample. In checking a fit of data to a normal or lognormal distribution, p is 2 if the mean and standard deviation are estimated from the sample; p is zero if they have been established from previous testing.

Table 2.3 (page 41) gives values of χ^2 for different values of α and of F. The function is also shown graphically in Figure 2.18.

Generally, the expected number of data falling in an interval (the frequency) should be a minimum of 5. If that is not achieved, then the interval is combined with an adjacent one. The intervals need not be of equal width.

The power of the chi-square test increases with the number of data. Ideally we ought to have a couple of hundred values. However, often so many are not available. Example 2.10 shows the procedure for the sample of 150 of Table 2.2.

Example 2.10. Earlier we concluded graphically (Figure 2.13) that the data of Table 2.2 were from a lognormal distribution. We test that assertion with the chi-square test.

TABLE 2.4
Table for χ^2 computation, first pass

Interval no.	Interval boundary	Observed	Expected	Chi squared
	2.698			
1		1.000 00	2.094 00	0.571 55
	2.751			
2		3.000 00	4.887 11	0.728 69
	2.804			
3		15.000 00	9.563 16	3.090 95
	2.858			
4		19.000 00	15.690 74	0.697 94
	2.911			
5		19.000 00	21.586 52	0.309 92
	2.964			
6		27.000 00	24.901 61	0.176 83
	3.017			
7		24.000 00	24.086 77	0.000 31
	3.070			
8		13.000 00	19.536 25	2.186 84
	3.123			
9		17.000 00	13.286 34	1.038 00
	3.176			
10		5.000 00	7.576 44	0.876 14
	3.230			
11		2.000 00	3.622 48	0.726 70
	3.283			
12		5.000 00	1.452 19	8.667 54
	3.336			

The logarithms of the data were first found in our computer routine, and the sample log mean and log standard deviation were found. These are used as estimates of the corresponding parameters for the underlying population. The hypothesis is then as follows:

H_0: The data of Table 2.2 are from a lognormally distributed population with log mean = 3.006942 and log standard deviation = 0.125696.

For 12 intervals, the interval boundaries are those shown in Table 2.4, along with the observed occurrences and expected occurrences (per a lognormal distribution) and the value of χ^2 computed by Equation (2.6). The last interval has an expected number of occurrences $E_1 = 1.45219$ and an observed number $O_1 = 5$. The result is an outsized χ^2 for that interval, and this is the reason that the intervals must be combined (i.e., the interval boundaries adjusted)—so that there are a minimum of 5 expected occurrences in any interval. The computer program does this automatically; the result appears in Table 2.5. The two intervals at the top and bottom of the table have expected values of the minimum of 5. Note that the chi-square value for the last interval is now reasonable.

Since the number of items in the computation is 10 and two parameters are being estimated (the population log mean and log standard deviation), the degrees of freedom are $F = 10 - 2 - 1 = 7$. From Table 2.3, for $\alpha = 0.005$, the calculated $\chi^2 = 20.3$. The

TABLE 2.5
Final interval arrangement and χ^2 computation for data of Table 2.2

Interval no.	Interval boundary	Observed	Expected	Chi squared
	2.698			
1		4.000 00	6.981 111	1.273 01
	2.804			
2		15.000 00	9.563 16	3.090 95
	2.858			
3		19.000 00	15.690 74	0.697 94
	2.911			
4		19.000 00	21.586 52	0.309 92
	2.964			
5		27.000 00	24.901 61	0.176 83
	3.017			
6		24.000 00	24.086 77	0.000 31
	3.070			
7		13.000 00	19.536 25	2.186 84
	3.123			
8		17.000 00	13.286 34	1.038 00
	3.176			
9		5.000	7.576 44	0.876 14
	3.230			
10		7.000 00	5.074 68	0.730 46
	3.336			

Chi squared = 10.38041

chi square for the sample is 10.38. Since this chi square is smaller, we accept the hypothesis of the lognormal distribution.

A chi-square test for goodness of fit to a normal distribution for these data results in $\chi^2 = 23.42$, and we would reject the hypothesis of a normal distribution.

Example 2.11. If you shoot craps, here is an analysis which may have relevance. A pair of dice are rolled 242 times, and the occurrences of the numbers are recorded, as shown in the first two columns of the table below. Are the dice honest? To put it more formally, should we accept a hypothesis that the outcomes are attributable solely to chance?

This problem involves discrete outcomes, whose probabilities are readily computed (no areas under the pdf curves need be found). Each die has 6 faces, so there are 36 possible outcomes (permutations). We can compute the possible ways of getting each number and hence the probability of each outcome. For example, 5 is obtained as shown below:

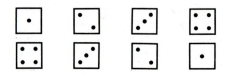

Out of 36 possibilities there are four for a 5, so the probability is 4/36 and the expected number of times for 5 to come up in 242 rolls becomes $4/36 \times 242 = 26.889$. We can now construct the following table to compute χ^2.

Number	Occurrences (observed) O_i	Probability of no.	Expected frequency E_i in 242 rolls	$\dfrac{(O_i - E_i)^2}{E_i}$
2	4	1/36	6.722	1.102
3	8	2/36	13.444	2.205
4	17	3/36	20.167	0.497
5	29	4/36	26.889	0.166
6	42	5/36	33.611	2.094
7	47	6/36	40.333	1.102
8	39	5/36	33.611	0.864
9	28	4/36	26.889	0.046
10	18	3/36	20.167	0.233
11	7	2/36	13.444	3.089
12	3	1/36	6.722	2.061
	$\sum = 242$			$\chi^2 = \sum = 13.459$

The number of terms in the χ^2 formula was 11. No parameters were estimated (they did not involve a mean or a standard deviation), so the degrees of freedom $F = 11 - 0 - 1 = 10$. From Table 2.3, at a significance level of $\alpha = 0.005$, the calculated $\chi^2 = 25.2$, and since the value we found is smaller, we accept the hypothesis that the rolls of the dice are attributable to chance, i.e., that the dice are not loaded.

KOLMOGOROV-SMIRNOV (K-S) TEST. This test is for goodness of fit to a specified distribution; i.e., the mean and standard deviation are known from previous history and will not be estimated from the sample.[1] It is a good test for fairly small samples of data, around 20 values or more, where the chi-square test could not be used. The K-S test is inapplicable if the specified distribution is discrete, as in the previous example.

The K-S test judges a fit on the basis of the difference between the values of the cumulative distribution function (cdf) of the sample values and the corresponding numbers from the cdf for the specified distribution. The maximum absolute difference is found and compared with critical values for a specified acceptance level to see if the hypothesis of a fit to the specified distribution should be accepted. The procedure is as follows:

1. The data are arranged in increasing order: $x_1, x_2, \ldots, x_n$.
2. The sample cdf is calculated as $cdf(x_i) = i/n$, where i commences with value 1 and proceeds to n, the number of sample values. The cdf thus progresses stepwise, with each step equal to $1/n$, from 0 to 1, as shown in Figure 2.19.

---

[1] An extension of the test does exist for the case in which the parameters must be estimated from the sample data. It is coded in some statistical software packages.

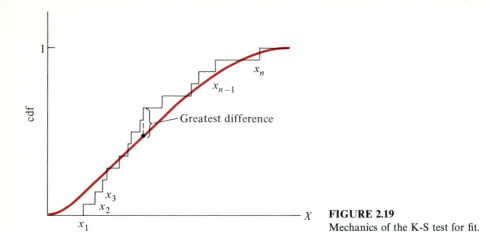

FIGURE 2.19
Mechanics of the K-S test for fit.

3. For each value of the variable of the sample, the specified cdf is calculated or found in a table. This is the curve in the figure.

4. The difference between items 2 and 3 for each sample value is found, and the maximum absolute value is recorded. This value is then compared with the critical value from Table 2.6. If it is less than the critical value, the hypothesis that the data are from the specified distribution is accepted.

Example 2.12. A part of an assembly has been manufactured in its present design for some time, and the extensive data which have accumulated indicate that the underlying population is normally distributed with the mean strength = 2516 newtons (N) and standard deviation = 172 N.

The quality-control department tests small batches of these parts from time to time, and they have recently obtained the following values of part strengths:

2190	2300	2327	2396	2288
2700	2703	2293	2355	2603
2398	2298	2407	2432	2685
2255	2324	2722	2490	2336
2776	2375			

Do these data fit a normal distribution with the historic mean and standard deviation? Table 2.7 shows the several steps outlined above. The maximum absolute difference between the sample cdf and the specified cdf (normal) occurs for data value 14: 0.373. From Table 2.6, we find $D_{n=22, \alpha=0.01} = 0.36$, and since the calculated value is greater than this figure, the hypothesis must be rejected. This would be a signal to the quality-control department that something is wrong on the line. The parameter $\alpha = 0.01$ means that about 1 percent of rejections like this one would be a mistake.

Although it is not necessary to the solution of the problem, a plot of the two cdf's is shown in Figure 2.20.

TABLE 2.6
Values of D for Kolmogorov-Smirnov test

No. of data	α level		
n	**0.10**	**0.05**	**0.01**
1	0.95	0.98	0.995
2	0.78	0.84	0.93
3	0.64	0.71	0.83
4	0.56	0.62	0.73
5	0.51	0.56	0.67
6	0.47	0.52	0.62
7	0.44	0.49	0.58
8	0.41	0.46	0.54
9	0.39	0.43	0.51
10	0.37	0.41	0.49
11	0.35	0.39	0.47
12	0.34	0.38	0.45
13	0.33	0.36	0.43
14	0.31	0.35	0.42
15	0.30	0.34	0.40
16	0.30	0.33	0.39
17	0.29	0.32	0.38
18	0.28	0.31	0.37
19	0.27	0.30	0.36
20	0.26	0.29	0.36
25	0.24	0.27	0.32
30	0.22	0.24	0.29
35	0.21	0.23	0.27
40	0.19	0.21	0.25
50	0.17	0.19	0.23
> 50	$\dfrac{1.22}{\sqrt{n}}$	$\dfrac{1.36}{\sqrt{n}}$	$\dfrac{1.63}{\sqrt{n}}$

Reprinted from F. J. Massey, Jr., "The Kolmogorov-Smirnov Test for Goodness of Fit," *Journal of the American Statistical Association*, vol. 46, no. 253, 1951, with permission of the American Statistical Association.

2.10   WEIBULL DISTRIBUTION

This distribution was formulated by W. Weibull[1] [34] in 1951. Its main use is to describe life phenomena, i.e., time to failure. In fatigue studies it is preferred by some over the lognormal distribution for that purpose. It is more versatile, in that the shape

---

[1] Waloddi Weibull (1887–1979) was the director of research at NKA Ball Bearing Co., Göteborg, Sweden (1916–1922), professor of mechanical engineering (1923–1941) and of applied physics (1941–1953) at the Swedish Royal Institute of Technology, and scientific adviser to Bofors.

TABLE 2.7
K-S procedure for Example 2.11

Datum no.	Value (step 1)	Sample cdf (step 2)	Normal cdf (step 3)	Absolute difference (step 4)
1	2190	0.045	0.029	0.016
2	2255	0.091	0.065	0.026
3	2288	0.136	0.092	0.044
4	2293	0.182	0.097	0.084
5	2298	0.227	0.102	0.125
6	2300	0.273	0.105	0.168
7	2324	0.318	0.132	0.186
8	2327	0.364	0.136	0.228
9	2336	0.409	0.148	0.261
10	2355	0.455	0.175	0.280
11	2375	0.500	0.206	0.294
12	2396	0.545	0.243	0.303
13	2398	0.591	0.246	0.345
14	2407	0.636	0.263	0.373
15	2432	0.682	0.313	0.369
16	2490	0.727	0.440	0.287
17	2603	0.773	0.694	0.079
18	2685	0.818	0.837	0.019
19	2700	0.864	0.858	0.006
20	2703	0.909	0.862	0.048
21	2722	0.955	0.884	0.070
22	2776	1.000	0.935	0.065

Maximum difference = 0.373

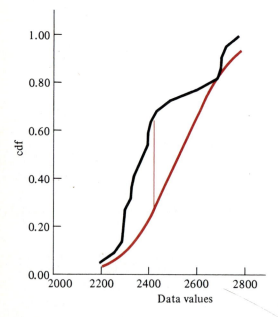

FIGURE 2.20
The cdf's for Example 2.12. Normal cdf is in color, and the vertical line shows the greatest absolute difference.

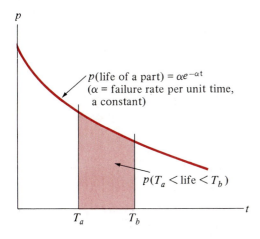

p(life of a part) = $\alpha e^{-\alpha t}$
(α = failure rate per unit time,
a constant)

$p(T_a < \text{life} < T_b)$

FIGURE 2.21
The exponential failure law applies when the failure rate is constant.

and position of the distribution may be varied by changing the parameters of the function.

Imagine a large fleet of automobiles, such as that of the postal service. If we ask how many windshields will suffer rock damage in any week, the answer is the same for any week of this year or next year; i.e., the chance of a windshield's being hit has nothing to do with how old it is or how many miles are on the car. The failure rate, sometimes called the *hazard function*, is constant. If the failure rate per unit time is α (so many windshields per week), then the probability of failure in time Δt is $\alpha \, \Delta t$. This leads to the pdf for the time to failure, i.e., the life of a windshield

$$\text{pdf}_{\text{windshield life}} = f(t) = \alpha e^{-\alpha t} \qquad t > 0$$

The form has caused this equation to be called the *exponential failure law*. It is shown in Figure 2.21. The probability of failure occurring between T_a and T_b (shown on the abscissa) is equal to the area under the curve between T_a and T_b. It follows that the probability of failure occurring before T_a is the area under the curve from 0 to T_a. Also the probability of survival to T_a is the complement of the area from 0 to T_a; or, what is the same, it is the area under the curve from T_a rightward (the probability of failure occurring *after* T_a).

Now suppose the failure rate is not constant in time. This occurs whenever anything wears, i.e., deterioration occurs. There are thousands of examples: humans, automobile tires, bearings, electric parts, pumps, etc. If the failure rate R is

$$R(t) = \alpha \beta t^{\beta - 1} \tag{2.7}$$

then the value of β determines how R varies in time. If $\beta = 1$, the rate is constant; $\beta < 1$ means a decreasing failure rate, $\beta > 1$ an increasing rate.

With the failure rate expressed in the form of Equation (2.7), the pdf for the time to failure is given by

$$\text{pdf}_{\text{time to fail}} = \alpha \beta t^{\beta - 1} e^{-\alpha t^{\beta}} \qquad t > 0 \tag{2.8}$$

This is the *Weibull probability distribution function*. The parameters α and β are positive constants. Figure 2.22 shows this pdf for $\alpha = 1$ and several values of β. The latter

pdf

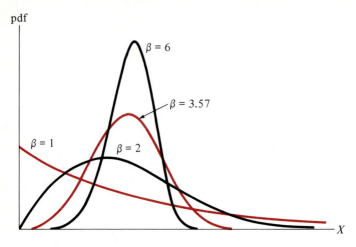

FIGURE 2.22
Weibull pdf's for $\alpha = 1$ and $\beta = 1, 2, 3.57, 6$.

determines the shape of the curve; $\beta = 1$ produces the exponential law, which follows from Equation (2.8) if that substitution is made. A close approximation to a normal distribution results when $\beta = 3.57$.

Relation (2.8) is the two-parameter Weibull pdf. There is a three-parameter form which accommodates a nonzero minimum value of the variable. In that form it is sometimes used for fatigue lives, rather than the lognormal distribution.

Since the Weibull distribution finds greatest use in time-to-failure applications, the cumulative distribution function (cdf) is very useful, since it represents the fractional number of failures up to any given time. The cdf is

$$\text{cdf}(t) = 1 - e^{-t^\beta / \alpha} \tag{2.9}$$

We will apply this relation to a sample to find the values of α and β. Before doing this, however, we must introduce another concept.

Median Ranks

Frequently in engineering we must make do with a limited number of data, because tests can be expensive. Suppose we had the results of 10 lifetime tests on a certain bearing type. If we arranged the lifetimes in increasing order, then we might say that 10 percent of bearings of this type would fail at a lifetime equal to or less than that of the first bearing to fail ($L_{10} \leq L_1$), 20 percent would fail at $L_{20} \leq L_2$, etc. These statements are unfortunately based on a single result, and they would be unlikely to apply in general to this bearing type, i.e., to the whole population. We need an estimate of what fraction of the population each bearing life of the sample represents.

Figure 2.23 might represent the lifetime pdf for the bearing type of our sample of 10. If we tested several batches of 10, the first bearing to fail would have a lifetime falling at different points, as shown in the figure. If a large number of batches of 10

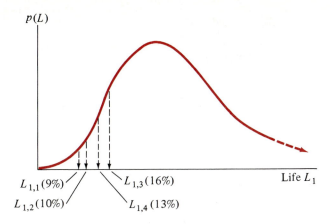

FIGURE 2.23
First bearing to fail in several batches of 10 would represent different probabilities. ($L_{1,1}$ = life of first bearing to fail in first batch of 10, etc.)

were tested, the pdf for L_1, a random variable, could be constructed. The median of those numbers is termed the *median rank* of the first bearing to fail. Similarly, median ranks for the other bearings in the sample could be found. The ranks depend on the sample size, naturally. Table 2.8 presents median ranks for sample sizes through 20.

The median rank MR may also be calculated approximately from

$$\text{MR} = \frac{j - 0.3}{n + 0.4}$$

where n is the sample size and j is the order number.

Other rankings are drawn from the distribution of each order of data: mean rank, 25 percent rank, 95 percent rank, etc. The median rank is most commonly used in engineering.

The application of median ranks is indicated when the sample size is small, around 25 to 30 or less.

Weibull Plots

Rewriting the cdf in Equation (2.9) and taking logarithms, we get

$$\frac{1}{1 - \text{cdf}(t)} = e^{t^\beta / \alpha}$$

$$\ln \frac{1}{1 - \text{cdf}(t)} = t^\beta / \alpha$$

$$\log \left[\ln \frac{1}{1 - \text{cdf}(t)} \right] = \beta \log t - \log \alpha \tag{2.10}$$

TABLE 2.8
Median ranks [74]

Order no. j	\multicolumn{10}{c}{Sample size n}									
	1	2	3	4	5	6	7	8	9	10
1	0.5000	0.2929	0.2063	0.1591	0.1294	0.1091	0.0943	0.0830	0.0741	0.0670
2		0.7071	0.5000	0.3864	0.3147	0.2655	0.2295	0.2021	0.1806	0.1632
3			0.7937	0.6136	0.5000	0.4218	0.3648	0.3213	0.2871	0.2594
4				0.8409	0.6853	0.5782	0.5000	0.4404	0.3935	0.3557
5					0.8706	0.7345	0.6352	0.5596	0.5000	0.4519
6						0.8909	0.7705	0.6787	0.6065	0.5481
7							0.9057	0.7979	0.7129	0.6443
8								0.9170	0.8194	0.7406
9									0.9259	0.8368
10										0.9330

Order no. j	\multicolumn{10}{c}{Sample size n}									
	11	12	13	14	15	16	17	18	19	20
1	0.0611	0.0561	0.0519	0.0483	0.0452	0.0424	0.0400	0.0378	0.0358	0.0341
2	0.1489	0.1368	0.1266	0.1178	0.1101	0.1034	0.0975	0.0922	0.0874	0.0831
3	0.2366	0.2175	0.2013	0.1873	0.1751	0.1644	0.1550	0.1465	0.1390	0.1322
4	0.3244	0.2982	0.2760	0.2568	0.2401	0.2254	0.2125	0.2009	0.1905	0.1812
5	0.4122	0.3789	0.3506	0.3263	0.3051	0.2865	0.2700	0.2553	0.2421	0.2302
6	0.5000	0.4596	0.4253	0.3958	0.3700	0.3475	0.3275	0.3097	0.2937	0.2793
7	0.5878	0.5404	0.5000	0.4653	0.4350	0.4085	0.3850	0.3641	0.3453	0.3283
8	0.6756	0.6211	0.5747	0.5347	0.5000	0.4695	0.4425	0.4184	0.3968	0.3774
9	0.7634	0.7018	0.6494	0.6042	0.5650	0.5305	0.5000	0.4728	0.4484	0.4264
10	0.8511	0.7825	0.7240	0.6737	0.6300	0.5915	0.5575	0.5272	0.5000	0.4755
11	0.9389	0.8632	0.7987	0.7432	0.6949	0.6525	0.6150	0.5816	0.5516	0.5245
12		0.9439	0.8734	0.8127	0.7599	0.7135	0.6725	0.6359	0.6032	0.5736
13			0.9481	0.8822	0.8249	0.7746	0.7300	0.6903	0.6547	0.6226
14				0.9517	0.8899	0.8356	0.7875	0.7447	0.7063	0.6717
15					0.9548	0.8966	0.8450	0.7991	0.7579	0.7207
16						0.9576	0.9025	0.8535	0.8095	0.7698
17							0.9600	0.9078	0.8610	0.8188
18								0.9622	0.9126	0.8678
19									0.9642	0.9169
20										0.9659

This is of the form

$$y = \beta x - \log \alpha$$

with

$$y = \log y' = \log \left[\ln \frac{1}{1 - \text{cdf}(t)} \right]$$

and

$$x = \log t$$

Thus a plot of y vs. x as defined above would be a straight line with intercept $-\log \alpha$ and slope β. The parameter β is known as the *Weibull slope*. The plot can be made

TABLE 2.9
Lifetime tests on a small sample lot of 10 roller bearings

		Ordered data			
Sample no.	Lifetime, h	Lifetime, 10^6 r†	Sample no.	Assigned median rank (cdf)	$y' = \ln \dfrac{1}{1 - \text{cdf}}$
1	846	6.8	10	0.067	0.069
2	164	8.4	6	0.163	0.178
3	461	26.9	5	0.259	0.300
4	146	47.3	4	0.356	0.440
5	83	53.1	2	0.452	0.601
6	26	59.9	7	0.548	0.794
7	185	71.9	8	0.644	1.033
8	222	118.3	9	0.741	1.351
9	365	149.4	3	0.837	1.814
10	21	274.1	1	0.933	2.703

† Revolutions of inner bearing ring (test speed = 5400 rpm).

on log-log paper, which is the reason for the mixture of logarithmic bases. Special Weibull paper is also obtainable, whose scale has the calculation required for y built in. The parameters α and β are determined from the plot, as shown in the following example.

Example 2.13. Table 2.9 shows the results of lifetime tests on a small sample lot of 10 roller bearings performed at the Split Ballbearing Co., to whom we are grateful for the data. The data were placed in order of life as shown, and the median ranks for a sample of 10 were taken from Table 2.8. The values of $y' = \ln [1/(1 - \text{cdf})]$ were calculated and then plotted vs. life on log-log paper (see Figure 2.24). The best straight line was drawn through the points. The resulting values of the parameters are $\alpha = 63.69$ and $\beta = 0.936$. The pdf then becomes

$$\text{pdf} = 59.61 t^{-0.064} e^{-63.69 t^{0.936}}$$

The failure rate from which this derives is Equation (2.7):

$$R = 59.61 t^{-0.064}$$

which decreases with time.

The time at which any given fraction of parts will have failed can be found by solving Equation (2.10) for t:

$$t = \left(\alpha \ln \frac{1}{1 - \text{fraction failed}} \right)^{1/\beta} \tag{2.11}$$

In the bearing business, the time at which 10 percent of bearings fail is used as an index of reliability. It is termed the L_{10} *life*. Thus

$$L_{10} = \left(\alpha \ln \frac{1}{0.9} \right)^{1/\beta} = (0.1054\alpha)^{1/\beta}$$

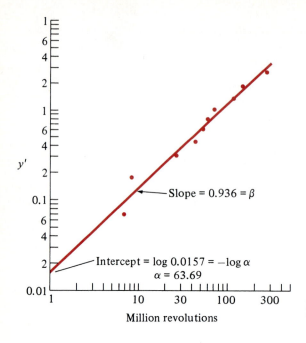

FIGURE 2.24
Weibull plot for Example 2.7.

The time for a given fraction failed can also be read off the Weibull plot at

$$y' = \ln \frac{1}{1 - \text{fraction failed}}$$

which, for 10 percent failed, is $y'_{L_{10}} = 0.1054$. The calculation of L_{10} for the example problem results in $L_{10} = 7.65E + 06$ revolutions (r), which is confirmed on the plot at $y' = 0.1054$.

If a data sample does not plot as a straight line as described above, either the three-parameter Weibull pdf is indicated or the data do not fit a Weibull distribution.

In lifetime testing, it is common to terminate tests in order to save time. The resulting samples are called *censored data* and require different handling from what we outlined in this synopsis. Reference 49 is quite readable on this topic.

PROBLEMS

Section 2.1

2.1. What is the probability of 5 *or more* successes in a series of 22 Bernoulli trials with a probability of success in any one of 0.4?
  Answer: 97.34%

2.2. Six people have shown up for our unisex dancing class. (That means it is all right for men to dance with men and women with women.) How many individual dances could we have without anyone having to dance with any given partner more than once?

2.3. Calculate the probability of 3 or fewer failures in a batch of 10 parts with probability ot failure = 0.3.

   Answer: 65.0%

2.4. A handful of 5 coins is tossed 100 times. How many times can we expect exactly 4 heads to turn up? How many times can we expect exactly 4 tails.?

2.5. Repeat Problem 2.4 for 4 or more heads and 4 or more tails.

   Answer: $p = 18.75\%$; 18 to 19 times

2.6. In 240 tosses of 2 dice, how many times can we expect to get a 7 or an 11?

2.7. In 145 tosses of 3 dice, how many times can we expect to get a 15?

   Answer: 6 or 7 (6.71 in many groups of 145)

Section 2.2

In the following problems, the histograms may be constructed so as to have interval boundaries to one figure beyond the data values or to have interval boundaries at whatever values result from dividing the span of the data by the number of intervals selected. Also, the problems may be done by hand or by computer.

2.8. The following are the lifetime results, in hours, of 90 fatigue tests on a part. Construct a histogram showing the distribution of these results. Provide scales for the number of data in an interval and for the fractional frequency of data in an interval.

335	278	433	596	226	211	214
163	219	449	217	268	207	456
283	293	215	446	334	414	271
352	352	137	411	434	290	297
269	389	219	231	274	190	188
222	310	314	251	224	233	225
200	235	240	182	307	294	283
201	633	201	277	304	303	548
179	432	266	143	365	292	227
152	423	200	334	289	650	344
244	305	282	166	243	275	161
171	270	285	487	388	232	170
175	178	447	129	439	181	

2.9. A company is in the business of melting steel scrap to make concrete reinforcing rod. The product must conform to the specifications of the American Society for Testing Materials (ASTM). Over the past year the company has accumulated the following values of S_{yt} for grade-60 rod (S_{yt} = 60 ksi minimum). Make a histogram of these data.

68.1	65.4	71.7	76.3	62.5	61.5	61.6	57.8	62.0	72.2
61.8	64.9	61.2	72.4	65.7	66.1	61.7	72.1	68.0	71.1
65.0	68.8	68.8	56.0	71.0	71.8	66.0	66.4	64.9	70.2
62.0	62.8	65.2	60.0	59.8	62.2	67.0	67.1	63.9	62.4
62.9	62.4	60.7	63.0	63.3	59.4	66.8	66.2	65.7	60.8
77.1	60.8	65.3	66.7	66.6	75.1	59.1	71.7	64.8	56.0
69.3	66.1	62.5	56.8	71.4	60.7	68.0	65.9	78.7	68.4
63.5	66.7	65.6	58.1	63.5	65.2	57.6	58.5	65.0	65.8
73.4	70.1	62.8	58.4	58.8	59.0	72.2	56.0	71.9	59.3
62.8	65.2	67.0	69.4	66.9	60.8	65.9	65.9	58.4	59.1

2.10. Construct a histogram for the data of Problem 2.11.

Section 2.3

2.11. The k values (spring rate) for a number of valve springs were recorded as follows. Construct the sample pdf for the data.

5008	5126	4995	5101	4800
5112	5005	4956	4965	5028
4895	5029	5018	4985	5176
5067	5035	5001	4977	5176
4958	5123	5076	5023	4993
5287	5089	5123	5223	5197
5127	5037	5098	5126	5167
5067	4995	4960	5080	5126
5127	5098	5098	4995	5017

2.12. Construct the sample pdf for the data of Problem 2.8.

2.13. Construct the sample pdf for the data of Problem 2.9.

Section 2.4

2.14. The bore size in millimeters on the crankshaft end of 10 randomly selected connecting rods is shown below. Find the sample mean, variance, median, and standard deviation.

2.15. Find the mean, median, variance, and standard deviation for the data of
(a) Problem 2.8
   Answer: $m = 273$, $\bar{x} = 289.689$, $s = 110.641$, $s^2 = 12\,240$
(b) Problem 2.9
(c) Problem 2.11
   Answer: $m = 5067$, $\bar{x} = 5059.22$, $s = 90.148$, $s^2 = 8127$

2.16. The table below lists the number of daily rejects from a manufacturing line. Find the mean, variance, standard deviation, and median of the data; construct a frequency histogram.

1	3	4	8	10	2	7	12
3	7	8	4	2	15	9	11
12	3	1	7	8	11	16	2
7	8	5	4	9	10	8	6
6	14	8	3	5	2	7	12
5	4	13	9	6	4	11	9
6	5	10	4	3	1	7	8
4	1	11	9	7	13	10	3

Sections 2.6 and 2.7

2.17. Calculate the total area under the normal pdf curve.

2.18. In Example 2.2, calculate the probability of having 2 to 9 successes. Check your answer against the results given in Example 2.4.

2.19. If the diameter of a crank produced in a line has been determined to be normally distributed, what percentage of the production will fall within 4 standard deviations of the mean?

   Answer: 99.94%

Section 2.8

2.20. For the data of the problems listed, construct a histogram for the logarithm of the data and find the log mean, log median, and log standard deviation. For what value x of the data is there a 50 percent chance of finding values greater than x?
(*a*) Problem 2.8   (*b*) Problem 2.9   (*c*) Problem 2.11

Section 2.9

2.21. Test the data of Problem 2.8 for a fit to a normal distribution.
   Answer: $\chi^2 = 16.268$. For $\alpha = 0.005$, we cannot accept the hypothesis.
(*a*) Use probability graph paper and estimate the mean and standard deviation from the plot. (Plotting so many points is rather tedious, so after ordering the data, use only every second one.)
(*b*) Use an appropriate analytical method with the mean and standard deviation estimated from the sample.

2.22. Test the data of Problem 2.9 for a fit to a lognormal distribution.
(*a*) Use probability graph paper and estimate the log mean and the log standard deviation from the plot. (After ordering the data, use only every other one.)
(*b*) Use an appropriate analytical method with the log mean and log standard deviation estimated from the sample.

2.23. Test the data of Problem 2.14 for a fit to a normal distribution with a mean of 80.000 millimeters (mm) and a standard deviation of 0.01 percent of the mean.
   Answer: Maximum absolute difference = 0.301. For an α level of 0.01, we accept the hypothesis.

2.24. A testing laboratory does 28-day compression tests on samples of concrete poured on a job. The results of a number of tests for a particular job are listed below. Do these data fit a normal distribution with a mean of 5000 psi and standard deviation equal to 5 percent of the mean?

5336	5314	5815	5014	5215	4886
4872	4651	5256	5139	4679	5008
4089	4840	5077	4589	4923	5333
4989	5015	4941	5182	3900	4676
4927	4740	4692	3903	4485	4782

Section 2.10

2.25. Nine assemblies are tested for lifetimes. They failed at 17.3, 5.5, 47.2, 230.0, 66.6, 101.8, 153.7, 149.3, and 409.1 hours (h). Find the Weibull slope and the value of α. What is the L_{25} life of these assemblies? Does the failure rate increase or decrease with time?
   Answers: $\beta = 0.817$, $\alpha = 55.25$, $L_{25} = 29.5$ h. Failure rate decreases with time.

CHAPTER
3

COMPUTER-AIDED DESIGN AND MANUFACTURE

The phrase *computer-aided* greatly understates the role of the computer in computer-aided design and manufacture (CADM). Computers of all kinds have become so ubiquitous that it would be very difficult to find any manufacturing operation that was not somewhere aided by computers. A new discipline is emerging—the union of design and manufacturing, aided by the computer. There is much to learn about this new discipline. This chapter is intended as an introduction.

It might be easier to understand CADM if you think about it as design and manufacturing without blueprints. This is a really radical departure from tradition, because the blueprint,[1] a copy of a handmade drawing, is, or was, central to our manufacturing technology and to the industrial revolution which created it.

Why were blueprints so important? Let us look at the way products are conceived and brought to the market. First there must be a concept. This may be in the form of a need definition, as discussed in Chapter 1. The need could be for something as frivolous as a Frisbee, which is not particularly needed but does look like something that people will buy. The concept is refined somewhat by a sketch, and then the sketch is redrawn to scale. This is where the blueprint came in.

Before blueprints, ideas went straight from the sketch to the machine shop. If the device being made included pieces which had to fit together and to do specific things after they were assembled, this was accomplished by literally cutting and trying. Raw material was measured, cut to shape, and trimmed and shaped to fit into

---

[1] The name *blueprint* came from the process once used to make full-size copies of drawings by placing sensitized paper in contact with the drawing, in sunlight. The resulting copy would be either blue lines on a white background or vice versa.

an assembly. The assembly was tested, changes decided upon, and the whole process repeated.

Drawings made to scale meant that many of the problems of fit and function could be solved in the design room. Blueprint copies meant that more than one person could work on the design of a complex machine and that accurate reproductions could be used—and soiled—in the machine shop, while the originals were kept safe and clean.

Thanks to quick and accurate copying, many people could be working on different phases of a development simultaneously. While one person kept the assembly drawing up to date, others made drawings of details, calculated stresses and heat flows, devised ways to make the product, and designed the jigs and fixtures needed. Calculations and scale drawings were checked for accuracy by the use of models subjected to stress and heat. Mock-ups checked that parts made according to the drawings really did fit together. Prototypes were then made to see if the resulting machine was the one intended by the designers. The use of blueprints thus made it possible to bring a complex product to market in a relatively short time by bringing many people together in a single enterprise. The process is challenging and exhilarating because of the many different activities going on at the same time.

The greatest challenge to the manager of such a project is the coordination of all these activities. Errors and delay can occur because, at each stage in this complex process, drawings and calculations must be put on paper in one location and then read or evaluated by others, who may be far away. Too often the manager's goal becomes that of minimizing errors and delay rather than making the best possible product at minimum cost. Of course, a well-managed enterprise will have the ability to detect errors. But their detection and correction cause delay and increase costs.

Modern computer-aided design places most of the activities described above inside a single computer. The computer usually communicates with several workstations. In this way, many designers have essentially instant access to the same information, or data base. Figure 3.1 is a graphical depiction of these activities.

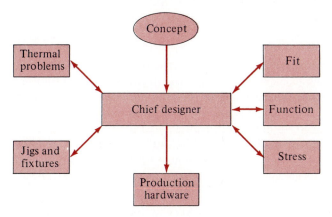

FIGURE 3.1
Flowchart of product development using computer-aided design and manufacture.

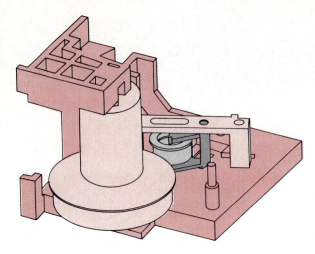

FIGURE 3.2
An assembly, with all parts to exact scale. (*Courtesy of Hewlett-Packard Co.*)

The original concept of a machine can be constructed to exact scale on the computer screen. The machine is constructed as an assembly. In this process, the individual pieces are also drawn to exact scale and position relative to one another, so that the fits and relative motions are determined as the design proceeds. Figure 3.2 shows a subassembly in an industrial plotter which was designed in this way. With all this information in one database, error is minimized.

The accuracy of a drawing done by computer is so different from that of a blueprint that it constitutes a genuine revolution. When drawings are made on paper, the precision of the image is limited by the eyesight of the drafter and the sharpness of the pencil. Thus drawings have always been less precise than the articles made from them. Precision is obtained by listing a dimension, which includes the permissible deviation, or tolerance, from ideal size.

Drawings made on the computer, however, are almost always *more* precise than the article can be made. Furthermore, two characteristics of CADM make it possible to use this precision:

1. Details can be expanded, without distortion, almost without limit.
2. The sizes, shapes, and other attributes of an object are kept in computer memory as numbers, or digits. Numerical instructions can be produced and transmitted directly to fabricating machines. Therefore, sizes, shapes and locations are always available to high precision. When stresses and/or heat flows must be calculated, powerful routines can be called up and applied directly to the part as shown on the screen. Fits and tolerances are easily checked because the designer can expand the picture on the screen to an almost unlimited degree. Instructions can be sent directly from the computer to machine tools to make models, prototypes, and finally production parts.

You might be able to appreciate the speed and flexibility of this process by reflecting on what happens when the part made by this paperless process is faulty and must be modified. No new drawings or fixtures or written instructions need to be made. The instructions given the computer are changed; the modified design is stored, and the new part is made immediately by instructions from computer to machines.

The foregoing is, to be sure, a rosy and somewhat breathless view of CADM. Computer-aided design and manufacture will never be a panacea. Computers instructed by people do make mistakes. There will always be questions of stress, heat transfer, and fit beyond practical computation. The important fact is that CADM is a new and fundamentally different way, a quantum jump in our ability, to bring quality products to market quickly and at minimum cost. It has become an essential part of every engineer's tool kit.

3.1 CADM ON THE MICROCOMPUTER

Computer-aided design was originally developed on the largest computers available; one of the first applications used a machine which was the sole occupant of a large building. It was capable of storing 1 000 000 bytes[1] of information. Its memory capacity and its calculation speed are now available on computers the size of a typewriter. These little machines became known as *microcomputers*, in reference to their capabilities. The name has become generic, but it now refers to physical size rather than capability. Indeed, the only limits to the miniaturization of the computer will be the size of the human finger and the resolving ability of the human eye.

Many programs are available for computer-aided design (CAD) on microcomputers. The first were two-dimensional in operation. That is, the designer was obliged to work as though she or he were drawing lines on a sheet of paper, using her or his knowledge of orthographic projection to construct auxiliary or pictorial views. The ability to work in three dimensions has been added to these programs in an evolutionary manner. The balance of this section will be illustrated with one of the most widely used of such programs.

A "two-dimensional" computer program will be most effectively used by exploiting its most powerful features:

1. The unlimited size of a "page".[2] This means that any number of views, to any scale, can be put on a single page.
2. The nearly unlimited number of pages. By assigning different parts of an assembly to different pages, they can be shown together or separately. They can be moved or their sizes adjusted for perfect fit.

---

[1] The memory of a computer is composed of millions of entities, each of which can exist in two states. One state is regarded as 0 and the other as 1. This unit of memory is called a *bit*. The memory required to store one letter or number, usually 8 bits, is called a *byte*. Computer memory is specified in kilobytes. One kilobyte is actually $2^{10} = 1024$ bytes.

The *word* is the size of the information unit which a particular computer can handle at one instant. Microcomputers use words of 4 to 32 bits; newer or larger computers work with words of 64 to 128 bits.

[2] A *page*, or *layer*, is analogous to a sheet of drawing paper, except that the computer page is perfectly transparent. Details on the bottom page are just as clear as those on the top.

FIGURE 3.3
A microcomputer work station.

At times it is helpful to show objects both separately and superimposed on one another. Examples are mechanical assemblies, the several floors of a building, or the many layers, each carrying a different circuit, of a printed electronic circuit board. This is done by assigning each detail to a different page. Each page may be shown in a different color. The computer will display as many layers as desired at the same time, and it can magnify the image by a factor of many thousands. Thus fits and tolerances can easily be determined and checked on the screen.

Figure 3.3 shows a typical microcomputer workstation. At the right is a digitizer[1] pad and "mouse." The mouse is equipped to sense movement relative to the digitizer pad by optical or electromagnetic means or through the motion of a rolling ball in its lower surface. As the mouse is moved relative to its pad, a cursor moves on the screen.

The original purpose of the digitizer was to enter the information on a conventional scale drawing into computer memory by tracing. This function can now be performed by machines which use copier technology to automatically convert an optical scan of the drawing to digital form.

A command can be given to the computer in several ways. It may simply be entered by name from the keyboard. Instead of the mouse, an electronic stylus can be used to choose an item from a tablet. The tablet is a very efficient way to use programs which are specialized for a particular technology, such as building design.

Probably the most widely used way for the designer to communicate with the computer is to use the digitizer merely to move the cursor on the screen. By pressing a button on the digitizer, a location or command is chosen.

---

[1] To *digitize* is to convert information to digital i.e., numerical, form.

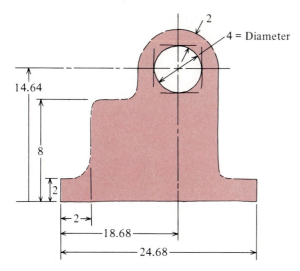

FIGURE 3.4
Sketch of a figure to be constructed on the computer screen.

Many of the features of a microcomputer-based CAD system can be demonstrated by constructing a simple figure, whose specifications are shown in Figure 3.4. Note that the size and location of some features of the part are specified by dimensions, while dashed lines indicate how the unspecified parts of the object may be finished. The dimensions do not have units such as inches or millimeters because, to the computer, they are simply numerical vectors. As vectors, they contain one number for each dimension. The user of the CAD program can assign any attribute desired to those numbers, as long as that attribute is appropriate for a vector.

The article we are going to design has a corner at its lower left. Everything about the analysis and construction of this article will be easier if its features can be located from that corner. The best way to arrange that is to place the lower left corner at the origin of our coordinate system, i.e., at the lower left corner of the screen. A problem now arises, because the article to be drawn is 25 units wide and 19 units high, while the *default*[1] size of the AUTOCAD page is only 12 units wide by 9 high. Also, the origin appears exactly at the edge of the screen—it is hard to see.

This is a good time to review the way in which any CAD program looks at a design. Think of the design as being drawn on an extremely large billboard. On a level surface in front of the billboard is a TV camera mounted on a remotely controlled vehicle so that the camera can be placed in front of any point on the billboard. The camera has a zoom lens of such power that it can either look at the entire billboard or zoom in to examine a flea. The image captured by the camera appears on the computer screen. The user specifies how big that "billboard" is to be with the command LIMITS and how much of it is seen on the screen with the command

---

[1] To speed up the operation of a computer-aided drawing program, nearly all variables are given their most usual values when the program is called up. These are known as the *default values*, and they are shown on the screen in the display between brackets.

ZOOM. In a practical sense, there is no limit to the size of the billboard—one of the sample drawings furnished with AUTOCAD shows the entire solar system. But greater size means more memory used. Therefore, the designer should specify drawing limits only large enough to encompass the design to be done.

In the present example, some space will be provided at the lower left corner by starting the limits at $(-2, -2)$ and similarly for the upper right corner with (30, 20). Then the command ZOOM is entered, and the response A (all) enlarges the field of view to include all the area within the limits.

When the program AUTOCAD is called into operation, a menu of commands appears on the screen. The command DRAW is selected or typed from the keyboard, and a submenu appears. While each program has its own structure of menus, the principles are the same; the user selects a general category and then progressively more specific commands. In this example, construction begins with the line which represents the bottom of the object, so LINE is chosen from the submenu.

As the designer uses a CAD program, he or she is provided with three pieces of information: (1) the commands given (in this case, the screen shows COMMAND: LINE), (2) the action expected (here the prompt "From Point:" appears), and (3) the available commands, shown in the menu. In other words, the command LINE has been accepted, and the designer is expected to designate the location of the starting point. The line will be started at the origin if 0, 0 is entered from the keyboard.

The end of the line is entered from the keyboard as 24.68, 0 and the top of the corner by 24.68, 2. This part of the construction is finished by entering the coordinates 22, 2 and finally pressing the ENTER key an extra time to turn off the LINE command. The construction so far is shown in Figure 3.5.

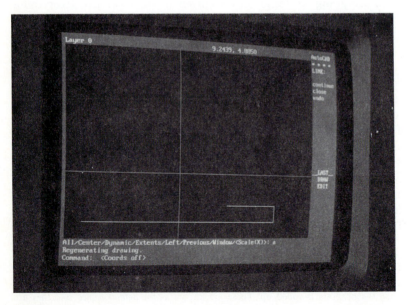

FIGURE 3.5
Part of a figure constructed by entering coordinates.

The specifications of Figure 3.4 call for a round hole 4 units in diameter with its center located 18.68 units to the right of the lower left corner and 14.64 units up. Since we have placed that corner at the origin, these are the coordinates of the center of the circle. From the menu **DRAW** and submenu **CIRCLE**, we choose CENter, DIA. The prompt "3P/2P/TTR/⟨Center Point⟩:" appears. It means that the location of the center point is expected, so 18.68, 14.64 is entered. The point appears, and the prompt changes to "D Diameter:" to which 4 is entered. To have the thickness of the material around the hole exactly 2 units, the command OFFSET is entered or chosen from the menu. We enter 2 for the offset distance, and we indicate the direction of offset by selecting a point outside the original circle. The result is a complete circle. To be able to trim that to the desired half-circle, horizontal lines are needed exactly through the center of the circle. The starting point of the line is placed exactly at the center of the circle by responding CEN to the "From:" prompt. A box appears at the cursor, which needs only to be placed somewhere on the circle. To make the line exactly horizontal, we switch to ORTHO mode by pressing function key F8. This causes all lines drawn to be either horizontal or vertical, so it is only necessary to use the mouse to select some point outside the circle and at approximately the same height, as shown in Figure 3.6.

Another line is constructed in the opposite direction. Now the starting point for a vertical construction line is established by entering INT and selecting one of the points where the outer circle crosses the horizontal lines. The other end of the line is placed near the base by using the mouse. Another line is drawn from the other intersection.

Finally, the unneeded half of the outer circle is removed with the command BREAK. The first response is F (I'm going to give you the first point) and then INT.

FIGURE 3.6
Construction of a horizontal line by using the ORTHO mode.

FIGURE 3.7
The "box" on the crosshair shows the area within which an intersection will be acquired.

FIGURE 3.8
Finished figure, except for corners, which will be replaced with fillets.

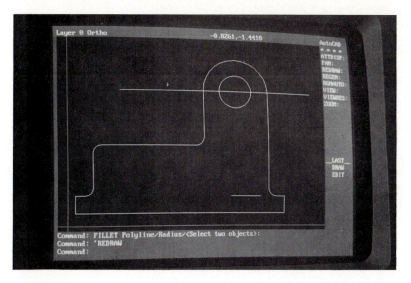

FIGURE 3.9
Finished figure.

The box on the crosshair is placed so that it covers the desired intersection and only that intersection. Then when the prompt calls for a second point, INT is again entered and the other intersection selected, as shown in Figure 3.7.

CAD programs regularly construct or erase circles in a counterclockwise direction. For that reason, the two points in Figure 3.7 were selected so that motion, and hence erasure, proceeded from the first to the last along the lower branch of the circle, the part to be erased.

Now we can place the vertical and horizontal surfaces on the left side of the figure, whose locations were specified by Figure 3.4. The lowest corner can be started with coordinates, and then the end of the horizontal line can be placed well to the right by using the mouse. The vertical line is started with coordinates 2, 1, and the upper end is located something more than 8 units up, while ORTHO keeps the line vertical. The upper horizontal line is placed in the same way. It is desirable that the lines overlap at the corners. Figure 3.8 shows the result.

The command FILLET is now used. To the first prompt we respond R, and to the prompt "Fillet Radius," 1. Then we invoke the command again, and we select two intersecting lines. The computer trims the lines and places an arc tangent to each line. The finished figure is shown in Figure 3.9.

3.2 FITTING PARTS TOGETHER

No machine will function properly unless its parts have the correct size and shape relative to each other. It is not possible to make an object of any exact size; if a number of objects are made with the same process, their sizes or dimensions will vary. In any given process, the amount of variation can be reduced, but the cost will increase accordingly. The designer thus must carefully plan the amount of variation that will be accepted in each part of an assembly, to strike the best possible balance between cost and function. This acceptable variation in dimension is known as the *tolerance*.

Two other definitions are needed. First, proper function requires that mating parts have some minimum space between them. This is known as the *clearance*. Second, the original design of a machine is done with dimensions that are ideal, as though clearance and tolerance were not necessary. These are known as the *nominal dimensions*.

> **Example 3.1.** A 0.5-in hole drilled by hand into a metal plate may have a precision of ± 0.002 in. That is, we can expect that the hole will have a diameter of not less than 0.498 or more than 0.502 in. If a pin is to fit into that hole with a minimum clearance of 0.001 in on the diameter, then the diameter of the pin cannot be greater than 0.497 in. If pins can be made, at reasonable cost, to an accuracy of ± 0.002 in, then the diameter of a pin may be specified as 0.495 ± 0.002 in. The maximum clearance C possible with these specifications is found with the largest hole and the smallest pin:
>
> $$C_{\max} = 0.502 - 0.493 = 0.009$$
>
> If a number of these parts are made and randomly assembled, then a few will have much more clearance than originally specified. This is known as *tolerance accumulation*. The condition is shown, greatly exaggerated, in Figure 3.10.

Tolerance accumulation, like the weather, cannot be avoided. A small increase in the complexity of parts that are to fit together can result in a large increase in tolerance accumulation, or in the number of ways in which the components might refuse to fit. Suppose, for example, that two pins are fastened to a baseplate. Another plate has two holes drilled in it big enough to fit over the pins. Suppose a large number of plates with holes are ordered from one supplier, and an equal number

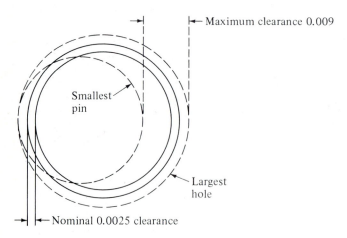

FIGURE 3.10
A simple example of tolerance buildup.

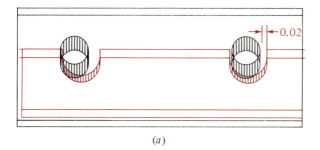

−0.02

(a)

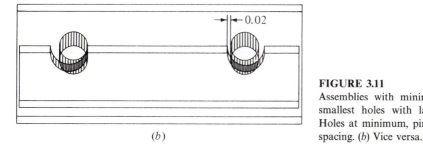

−0.02

(b)

FIGURE 3.11
Assemblies with minimum clearance; smallest holes with largest pins. (a) Holes at minimum, pins at maximum spacing. (b) Vice versa.

with pins from a different vendor, located far away. The two orders are received and dumped into a single pile. We need to know whether a set of holes chosen at random will fit over a set of pins which has also been chosen at random. There are two ways in which the two parts may fit together with minimum clearance. The outside of the pins may approach the outside of the holes (Figure 3.11a), or the inside of the pins may approach the inside of the holes (Figure 3.11b). This situation is examined in detail in the following example.

Example 3.2. Suppose that the specifications for the parts discussed above call for a nominal distance between centers of 6 in and a nominal diameter for the holes of 1 in. Suppose also that the tolerance for that center distance, as well as for the diameters of both holes and pins, is ± 0.02 in. The minimum clearance between hole and pin is also 0.02 in. We need to determine the nominal size and center distance for the pins.

Tolerance is usually understood to be applied to the diameter, and so the diameter of the smallest hole to pass inspection will be 0.98 in, and the largest, 1.02 in. For round parts, clearance also is specified on the diameter. In this example, a clearance of 0.02 in has been specified. Since the smallest acceptable hole diameter is 0.98 in, the largest pin that will fit must have a diameter d no larger than

$$d_{\text{max}} = 0.98 - 0.02 = 0.96 \text{ in}$$

The nominal diameter of the pin is thus

$$d_{\text{nom}} = 0.96 - 0.02 = 0.94 \text{ in}$$

and the smallest acceptable pin will have a diameter of

$$d_{\text{min}} = 0.94 - 0.02 = 0.92 \text{ in}$$

Combining the smallest pin with the largest hole will yield a maximum possible clearance of

$$C_{max} = 1.02 - 0.92 = 0.10 \text{ in}$$

Inevitably, at least a few assemblies will have 5 times the specified clearance of 0.02 in.
        Our overriding concern is whether the two parts will always fit together. Checking the condition of Figure 3.11b, we see that the combination of the largest pins and the greatest center distance produces the greatest dimension to the outside of the pins:

$$6.02 + 0.96 = 6.98 \text{ in}$$

This must fit into the combination of smallest hole and smallest center distance:

$$5.98 + 0.98 = 6.96 \text{ in}$$

Clearly, it is likely that some of the parts will not fit their randomly drawn mates.

We can adjust dimensions with some simple algebra to guarantee that all parts will fit together. The requirement here is that two combinations of parts fit together, which provides two equations. Therefore, two unknowns can be solved for. These can be two dimensions or two tolerances or one of each.

Example 3.3. The holes and pins of Example 3.2 are to be dimensioned and toleranced so that any combination will fit together. The nominal center distance and diameter of the holes are again set at 6 and 1 (all dimensions in inches), but to make the example easier to follow, the tolerances are changed to ± 0.04 on the center distance and ± 0.02 on the holes. For the pins, the tolerances are ± 0.01 on the diameter and ± 0.03 on the center distance. The minimum clearance is 0.02. We need to find the nominal center distance and diameter of the pins.
        First we examine the extreme condition for assembly shown in Figure 3.11a, which is that the pins which are largest and farthest apart (P_{maxmax}) must fit the closest together into the smallest holes (H_{minmin}). This can be expressed mathematically, with L and d representing the unknown nominal center distance and diameter of the pins:

$$6 - 0.04 + 1 - 0.02 - 0.02 = L + 0.03 + d + 0.01$$

$$L + d = 6.88$$

Figure 3.11b shows the smallest center distance and largest pin (P_{minmax}) fitting into the largest center distance and smallest hole (H_{maxmin}):

$$6 + 0.04 - (1 - 0.02) + 0.02 = L - 0.03 - (d + 0.01)$$

$$L - d = 5.12$$

The two equations are solved for the nominal dimensions:

$$L = 6.0 \qquad d = 0.88$$

These nominal dimensions will ensure that all combinations of pieces can be assembled to each other. The maximum possible clearance has increased to $1.02 - 0.87 = 0.15$.

When the problem of ensuring desired fits is more complex than our 2×2 example, and it nearly always is, we can see the power of CAD. This power results because individual parts can be put on separate layers, which appear as though they

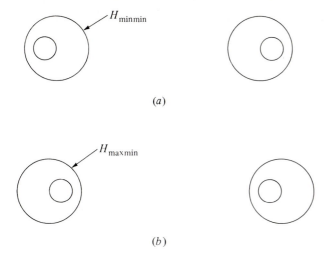

FIGURE 3.12
(a) One-half the minimum clearance and one-half the tolerance have been laid off to locate the outer extremities of the pins. (b) Same construction is used to locate the inner extremities of the pins.

were drawn on perfectly transparent sheets of plastic. Any combination of layers can be shown at one time and can be moved relative to one another. In other words, the designer does not have to calculate or guess at sizes; he or she can determine nominal sizes by assembling the parts on the screen.

Example 3.4. We will solve the problem of Example 3.3 on the computer screen. The method is suggested by a look at Figure 3.11, which shows that both extremes of fit involve the minimum hole diameter and the maximum pin diameter. That is, we can do all the construction with only one diameter for holes and one for pins.

We start the construction with holes of minimum size and center distance on the layer H_{minmin}. The extremities of the pins at condition P_{maxmax}, as shown in Figure 3.11a, can be located by measuring in an amount equal to one-half the clearance on each side. By moving in an additional amount equal to one-half the tolerance on the center distance of the pins, the outside of the pins at the nominal center distance is located. Each pin is represented by a small circle in Figure 3.12a, because the pin diameter is not known yet.

Next the condition of Figure 3.11b is satisfied. On the layer H_{maxmin}, holes of maximum center distance and minimum size are placed. The construction is similar to Figure 3.12a, except that the half-clearance and half-tolerance are laid off outward from the inner extremities of the holes.

The maximum-size pins can now be constructed by laying off construction lines tangent to the small circles of Figure 3.12, as shown in Figure 3.13. These are images of the pins at nominal spacing and maximum size, so their image is placed on layer P_{nommax}. The other layers are turned off, and the automatic dimensioning capability is used to measure the center distance and diameter.

With the dimensions of each member determined, the extreme combinations of center distance and size could be constructed, and each designated as a single entity.

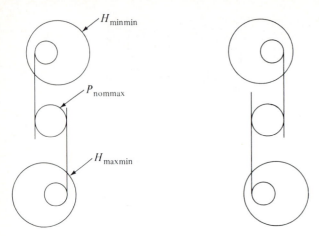

FIGURE 3.13
Construction lines are used to define the largest pins at nominal spacing.

By assigning each to a separate layer, they can be moved relative to one another and turned on and off at will to show all the possible ways in which the parts could fit together. Designating a figure on the screen as a single entity requires use of the command BLOCK, which is discussed in the next section.

3.3   KINEMATICS WITH CAD

This section is an example of the use of CAD to show the assembly and range of motion of a mechanism. The motion will be shown by constructing the mechanism in a different position on each of several different layers. The display can be animated by having the computer turn on and off in sequence.

A popular type of casement window is shown in Figure 3.14. It uses an arrangement of links to control the motion of the window sash so that, just before the window is fully closed, the entire sash is parallel to its seat. The last little bit of motion is thus perpendicular to the weather strip, compressing it rather than sliding on it. At the other extreme of motion, when the window has been moved out perpendicular to the frame, it has also moved to the left so that both sides of the glass can be reached for washing.

A schematic sketch of the mechanism is shown in Figure 3.15. The link AE (link 1) pivots around point A, which is fixed to the frame. The hinge C, which joins link 2 and link 3, slides on a track represented by the line from A through C. As long as link 2 is the same length as ED, and link 3 the same as BE, a parallelogram will be defined by $BCDE$. Thus, as AE rotates in toward the window frame, CD will remain parallel to AE, and the window will be moved perpendicular to its seat.

The objective of this exercise is to verify the design of this mechanism by showing it in every possible configuration. To do this, first each part of the window mechanism will be drawn as a separate unit called a *block*. This is necessary because, while we have no difficulty in seeing an object on the screen as an entity, the computer

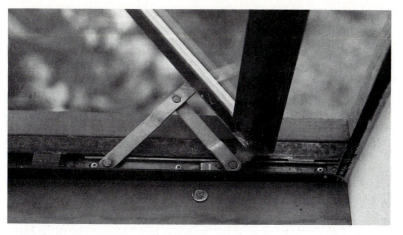

FIGURE 3.14
Photograph of a casement window mechanism.

knows the object only as a number of separate lines and arcs that happen to be in the same place. When a collection of lines, arcs, and points is designated as a *block*, it is carried in memory as a single entity. As a single entity, the block can be moved relative to other entities, or copies placed at any chosen location, to any scale.

With each link of the mechanism designated as a block, the links are assembled in a different orientation on each layer. The construction of all the blocks can be illustrated with the creation of link 1. The pivot at *A* is represented by a small circle drawn at the origin. Then pivots *B* and *E* are represented by circles at appropriate spacing along the horizontal, i.e., with the coordinate $y = 0$. Lines and the label "1" are placed in the usual way.

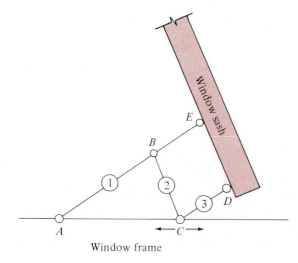

FIGURE 3.15
Schematic of a mechanism similar to that of Figure 3.14.

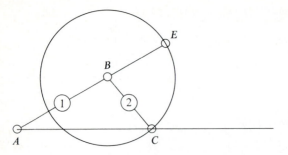

FIGURE 3.16
Second link of the casement mechanism being placed.

The block (link 1 here) can be inserted at any orientation and scale desired. The *insertion base point* is given as (0, 0). This means that when the block is inserted into an assembly, the center of pivot *A* will be placed at the designated point.

The other details—links 2 and 3, sash, and frame—are created in the same way. The insertion base points of these blocks will be *B* (for link 2), *C* (for link 3), and *E* (for sash).

The assembly technique will be illustrated with two positions. In the first, *AE* has rotated through a counterclockwise or positive angle of 30°. The construction starts with the insertion of link 1 with insertion point at *A* and an angle of +30°. Now the position of pivot *C* must be located. This is done with a traditional kinematics technique. First a line is laid off horizontally from *A* to represent the slider guide. Then a circle is placed with center at *B* and radius equal to the length of *BC*. The intersection of the circle and the line is the location of *C*. Link 2 is then placed with the insertion point at *B*; i.e., the computer is instructed to place the insertion point at the center of the circle at *B*. When the computer asks for an angle, it is told to find the intersection of the circle and the line. The result is shown in Figure 3.16.

Point *D* is similarly located at the intersection of arcs from *C* and *E*, and link 3 is inserted. Finally, the sash is placed with insertion point at *E*, and an angle is determined by selecting point *E*. The completed assembly is shown in Figure 3.17.

This mechanism will reach the limit of its motion when link 2 is perpendicular to line *AC*. Thus, the assembly for this position starts with a line parallel to *AC* at

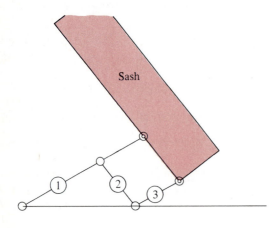

FIGURE 3.17
Complete linkage.

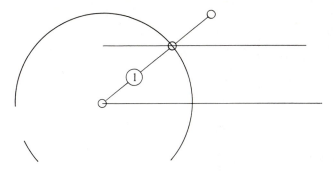

FIGURE 3.18
Construction to assemble the mechanism in extreme position.

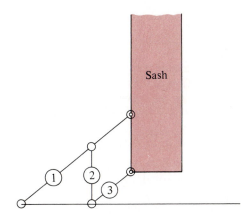

Sash

FIGURE 3.19
The mechanism in full-open position.

a distance equal to the length of link 2. This intersects an arc with center at the origin and radius equal to *AB*, as shown in Figure 3.18. Link 2 is then inserted with an angle of $-90°$, and the rest of the assembly is put together as before. Figure 3.19 shows the result.

To complete a kinematic analysis, additional assemblies are made at angles of $0°$ and $15°$. They may be examined by manually turning the layers on and off, or each layer may be captured as a "slide." Instructions placed in a separate text file through a program such as EDLIN or WORDSTAR control the order in which "slides" are turned on and how long each is left on the screen.

3.4 A THREE-DIMENSIONAL CAD UTILITY

The bulk of engineering design continues to be done in a two-dimensional format, if for no other reason than that most designers learned their trade on paper. There are other good reasons: Many of the objects that we design are, in essence, two-dimensional, many details are best seen in two dimensions, and limited computer power is most efficiently used in a two-dimensional format.

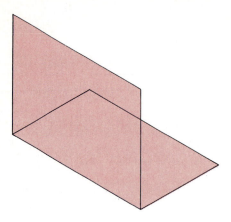

FIGURE 3.20
Isometric view of the bottom and front surfaces of the block.

Several CAD programs have been conceived as fundamentally three-dimensional in their operation. An example is CADKEY. The structure of CADKEY can be demonstrated by constructing a simple block in two ways. In the first method, the block is described by entering the coordinates of its corners. Normally, the coordinates are related to the computer screen, with x increasing to the right, y increasing toward the top, and z increasing toward the viewer. These are called *view coordinates*. For this exercise, "world" coordinates, fixed to the object, will be used. In reference to the top or plan view, the two coordinate systems are identical.

The top view of the block is a simple rectangle. We set the first corner of the block at coordinates (2, 2, 2). The rest of the top view is created by lines to the other corners which are at (7, 2, 2), (7, 5, 2), and (2, 5, 2). The construction so far appears to be the top view of a box, but it is in fact just the bottom of the box. Like the legendary Powder River, it has no depth.

For the next step, we change to the front view. The screen shows only a line because, as mentioned above, the construction has no depth so far. The front surface of the block is then created by constructing lines to the coordinates of the two top corners. To see what the object so far constructed really looks like, we switch to the isometric view shown in Figure 3.20. Two points have been placed, by using coordinates as before, to represent the two remaining corners of the box. The remaining lines are filled in point to point. The result is called a *wire frame model*, because all the lines constructed are visible. It can be made to look more realistic by manually changing those lines judged to be hidden from solid to dashed. The completed isometric view is shown in Figure 3.21.

The second method of constructing a figure is to move a shape through space. The operation is usually called *sweeping*, or *extruding*. Again, we start in the top view and construct the bottom of the box by commanding the computer to generate a rectangle at a working depth of 2 units. This is equivalent to requiring that the z coordinate of all parts of the rectangle be 2. The box is then generated by the command to transform or "sweep" the rectangle through the vector (0, 3, 0). The result is identical to that achieved by the first method.

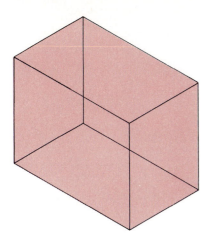

FIGURE 3.21
Isometric view of the complete block. Hidden lines have been replaced by dashed lines.

Finally, we give a short illustration of how new auxiliary views can be defined and objects extruded in any direction. A new surface is defined on the block of Figure 3.21 by constructing new lines from the rear top corner to the front top corner and from each to a point 1 unit down from the top right corner. The lines no longer needed are removed, to produce Figure 3.22.

Our next step is to define a new view which will show that plane in true size. Two points are chosen to define the x axis (view coordinates) of the new view and the direction of the y axis. The new view is displayed on command (Figure 3.23). New shapes constructed while this view is active will normally be parallel to the surface we used to define the view. New shapes can be put *on* that face by defining the working depth as that of any line on the face. New shapes can then be "grown" perpendicular to the face, as shown in Figure 3.24, which is an isometric view. In this way, complex objects can be generated with absolute control of shape and size.

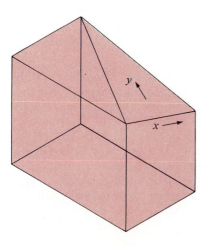

FIGURE 3.22
An oblique surface has been defined.

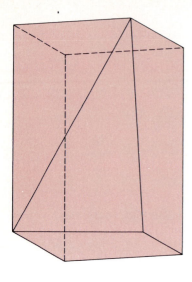

FIGURE 3.23
A new view showing the true size of the oblique surface.

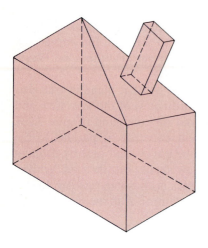

FIGURE 3.24
A rectangle drawn on the oblique surface has been swept outward to generate a prism.

PROBLEMS

Section 3.1

3.1. Computers work with binary arithmetic, since each entity within the machine must be in one of two states: 0 or 1. Each digit in binary arithmetic represents a power of 2; thus 10211 in binary means (reading from left to right) $1(2^4) + 0(2^3) + 2(2^2) + 1(2^1) + 1(2^0)$, or 27. Express the numbers 13, 21, and 35 in binary form. What is the largest number that can be expressed in 4 bits?

3.2. Construct the principal view shown on your computer. The figure is *not* shown to scale. Dimensions are in centimeters. (See Fig. P3.2 on page 79.)

3.3. If your CAD utility has a suitable analysis capability, construct the cross section of an 8×13.75 lb/ft channel, and evaluate its area and moments of inertia, using both the CAD utility and hand calculation.

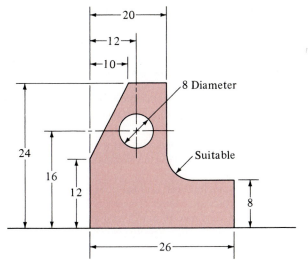

FIGURE P3.2
A two-dimensional figure.

3.4. Sometimes it makes more sense to fabricate a beam onsite rather than order a standard shape. Such a beam has been proposed as a $\frac{1}{2}$-in plate 10 in high, with $6 \times \frac{1}{2}$-in plates welded to the top and bottom. Evaluate its properties using both your CAD facility, and hand calculation.

3.5. In this problem, you will accurately construct a gear tooth. First place the pitch circle with a diameter of 6 in, with a vertical line through its center. From the intersection of the line with the top of the circle, place a line at the pressure angle of 20°, or 70° from the vertical. The base circle of the gear is tangent to this line.

The side of the tooth will be an involute curve, which is the path followed by the end of a piece of string unwrapped from a round object. You will simulate this by producing multiple images of a set of dots laid along a line tangent to the top of the base circle. Each image is rotated from its neighbor by the same small angle. The space between dots is equal to the arc measured on the base circle by the small angle chosen (Figure P3.5).

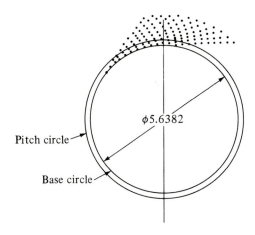

FIGURE P3.5
Construction for an involute gear tooth.

Lay off a line from the center of the pitch circle 5° right of the vertical. Measure the distance between the intersection of the vertical line, and this one, with the base circle. Place a very small circle at the intersection with the vertical line. Array 24 copies of that circle in one horizontal line at the distance just measured. Finally array the set of 24 around the center of the pitch circle, at a spacing of 5 degrees. The shape of the involute curve should be obvious.

Refer to your reference to finish the side of the tooth, and then mirror that to make the complete tooth. Evaluate its properties using both your CAD facility and hand calculation.

3.6. With the gear tooth of Problem 3.5 or one furnished, make it a block or detail and then reproduce it to make the complete 6-in pitch diameter, diametral pitch 2 gear. Show a suitable hub, center hole, and keyway.

3.7. Figure P3.7 is a layout of an overhead cam, follower, and valve for an automotive engine. The cam profile is approximated by circular arcs. Draw this assembly, using the following dimensions: camshaft diameter, 25 mm; cam rise, 14 mm. The diameters of the follower, valve shaft, and valve head are 30, 8, and 27 mm, respectively. The height from the cam centerline to the flat valve face is 100 mm. Choose other dimensions etc. for appearance.

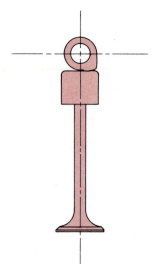

FIGURE P3.7
Layout of an overhead cam and valve.

Section 3.2

3.8. Design a device that would slide on the same flat surface as the object of Problem 3.2, carrying a pin to fit into the hole. The smallest clearance between the tightest combination of pin and hole must be 0.3 mm. The tolerance is ± 0.4 mm on the location and ± 0.5 mm on the diameter of both hole and pin. Show, by illustrating the extreme possibilities on the screen or plotter, that your design will work.

3.9. Consider the parts of Example 3.3, but with a center distance for both holes and pins of 8 ± 0.02. The nominal diameter of the holes is 1.07, and that of the pins is 1.00. The tolerance on both diameters is ± 0.01. Determine the minimum clearance in this way: Construct an extreme condition for the holes, make it a block or detail, and insert it on layer H_{minmin}. Repeat for the other extreme condition for holes and the two extreme conditions for pins.

Check the minimum clearance by turning on the appropriate layers and moving either holes or pins so that they touch. Then measure the minimum clearance for each extreme condition.

Section 3.3

3.10. A proposed "double wishbone" front-suspension system for a light truck is shown in Figure P3.10. The proposed range of motion of the top link is 30° up and 20° down. Evaluate the change in the camber, i.e., the inclination of the wheel from the vertical, from the lower to the upper extreme position.

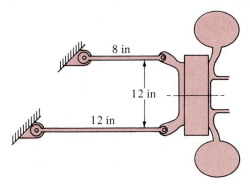

FIGURE P3.10
Sketch of an automotive suspension.

3.11. A window mechanism like that discussed in the text is shown in Figure P3.11. Note that the fixed pivot is located in a different place and the lengths of the links are not the same. If point A just matches the edge of the frame with the window closed, determine how much gap will exist with the window sash open at 45° to the frame.

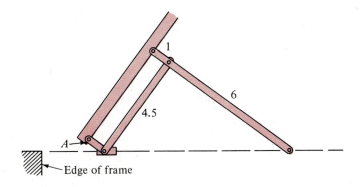

FIGURE P3.11
Schematic of a window mechanism.

3.12. Refer to the valve mechanism of Problem 3.7. It is for a four-stroke-cycle engine, which means that the cam will rotate once for every two rotations of the engine. The total valve lift is 14 mm. Construct the cam so that it opens the valve with uniform acceleration during 120° of *engine* rotation, holds it open for 20°, and allows the valve to close, also with constant acceleration, for 120°.

Section 3.4

3.13. Draw the object shown in Figure P3.13 in three views and isometric, using the capabilities of your CAD program.

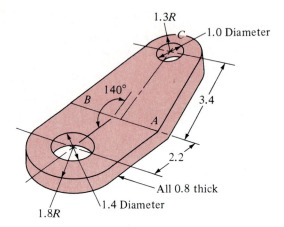

FIGURE P3.13
An adapter plate.

3.14. If your CAD program can do it, construct a view of the object sketched in Figure P3.13 which shows the face *ABC* as true size. On that face, construct a square prism 3 units high.

CHAPTER

4

REVIEW OF FORCE, MOMENT, AND STRESS

In this chapter the principles of static and dynamic equilibrium are reviewed. Mohr's circle of stress, an application of statics principles, is discussed in some detail because of its importance in failure theories. Stresses, deflections, and other calculations for the mechanical components treated in the text are reviewed in specific chapters.

4.1 FORCE AND MOMENT VECTORS

Forces

Figure 4.1 shows a person pushing at the back of a stuck car. The force has size and direction, which make it a vector. It also has a point of application. For a rigid body, the force may be applied anywhere along its line of action without affecting the external reactions, which in this case would be the forces of the ground on the wheels. Thus, those forces would be the same if, say, the person pulled equally in the same direction at point b at the front of the car, rather than pushed at point a.

Vectors may be resolved into components. In the illustration, the person's force is the equivalent of the components F_x and F_y. (The axes need not line up with the paper.) F_z, perpendicular to the paper, is zero in this case.

Moments

The *moment* of a force about a point is defined as $\mathbf{M} = \mathbf{r} \times \mathbf{F}$, as shown in Figure 4.2, where $\mathbf{r}$ is the vector from the point to any position on the line of action of $\mathbf{F}$. The magnitude of $\mathbf{M}$ is $rF \sin(\mathbf{r}, \mathbf{F})$, where $(\mathbf{r}, \mathbf{F})$ is the angle between $\mathbf{r}$ and $\mathbf{F}$. The angle

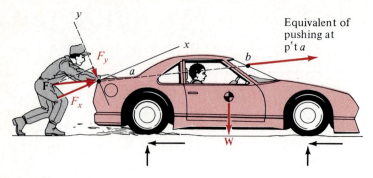

FIGURE 4.1
The effect of a force is the same if it is applied anywhere along its line of action.

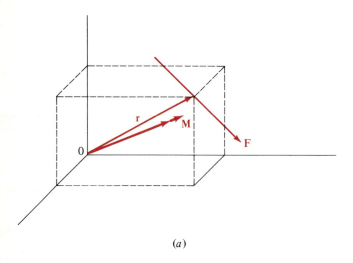

(*a*)

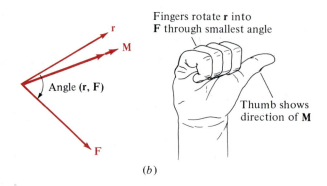

(*b*)

FIGURE 4.2
Finding the moment vector. (*a*) Vector representation of a moment about a point 0. (*b*) The magnitude of the moment is $rF \sin (\mathbf{r}, \mathbf{F})$. Also the right-hand rule is shown.

is most easily seen by moving the vectors along their lines of action so as to make their tails coincident, as shown in Figure 4.2b. If the angle is 90°, the sine is 1, so the easy way to find the size of the moment is to make **r** perpendicular to **F**. The moment is then the force times the distance perpendicular to it from the point.

The direction of the moment vector **M** is determined by the right-hand rule, as shown in Figure 4.2b. The fingers indicate the rotation which brings **r** into coincidence with **F** through the smallest angle; they also show the direction of the moment. The thumb indicates the direction of the **M** vector, which is perpendicular to the plane determined by lines **r** and **F**. A line in the direction of the moment vector through the point about which the moment is taken is called the *axis of the moment*; i.e., the force tends to produce motion about that axis.

In determinant form, the moment vector is

$$\mathbf{M} = \mathbf{r} \times \mathbf{F} = \begin{vmatrix} \mathbf{i} & \mathbf{j} & \mathbf{k} \\ r_x & r_y & r_z \\ F_x & F_y & F_z \end{vmatrix}$$

In much engineering work, forces and dimensions can be shown on a single plane, and then the components r_z and F_z are zero, so that only the $\mathbf{k}(z)$ component remains:

$$\mathbf{M} = (r_x F_y - r_y F_x)\mathbf{k}$$

Since $+\mathbf{k}$ is usually toward us, out of the paper (Figure 4.3a), the moment is positive counterclockwise about the z axis and is equal, as we see above, to the counterclockwise moments of the components of F about that axis. If the result is negative, the moment is in the other direction, i.e., clockwise. This computation of moment is the same as the product of the size of the force and of the perpendicular distance from the z axis to the line of action of the force. Since this is a simple calculation, the formality of a vector computation is usually unnecessary in two-dimensional problems.

When the moment vector is perpendicular to the paper, the tip of the arrow is shown within a circle if the vector is aimed outward, as seen in Figure 4.3b. The tail is shown as a cross within a circle when the vector is aimed inward. In the side or perspective view, we use a double-headed arrow in this book to avoid confusion with force vectors.

Curved arrows are often used to represent moment vectors, illustrated in Figure 4.3b. The addition of such vectors poses no particular problem if they all lie in the same plane. When the moments are in different planes (vectors of different directions, such as 10**i** and 13.5**k**), the moments must be represented by their three components $a\mathbf{i} + b\mathbf{j} + c\mathbf{k}$ and are then added just as any other vectors are.

Another term for moment is *torque*. Usually *moment* is used when the result is bending, as with a beam, and *torque* is used in the sense of twisting, as of a shaft.

Couples

A *couple* consists of two equal and opposite parallel forces. The sum of the moments of the forces about any point in the plane of the forces is the same, as illustrated in

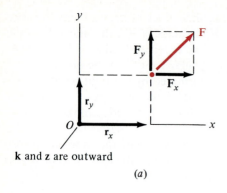

k and z are outward

(a)

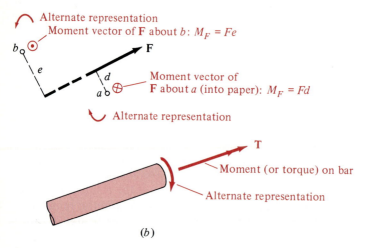

(b)

FIGURE 4.3
Moments in two-dimensional problems. (a) Moment of $\mathbf{F}$ about 0 is $(r_x F_y - r_y F_x)\mathbf{k}$. (b) Moment representations.

Figure 4.4. The point of application of a couple is thus immaterial; the vector representing it is called a *free vector*. Since the forces of a couple are equal and opposite, the application of a couple results in no net force; only a moment is applied.

A common use of a couple is the cruciform lug wrench for vehicle wheels, shown in Figure 4.5a. The mechanic pulls up on one end of the handle while pushing down with more or less the same force on the other end. The mechanic exerts a moment equal to the size of one of the forces times the distance between his or her hands. Imagine now a bar welded to the lug nut and with another nut welded to it, as shown in Figure 4.5b. If the mechanic applies the couple to the second nut, the effect is the same as if she or he applied the wrench directly to the lug nut itself.

Force-Moment Equivalents

Sometimes it simplifies problems to replace a force by an equal and parallel force acting through another point, plus a moment, as shown in Figure 4.6. Note that the

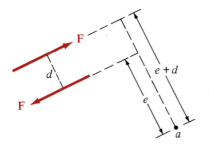

$\mathbf{M}_a = \mathbf{F}e - \mathbf{F}(e+d) = -\mathbf{F}d$ (independent of distance e)

FIGURE 4.4
The moment of a couple is independent of any point. It is equal to the force times the distance separating the lines of action.

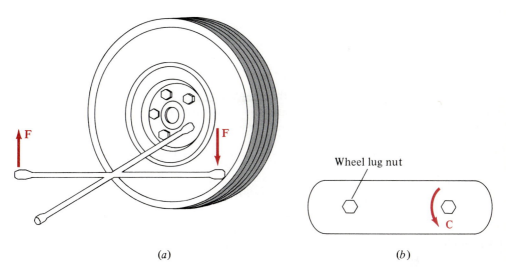

(a)   (b)

FIGURE 4.5
Example of couple application. (a) Cruciform lug wrench. If **F**'s are equal, a couple equal to **F**d is applied. (b) If an extension were used, the location of the applied couple would be immaterial.

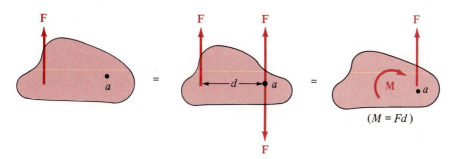

$(M = Fd)$

FIGURE 4.6
Replacement of a force by a force at another place plus a moment.

moment is in the same direction and is equal to the moment of the original force about the new point a.

Occasionally the reverse is useful—a moment plus force may be replaced by a single, equal, parallel force acting on a line which is removed from the original force line of action by $d = M/F$. The mechanics are the same as in Figure 4.6, read from right to left. The formality of the process is unnecessary if one notes that the single force has the same moment about the original point a as the moment being removed.

4.2 STATIC EQUILIBRIUM

General Three-Dimensional Case

Newton's laws of motion for a body, or assemblage of bodies, of constant mass are

$$\sum \mathbf{F} = m\mathbf{a}_{CM} \tag{4.1a}$$

and

$$\sum \mathbf{T}_{CM} = I_{CM}\alpha \tag{4.1b}$$

The term $\mathbf{a}_{CM}$ is the acceleration of the center of mass (CM), and α is the angular acceleration. If the accelerations are zero (the system is at rest, i.e., static, or is moving at constant linear and angular velocity), these expressions become the laws of static equilibrium:

$$\sum \mathbf{F} = 0 \qquad \sum \mathbf{M} = 0 \tag{4.2}$$

In the usual xyz (cartesian) coordinates, relations (4.2) become

$$\sum F_x = \sum F_y = \sum F_z = 0 \qquad \sum M_x = \sum M_y = \sum M_z = 0 \tag{4.2a}$$

The moments may be taken about any point at all.

Example 4.1. Figure 4.7 is a sketch of a gear on a shaft which drives a load at constant speed. Bearing A is a radial bearing, while the bearing at B can take an axial (thrust) as

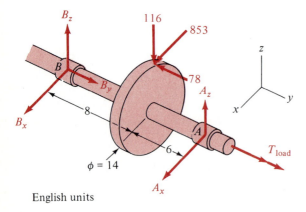

English units

FIGURE 4.7
Gear-shaft unit for Example 4.1.

well as a radial load. The bearing loads at A and B are needed, so the bearings can be selected.

A convenient xyz coordinate system is chosen, as shown in the figure. The forces acting on the system are

$$\mathbf{F}_{gear} = 853\mathbf{i} - 78\mathbf{j} - 116\mathbf{k}$$

$$\mathbf{F}_A = A_x\mathbf{i} + 0\mathbf{j} + A_z\mathbf{k}$$

$$\mathbf{F}_B = B_x\mathbf{i} + B_y\mathbf{j} + B_z\mathbf{k}$$

$\mathbf{F}_A$ and $\mathbf{F}_B$ are the forces exerted *on* the shaft *by* the bearings. We seek the opposite forces, which are, of course, of the same magnitude.

Summing the force components in the three directions, we get

$$A_x + B_x + 853 = 0 \qquad (1)$$

$$0 + B_y - 78 = 0 \qquad (2)$$

$$A_z + B_z - 116 = 0 \qquad (3)$$

Using B as a convenient point for the moment summation, we see that the vector from B to the point of application of the forces on the gear is $\mathbf{r} = 0\mathbf{i} + 8\mathbf{j} + 7\mathbf{k}$. The moment then is

$$\mathbf{M}_B = 0 = \begin{vmatrix} \mathbf{i} & \mathbf{j} & \mathbf{k} \\ 0 & 8 & 7 \\ 853 & -78 & -116 \end{vmatrix} + \begin{vmatrix} \mathbf{i} & \mathbf{j} & \mathbf{k} \\ 0 & 14 & 0 \\ A_x & 0 & A_z \end{vmatrix} + T_{load}\mathbf{j}$$

which gives

$$\mathbf{i}: \qquad -928 + 546 + 14A_z = 0 \qquad (4)$$

$$\mathbf{j}: \qquad 5971 + T_{load} = 0 \qquad (5)$$

$$\mathbf{k}: \qquad -6824 - 14A_x = 0 \qquad (6)$$

Thus we have a total of six equations in six unknowns: A_x, A_z, B_x, B_y, B_z, and T_{load}. The solutions are

$$A_x = -487.4 \qquad \text{from (6)}$$

$$A_z = 27.3 \qquad \text{from (4)}$$

$$B_x = -365.6 \qquad \text{from (1)}$$

$$B_y = 78.0 \qquad \text{from (2)}$$

$$B_z = 88.7 \qquad \text{from (3)}$$

$$T_{load} = -5971 \qquad \text{from (5)} \quad \text{(direction opposite to that shown)}$$

The loads on the bearings are then

$$A_{rad} = \sqrt{487.4^2 + 27.3^2} = 488.2 \text{ lb}$$

$$B_{rad} = \sqrt{365.6^2 + 88.7^2} = 376.2 \text{ lb}$$

$$B_{thrust} = 78.0 \text{ lb}$$

Two-Dimensional Equilibrium

In the common problem where all the significant forces are in the same plane, F_z is zero as are M_x and M_y, and the equilibrium conditions reduce to

$$\sum F_x = \sum F_y = \sum M_z = 0 \tag{4.3a}$$

Frequently the subscript z is omitted from the moment requirement.

The summations $\sum F_x = 0$ and $\sum F_y = 0$ are often satisfied most easily with a force-vector polygon, whose closing requires that the x and y components total zero. If greater precision is required than what can be obtained graphically, it is still a good idea to draw the polygon to show the components of forces which are to be summed and to give an idea of the ultimate result.

Two alternative equivalents of Equations (4.3a) are sometimes useful:

$$\sum F_x = 0 \qquad \sum M_a = \sum M_b = 0 \tag{4.3b}$$

Points a and b may not lie on a line parallel to the y axis.

$$\sum M_a = \sum M_b = \sum M_c = 0 \tag{4.3c}$$

Points a, b, and c may not lie on a single line.

Useful Corollaries

Two corollaries of the equilibrium laws often save time:

1. For a body subjected to forces at two points only (no moments), the resultants of the forces must be equal and opposite in direction ($\sum F = 0$), and they must also be collinear ($\sum M = 0$). This is shown in Figure 4.8a. (Significant body forces must be included: gravity and dynamic. The latter are discussed in the following section.)
2. For a body subjected to three forces (no moments), the forces must be coplanar ($\sum F = 0$) and their lines of action must intersect at a common point ($\sum M = 0$). This is illustrated in Figure 4.8b. The common point is at infinity if the three forces happen to have the same direction. (The coplanar requirement in this case results from $\sum M = 0$.) An example is shown in Figure 4.8c.

Redundancy

With the six equations of the three-dimensional laws (4.2a), six unknowns may be found. The three equations of the two-dimensional set (4.3a) allow the solution of three unknowns. A simple example of a statics problem with more unknowns than equations is a beam resting on three supports, such as that shown in Figure 4.9; the forces are coplanar. This beam has more supports than it needs—any two would hold it up. In other words, it is supported by more than what is required by the laws of static equilibrium. That is called *redundancy*.

A horizontal summation of forces produces $0 = 0$; there is force balance in this direction. The force summation vertically shows

$$R_1 + R_2 + R_3 = P$$

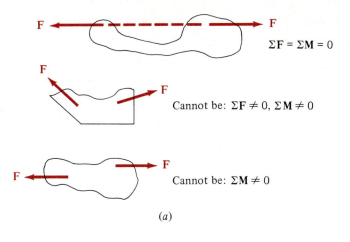

$$\Sigma F = \Sigma M = 0$$

Cannot be: $\Sigma F \neq 0, \Sigma M \neq 0$

Cannot be: $\Sigma M \neq 0$

(a)

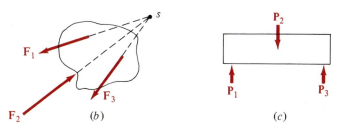

(b)

(c)

FIGURE 4.8
Applications of static laws. (a) If a body is subjected to forces (only) at two points, their resultants must be equal and collinear. (b) If a body is subjected to three forces only, they must be coplanar and their lines of action must intersect at a common point. (c) If the three forces are parallel, the intersection is at infinity.

and a moment summation about, say, the left end gives

$$0.5LR_2 + LR_3 = 0.8LP$$

There are two equations in the three unknowns R_1, R_2, and R_3. Any two can be found in terms of the third, but the three values cannot be obtained in terms of the load only. To use other axes for the force summation is only to sum components of

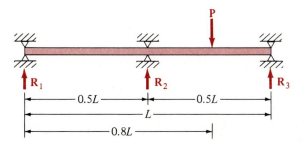

FIGURE 4.9
Beam with redundant support.

the forces, a rewrite of the earlier result. To sum moments about some other point is simply to rewrite the moment equation. An additional relation is necessary. It is supplied in this case by the requirement that the three support points fall on a straight line, i.e., there is no deflection of the beam in the center. We deal with the computation of deflections in Chapter 14.

Component Sectioning

In machine-component analysis, the element is often shown cut to facilitate the computation of stresses. Two examples are displayed in Figure 4.10. One part of the cut element is to be examined; the other part is set aside. But the effect of the latter on the part being analyzed is accounted for on the free-body diagram by showing

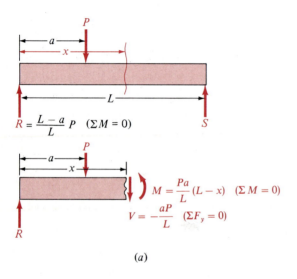

$$R = \frac{L-a}{L}P \quad (\Sigma M = 0)$$

$$M = \frac{Pa}{L}(L-x) \quad (\Sigma M = 0)$$

$$V = -\frac{aP}{L} \quad (\Sigma F_y = 0)$$

(a)

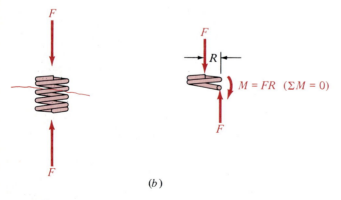

$$M = FR \quad (\Sigma M = 0)$$

(b)

FIGURE 4.10
The laws of statics apply to parts of machine elements. (a) Section cut from simply supported beam. (b) Section cut from compression spring.

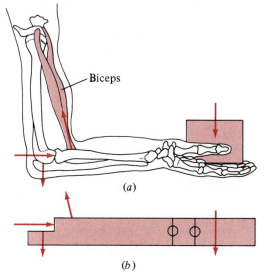

(a)

(b)

FIGURE 4.11
An assemblage of bodies may be isolated for static analysis. (a) Arm holding up a weight. (b) Free-body diagram of forearm and hand.

the forces and/or moments which the part exerts internally, as you see in the figure. The laws of static equilibrium apply now to the cut part and make possible the computation of the internal forces and/or moments which exist, as illustrated in the figure. (In considering the whole body, the internal forces and moments do not influence the static equilibrium, since they occur in pairs of equals and opposites, i.e., they total zero.)

Application to Assemblages

As intimated earlier, the laws of static equilibrium apply to assemblages of bodies as well as to single bodies. Figure 4.11 shows an example, the human arm mechanism. A free-body diagram of the forearm with hand, a sizable number of articulating parts, is shown in Figure 4.11b. At equilibrium, all the forces and moments on this assemblage must total zero. The forces (and moments) acting between the parts and the various muscle forces occur in pairs of equal opposites and thus do not influence the equilibrium.

4.3 DYNAMIC FORCE EQUILIBRIUM

A body under the influence of unbalanced forces and/or moments (i.e., the laws of static equilibrium are not satisfied) will accelerate according to Newton's laws of motion, expressions (4.1), which we repeat here:

$$\sum \mathbf{F} = m\mathbf{a}_{CM} \tag{4.1a}$$

$$\sum \mathbf{T}_{CM} = I_{CM}\boldsymbol{\alpha} \tag{4.1b}$$

In (4.1a), $\sum \mathbf{F}$ is the resultant of all forces acting on the body. The law is valid no matter where the force is applied; i.e., there is no implication that the force must

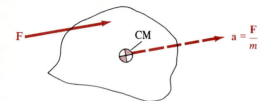

FIGURE 4.12
The acceleration of the center of mass of the body will be in the direction of **F** and equal to F/m.

be applied at the center of mass. The acceleration of the center of mass is in the direction of the force and at a rate equal to F/m. This is illustrated in Figure 4.12. Here the force resultant does not pass through the center of mass. Thus an instant later, besides the acceleration in the direction of F, this body will have rotated, as we discuss below.

In Equation (4.1b), $\Sigma \mathbf{T}$ is the sum of all moments and of the moments of all forces about the center of mass; I_{CM} is the mass moment of inertia about the mass center, and α is the angular acceleration. Angular acceleration requires no point reference; i.e., we do not indicate the angular acceleration to be about the center of mass or about some other point. For example, in Figure 4.13, the angular acceleration (and/or angular velocity or position) of the bar about the pivot is the same as it is about its center of mass. It is a free vector.

Newton's laws apply to rigid bodies as well as to interconnected assemblages of bodies (e.g., a piston/rod/crank assembly) and to unconnected assemblages (e.g., a group of stars). The laws are snapshots of the action—they apply to the instant for which they are written. A moment later the position, velocity, indeed the configuration of the body (bodies), will in general be different. Our concern is usually with rigid bodies.

The analysis is facilitated if we move the right sides to the left in Newton's laws (4.1) and write

$$\mathbf{F} - m\mathbf{a}_{CM} = 0 \tag{4.1a'}$$

$$\mathbf{T}_{CM} - I_{CM}\alpha = 0 \tag{4.1b'}$$

Thus, if we include the vectors $-m\mathbf{a}_{CM}$ and $-I_{CM}\alpha$ on the free-body diagram, we have only to sum all the force and all the moment vectors to zero. The problem becomes, in essence, a statics problem. The summation of moments may be taken about any convenient point. If this is not the center of mass, the moment of the inertia force $-m\mathbf{a}_{CM}$ must be included.

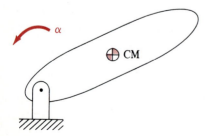

FIGURE 4.13
Angular acceleration is not referenced to a particular point.

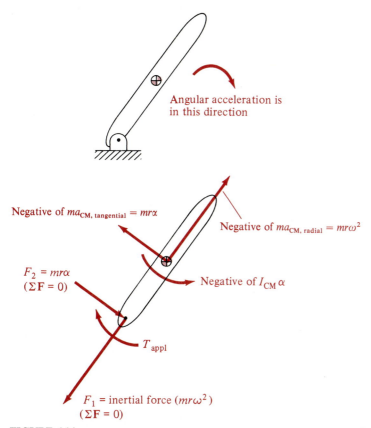

Angular acceleration is in this direction

Negative of $ma_{CM,\,tangential} = mr\alpha$

Negative of $ma_{CM,\,radial} = mr\omega^2$

$F_2 = mr\alpha$
$(\Sigma F = 0)$

Negative of $I_{CM}\alpha$

T_{appl}

$F_1 =$ inertial force $(mr\omega^2)$
$(\Sigma F = 0)$

FIGURE 4.14
Forces and torques on a body rotating about an axis not through its center of mass.

The conversion of a dynamics problem to one of statics as expressed above is known as *d'Alembert's*[1] *principle*. The term $-m\mathbf{a_{CM}}$ is known as the *d'Alembert force* or the *inertial force*, and $-I_{CM}\boldsymbol{\alpha}$ is called the *d'Alembert torque* or the *inertial torque*.

A simplification of expression (4.1b′) obtains for the common case of a body rotating about an axis not through its center of mass, as in Figure 4.14. The free-body diagram is shown in the figure, with the inertia force and inertia torque included as outlined above. Summing torques about the pivot, we get

$$T_{appl} - I_{CM}\alpha - mr^2\alpha = 0$$

or

$$T_{appl} = (I_{CM} + mr^2)\alpha = I_{pivot}\alpha$$

The moment summation could have been made about the center of mass rather than the pivot. In that case, however, the force $\mathbf{F}_2$ must be included, with the same result.

---

[1] Jean le Rond d'Alembert (1717–1783) was a French mathematician and philosopher. Abandoned as an infant, he was raised by a poor glazier. D'Alembert contributed new methods for the solution of ordinary partial differential equations and developed a criterion for determining the convergence of series. His principle for the solution of dynamics problems was part of his *Traité de Dynamique* (1743). He produced scholarly writings in music, letters, and philosophy.

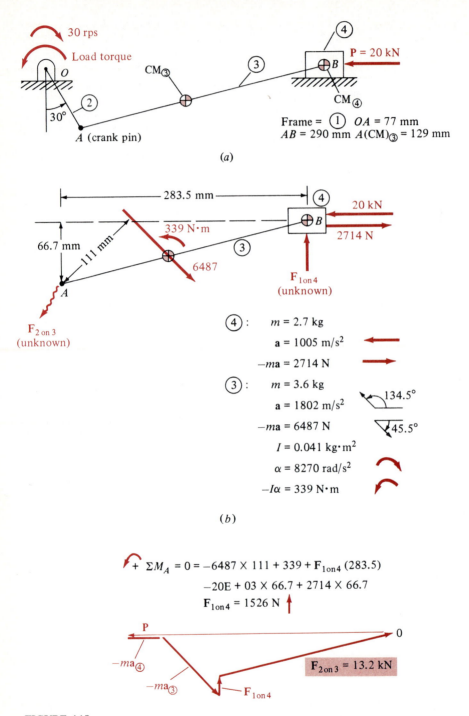

FIGURE 4.15
Dynamic force analysis of diesel piston and crank in Example 4.2. (a) Schematic of diesel piston, rod, and crank. (b) Free-body diagram of piston and piston rod. (c) Force solution. The force on the crank pin is opposite in direction to $F_{2 \text{ on } 3}$.

Example 4.2. Figure 4.15 is a schematic of a piston, rod, and crank of a diesel engine. At the instant shown, the combustion produces a force of 20 000 N on the piston. The crank is turning clockwise. The force acting on the crank pin A is needed.

The piston and piston rod are shown as a free body in Figure 4.15*b*. Kinematic analysis of the mechanism yields the accelerations listed, along with the inertia forces and torques, which are then shown as vectors on the free-body diagram.

There are three unknowns appearing on that diagram: the size of $F_{1\ on\ 4}$ and the direction and size of $F_{2\ on\ 3}$ (or two components of that force, which is the same). All three of the statics equations are required for solution.

A summation of moments about A will yield the size of $F_{1\ on\ 4}$ directly, Figure 4.15*c*. Note that a summation about any other point would involve the two unknowns associated with $F_{2\ on\ 3}$.

The force summations are accomplished most easily with a vector polygon, drawn in Figure 4.15*c*, with the desired force indicated. To get greater precision, the summations can be done analytically, producing the same result. This is left as an exercise in Problem 4.6. It is, as we said earlier, a good idea to draw the polygon in any case.

4.4 STRESS ANALYSIS—ANALYTICAL

All design work requires that the part or assembly successfully endure the loads and consequent stresses placed on it. This review aims at determining the critical stresses existing at a point and their orientation.

Simple Tension Member Stress Analysis

Consider the tension member, shown in Figure 4.16*a*, which is subjected to the axial forces P only. To find the force on a section *a-a* perpendicular to the applied force, we draw a free-body diagram of the upper or lower part of the bar. Equilibrium of

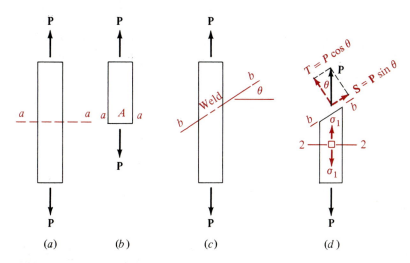

FIGURE 4.16
Analysis of stresses resulting from a uniaxial load.

this part, shown in Figure 4.16b, requires that the total force on the cut surface be a tension P. If that force is spread uniformly over the surface, then a tension stress $\sigma = P/A$ exists at any point on the surface. This is also called a *normal stress*, because the vector which represents the stress is perpendicular, or normal, to the surface. Since there is no force parallel to the surface, i.e., no shearing force, the shear stress τ on *a-a* is zero.

Now suppose that the bar of Figure 4.16 has been welded together at some angle θ to its axis (Figure 4.16c). To determine what stresses the weld is subjected to, we draw a free-body diagram of the lower part, illustrated in Figure 4.16d. The force on section *b-b* must again be equal to P. It can be resolved into a tension T and a shear S, as shown:

$$T = P \cos \theta \qquad S = P \sin \theta$$

The tension stress $\sigma_{b\text{-}b}$ and the shear stress $\tau_{b\text{-}b}$ are found by dividing T and S by the area of section *b-b*, $A/\cos \theta$:

$$\sigma_{b\text{-}b} = \frac{T}{A/\cos \theta} = \frac{P}{A} \cos^2 \theta$$

$$\tau_{b\text{-}b} = \frac{S}{A/\cos \theta} = \frac{P}{2A} \sin 2\theta$$

(4.4)

Of course, P/A is the tension stress within the bar in the P direction. Note that as the angle θ is reduced toward zero, the tension stress $\sigma_{b\text{-}b}$ approaches P/A and the shear stress $\tau_{b\text{-}b}$ approaches zero, as we know they must. We also see from Equations (4.4) that $\sigma = P/A$ (for $\theta = 0$) is the greatest tension stress for any θ.

The term used for the extreme normal stresses (maximum and minimum) is *principal stresses*, and the planes on which they act are called *principal planes*. The direction of a principal stress is called a *principal direction*. We also see that $\theta = 90°$ defines a plane parallel to the applied force P, and as the equations indicate, the normal and shear stresses on this plane are zero: $\sigma_{b\text{-}b} = \tau_{b\text{-}b} = 0$. This element is shown in Figure 4.16d. It turns out that principal stresses are always found on planes which are orthogonal, i.e., at 90° to each other, and the shear stress on these planes is zero. In a three-dimensional solid, there will be three principal stresses. The usual nomenclature is $\sigma_1, \sigma_2, \sigma_3$, in descending order of algebraic size. (If, for example, the three stresses are 2, 6, and -26, then $\sigma_1 = 6$, $\sigma_2 = 2$, and $\sigma_3 = -26$.)

Turning to the second of equations (4.4), we see that the maximum value of the shear stress $\tau_{b\text{-}b}$ can be computed through the usual device of setting $d\tau/d\theta = 0$, or by simply observing that $\sin 2\theta$ has its greatest value when $2\theta = 90°$, or $\theta = 45°$. Substituting this in the second of Equations (4.4) yields

$$\tau_{\max} = \frac{\sigma_1}{2} \sin 90° = \frac{\sigma_1}{2}$$

The first of these results is always valid—the maximum shear stress occurs on planes at 45° to the principal directions. The second result ($\tau_{\max} = \sigma_1/2$) is true enough for this uniaxial-load situation; however, $\tau_{\max}$ is really $(\sigma_1 - \sigma_2)/2$, with σ_2 being the stress on the other principal plane, which here happens to be zero. This $\tau_{\max}$ is the largest shear stress in the two-dimensional model we are discussing. We will account for the third dimension later.

The magnitudes of the principal stresses, and of the maximum shear stress, which is easily calculated once the principal stresses are known, are very important in design. They are the quantities to watch in predicting failure.

Principal-Stress Determination—Analytical Background

Most engineering problems can be handled as plane-stress problems; i.e., the vectors representing the stresses lie in a single plane, as in Figure 4.17. Such a sketch shows the edges of the planes on which the stresses act. The planes are perpendicular to the paper, in other words, and the problem is a two-dimensional statics problem.

In a three-dimensional problem, a tension stress in the third dimension σ_z would act on the face which is on the surface of the paper. There could also be stresses in the z direction acting on the x and y planes. (The planes are denominated according to their outward normals.) They would be the shear stresses τ_{xz} and τ_{yz}. These and σ_z are absent in plane stress. When they are present, the third equation of statics, $\Sigma F_z = 0$, must be used. In the plane-stress problem, the stress $\sigma_z = 0$ is a principal stress, even though it is zero, because there are no shear stresses on that plane.

The shear stresses on the x planes are equal and opposite, as are those on the y planes. Further, the x and y plane shear stresses are equal to each other. This is dictated by the three equilibrium requirements for two-dimensional statics discussed in Section 4.2. In more advanced stress analysis, a differentiation is made in the notation of these shear stresses; the shear stress on the face whose normal is in the $\pm y$ direction is called τ_{yx}, meaning the shear stress on the y face in the x direction. On the x face it would be τ_{xy}. Since $\tau_{xy} = \tau_{yx}$, we ignore the distinction here. We have

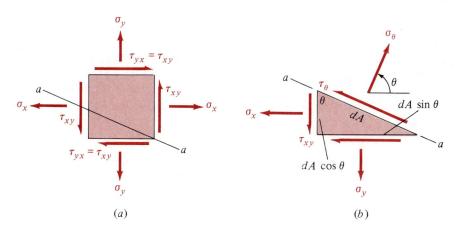

(a)   (b)

FIGURE 4.17
Determination of stresses for general orientation. (a) Stresses on element with xy orientation. (b) Free-body diagram for determining stresses on plane with orientation θ.

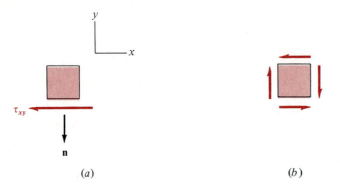

FIGURE 4.18
Shear-stress sign convention. (a) τ_{xy} is positive because the outward normal **n** points in the $-y$ direction and τ_{xy} points in the $-x$ direction. (b) Negative shear stresses.

followed a common sign convention for the shear directions, however. The shear stress is positive if the outward normal of the face and the shear stress both point in their positive or negative directions. The shear stress is negative if the outward normal of the face points in the positive direction and the shear stress points in the negative direction, or vice versa. For example, in Figure 4.18a, the shear stress τ_{xy} shown is positive. The outward normal is in the $-y$ direction, so that a positive shear stress points in the $-x$ direction. Negative shear stresses are indicated in Figure 4.18b. You can see that once the direction of one shear stress is established, the others on the element are automatically fixed.

Expressions for stress as a function of element orientation can be found by the process leading to Equations (4.4). Then, through differentiation, it is possible to determine maximum values and their orientations. It is simplest to take the surface a-a through a corner, to eliminate one vertical side. The resulting free-body diagram is shown in Figure 4.17b. Note that we are measuring angle θ counterclockwise from the x axis to the outward normal of the face a-a (the normal has the same direction as σ_θ).

The unknowns are the normal stress σ_θ and the shear stress τ_θ. Two equations of static equilibrium are available to determine them: $\Sigma F_x = 0$ and $\Sigma F_y = 0$. It is easiest to take these summations in the τ_θ and σ_θ directions. Note that *forces* must be summed, which means that the stresses shown must be multiplied by the areas on which they act. The latter are indicated in the figure. The stresses result directly and are

$$\sigma_\theta = \frac{\sigma_x + \sigma_y}{2} + \frac{\sigma_x - \sigma_y}{2} \cos 2\theta + \tau_{xy} \sin 2\theta$$

$$\tau_\theta = -\frac{\sigma_x - \sigma_y}{2} \sin 2\theta + \tau_{xy} \cos 2\theta$$

(4.5)

The stresses on the adjacent sides of an element with this orientation are found by substituting $\theta + 90°$ for θ.

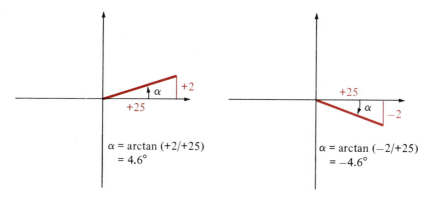

FIGURE 4.19
Tangent relations.

The maximum and minimum values of σ_θ are found by setting $d\sigma_\theta/d\theta$ to zero in the usual way and similarly for τ_θ. The results are

$$\sigma_{1,2} = \frac{\sigma_x + \sigma_y}{2} \pm \sqrt{\left(\frac{\sigma_x - \sigma_y}{2}\right)^2 + \tau_{xy}^2}$$

$$\tau_{max} = \pm\sqrt{\left(\frac{\sigma_x - \sigma_y}{2}\right)^2 + \tau_{xy}^2}$$

$$\tan 2\theta_{\sigma_{1,2}} = \frac{2\tau_{xy}}{\sigma_x - \sigma_y}$$

These expressions are built into some hand calculators. The results do not differentiate between σ_1 and σ_2. In the last equation of the group, $\tan 2\theta_{\sigma_{1,2}}$ yields the two angles between the x axis and the normals to the faces on which the principal stresses act (Figure 4.17). The shear stress τ_{xy} in the numerator of the tangent expression must have the proper sign (Figure 4.17). Both $\tan \alpha = -2/+25$ and $\tan \alpha = +2/+25$ yield different values of angle α, that is, orientations, as depicted in Figure 4.19. Actually, the signs of the numerator *and* denominator in the tangent (and cotangent) must be specified, because they determine the quadrant of the angle, as seen in Figure 4.20. It is not sufficient, for example, to write $\tan \alpha = -6$. This could

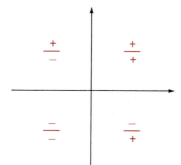

FIGURE 4.20
The signs of both the numerator and the denominator in the tangent relation must be specified.

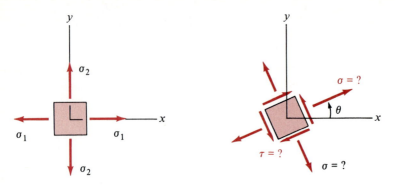

FIGURE 4.21
Starting with principal stresses, the object is to find the stresses on an element with orientation θ.

be $+6/-1$ or $-6/+1$, and the value of α would be quite different—180° different, as you see in the figure. Calculators and computers generally will not make the above distinction for you.

4.5   MOHR'S CIRCLE

A quick and reliable method to find stresses on different orientations, which is at the same time an excellent way to recall the equations, is a graphical scheme bearing the name of the German engineer Otto Mohr,[1] though it was not developed by him. It is understood most easily by starting with a situation where the x and y planes are planes of principal stress, as sketched in Figure 4.21, and then finding the stress situation on an element with orientation θ. Because x and y are the principal directions, τ_{xy} is zero, and Equations (4.5) are somewhat simpler:

$$\sigma_\theta = \frac{\sigma_1 + \sigma_2}{2} + \frac{\sigma_1 - \sigma_2}{2} \cos 2\theta$$

$$\tau_\theta = \frac{\sigma_1 - \sigma_2}{2} \sin 2\theta$$

(4.6)

The trigonometric terms of relations (4.6) suggest the projections of the radius of a circle, and the constant term of the σ_θ relation suggests displacing the circle's center, as seen in Figure 4.22.

---

[1] Otto Mohr (1835–1918) was a graduate of the Hannover Polytechnic Institute. His early work was in structural design for the railroads in northern Germany, but his forte seemed to be in the engineering theory of structures, in which he spent the last half of his career as a professor in Stuttgart and Dresden.

　　Mohr developed a simple method for establishing the deflection curve of a loaded beam and refined and expanded the graphical method of determining stresses for different orientations at a point, which we know as Mohr's circle. The representation in two dimensions by a circle was not Mohr's discovery, but rather that of Culmann, published in 1866, some 16 years prior to Mohr's paper. Mohr also developed a failure theory, which still has validity, discussed in Chapter 7.

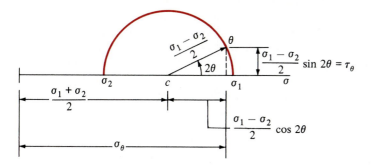

FIGURE 4.22
The terms of expression (4.6) suggest a certain geometry.

With σ_1 and σ_2 laid off on the abscissa axis and used as the diameter of a circle, the center and the radius correspond to $(\sigma_1 + \sigma_2)/2$ and $(\sigma_1 - \sigma_2)/2$, respectively, in Equations (4.6). If now a radius is laid off at an angle 2θ with the abscissa axis, the horizontal projection of the point labeled θ results in σ_θ, as shown in Figure 4.22, while the vertical projection is τ_θ. Now, examine Figure 4.23. Mohr's circle is shown in Figure 4.23a and the principal-stress element in Figure 4.23b. (Although we normally orient the principal element with the axes of the Mohr-circle diagram, σ_1 is

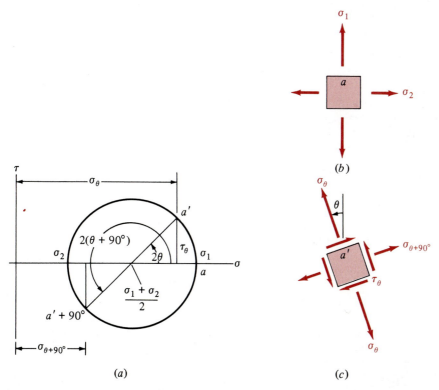

FIGURE 4.23
Mohr's circle of stress. (a) The circle. (b) Principal element. (c) Stresses on element with orientation θ.

not of necessity in line with the σ axis, as in the figure.) If the element is now rotated counterclockwise through angle θ, as we have been discussing, the side with σ_1 on it originally, marked a, moves to a', as indicated on the circle and on the rotated element in Figure 4.23c. If the rotation is continued to $2(\theta + 90°)$, the point labeled $a' + 90°$ is reached (the other end of the diameter commencing at point a'). The projections become the stresses σ and τ on the other pair of sides of the rotated element, as shown in Figure 4.23c. The following is a numerical example, with one of the principal stresses being negative.

Example 4.3. We have a stress element oriented with the directions of the principal stresses, which are 6 ksi of tension and 10 ksi of compression, as shown in Figure 4.24a. Now σ_y is compression and therefore negative. Since 6 is algebraically greater than -10,

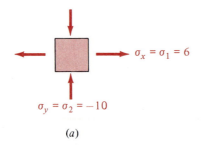

$$\sigma_x = \sigma_1 = 6$$
$$\sigma_y = \sigma_2 = -10$$

(a)

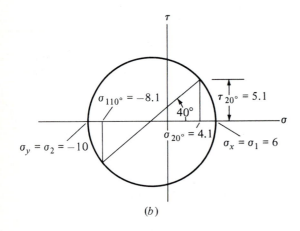

$\sigma_{110°} = -8.1$
$\tau_{20°} = 5.1$
$40°$
$\sigma_{20°} = 4.1$
$\sigma_y = \sigma_2 = -10$
$\sigma_x = \sigma_1 = 6$

(b)

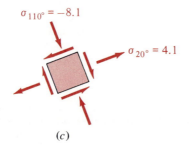

$\sigma_{110°} = -8.1$
$\sigma_{20°} = 4.1$

(c)

FIGURE 4.24
Analysis for Example 4.3. (a) Original element. (b) Mohr's-circle analysis. (c) Stresses on element at 20° counterclockwise.

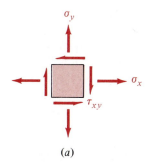

(a)

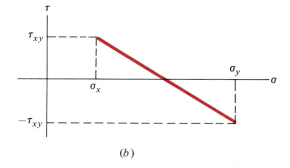

(b)

FIGURE 4.25
Shear-stress sign convention for Mohr's circle. (a) The clockwise shear stresses on x faces are *Mohr-positive* shear stress; those on y faces are *Mohr-negative*. (b) Plot of stresses.

$\sigma_1 = 6$ ksi and $\sigma_2 = -10$ ksi. Determine the stresses for an orientation 20° counterclockwise from the xy axes.

We draw Mohr's circle (Figure 4.24b) and lay off a diameter at $2\theta = 2 \times 20 = 40°$ counterclockwise from the abscissa axis. The stresses are measured as indicated, and we end up with the element sketched in Figure 4.24c. The directions of the shear stresses require some discussion.

Shear-Stress Directions

In Example 4.3, in which the original stress element was a principal element, the shear-stress size on a rotated element was easily found. We did not, however, have the information for determining the directions of those stresses. Similarly, if you start with a general element as in Figure 4.17, there is no problem laying off σ_x and σ_y for a Mohr's-circle diagram, but should the shear stresses be plotted up or down? Doing it either way leads to the same circle and the correct principal stresses, but the orientation will be different.

The above difficulty can be overcome with a sign convention. One is to call shear stresses which point clockwise *Mohr-positive* and those which point counterclockwise *Mohr-negative*.[1] In the sketch of Figure 4.25a, the shear stresses on the x faces are

---

[1] This is opposite to the sign convention for shear stresses cited in Section 4.3, which is why we attach the modifier *Mohr* to the terms *positive* and *negative*. Alternatively, the shear stresses can be plotted as positive downward, which is done in some texts.

Mohr-positive. Thus on the Mohr's-circle diagram, τ_{xy} is plotted up from σ_x. On the y faces the shear stresses are counterclockwise, hence are plotted down from σ_y, as shown in Figure 4.25b. With this convention, you can see that the shear stresses were correctly shown in Example 4.3 (Figure 4.24).

Maximum Shear Stresses in the Loading Plane

Since shear stresses are plotted on the vertical axis of the Mohr's-circle diagram, their maximum value is the radius of the circle $\tau_{max} = (\sigma_1 - \sigma_2)/2$, as shown in Figure 4.26a. The orientation of the element on which they act is $\theta = 45°$ ($2\theta = 90°$) from the principal directions. The normal stresses on that element are the same in both directions and are equal to the average of the principal stresses, i.e., the σ value at the center of the circle, as seen in the figure. The original side of the principal element with stress σ_1 marked a (Figure 4.26a and b) has rotated to a', as shown in Figure 4.26a and c. The shear stress on it is seen to be Mohr-positive, clockwise on side a'.

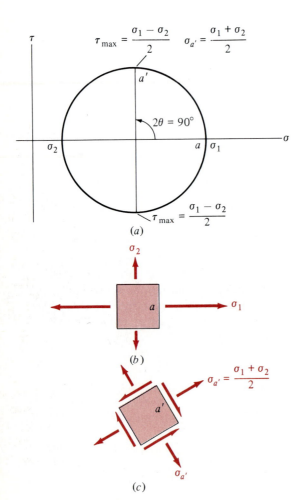

FIGURE 4.26
Maximum shear stresses. (a) Mohr's circle, showing maximum shear stresses (b) Principal element. (c) Element with maximum shear stresses.

The shear stresses so determined are the greatest existing in the σ_1-σ_2 plane, i.e., in the plane of loading, which we often call the xy plane. We will see below that the maximum shear stress in the xz or yz plane may be larger.

Element Orientations on Mohr's Circle

A convenient scheme for showing element orientations on a Mohr's-circle diagram is developed in Figure 4.27. Two diameters of the circle are shown. One, ab, lies along the abscissa axis, on which the principal stresses are located. The other, cd, is associated with xy axes which make an angle θ with the principal axes. The angle between the two diameters is 2θ, as required by the Mohr's-circle relations. An isosceles triangle is defined by aOd, and each of its equal angles is $2\theta/2 = \theta$. Thus, line ad is located at angle θ from the principal directions, and the xy axes, aligned with ad, have the proper orientation with respect to the principal directions. Line bc is parallel to ad, and lines bd and ac are perpendicular to ad. Hence an xy system aligned along any one of these lines will have the same orientation with respect to the principal axes, as shown in the figure. Any one of the four lines can be selected to show the

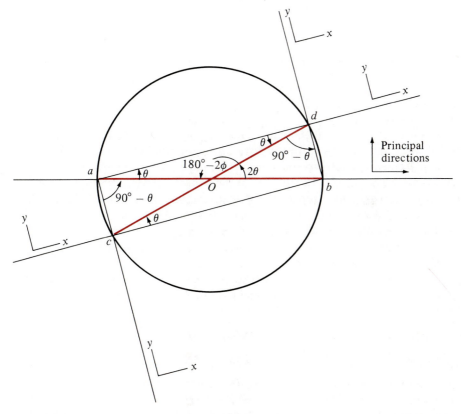

FIGURE 4.27
Equivalent xy orientations.

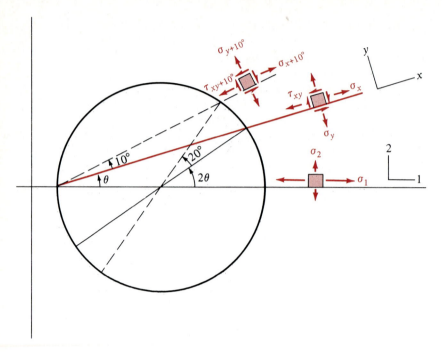

FIGURE 4.28
Element orientations shown on Mohr's-circle diagram.

stress element; note that they connect the ends of the two circle diameters in all four possible combinations.

An element is depicted in the above manner in Figure 4.28. Note that the xy axes, which we usually attach to the stresses on the original element, will not be in their usual placement parallel to the sides of the paper. The principal stresses will, however, always be oriented with the axes of the Mohr's-circle diagram. In showing the situation, you can either repeat the principal-element sketch on the original problem description or sketch the physical part itself or the xy axes on the Mohr's-circle figure. With this scheme of element representation, any orientation can be clearly represented. In Figure 4.28 the principal element is shown as well as an element 10° counterclockwise from xy.

Example 4.4. At some point in a structural member, the stresses (in pounds per square inch) are found to be those shown in Figure 4.29a. Determine the principal stresses and directions and the stresses on an element at 80° clockwise from xy.

Since the shear stresses on the x faces are clockwise, they are Mohr-positive; hence $(-15, 20)$ is one end of the Mohr's-circle diameter for this element. Similarly, the shear stresses on the y faces are Mohr-negative, and the other end of the diameter is at $(45, -20)$. The Mohr's circle is thus established, as shown in Figure 4.29b.

Again, note that the xy orientation is not aligned with the axes of the Mohr's-circle diagram. The Mohr's-circle axes are associated with the principal-stress directions. To avoid confusion, the entire member is sketched with each element shown.

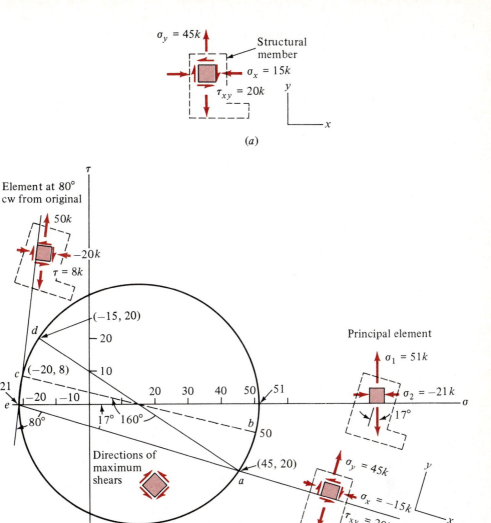

FIGURE 4.29
Analysis for Example 4.4. (*a*) Known stresses. (*b*) Mohr's circle.

The values of the principal stresses are picked off the chart or computed from the geometry: $\sigma_1 = 51\,000$ and $\sigma_2 = -21\,000$.

To determine the situation for an element rotated 80° clockwise from the original, a line is drawn from point e in Figure 4.29b at 80° to the original orientation. The new stress element is sketched. Notice that the part itself is shown in its original orientation; only the stress element has been rotated. The circle diameter corresponding to the new element is laid off from the intersection c of the line just drawn and the circle. The stresses $\sigma = -20$ and $\tau = +8$ can be read at point c, and $\sigma = 50$ at point b. The directions of the shear stresses are determined according to whether they are Mohr-positive or Mohr-negative.

Sometimes the point labeled c, at the end of the diameter determining the stresses on the element, is hard to locate exactly because of the shallow angle of intersection of the circle and the orientation line. You can solve that problem by measuring off $2 \times 80 = 160°$ from the original diameter ad, as shown, or by rotating the element $10°$ counterclockwise, instead of $80°$ clockwise—the same orientation results.

Example 4.5. At a point at the surface of a flat part, the stress along the x axis is known, as shown on the element sketched in Figure 4.30a. Deduce the tension stress in the y direction on the element and the shear stresses.

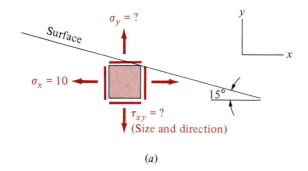

(a)

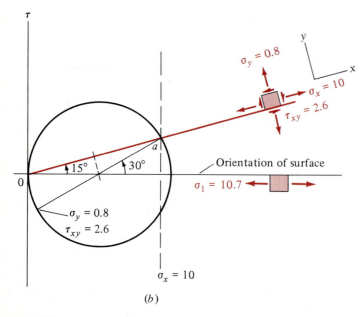

(b)

FIGURE 4.30
Example 4.5. (a) The problem. (b) The solution.

On the surface there is no tension stress and no shear stress. (Nothing is pushing, pulling, or rubbing it.) If the element were oriented along the surface, there could be only a tension (or compression) stress normal to the surface. In other words, an element oriented with the surface will be a principal element (no shear stress), and the xy orientation is 15° counterclockwise from it.

One end of the Mohr's-circle diameter corresponding to the principal stresses is at the origin, since one principal stress is zero. From that point we draw a line at 15° counterclockwise from the σ axis. This corresponds to the x axis, as shown in the figure. Since $\sigma_x = 10$, a vertical at that level locates point a, which will be on the circle. Now you are stuck until you recognize that Oa is a chord of the circle, and the center will thus be along the perpendicular bisector of the chord. With the center located, the circle can be drawn, and the problem is solved. Another way to locate the center is to draw in the diameter from a at an angle $2 \times 15 = 30°$ with the·σ axis.

Inferring Principal Directions from Element Distortion

The use of a shear-stress sign convention can be obviated by considering the distortion of the stress element, as follows.

Consider the problem of laying out the Mohr's circle so that the principal stresses are located correctly with respect to the known stresses, and assume that we have forgotten how the sign convention works, or that we wish to double-check our work. Figure 4.31a shows an element subjected to a general stress situation, including shear stresses. We take σ_x to be greater than σ_y, as indicated by the length of the vectors. We can draw the Mohr's circle and get the size of the larger principal stress, but will it act in a direction clockwise from the x axis, or counterclockwise? We can solve the problem by a conceptual experiment. First, consider what would happen if there were no shear stresses, as shown in Figure 4.31b. Then σ_x would clearly be the maximum principal stress.

If we draw the element again without the tensile stresses (Figure 4.31c), we see that the distortion produced by the shear stresses results in an extension along the 45° axis 1-1. Hence, the larger principal stress in this case is in that direction.

Putting the pieces together, we deduce that the 1-1 principal direction for the stress situation of the element in question (a in the figure) lies between the direction shown in Figure 4.31b and that shown in Figure 4.31c. That is, the 1-1 principal direction is located clockwise something less than 45° from the x axis, as shown in Figure 4.31d.

Example 4.6. Refer to Example 4.4, for which the original stress element is shown in Figure 4.29a. Figure 4.32a shows that the principal axis 1 would be along y if there were no shear stress and at 45° to y if there were only shear stress. The actual 1-1 axis is between the two, as indicated, <45° counterclockwise from y. The diameter of the Mohr's circle and the xy element (the original) can therefore be located with this information, as shown in Figure 4.29b, such that a counterclockwise rotation of about 17° of the element brings it to the principal orientation. The Mohr's-circle diameter of the original element is ad. If we had picked the other diameter with the same magnitudes of σ's and τ, but disposed the other way, as shown in the sketch of Figure 4.32b, then the 1 direction would be clockwise from the y direction, which we know to be wrong.

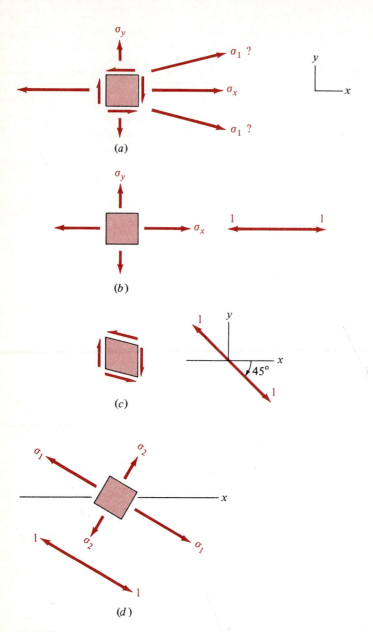

FIGURE 4.31
Determining 1-1 direction. (*a*) Stress element and possible directions of σ_1. (*b*) The 1-1 principal direction, considering only tension stresses. (*c*) The 1-1 principal direction, considering only shear stresses. (*d*) Actual orientation of 1-1 principal direction.

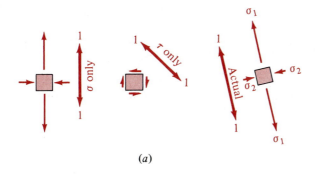

(a)

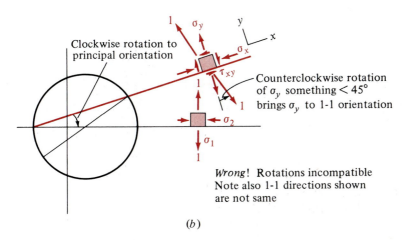

(b)

FIGURE 4.32
Determination of element orientation for Example 4.6. (a) Determination of principal directions. (b) Incorrect orientation of element.

Inferring Shear-Stress Directions from a Principal Element

Let us assume that the principal stresses at a point are known, and to vary the problem, we take them both to be compressive, as shown in Figure 4.33. Recall that the larger of the two in algebraic terms is σ_1. For example, if the stresses are -10 and -2, then $\sigma_1 = -2$. Mohr's circle is entirely to the left of the τ axis. The question is, What are the stresses on an element with orientation $10°$ clockwise from the principal element? At $10°$ clockwise from the line σ_1-σ_2 we lay off the line σ_1-a and draw the element. The σ_2 sides of the element have rotated to those labeled $\sigma_{2+10°}$, and the end of the circle diameter moves from σ_2 to a, where the normal stress on those sides is then measured as the abscissa value of point a. The stress on the $\sigma_{1+10°}$ sides is found, similarly, at b. The size of the shear stress is the ordinate value of a or of b.

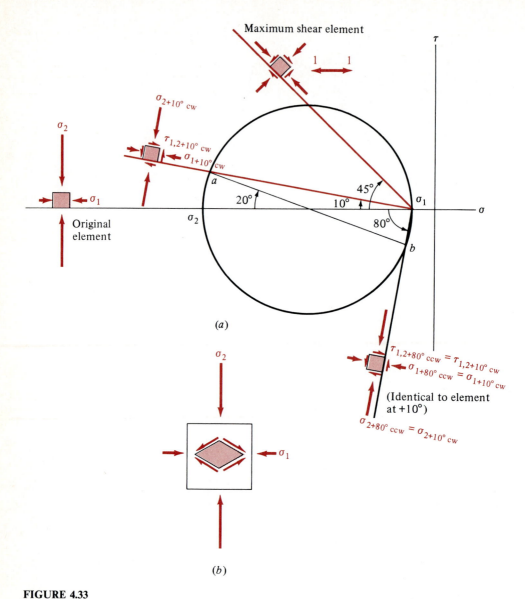

FIGURE 4.33
Analysis of problem, starting with principal stresses. (*a*) Mohr's circle. (*b*) Distorted element is at 45° to principal directions and the shear stresses causing the distortion. Element is also shown on Mohr's circle.

The element with maximum shear is at 45° to the principal directions; it is shown at *b* in the figure. Its distortion is a contraction in the direction of the larger compressive stress σ_2. Thus the shear stresses have the directions indicated. Since all orientations within $\pm 45°$ of the maximum-shear element have the same sign for shear, being on the same side of the σ axis, the directions of the shear vectors on the element at 10° are now known.

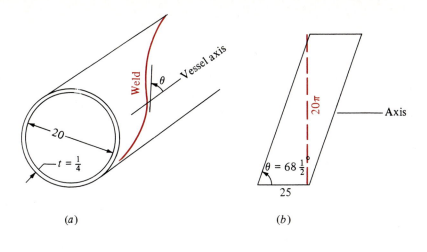

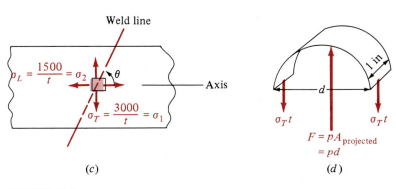

FIGURE 4.34
Pressure-vessel weld analysis for Example 4.7. (*a*) Welded vessel. (*b*) One turn unwrapped. (*c*) Side view, showing longitudinal and transverse stresses. There are no shear stresses, hence these are principal stresses. (*d*) Calculation of σ_T.

The problem may be worked equally well with an 80° counterclockwise rotation, shown in Figure 4.33*a*, rather than 10° clockwise, for the resulting orientation is the same. You should verify this.

Example 4.7. A cylindrical thin-wall pressure vessel is formed by spiral welding, as shown in Figure 4.34*a*. The weld makes one wrap in 25 in. The internal pressure is 300 psi, and since the vessel is closed at the ends, the pressure acts longitudinally as well as radially. What are the normal and shearing forces carried per inch of weld?

If you imagine a turn of the spiral of the vessel unwrapped, as shown in the sketch in Figure 4.34*b*, the angle that the weld makes with the vessel axis is

$$\theta = \arctan \frac{20\pi}{25} = 68\tfrac{1}{2}°$$

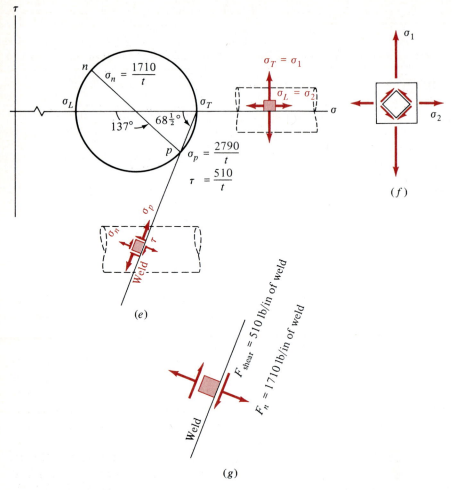

FIGURE 4.34 (Continued)
(e) Mohr's circle. (f) Determination of shear directions. (g) Required forces on weld.

First, we consider a stress element aligned with the axis of the cylinder. The axes of the element are axes of symmetry of the cylinder; the element will not be subjected to shear stress. It is, in other words, a principal element.

There are two stresses[1] to deal with: the longitudinal stress (subscript L) and the transverse stress (subscript T), as shown in Figure 4.34c. The latter is frequently termed the *hoop stress*. The longitudinal stress is

$$\sigma_L = \frac{F_1}{A_{\text{cross section}}} = \frac{p\pi d^2/4}{\pi dt} = \frac{pd}{4t} = \frac{1500}{t}$$

---

[1] There is actually a third stress, in the radial direction; but in thin-walled vessels it can be neglected, so this is a two-dimensional problem.

The thickness t is left as a symbol rather than its actual value, because it will cancel later when we compute the force.

The transverse stress is the transverse force in a unit of vessel length divided by the cross-sectional area in a unit length, as shown in Figure 4.34d. Thus

$$\sigma_T = \frac{pd}{2t} = \frac{300 \times 20}{2t} = \frac{3000}{t}$$

There are no shear stresses on this element, which makes σ_L and σ_T principal stresses, as we observed earlier. The element is therefore shown on the σ axis of Mohr's circle in Figure 4.34e. The stress situation for an element on the weld line ($\theta = 68\frac{1}{2}°$ counterclockwise) is also shown on that diagram. The direction of the shear stresses is determined by examining the distortion of an element at 45° to the principal directions, as indicated in Figure 4.34f. The stresses are read off the circle at points p and n, and the normal and shearing forces per inch of weld are obtained by multiplying the stresses by the cross-sectional area of 1 in of weld, which is $1 \times t$. The result is shown in Figure 4.34g.

Maximum Shear Stress (Three Dimensions)

Suppose we have the general two-dimensional stress situation shown in Figure 4.35a. We draw Mohr's circle and find the principal stresses are on planes at an angle clockwise from the given planes, shown in Figure 4.35b.

Now, recall that the stress analysis performed by Mohr's circle involves rotation of stress elements about a principal axis. In the free-body diagram from which we developed Mohr's circle (Figure 4.17b), there were no shear stresses in the plane of the paper, that is, $\tau_{xz} = \tau_{yz} = 0$; hence the stress $\sigma_z = 0$ was a principal stress, and the z axis was the third principal direction. When we rotated the two-dimensional view of the element, we were rotating it about the z axis. We did this in Figure 4.35 in going from the original orientation at a to the principal orientation at b.

The principal element in Figure 4.35b is shown in isometric view in Figure 4.35c. If you view it along the σ_2 axis (2 axis), what you see is shown in Figure 4.35d. It looks like a plane-stress situation in σ_1 and σ_z. A Mohr's-circle analysis rotating the element in Figure 4.35d about the 2 axis results in a circle whose diameter extends from σ_1 to σ_3, the latter being at the origin ($\sigma_z = \sigma_3 = 0$). In like manner, the circle for the σ_2-z (z = principal direction 3) plane is drawn.

Elements in the 1-3 and 2-3 planes can be analyzed as we have done with Mohr's circle earlier. Of course, elements in the 1-2, 1-3, and 2-3 planes do not exhaust the possible orientations. Although elements not in these planes can be investigated analytically, the stresses on them cannot be found graphically with Mohr's circle. What can be said, however, is that the stresses fall within the largest of the three circles and outside the smaller ones.

We will soon be interested in the largest existing shear stresses, which we know are at 45° to the principal directions. For the situation of Figure 4.35, then, the largest shear stresses are in the 1-3 (1-z) plane (largest radius), i.e., the plane shown in Figure 4.35d. This is not the plane of loading, which is the 1-2 plane. The maximum shear stresses appear as sketched in Figure 4.35e. Their magnitude is the radius of the 1-3 circle: $\tau_{max} = \sigma_1/2$.

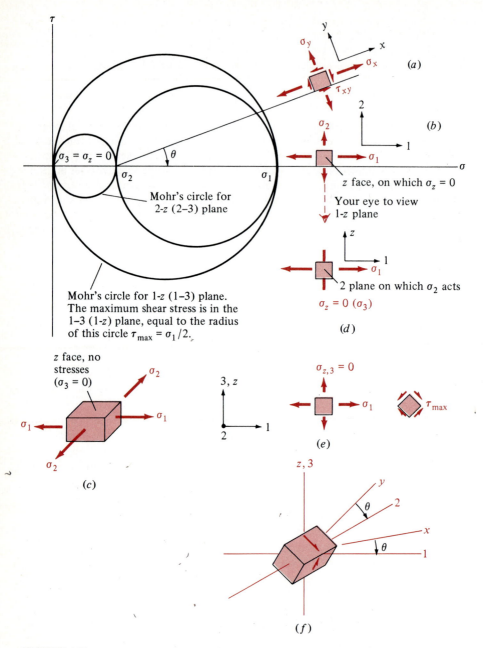

FIGURE 4.35
Determination of maximum shear in two-dimensional loading. (a) Original element. (b) Principal element in loading plane. (c) Isometric of principal element. (e) Maximum shear-stress orientation. (f) Perspective of element with maximum shear stress.

To visualize the orientation of the planes of maximum shear stresses, the element (at a in Figure 4.35) must first be rotated about the z axis through a clockwise angle θ to the 1-2 directions (b in the figure). A 45° rotation (either direction) about the 2 axis (Figure 4.35e) produces the orientation of maximum shear stress. A perspective view is shown in Figure 4.35f.

In Figure 4.35, σ_1 and σ_2 were both positive, and it turned out that the largest circle was that of the 1-3 (z) plane. As σ_1 and σ_2 are not always positive, we need to be more general. It is common in stress analysis to label the three principal stresses $\sigma_1, \sigma_2, \sigma_3$, in decreasing order of size, as stated earlier. Then the Mohr's circles always appear as sketched in Figure 4.36a, where the advertised order has been observed. For two-dimensional loading, one of these stresses will be $\sigma_z = 0$, and the vertical τ axis will pass through it. There are thus three possibilities, as shown in Figure 4.36b. Whichever the case, the largest circle has the diameter $D = \sigma_1 - \sigma_3$, and the maximum existing shear stress is thus

$$\tau_{\max} = \frac{D}{2} = \frac{\sigma_1 - \sigma_3}{2}$$

It acts on planes at 45° to the 1-3 axes.

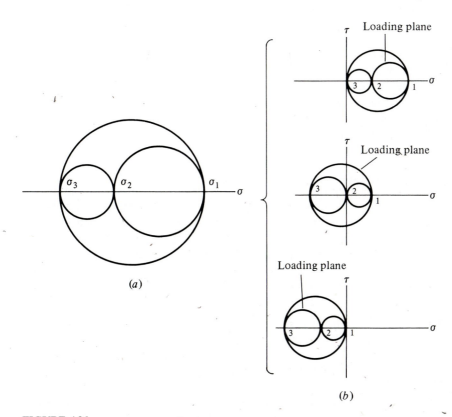

(a)

(b)

FIGURE 4.36
Possible Mohr's-circle arrangements for two-dimensional loading. (a) General three-dimensional stress circles. (b) Two-dimensional ($\sigma_z = 0$) possibilities.

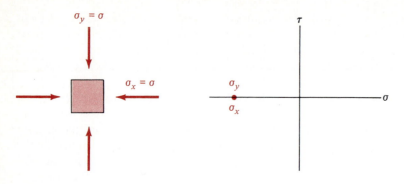

FIGURE 4.37
Equal principal stresses.

Hydrostatic Pressure

Suppose $\sigma_x = \sigma_y = \sigma$ and there is no shear, as depicted in Figure 4.37. Now σ_x and σ_y are principal stresses, and Mohr's circle reduces to a point, i.e., the radius is zero, as shown. The analysis of such a circle leads to the conclusion that the normal stress is the same in any direction. If we imagine that the stress in the third dimension is also the same, that is, $\sigma_z = \sigma$, then the Mohr's circles in the xz and yz planes would be points as well, and the stress, i.e., pressure, would not vary, no matter what direction we looked in. That is an elementary principle of physics; it is called *hydrostatic pressure*, which means "under still water."

We also note that since the radius of the Mohr's circle(s) is zero, there is no shear on any orientation. That is an interesting fact, and it plays a part in material analysis and the derivation of one theory of material failure.

4.6 OTHER USES OF MOHR'S CIRCLE

The same types of equations represented by Mohr's graphical approach appear in other branches of engineering and science, and similar analyses can be carried out and are extremely useful. The finding of principal moments of inertia or the reverse problem of transferring moments of inertia to inclined axes is a case in point, important in beam theory and in the dynamics of rotating bodies. Mohr's circle will also be found in Chapter 16, where it is used to approximate the modulus of elasticity of composite laminae.

PROBLEMS

Section 4.1

4.1. A 30-lb traffic sign is cantilevered over the street, as shown in Figure P4.1. A high wind produces a force on the sign of 17 lb. Determine the moment at point O at the base of the sign. Assume that the force acts at the centroid of the sign.

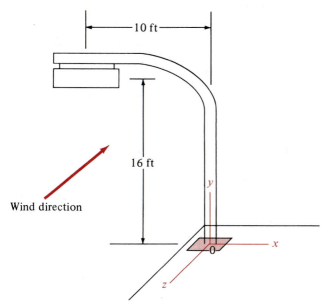

FIGURE P4.1

4.2. Figure P4.2 shows an A frame used to hoist engines out of cars. The engine has a mass of 160 kg. Determine the forces at *A, B, C,* and *D*.

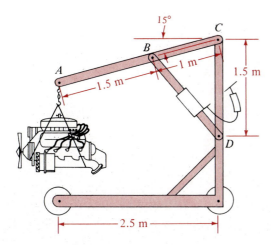

FIGURE P4.2

Section 4.2

4.3. The pilot of a plane brings the nose wheel down rather hard, resulting in a vertical force of 600 lb at the wheel, as shown in Figure P4.3 (page 122). Find the forces and/or moments at section *a-a*.

4.4. The C clamp shown in Figure P4.4 (page 122) is tightened to produce a force of 250 N on the parts being glued. Find the forces and/or moments at the center of sections *a-a* and *b-b*.

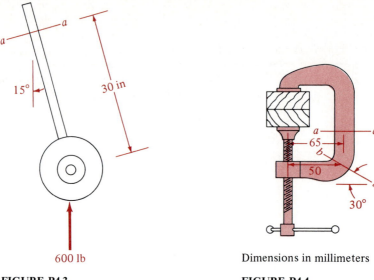

FIGURE P4.3   FIGURE P4.4

Dimensions in millimeters

4.5. A plumber applies a force of 20 lb downward to remove a section of pipe, as shown in Figure P4.5. Determine the internal forces and/or moments at point *a* of the pipe.

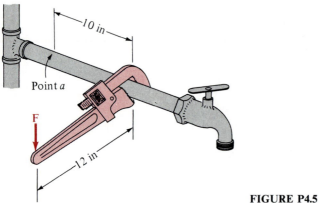

FIGURE P4.5

4.6. Perform the force summation of Example 4.2 analytically, showing the result to be the same as that from the vector polygon in Figure 4.15c.

Sections 4.4 and 4.5

4.7. Determine the principal stresses and their orientation as well as the maximum shear stresses in the plane of loading for the situations listed below, indicating the results on properly oriented elements.

(a) $\sigma_x = 500$, $\sigma_y = 300$, $\tau = 200$ MPa

   Answer: $\sigma_1 = 625$ (30° counterclockwise from x), $\sigma_2 = 175$, $\tau_{max} = 225$

(b) $\sigma_x = 500$, $\sigma_y = 300$, $\tau = -200$ MPa

(c) $\sigma_x = -100$, $\sigma_y = 0$, $\tau = 20$ ksi
   Answer: $\sigma_1 = 4$, $\sigma_2 = -104$ (11° clockwise from x), $\tau_{max} = 54$
(d) $\sigma_x = \sigma_y = -20$, $\tau = 10$ ksi
(e) $\sigma_x = -400$, $\sigma_y = 100$, $\tau = -50$ MPa
   Answer: $\sigma_1 = 105$, $\sigma_2 = -405$ (6° counterclockwise from x), $\tau_{max} = 255$

4.8. At a point in a structure the stresses below are known to exist. Find the stresses for the orientations indicated. Show the stresses on a properly oriented element.
   (a) $\sigma_x = -500$, $\sigma_y = 100$, $\tau = -100$ MPa
      (1) 25° clockwise
         Answer: $\sigma = -85$, -310; $\tau = 290$
      (2) 15° clockwise
   (b) $\sigma_x = 40$, $\sigma_y = 20$, $\tau = -10$ ksi
      (1) 15° clockwise
      (2) 75° counterclockwise
   (c) $\sigma_x = -15$, $\sigma_y = -35$, $\tau = -15$ ksi
      (1) 80° counterclockwise
      (2) 15° clockwise
   (d) $\sigma_x = -100$, $\sigma_y = 350$, $\tau = 0$ MPa
      (1) 45° counterclockwise
         Answer: $\sigma = 125$, 125; $\tau = 225$
      (2) 70° clockwise
   (e) $\sigma_x = 80$, $\sigma_y = 80$, $\tau = -40$ MPa
      (1) 20° clockwise
      (2) 40° counterclockwise
   (f) $\sigma_x = 10$, $\sigma_y = 30$, $\tau = 5$ ksi
      (1) 70° clockwise
      (2) 20° counterclockwise
         Answer: $\sigma = 15$, 25; $\tau = 10$
   (g) $\sigma_x = 80$, $\sigma_y = -40$, $\tau = 0$ ksi
      (1) 15° counterclockwise
      (2) 90° counterclockwise
   (h) $\sigma_x = \sigma_y = 0$, $\tau = 30$ MPa
      (1) 35° clockwise
         Answer: $\sigma = -275$, 27.5; $\tau = 10$
      (2) 50° clockwise

4.9. Two pieces of material are keyed together and have stresses acting on the assembly, as shown in Figure P4.9. The keys resist shear, and the pieces themselves resist compression,

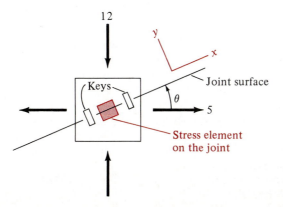

12

y

x

Joint surface

Keys

θ

5

Stress element
on the joint

FIGURE P4.9

but a state of tension across the joint would cause separation. What angle θ, or range of θ, can be used for the joint?

4.10. In Problem 4.9, suppose the joint is glued and that the glue can tolerate a tension stress of 2. What, then, are the limits on angle θ?

   Answer: $\pm 65°$

4.11. A solid steel shaft is subjected simultaneously to an axial load and a torque (Figure P4.11). Find the principal stresses and the maximum shear stresses in the plane of loading and show them on properly oriented elements.

   (a) Shaft diameter = 4 in, $F = +131\,000$ lb, $T = +65\,500$ lb·in.

   (b) Shaft diameter = 100 mm, $F = +5.0E + 05$ N, $T = -7.0$ kN·m

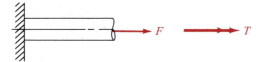

FIGURE P4.11

4.12. Sketch the directions of the shear stresses on the elements shown in Figure P4.12. Also sketch the element with maximum shear and indicate the directions of the stresses.

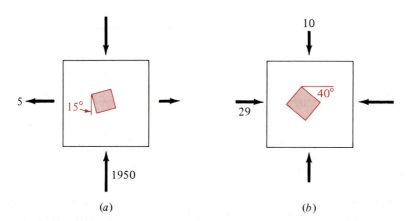

(a)   (b)

FIGURE P4.12

4.13. For the elements shown in Figure P4.13 (page 125), which are all principal elements (no shear), the relative sizes of the stresses are indicated by the vector lengths. Show the directions of the shear stresses on an element at 45°, and sketch the element in the deformed state, i.e., show the deformation caused by the shear stresses.

4.14. Find the shear and tension per unit length of weld in Figure P4.14 (page 125).

   (a) $F = 50\,000$ lb, $P = 0$. Dimensions are in inches.

      Answer: Shear = 7300 lb/in

   (b) $F = P = 1.0E + 0.5$ N. Dimensions in centimeters.

      Answer: Tension = 3.0 kN/mm

4.15. In Problem 4.14a, what value of F ($P = 0$) will result in 6000 lb/in of weld in tension? What is the resulting shear per inch of weld?

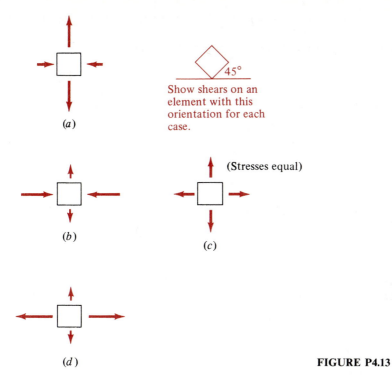

(a)

Show shears on an
element with this
orientation for each
case.

45°

(b)

(c)

(Stresses equal)

(d)

FIGURE P4.13

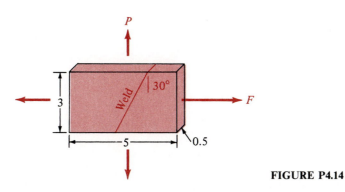

P

Weld

30°

F

3

5

0.5

FIGURE P4.14

4.16. In Problem 4.14*b* with *F* remaining as indicated, what value of *P* will result in a shear of 700 N/mm of weld? What is the resulting tension per millimeter of weld?
   Answer: $P = 86$ kN, tension = 2.93 kN

4.17. For the pipe of Example 4.7, if the weld stress in tension is not to exceed 15 ksi and the wall thickness is 0.25 in, what pressure will be permitted?

4.18. For the pipe of Example 4.7, what wrap angle θ is permitted if the weld stress in tension is not to exceed 7 ksi and the wall thickness is 0.375 in?
   Answer: 30°

4.19. Find the maximum shear stress and indicate the plane in which the stress is acting for

    (*a*) Problem 4.7

        Answer: (*a*) 3121.5, 45° to 1-*z*;   (*c*) 54, 45° to *xy*

    (*b*) Problem 4.8

    (*c*) Problem 4.11

        Answer: (*a*) 7.4, 67.5° clockwise from axis

    (*d*) Problem 4.12

        Answer: (*b*) 19.5, 45° to forces

CHAPTER 5

FACTORS OF SAFETY

Despite the occasional rhetoric of politicians and journalists, nearly everyone knows that the future can only be estimated. History is full of instances where a dike, for example, has been designed and built to withstand the worst flood in the preceding century, only to be immediately hit by an even worse deluge. If the maximum anticipated load on some machine part is, say, 20 000 lb, engineers must design for a greater load than that, to allow for the unforeseeable calamity. They do this by multiplying the load by a factor of safety (FS). Thus, if the factor of safety is 1.3, engineers aim for a strength of 26 000 lb. The extra 6000 lb of strength is built in "to be safe." There may also be uncertainties in the strength of the material chosen for the part, and these will be handled in a similar fashion. Other factors such as usable life or stiffness may be important in any given case. Factors of safety are applied to estimates of performance (gas mileage, load-carrying capacity, etc.) to ensure that advertised capabilities can be realized. The list of ways in which prudence tempers optimism is nearly endless.

Some factors of safety are chosen simply because experience has shown that they work. Factors of safety can also be calculated on a statistical basis and are then based on the mathematics of probability.

5.1 DEFINITIONS

Factor of Safety

Consider the simplest of loaded members, the tension bar in Figure 5.1. The stress in the bar is $\sigma = P/A$, or $\sigma = kP$, where k is a proportionality constant equal to $1/A$. The expression simply states that stress is proportional to load, i.e., doubling the

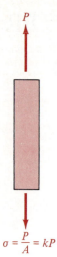

$\sigma = \dfrac{P}{A} = kP$ **FIGURE 5.1**
Stress is proportional to load in many members.

load doubles the stress. Because of uncertainties in the load, we may wish to apply a factor of safety FS_P to load P. This is the situation described in the introductory paragraph. Then

$$\sigma = k(FS_P)P \tag{5.1}$$

We require a material with a strength at least equal to Equation (5.1): $S = k(FS_P)P$. There will be uncertainties in our knowledge of the strength of the material chosen, so we divide our estimate of the strength by a factor of safety:

$$\frac{S}{FS_S} = k(FS_P)P \tag{5.2a}$$

Now the factors FS_S and FS_P may be put together on one side or the other of Equation (5.2a):

$$S = k(FS_P)(FS_S)P \tag{5.2b}$$

or

$$\frac{S}{(FS_P)(FS_S)} = kP \tag{5.2c}$$

and they may be lumped:

$$S = k(FS)P \qquad \text{or} \qquad \frac{S}{FS} = kP$$

The lumped factor is sometimes called an *overall factor of safety*. Solving for FS gives

$$FS = \frac{S}{kP} = \frac{\text{material strength}}{\text{computed stress}} \tag{5.3}$$

This is the usual definition of factor of safety.

In expression (5.2a), S/FS_S is often called a *design* or *working stress*. That is, the stress in the part is to be kept to that level or below. It allows for a margin of safety so that we do not design right up to the failure load. Similarly, $(FS_P)P$ is called the *permissible load*.

The outline above applies equally well to any part in which stress is proportional to load: tension, bending, torsion. Of course, the constant k is different, and P may be a moment or a torque, rather than a force. There are situations in which stress is not proportional to load and where the stresses derive from more than one load. We discuss these in Section 5.4.

Margin of Safety

The margin of safety (MS) indicates the amount by which the design (in terms of load or of stress) exceeds the load level. It may be expressed in nondimensional form as

$$MS = FS - 1 \qquad \text{(dimensionless)}$$

or in terms of strength or of load:

$$MS_{load} = \text{failure load} - \text{actual load}$$

$$= MS \times \text{actual load} \qquad \text{(units of load)}$$

$$MS_{strength} = \text{failure stress} - \text{computed stress}$$

$$= MS \times \text{computed stress} \qquad \text{(psi or Pa)}$$

5.2   STATISTICAL FACTOR OF SAFETY

Especially in cases of high production, quality is often expressed in statistical terms. A manufacturer may design for 5 percent failures, or 2 percent, or may aim for zero failures. The last is a typical requirement of hardware for space flights. A zero failure rate, by the way, requires not perfection, but rather ruthless elimination of components whose performance is below acceptability. For most manufactured products, however, there exists a balance between cost, which is usually associated with reliability, and what the market is willing or able to pay. The acceptable percentage of failures of a product is determined by this balance.

Two factors influence failure: the robustness of the product and the severity of the service conditions. There is, for example, a variation in the quality of apparently identical washing machines from any single manufacturer, and there is most certainly a variation in the treatment that the machines receive in service. Failure occurs when a machine happens to get rougher treatment than it can take, and the service time between failures is therefore associated with this event. The factor of safety is computed or estimated so as to maintain the life between failures at some acceptable minimum value.

In Chapter 2 distributions of variables were discussed. In this instance, there will be a distribution of stress, resulting from the different loads a part sees in service, and there will be a distribution of material strengths. If we are to deal with these parameters as distributions, we must refine the definition of factor of safety slightly:

$$FS = \frac{\text{mean strength}}{\text{mean computed stress}} \tag{5.4}$$

In the above definition, the strength is the level of stress defining failure. This could be the yield strength, ultimate strength, fatigue limit, or whatever property is appropriate. Computed stress is the engineer's best judgment of the stress level in the part. The object will be to determine the value of FS needed to keep failures to an acceptable level.

No Failures

Let us calculate first the required factor of safety if the goal is no failures. We base the factor of safety on the stress resulting from the highest conceivable load and the lowest conceivable material strength. The situation in terms of probability density distributions is shown in Figure 5.2. The curves are assumed to have finite tails. Thus, if the distributions are normal, the contributions of the curves outside the cut-off are neglected. ($0.27/2 = 0.135$ percent on each end, if the curves are truncated at 3 standard deviations. See Figure 2.8.)

The key to no failures is the separation of the two curves. To achieve this, $\bar{\sigma} + \Delta\sigma \leq \bar{S} - \Delta S$. Remembering the definition $FS = \bar{S}/\bar{\sigma}$, after a little algebra we get

$$FS \geq \frac{1 + \Delta\sigma/\bar{\sigma}}{1 - \Delta S/\bar{S}} \tag{5.5}$$

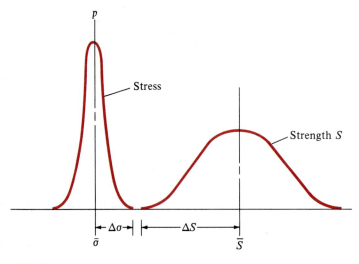

FIGURE 5.2
Orientation of stress and strength distributions for no failures.

The term $\Delta\sigma/\bar{\sigma}$ is the tolerance on stress, expressed as a fraction of the mean. Similarly, $\Delta S/\bar{S}$ is the tolerance on strength.

Relation (5.5) implies no failures, as long as the tolerances are in fact as advertised. Note that the calculation does not depend on the actual values of the mean strength and stress, only on the fractional tolerances of those parameters. We also note that the strengths and/or stresses need not be normally distributed. The basic requirement is that all strengths be greater than all stresses, as depicted in the figure; that principle can be applied even with nonnormal, nonsymmetric distributions.

Example 5.1. The mean value of the load on a part we estimate to be $\bar{F} = 95\,000$ N, and we believe it will vary $\pm 10\,000$ from that figure. Stress being proportional to the load in this piece, we then know that

$$\frac{\Delta\sigma}{\bar{\sigma}} = \frac{10\,000}{95\,000} = 0.105$$

Our quality-control (QC) department states it can give us strengths within 15 percent, so

$$\frac{\Delta S}{\bar{S}} = 0.15$$

What factor of safety should be used to avoid failures altogether? From expression (5.5)

$$FS \geq \frac{1 + 0.105}{1 - 0.15} = 1.3$$

Failures to Be Held to Preset Level

If the distribution curves (assumed normal) for strength and stress overlap, then we have a probability of failure. We need to compute that. The situation is depicted in Figure 5.3. The region of overlap is where the problem lies. We ask, What is the probability of finding strengths in the elemental range dS shown in the figure and, at the same time, finding stresses which are greater? Then we integrate to include all elements dS, that is, from $-\infty$ to $+\infty$.

$$p(S \text{ in } dS) = p_s\, dS$$

$$p(S \text{ in } dS \text{ and } \sigma > S) = p_S\, dS \int_{\sigma=S}^{\infty} p_\sigma\, d\sigma$$

$$p(\text{failure}) = \int_{S=-\infty}^{+\infty} p_S \left(\int_{\sigma=S}^{+\infty} p_\sigma\, d\sigma \right) dS$$

There is a better way. Instead of looking at two distributions, we examine the single distribution of strength minus stress. To determine that curve, we need to use the following theorems, which we accept as true:

1. If two variables are normally distributed, then the difference of the variables is also normally distributed.

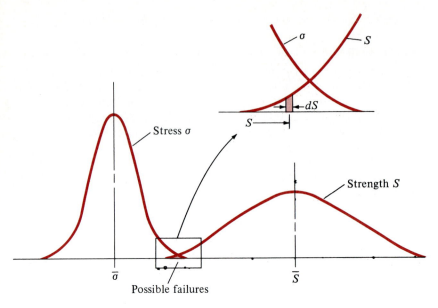

FIGURE 5.3
Stress and strength distributions where some failures are expected.

2. The mean of the difference distribution is the difference of the means of the individual distributions:

$$(\overline{S - \sigma}) = \bar{S} - \bar{\sigma}$$

3. The variance of the difference distribution equals the sum of the variances of the individual distributions:

$$SD_{S-\sigma}^2 = SD_S^2 + SD_\sigma^2$$

(Because we are using the symbol σ for stress, we adopt SD for the standard deviation of the population.)

The distribution $S - \sigma$ is shown in Figure 5.4. The area where $S - \sigma$ is negative represents the probability of failure, i.e., the fraction of failures which can be expected to occur. To compute that, we make the change to the standardized variable, as described in Section 2.7:

$$Z = \frac{(S - \sigma) - (\bar{S} - \bar{\sigma})}{SD_{(S-\sigma)}}$$

The distribution of Z is shown in Figure 5.5. The value of Z corresponding to the right-hand boundary of the failure area in Figure 5.5 is obtained by placing $S - \sigma$

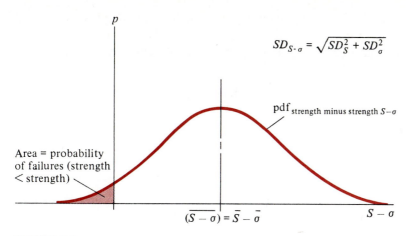

FIGURE 5.4
Distribution of strength minus strength.

equal to zero, so then

$$Z_f = \frac{-(\bar{S} - \bar{\sigma})}{\mathrm{SD}_{S-\sigma}} = \frac{\bar{\sigma} - \bar{S}}{\sqrt{\mathrm{SD}_S^2 + \mathrm{SD}_\sigma^2}} \qquad (5.6)$$

We can put this expression in terms of the factor of safety (FS) by using its definition:

$$\mathrm{FS} = \frac{\bar{S}}{\bar{\sigma}} = 1 - \frac{Z_f\sqrt{\mathrm{SD}_\sigma^2 + \mathrm{SD}_S^2}}{\bar{\sigma}} \qquad (5.7)$$

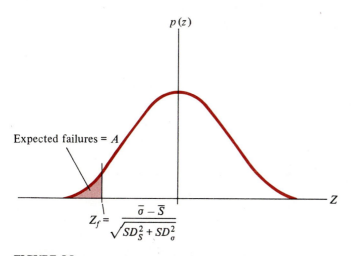

FIGURE 5.5
Distribution of standardized variable shows failure region corresponding to that in Figure 5.3.

A more useful relation results if we recall that 99.73 percent of all values fall within 3 standard deviations of the mean and neglect the other 0.27 percent. We then substitute $\Delta S/3 = SD_S$ and $\Delta\sigma/3 = SD_\sigma$. Putting the $\bar{\sigma}$ on the right-hand side of expression (5.7) under the radical and using a little algebra lead to

$$FS = 1 - \frac{Z_f\sqrt{(\Delta S/\bar{S})^2(FS)^2 + (\Delta\sigma/\bar{\sigma})^2}}{3} \tag{5.8}$$

from which

$$[9 - Z_f^2(\Delta S/\bar{S})^2](FS)^2 - 18FS + 9 - Z_f^2\left(\frac{\Delta\sigma}{\bar{\sigma}}\right)^2 = 0$$

This equation may be solved for the FS in terms of known strength and loading ($\Delta S/\bar{S}$ and $\Delta\sigma/\bar{\sigma}$):

$$FS = \frac{18 + \sqrt{324 - 4[9 - Z_f^2(\Delta S/\bar{S})^2][9 - Z_f^2(\Delta\sigma/\bar{\sigma})^2]}}{2[9 - Z_f^2(\Delta S/\bar{S})^2]} \tag{5.9}$$

The square root in this expression technically has a $\pm$ sign in front of it, but we will see in the example following that the negative square root is invalid in these problems.

We need to know Z_f as it relates to the shaded area under the curve in Figure 5.5, i.e., to the probability of failure. This is found with the standardized probability table, Table 2.1. The results for values of usual interest are displayed in Table 5.1. The bottom of the table gets somewhat beyond respectable design (5 percent failures), but it is included anyway.

To find the expected failures if the safety factor and the fractional tolerances on strength and stress are known, expression (5.8) is solved for Z_f:

$$Z_f = \frac{3(1 - FS)}{\sqrt{(\Delta S/\bar{S})^2(FS)^2 + (\Delta\sigma/\bar{\sigma})^2}} \tag{5.10}$$

and the expected failure rate is then found from Table 5.1.

TABLE 5.1
Values of the standardized variable for given failure levels

Failures, %	Z_f
0.1	−3.090
0.25	−2.808
0.5	−2.576
1.0	−2.326
1.5	−2.170
2.0	−2.054
2.5	−1.960
3.0	−1.880
4.0	−1.751
5.0	−1.645

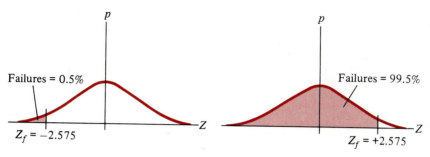

FIGURE 5.6
Two solutions for Z_f.

Example 5.2. For the case of Example 5.1, what factor of safety should be used to hold the failure rate to 0.5 percent?

Entering expression (5.9) with these numbers (and putting a $\pm$ sign before the square root), we have

$$FS = \frac{18 \pm \sqrt{324 - 4[9 - 2.576^2(0.15^2)][9 - 2.576^2(0.105^2)]}}{2[9 - 2.576^2(0.15^2)]}$$

$$= 1.017 \pm 0.159 = 1.176, 0.858$$

Obviously the value we want is 1.176, but we ought to be curious about the other result. Note in Equation (5.9) that Z_f appears only as a square. Thus, if we solve that relation for $Z_f = -2.575$, we will get a result for $Z_f = +2.575$. On the distribution curve for Z, you will thus see that this would be $1 - 0.5 = 99.5$ percent failures, as illustrated in Figure 5.6. Most of the stress distribution would be above the strength distribution; i.e., the mean stress is well above the mean strength. In fact, by applying the definition of safety factor, $S = 0.858\bar{\sigma}$ (99.5 percent failures). The positions of the distribution curves shown in Figure 5.3 are reversed for this case.

Example 5.3. In Example 5.2, if a factor of safety $FS = 2$ is chosen, what failure rate is to be expected?

For this calculation, we use Equation (5.10):

$$Z_f = \frac{3(-1)}{\sqrt{0.15^2(4) + 0.105^2}} = -9.44$$

This is over 9 standard deviations removed from the mean ($Z = 0$) and represents a situation where the strength distribution is well above the stress distribution, the case we discussed earlier in the paragraph under "No Failures." In fact, we computed in Example 5.1 the value of FS necessary to have zero failures: $FS = 1.3$. Thus, we should have known that a factor of safety of 2 would also yield no failures.

Let us revise the problem and ask what failure rate results for $FS = 1.1$. Using relation (5.10) again, we have

$$Z_f = \frac{3(0.1)}{\sqrt{0.15^2(1.1) + 0.105^2}} = -1.586$$

Turning to Table 5.1, we find that this corresponds to something more than 5 percent failures.

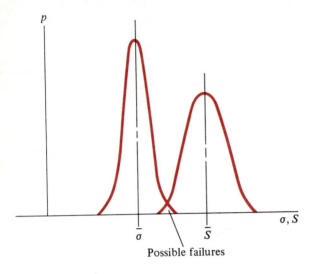

FIGURE 5.7
When the distributions are narrow (small standard deviation), small shifts cause large changes in the overlap, i.e., possible failures.

We have computed several results for the same problem. Here they are, tabulated:

FS	1.1	1.176	≥1.3
Failures	>0.05	0.005	0

It is significant that if we raise the value of FS from 1.1 to 1.3 (20 percent change), we reduce the failures from more than 5 percent to none at all (within the limitations we discussed in connection with Figure 5.2). A change in FS from 1.1 to 1.176 is even more striking: the failures are reduced from more than 5 percent, to 0.5 percent. Why this sensitivity to FS? The answer is found in the shapes of the distribution curves. Recall that the tolerance on stress was 0.105 and that for strength 0.15. Thus the distribution curves are relatively quite narrow, and small shifts in their positions cause large changes in the area overlap, as seen in Figure 5.7.

For the numbers of this example, one would certainly opt for FS = 1.3, with the prospect of no failures, in preference to, say, FS = 1.1, with more than 5 percent failures, or even FS = 1.2, with about 0.5 percent failures.

Let us look at some other combinations of stress and strength tolerances. Table 5.2 shows some results for a strength tolerance of 10 percent. The striking point of the table is the fact that with 10 percent quality control, which is quite loose, and a load variation of 30 percent, a factor of safety of only 1.34 is needed to hold the rate of failure to 1 in 1000 (0.1 percent). In the environment of an even more lenient quality control of 15 percent, with load variation of 100 percent, a factor of safety of 1.80 will hold the failure rate to 1 percent.

These numbers seem quite low in comparison to those usually found in practice. Why is this? First, the analysis assumes that the load and strength distributions are normal. They may, in fact, be skewed, nonsymmetric. Second, it is assumed that the load and strength distributions are known with certainty. This may be true for

TABLE 5.2
Factor of safety for quality control = 10 percent (loads and strengths normally distributed)

Failures, %	Load variation, %		
	10	20	30
0.1	1.16	1.24	1.34
2	1.10	1.16	1.22
5	1.08	1.13	1.18

parts in a tightly run operation, but there will always be many devices made with little or no organized quality control. Production testing may be done in effect by the customer. And regardless of the care taken in production, the loads to be encountered are often known only vaguely.

The above discussion and conclusions were based on distributions of strength and of stress which are normal. Static strengths follow such patterns, but fatigue strengths are lognormal. An analysis can be pursued along the same lines with the assumption of a normal distribution of stresses and a lognormal distribution of strengths, but the complexity is greater than the accuracy of the input—it is not worth it. We saw in Chapter 2 that one cannot practically differentiate between a normal and a lognormal distribution for standard deviations equal to about 25 percent of the mean and less. To the extent that a fatigue strength distribution satisfies this limitation, it can be considered normally distributed and the conclusions of this paragraph can be applied.

Finally, we should remember that the above calculations of the factor of safety assumed a tolerance on load and strength of 3 standard deviations. That is, the tails of the distribution curves beyond 3 standard deviations were neglected. In Section 2.6, we saw that this meant that 0.27 percent of the values in each distribution were neglected. In calculating safety factors for very low failure rates, the neglected tails of the distributions can be significant. One could use a tolerance of, say, 4 standard deviations, excluding then only 0.006 percent of the values. The 3 in the denominator of expression (5.8) would then be replaced by a 4, and expressions for FS and Z_f follow. As we have seen, however, in such cases it is better to calculate FS on the basis of no failures.

5.3   SOME PRACTICAL GUIDELINES

In practice, factors of safety are obtained from several sources.

First, in a manufacturing business, factors of safety are often set by experience in the design and manufacture of a particular product. If possible, a program of testing of the complete product is used to simulate the customer's experience. Factors of safety are then applied to keep product failure within tolerable limits. Usually the process is a trade-off between the cost of reducing failures and the price that can be realized for the product.

Second, where the public safety is involved, factors of safety or permissible failure rates are fixed by law or by a society code to which the manufacturer subscribes. Examples are boiler codes, pressure piping codes, and in recent years an entire set of very restrictive codes dealing with nuclear power installations. Naturally, as the science of design develops in a given technology, the necessary factors of safety decrease.

Third, engineers faced with a new problem must fix the factor of safety themselves. They must consider the uncertainties of the application and the other aspects, such as the public safety. Vidosic [5] suggests the following guidelines.

1. FS = 1.25 to 1.5 where component performance is well known, as from extensive testing, and loads are well known, as from experience. This range is used extensively where there is a necessity to minimize weight. Aircraft and now automobiles are the best examples.
2. FS = 1.5 to 2 for well-known materials under reasonably constant environmental conditions, subjected to loads and stresses that can be determined readily.
3. FS = 2 to 2.5 for average materials used in ordinary environments and subjected to loads and stresses that can be evaluated.
4. FS = 3 to 4 for untried materials used under average conditions of environment, load, and stress. This applies where stresses or environments are uncertain but the materials are better known and where safety is an overriding concern and extra cost or bulk is acceptable. Nuclear power plants, tramway or elevator cables, components of ski lifts are examples.

One may turn to handbooks in the hope of finding numbers considered suitable in specific problems. Some frustration results, for the factor may be implicit in a successful formula, or it may be stated to be the designer's decision based on the uncertainties of the problem. *Kent's Mechanical Engineers' Handbook* [6] is specific only to the extent of stating, for example, that for repeated shock loading the factor should be "not less than 10," or if strains are of importance, the factor should be very large, "possibly as high as 40." We point this out not by way of criticism of such handbooks, but rather to emphasize that factors of safety are very much the outcome of experience with successful (and unsuccessful) designs of particular engineering products.

The designer must be prudent. Designers and builders are now regularly taken to court for negligence. Where there is some question about the adequacy of the FS and the cost of a greater factor is not excessive, the greater factor should be used.

5.4   OTHER CONSIDERATIONS

Parameters Other than Stress Are Sometimes Important

Our discussion of factors of safety has, until now, implicitly assumed that failure occurs by yielding or fracture of the material under load. Failure in the broader sense is failure to perform the function for which the part was designed. Excessive deflection of a gear shaft under load, for example, may cause interference in a gear

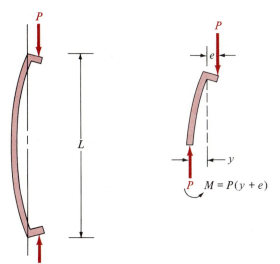

FIGURE 5.8
Eccentrically loaded column. Stress is not proportional to load.

set or improper clearance on a magnetic disk drive. The stiffness of the shaft, in other words, can be very important. It follows naturally that we would apply a factor of safety to our stiffness calculation, as well as to the calculated strength in yielding or fracture. Similar problems arise in many situations where some aspect of design in addition to element integrity is critical.

Stress Not Proportional to Load

In many circumstances, the stress in a part is not proportional to the load. A simple example is an eccentrically loaded column, as shown in Figure 5.8. A free-body diagram shows that the maximum moment occurs at the center of the column. The stress $\sigma = Mc/I$ is proportional to the moment, which in turn depends on F and y:

$$\sigma = kM = kP(y + e)$$

When the force increases, so does the moment arm y; hence the stress rises faster than the force. If an overall factor of safety is to be used, it should be applied to P. Of course, separate factors reflecting the uncertainties in material strength and in the load may be used.

Combinations of Loadings

The factor of safety to be applied ought to be determined by how accurately the loads are known and how accurately we can calculate the stresses resulting from those loads.

When several loads are to be applied to a part, we may have to deal with different amounts of uncertainty. For example, consider a simply supported beam subjected to a precisely known concentrated load of 100 lb and a distributed load of

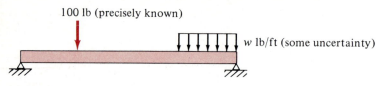

FIGURE 5.9
Example of beam loading.

FIGURE 5.10
Example where moments and hence stresses and deflections due to loads (F and P) are not independent.

approximately w lb/ft, as shown in Figure 5.9. In this simple beam, as long as deflections are small compared to the length of the beam, we can use superposition. The stresses and deflections resulting from the two loads can be calculated as the sum of the stresses and deflections which would result from each of the loads by itself. In other words, we can treat the loads as independent loads, with the effect of each being independent of the other.

If a factor of safety were to account for uncertainties in loading only, it would clearly be applied to the imperfectly known distributed load w only, and not to the precisely known 100-lb load. The factor of safety is intended, however, to account also for uncertainties of strength and imprecision of calculation. These apply to both loads. A sensible approach would be to choose an FS appropriate to the uncertainties in material strength and to apply it to the strength. The 100-lb load being known precisely, no FS need be applied to it. A factor of safety would be applied to w, reflecting the uncertainties in the distributed load.

As a designer, you should approach these cases of multiple loads with some caution, for the stresses and deflections of the one loading may depend on the other loading, and then superposition may not be used. For example, instead of the beam with two lateral loads of Figure 5.9, we might have a beam with lateral loads plus an axial load, as sketched in Figure 5.10. The moments (and hence stresses and deflections) resulting from these two loads are not independent. The moments due to the axial load P clearly depend on the deflections resulting from the lateral loads F_1 and F_2. You cannot calculate the effects independently and then add them. The factors of safety should be applied after the strength relation has been derived.

PROBLEMS

5.1. The stiffness of a gear shaft is very important. Suppose (as suggested in Section 5.4) we set out to apply factors of safety to both the applied load and the allowable deflection. Does the FS on deflection in effect apply an additional FS to stress?

5.2. We want to keep the deflection of a tension member within a certain limit. To be sure, we apply an FS to that limit. Also, we do not want failure, so we apply an FS to S_y. For good measure, we apply an FS to the steady axial load. Discuss the redundancy here.

5.3. The scatter of strength on the plot of the fatigue test results on a material at 10^8 cycles is ± 20 percent of its mean value. The service load varies ± 50 percent from its mean value. We can tolerate no more than 0.5 percent failures. What should be the FS on a statistical basis?

  Answer: 1.50

5.4. What is the significance of an FS of less than 1?

5.5. We wish to design a tension member to carry a load averaging 10 000 lb. This will vary from a high of about 11 000 to a low of about 9000 lb. Assume that these variables are normally distributed. Our quality control on strength is ± 5 percent.

  (*a*) Determine what FS we should use if we can tolerate 2 percent failures.

  Answer: 1.08

  (*b*) Determine what FS we should use if no failures are the requirement.

  Answer: 1.16

5.6. Show the joint distribution of load and strength if a 50 percent failure rate is anticipated. What is the FS?

5.7. A machine element is designed for a load which varies with a normal distribution from 10 000 to 16 000 lb. For 0.1 percent tolerable failures and an FS of 1.3, what should be the average strength and the tolerance on the strength?

  Answer: 16 900 lb, 2310 lb

5.8. Artillery shells are intended to *fail* when the charge ignites, and similar situations exist in the design of safety links, fuses, and blowout plugs. Sketch the load and strength probability density curves for such items.

5.9. Quality control in a plant is ± 4 percent. The design strength of a bar is 20E + 04 N. The FS being used is FS = 1.1, and the loading in the bar is expected to vary ± 3 percent. What failure rate can be expected?

  Answer: Less than 0.5 percent

5.10. Refer to the beam example of Figure 5.9. Assume the precision of the material specifications is described by a standard deviation of 5 percent of the mean value, and the load is known to a standard deviation of 12 percent of its mean value. Choose appropriate factors of safety.

CHAPTER

6

STATIC (MONOTONIC) MEASUREMENT OF MATERIAL PROPERTIES

The most important factor in the choice of a material for design is its mechanical properties. They are readily available from handbooks and vendors; some are given in this text in Appendices C, D, and F. You, as an engineer, should know how these properties are obtained, in order to evaluate their significance for a particular problem. The common static tests and the properties deriving from them are described in this chapter. In Chapter 8 we discuss properties obtained under dynamic conditions (fatigue).

6.1 TENSION TEST

The tension test, because of its simplicity, has become the traditional source of common material property data. Naturally, the properties deriving from it have direct applicability to many machine members loaded in tension, but the results, as we will see, can be applied to other types of loading as well.

Figure 6.1 is a photograph of a universal tension machine (*universal* means "for tension or compression"). The load is applied by a double-acting hydraulic cylinder under the lower specimen grip. The cathode ray tube (CRT) displays data during and after the test, and the printer-plotter provides a permanent record. The load is measured by an electronic load cell, making computerized recording and test

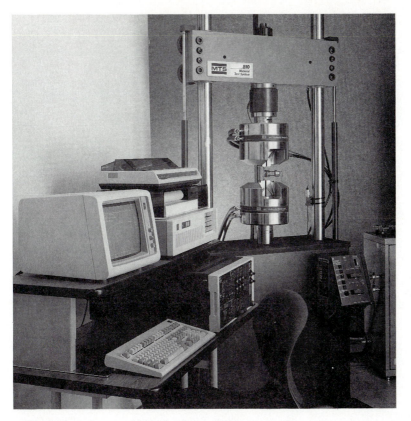

FIGURE 6.1
Universal testing machine, with CRT and loading control. The specimen in this machine has just been broken. (*Courtesy, MTS Systems Corp.*)

control possible. Such machines thus have the capability of running preprogrammed tests, i.e., maintaining loads for given times, producing selected rates of load application or of strain, cycling stress or strain, etc.

Test specimens vary, depending on the material and use. Naturally, flat specimens are usually used in testing sheet or thin plate and round specimens in testing rod. The standard round specimen is shown in Figure 6.2, screwed into testing-machine grips. These specimens are 0.505 in in diameter, a figure chosen to give a convenient cross-sectional area of 0.2 in[1].

Since the deflection of most engineering materials under load is quite small, the measurement of this quantity is delicate. Shown in Figure 6.2 is an extensometer, employing a lever to multiply the deflection and a linear differential transformer[1] to measure it. Electric resistance gages (strain gages) are frequently employed, and

---

[1] This instrument is simply a transformer with a movable core. The output is very sensitive to core position.

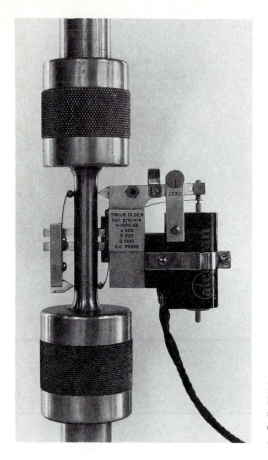

FIGURE 6.2
Standard specimen mounted in grips with extensometer attached. (*Photograph courtesy of Tinius Olsen Testing Machine Co., Inc., Willow Grove, Pa.*)

optical schemes are occasionally used. The extension is measured over a specified length of the specimen, termed the *gage length*.

The data taken during a tension test are simultaneous measurements of the load and specimen elongation. Stress is inferred from the former and strain from the latter. Interpretation must now be given to the results.

Engineering Stress and Strain[1]

The concept of stress was discussed in Chapter 4. In calculating the tensile stress in a tension test, load/area (P/A), the question is what area A to use, for the specimen's cross section diminishes as the stress increases during the test. A is commonly taken

---

[1] These concepts are due to Baron Augustin L. Cauchy (1789–1857), a French mathematician. Cauchy practiced engineering for a time until, in his midtwenties, Laplace and Lagrange persuaded him to devote his time to mathematics. He was the first to give a rigorous proof of Taylor's theorem; Cauchy developed the wave theory of optics and contributed the concepts of stress and strain to elasticity.

FIGURE 6.3
Necked-down tension specimen.

as A_0, the original (unloaded) cross-sectional area, so the stress is then

$$\sigma = \frac{P}{A_0}$$

This stress is called simply the *stress* or *engineering stress* or *nominal stress*, meaning stress in name. These terms differentiate it from the *true stress*, or load over actual existing cross-sectional area, discussed later.

Engineering or nominal strain is, in similar fashion, the length change in an original unloaded length L_0:

$$\varepsilon = \frac{\Delta L}{L_0}$$

where L_0 is a portion of the central part of the specimen, termed the *gage length*, as indicated earlier. At a certain stage in the test of some ductile materials, the specimen necks down, and the length change becomes concentrated in a limited region of the specimen, as seen in Figure 6.3. Hence engineering strain depends on the gage length L_0 used. The standard tension specimen has a gage length of 2 in.

Material Behavior and Failure Modes

Two general types of material behavior are distinguished in the tension test: ductile and brittle. Ductile behavior is characterized by an ability to deform plastically, as with wet clay. The plastic deformation is the outward manifestation of slip of adja-

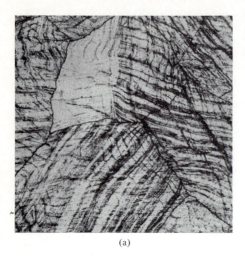

(a)   (b)

FIGURE 6.4
(a) Slip in steel (1000X, etched). (b) Cup and cone rupture of ductile material.

cent layers of material within the grains, a shear phenomenon, as shown in Figure 6.4a. Recall that the maximum shear stress occurs on planes at 45° to σ_1. The locus of such planes, with σ_2 and σ_3 equal to zero, is a cone, which is the form of ductile rupture, as seen in Figure 6.4b.

There is little, if any, plastic deformation with brittle behavior, such as with dry clay and glass. In these materials the resistance to slip is greater than the resistance to separation, hence rupture occurs on a plane normal to the tension stress. When the slip and separation resistances do not differ substantially, it becomes hard to say whether the behavior is ductile or brittle. A permanent elongation of about 5 percent is generally taken as the criterion of definition.

Most structural steel, brass, and aluminum are ductile under the slow loading and room-temperature conditions of the usual tension test. These materials are commonly called *ductile materials*. However, understand that an ordinarily ductile material may exhibit brittle behavior if the temperature is lowered and/or the loading rate is increased sufficiently. A certain combination of stresses will also cause brittle behavior; we discuss that later. Loads which vary in time (fatigue—Chapters 8 and 9) and loads applied for a long time at elevated temperatures (creep) will also cause brittle-appearing failures.

Most cast irons, hardened steels, ceramics, and concrete are usually brittle. However, brittle materials may be made to act in a ductile manner by increased temperatures and low rates of loading.

Stress-Strain Diagram

A typical stress-strain pattern of a tension test on a quite ductile material (soft steel or brass) is shown in Figure 6.5. You will see that until about one-half the maximum

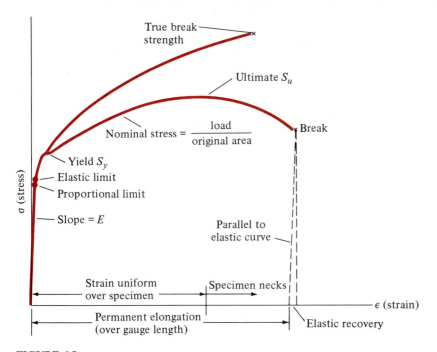

FIGURE 6.5
Tension stress-strain relation for material with observable yield (soft steel or brass). (Strain is exaggerated in the elastic region.)

load is reached, the strain is very small compared to what happens subsequently. With a strain scale like the one in the figure, you would barely distinguish the curve from the y axis. (We have exaggerated the strain in this region.) During this period the strain is uniform along the specimen—there is no localized deformation. The stress σ is observed to be proportional to strain ε (straight line through the origin). This is expressed in Hooke's[1] law:

$$\sigma = E\varepsilon$$

---

[1] Robert Hooke, an Englishman, lived from 1635 to 1703. At about age 18 he went to Oxford and there worked with Robert Boyle (Boyle's law) and also became interested in astronomy. At age 27 Boyle helped Hooke locate at the Royal Society as curator of experiments. Hooke became interested in microscopy and wrote a book on the subject. He also offered new insights into the nature of light. When he was 31, Hooke presented a paper to the Royal Society concerning the gravitational attraction of stars and the laws of their motion. Hooke was a contemporary of Newton (1642–1727); the two detested each other.

    Hooke is best known among engineers for *Hooke's law*, which states that the force on an elastic body is proportional to the deformation it causes. He also experimented with wood beams, noting that the fibers on the convex side were stretched and those on the concave side were compressed.

The proportionality constant E measures the stiffness of the material and is known as *Young's*[1] *modulus* or more commonly the *modulus of elasticity*.

Beyond a certain point, termed the *proportional limit*, the straight-line relation ceases. In the region up to the proportional limit and slightly beyond it, the strain will return to zero if the load is removed; the specimen regains its original size and shape. The upper limit for this to occur is called the *elastic limit*. In engineering steel, the proportional and elastic limits are very close together. Most engineers are not concerned with the distinction; moreover, these limits have little use in engineering.

Shortly above the elastic limit, the material extends rapidly, and the load may drop somewhat. This is known as the *yield point*, and the corresponding stress is called the *yield strength* S_y (sometimes with an additional subscript t, to denote tension). For the material of Figure 6.5, this point is easily discerned. Beyond the yield, the engineering stress rises again and reaches a maximum, termed the *ultimate tensile strength* S_u or S_{ut}. Localized deformation appears at this stage; the specimen necks down, as shown in Figure 6.3. While the true stress (load divided by actual cross-sectional area) continues to increase, the area decreases faster, so the load, and hence the engineering stress, falls off until the piece breaks. Rupture is thus spontaneous, unless the load is reduced. The ultimate tensile stress is sometimes called the *point of instability* for this reason.

If a specimen is loaded beyond the elastic limit, say point B in Figure 6.6a, and then the load is removed, the stress returns to zero along a line parallel to the proportional part of the curve, as shown in the figure. Thus a certain part of the strain is recovered; it is called the *elastic strain*. The remainder, the *plastic strain*, is permanent deformation, or set. This is sketched in the figure, along with the permanent deformation at rupture. The latter is frequently expressed as the reduction

---

[1] Thomas Young (1773–1829), an Englishman, mastered the classics at an early age and was able to teach them while he was in his teens. He completed a degree in medicine in Germany at age 23. Young's scientific contributions started in physics and progressed to engineering mechanics, and it was in these fields, rather than in medicine, that he did his most notable work. At age 29 he was elected a member of the Royal Society, to whose members he had already read several papers on acoustics and light, the earliest at age 20.

To mechanical engineers, Young's name is familiar in *Young's modulus* E, the ratio between stress and strain below the elastic limit. Young actually defined an associated property, the specific modulus, or the amount that a column of a material will shorten under its own weight, which he published in 1807. At that time the concepts of stress and of strain were not yet known; our present-day definition of Young's modulus is due to Henri Navier (1785–1836), a French professor of civil engineering and author of numerous works on elasticity.

Young made a number of observations to which the names of others have been attached. He observed the Poisson effect, i.e., when a material is stretched in one direction, there is a concomitant diminution in the other dimensions. He observed that yielding of materials limited the useful load they could bear. He concluded that impact loading was more severe than the same force applied statically. Young also studied the buckling of columns, offering the solution to the problem of eccentric loading and of columns with initial curvature. He observed that buckling failure was restricted to columns with what we term *slenderness ratios* greater than some limiting value, below which they fail by crushing, rather than by buckling.

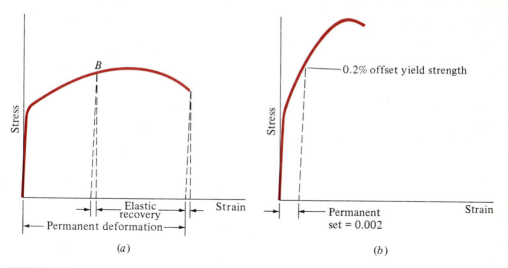

FIGURE 6.6

(a) Unloading-reloading action for soft steel. (Strain is exaggerated in the elastic region.) (b) Heat-treated or cold-rolled steel with definition of yield point.

in area (RA):

$$RA = \frac{A_0 - A_f}{A_0}$$

where the subscript f indicates fracture. This is often written as a percentage reduction in area and is commonly published as a measure of ductility.

If, after release of the load from point B, the load is reapplied, the stress follows the proportional line back up to point B and then the stress will trace out the remainder of the curve, as if the unloading-loading cycle had never taken place. The proportional limit, elastic limit, and yield point have been raised in the process. This is essentially what happens in cold rolling, and the ultimate strength is increased as well, as shown in Figure 6.6b. Heat-treated steels show the same pattern. The material is clearly less ductile; there is no clear yield point—the yielding occurs gradually in the plastic region. This has led to the definition of the yield as the stress required to produce a given permanent deformation, termed *offset*. The most commonly used figure is 0.2 percent. The stress is determined at the intersection of the stress-strain curve and a line through $\varepsilon = 0.002$ parallel to the elastic line, as in Figure 6.6b. Note that for a material which shows a clear point of yielding, this definition places the yield stress there. Generally the usefulness of materials is at stress levels below the yield.

The properties discussed above are published for nearly all engineering materials. Short lists can be found in Appendices C and D.

The area under the stress-strain curve up to the elastic limit represents recoverable energy per unit volume of material. That is known as the resilience. The area under the curve up to the fracture point is the energy per unit volume necessary to rupture the specimen and is called the *toughness*.

True Stress and Strain

TRUE STRESS. As earlier indicated, the *true stress* is simply the load divided by the area at the instant of the reading:

$$\sigma_t = \frac{P}{A}$$

The measurement of true stress is accomplished in the tension test by measuring the diameter, or other cross-sectional dimension(s), of the piece simultaneously with the load. When the piece necks down, these measurements are naturally made at the narrowest point.[1]

The *true fracture strength* is the true stress at rupture and is given the symbol σ_f. Some values for steel and aluminum are shown in Appendix C.

TRUE STRAIN. Since true stress is measured at the narrowest point of the specimen, an expression for true strain at that point is desirable. It can be developed by letting the change in strain in the straight specimen be equal to the length change divided by the length existing at the instant of measurement:

$$d\varepsilon_t = \frac{dL}{L}$$

which upon integration becomes

$$\varepsilon_t = \ln \frac{L}{L_0}$$

With deformation localized in the neck, this is no better than the nominal strain. That problem can be solved by dealing with the specimen dimensions at the critical point, the neck. Imagine a specimen with uniform cross section equal to the minimum diameter d. Since plastic deformation is very nearly a constant-volume process, we write

$$\frac{\pi}{4} d^2 L = \frac{\pi}{4} d_0^2 L_0$$

whence

$$\frac{L}{L_0} = \left(\frac{d_0}{d}\right)^2$$

Substituting in the expression for ε_t, we get

$$\varepsilon_t = 2 \ln \frac{d_0}{d}$$

---

[1] In actual fact, the stress situation when the specimen necks down becomes somewhat complex. It is triaxial, i.e., three-dimensional, and causes the final rupture to occur in tension—the flat part of the cup-and-cone fracture. The stress during necking, as computed by the load over the existing area, can be corrected to give the actual stress, but this is not of particular concern to design engineers. It is of interest principally to researchers in failure phenomena.

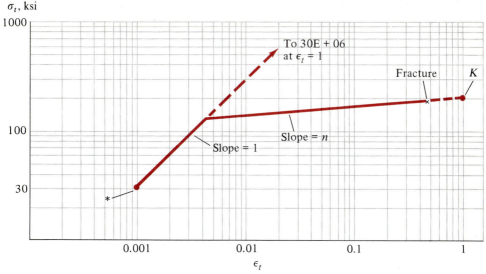

*30 ksi at ϵ_t = 0.001 corresponds to 30E + 06 at ϵ_t = 1

FIGURE 6.7
True stress–true strain plot (approximate for alloy steel). For materials having a marked yield region, a flat transition between the two lines occurs.

The *true fracture strain* or *ductility* is the true strain at rupture and is designated ε_f. If the ratio d_0/d in the expression for ε_t immediately above is expressed in terms of the reduction in area, then the true fracture strain becomes

$$\varepsilon_f = \ln \frac{1}{1 - \text{RA}}$$

Some values of ε_f are shown for steel and aluminum in Appendices C and D.

True stress–true strain plots on log-log paper result in straight-line segments, intersecting near the yield point, as shown in Figure 6.7. Below the yield (really the proportional limit)

$$\sigma_t = E\varepsilon_t$$

or
$$\log \sigma_t = \log E + \log \varepsilon_t$$

This is a straight line with intercept $\log E$ (at $\log \varepsilon_t = 0$, that is, $\varepsilon_t = 1$) and a slope of 1 (independent of the material).

Above the yield, the straight-line expression is

$$\log \sigma_t = n \log \varepsilon_t + \log K \tag{6.1}$$

The intercept is $\log K$, and the slope is n. By taking antilogs, Equation (6.1) gives

$$\sigma_t = K\varepsilon_t^n$$

Since the elastic part of the strain is very small compared to the plastic part, ε_t can be replaced with ε_{tp}, the true plastic strain:

$$\sigma_t = K\varepsilon_{tp}^n \tag{6.2}$$

where K is known as the strength coefficient and n as the strain-hardening exponent. Values for some steels and aluminums are shown in Appendix C.

RELATION BETWEEN TRUE AND ENGINEERING STRESS AND STRAIN BELOW THE POINT OF INSTABILITY. For true stress, recall that $\sigma_t = P/A$, where A is the actual area at the time the load is measured. The volume of material remains nearly constant, so

$$A = A_0 \frac{L_0}{L} \quad \text{and} \quad \sigma_t = \frac{PL}{A_0 L_0}$$

And since the engineering stress

$$\sigma = \frac{P}{A_0}$$

then

$$\sigma_t = \sigma \frac{L}{L_0} \tag{6.3}$$

The true strain was found to be

$$\varepsilon_t = \ln \frac{L}{L_0}$$

The engineering strain is defined as

$$\varepsilon = \frac{L - L_0}{L_0} = \frac{L}{L_0} - 1$$

whence

$$\frac{L}{L_0} = \varepsilon + 1$$

Then
$$\varepsilon_t = \ln(\varepsilon + 1) \quad \text{(below point of instability)} \tag{6.4}$$

We earlier stated that below the yield point the differences between the engineering and true stresses and strains are insignificant. For most metals the engineering strain at yield is 0.005 or less, so

$$\frac{L}{L_0} = 1.005$$

Then from (6.3) $\qquad \sigma_t = 1.005\sigma \simeq \sigma$

and from (6.4) $\qquad \varepsilon_t = \ln(\varepsilon + 1) = \ln 1.005$

$$= 0.00499 = 0.9975\varepsilon \simeq \varepsilon$$

Also the value of the (true) modulus of elasticity is

$$E_t = \frac{\sigma_t}{\varepsilon_t} = \frac{\sigma L}{L_0 \ln (\varepsilon + 1)}$$

$$= \frac{1.005\sigma}{0.9975\varepsilon} = 1.0075 \left(\frac{\sigma}{\varepsilon}\right) \simeq \frac{\sigma}{\varepsilon} = E$$

Hence, there is usually no need to measure true stress and strain to determine the true modulus of elasticity.

True and nominal stresses and strains differ substantially only in the plastic region, and engineering materials are seldom used in static applications at that level. (Building frames are sometimes designed into the plastic range. Also when the stresses on an element vary in time, in some instances they may be above the yield locally. This is discussed in Chapter 9.) Thus, engineering computations are carried out on the basis of nominal or engineering stress and strain. Although the ultimate strength is in the plastic region for ductile behavior, it is a universally published figure. Several other properties correlate with ultimate strength, so that they may be estimated from it.

Brittle Behavior

The action of brittle materials in the tension test is not as complex as that for the ductile case. There is little, if any, plastic flow, which is really the definition of brittleness. Typical stress-strain diagrams for such behavior are shown in Figure 6.8. Some are quite straight to fracture. In others the plot is curved, but we can usually define with reasonable accuracy a value of E for use in engineering calculations. In none of them is it possible to discern what could be called a yield or elastic limit. All we have is the ultimate strength, and this is considerably greater in compression than in tension.

Just about any ductile material will become brittle at a low enough temperature and/or if it is strained rapidly enough, as mentioned earlier. Triaxial tension also causes ductile materials to behave in a brittle manner because even though the three principal stress levels may be quite high, the maximum shear stress $(\sigma_1 - \sigma_3)/2$ may be relatively small. (It could, in fact, be zero, if the principal stresses are equal, as

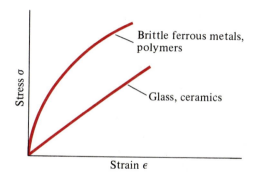

FIGURE 6.8
Stress-strain tension curve shapes for brittle materials (not to same scales).

FIGURE 6.9
Torsion-testing machine. (*Photograph courtesy of Tinius Olsen Testing Machine Co., Inc., Willow Grove, Pa.*)

discussed in Section 4.5.) Since yielding is a shear phenomenon, it cannot take place, and failure is by tensile rupture—brittle behavior. For this reason, sharp notches perpendicular to the tension direction are particularly bad as well as abrupt changes in cross sections.

6.2 TORSION TEST

The torsion test is not as common as the tension test. The results are not as useful, and the test has several limitations. Figure 6.9 is a photograph of a torsion machine.

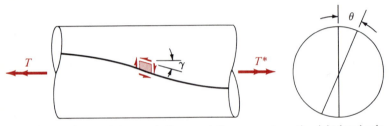

*The double-headed arrow indicates twisting moment per the right-hand rule.

FIGURE 6.10
Schematic of torsion test.

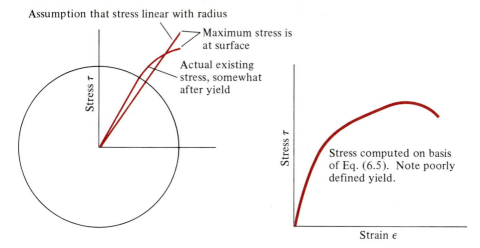

FIGURE 6.11
Stresses in torsion test.

The load (a moment) is applied, measured, and recorded graphically against the twist angle at the console unit on the right. The carriage with grip on the left is movable to accommodate specimens of different lengths. The usual specimen is a round bar with appropriate ends, usually squared, for insertion in the machine grips. A sketch of the bar under load is shown in Figure 6.10. A longitudinal element will have assumed the shape of a helix, and a strained element on the surface with stresses appears as in the figure. The theory underlying computation of these stresses is based on the assumption that diameters of cross sections remain straight, as shown at the right in Figure 6.10. It follows that shear strain (the angular distortion of an element γ, as shown) varies from zero at the center linearly to a maximum at the surface, and the shear stresses will do likewise as long as stress is proportional to strain, i.e., $\tau = G\gamma$. The proportionality constant G is called the *shear modulus* or the *modulus of rigidity*. It is analogous to the modulus of elasticity E for tension, and the two are related by $E = 2G(1 + v)$, a result from the theory of elasticity. Here v is Poisson's ratio, about 0.3 for most metals.[1]

The linear variation of nominal shear stress with distance from the center is shown in Figure 6.11. The nominal shear stress is greatest at the surface:

$$\tau_{\text{surf}} = \frac{16T}{\pi d^3} \tag{6.5}$$

With the stress varying from zero at the center to a maximum at the surface, yielding occurs progressively from outside inward, rather than all at once across the section,

---

[1] When a material is stretched in one direction, its dimensions in the orthogonal directions diminish. If x is the loading direction, then Poisson's ratio is $-\varepsilon_y/\varepsilon_x$. Simeon Denis Poisson (1781–1840), a French mathematician, succeeded Fourier as full professor at the Ecole Polytechnique at age 26. Poisson is best known for his work on definite integrals, electromagnetic theory, and probability. At 31 he was elected to the Académie Française, one of France's highest honors.

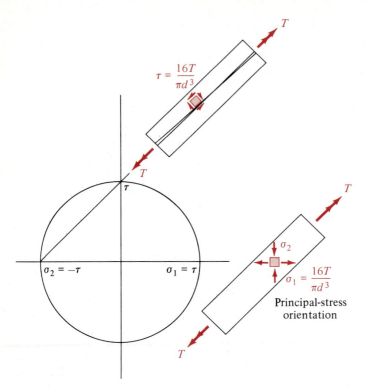

FIGURE 6.12
Mohr's circle for torsion test.

and a yield point and shear yield strength S_{sy} are thus not easily identified. The true stress shortly after yield is sketched in Figure 6.11. Since both the true and the nominal stresses must result in the same applied moment, the true stress is higher in the interior and lower at the surface than that calculated by Equation (6.5). Thus, the yield or ultimate shear stress, computed on the basis of Equation (6.5), will be greater than the actual existing stress. The material property so recorded is therefore nonconservative, except when it is used for computing shaft twisting strengths.

The stress element of Figure 6.10 is in pure shear, i.e., there are no normal stresses σ. The resulting Mohr's circle is shown in Figure 6.12. At 45° to the original element, i.e., at 45° to the axis of the piece, the principal stresses are found to be $\sigma_1 = \tau$ and $\sigma_2 = -\tau$, as shown in the figure. The tensile principal stress σ_1 will lead to a tension fracture for brittle materials. Thus most cast-iron breaks are on a 45° helix. You can show this with an ordinary piece of classroom chalk. If the material is ductile, however, the break will be square across, in shear. This is found with most steels, brass, and aluminum.

6.3   HARDNESS TESTS

Hardness tests measure the resistance of a material to penetration. Because hardness correlates well with material strength and the tests are easy and inexpensive to per-

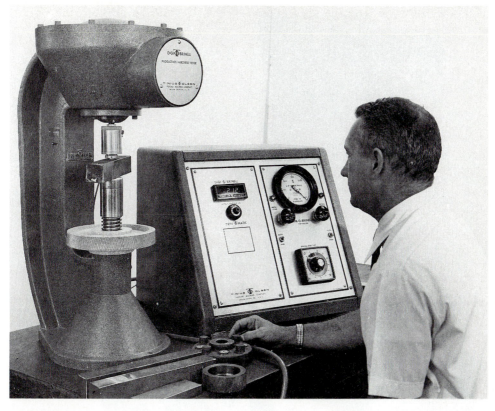

FIGURE 6.13
Brinell hardness tester. (*Photograph courtesy of Tinius Olsen Testing Machine Co., Inc., Willow Grove, Pa.*)

form and do not destroy the specimen, they are very useful for control purposes. The equipment can be placed on a table, and portable handheld machines are available, making for great flexibility.

The Brinell tester uses a 10-mm steel ball, which is pressed into the specimen by a force of 3000 kg (now 29 420 N). The Brinell hardness number is equal to the kilogram load divided by the area of the impression in square millimeters. Today that computation is done automatically and the result displayed digitally, as shown in Figure 6.13. The machine senses the depth of the penetration to get the area of the crater. Brinell hardness numbers H_B for steel range from 100 for low-carbon steel to 600 or more for high-strength alloy steels. These numbers have been found to relate well to the ultimate strength of steels:

$$S_{ut} \simeq \begin{cases} 500H_B \text{ psi} \\ 3.45H_B \text{ MPa} \end{cases} \tag{6.6}$$

The Rockwell machine is more widely used because it is smaller and somewhat easier and faster to operate. There are several scales; B and C are the ones most commonly used with metals. The readings are given as 75B, say, or 63C. The B scale uses a 100-kg (now 981-N) load on a $\frac{1}{16}$-in steel ball penetrator. Soft steels and

FIGURE 6.14
Standard Rockwell hardness tester. The weights for applying the load are discernible at the rear, and a multiplying lever is seen at the top. To the right of the instrument are several specimen platforms and another penetrator. The cases contain test pieces for calibration. A test piece is shown in the instrument. (*Courtesy, Clark Instruments.*)

brass ($\sim$80B) are generally read on this scale. The C scale employs a diamond cone with a load of 150 kg (now 1480 N). Most steels are read on Rockwell C (medium-strength steels $\sim$35C, high-strength steels $\sim$50C and up). In both scales a preload to set the specimen is used, and the hardness is obtained directly from a dial. Figure 6.14 shows a standard Rockwell machine. The machine in Figure 6.15 provides digital readout.

The relation of one hardness-test number to another is not exact; it depends on the material. That between Rockwell C, Brinell, and two other less frequently used tests for steel is depicted in Figure 6.16. The figure also relates hardness to ultimate tensile strength.

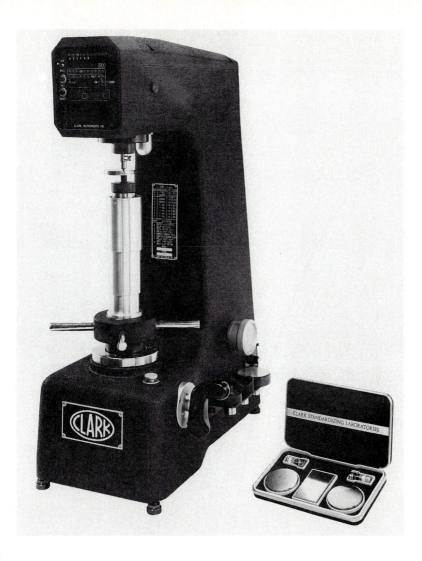

FIGURE 6.15
Rockwell hardness tester with digital readout. (*Courtesy, Clark Instruments.*)

6.4   OTHER STATIC TESTS

A number of other tests are commonly made in materials testing laboratories. Since the properties deriving from them have little applicability to the types of problems developed in this text, the tests are only mentioned. The following list is not exhaustive.

1. *Creep test.* Materials under high loads, and especially at high temperatures, will exhibit continuously increasing strain. This phenomenon is important in such things as turbine blades.

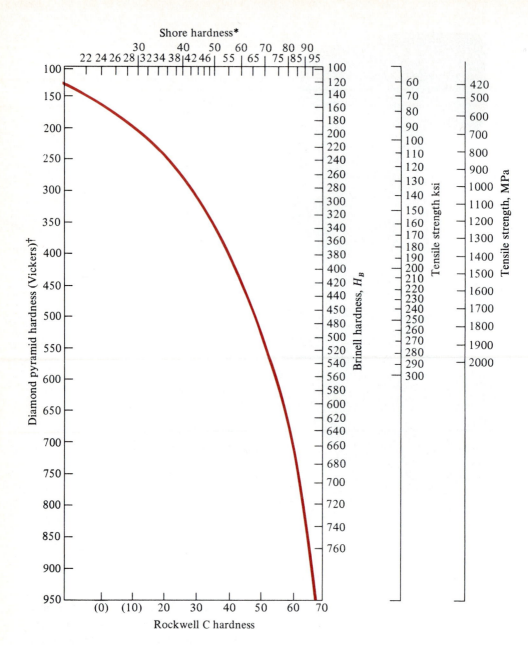

*Allows diamond-tipped mass to fall on specimen. Hardness read from the rebound height.

†Penetration test using small diamond-tipped pyramid indentor.

FIGURE 6.16

Approximate conversions between hardness scales and tensile strength of steel. (*Courtesy, International Nickel, Inc.*)

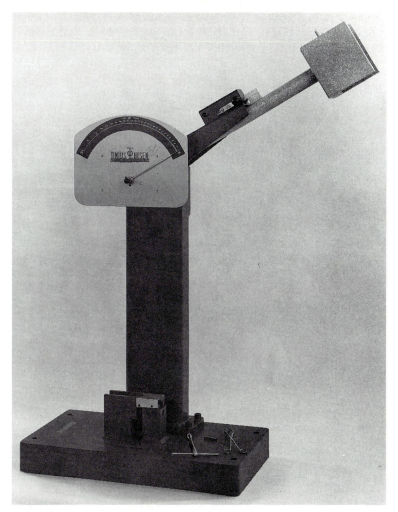

FIGURE 6.17
Charpy impact tester. (*Courtesy of Tinius Olsen Testing Machine Co., Inc., Willow Grove, Pa.*)

2. *Impact test.* The most common machine is shown in Figure 6.17. Typically, a notched specimen (notched so that it will break rather than fold over) is broken by being struck by a pendulum; the energy absorbed by the specimen is measured as that lost by the pendulum. The objective of the test is the study of shock resistance, i.e., energy absorption capability, particularly as a function of temperature. We mentioned in the tension-test discussion that ductile materials, if sufficiently cooled, become brittle. The temperature at which this happens is found by plotting energy absorption vs. temperature. At the critical temperature a sharp drop occurs.

3. *Environmental tests.* Since many engineering materials are used in corrosive atmospheres, tests attempting to duplicate a particular situation are frequently made.

In this category one may also include such extremes as nuclear bombardment, which alters material properties at the atomic level, and water, which erodes metal from elements like pump blades and casings.

PROBLEMS

Section 6.1

6.1. With brittle behavior in a tension test, why do we not see a yield point?

6.2. What accounts for the falling off of nominal stress after the ultimate strength?

6.3. When and why are nominal and true stresses nearly the same? When and why are they quite different?

6.4. Define the proportional limit and the elastic limit.

6.5. How is the modulus of elasticity measured?

6.6. What can one substitute for a yield strength for brittle materials?

6.7. Figure P6.7 shows a load vs. strain tension trace made on a specimen of 4140 normalized steel. The diameter of the specimen was 0.502 in. Because the extensometer might be damaged when the specimen broke, it was removed at $\varepsilon = 0.85$, interrupting the strain record. The load record, however, continued up to a maximum and then decreased. The

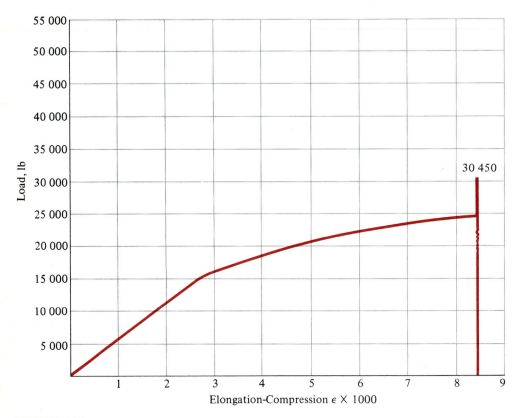

FIGURE P6.7

small wiggle just above 20 klb is where the break occurred.

(a) Find the modulus of elasticity for this material.

(b) Since there is no clear yield, find the 0.2 percent offset yield strength.

(c) What was the ultimate strength?

6.8. The data in Table P6.8 show measurements on a specimen of aluminum. Plot true stress vs. true strain and find

(a) The strength coefficient K

(b) The strain-hardening exponent n

(c) E_{true}

(d) E_{eng}

(e) True break strength

(f) Ultimate tensile strength

(g) Percentage reduction of area

TABLE P6.8

Load	d	Extension × 1000
0	0.5062	0.91
1000	0.5062	1.93
2000	0.5061	2.92
3000	0.5060	3.95
4000	0.5060	4.93
5000	0.5059	6.94
6000	0.5058	6.79
6800	0.5057	7.55
7200	0.5056	
8000	0.5054	
8200	0.5035	
8320	0.5010	
8440	0.5005	
8600	0.4945	
8840	0.487	
8460	0.454	
8060	0.430	
7400	0.400	

Section 6.2

6.9. Explain and account for the types of breaks encountered in the torsion test.

6.10. Why is the yield in a torsion test poorly defined?

Section 6.3

6.11. An alloy steel has a hardness $H_B = 460$. Estimate the strength of the steel in pounds per square inch. (What strength is it?)

6.12. A medium-carbon steel has a hardness $H_B = 500$. Estimate the strength in megapascals.

6.13. A steel has a hardness of Rockwell 40C. Estimate its tensile strength in kilopounds per square inch.

CHAPTER
7

PREDICTION OF STATIC FAILURE

Except for safety devices, military ordnance, and a few other items which are intended to fail, a large part of design is the art and science of avoiding material failure under service loadings. This implies the necessity of predicting failure conditions, so as to avoid them. Even in the design of devices whose function is to fail, the same information is essential.

Failure theories typically seek to correlate a calculated level of stress or energy with material properties that have been measured in a standard test, such as the tension test. One of the classic theories dates from Coulomb in the eighteenth century, and the majority of the others of note are from the nineteenth century. The current effort is concentrated on fracture mechanics, which is based on the energy exchanges involved in the propagation of cracks. This is very important when the presence of flaws is highly likely or must be assumed.

7.1 FAILURE THEORIES, DUCTILE BEHAVIOR

Maximum-Shear-Stress Theory

This theory, also known as the *Coulomb*[1] *theory*, asserts that failure occurs by yielding when the maximum shear stress reaches a critical level. That critical level is most easily determined in a tension test. In that test the load is uniaxial, and the maximum shear is one-half the tension stress, as we saw in Section 4.4. Thus the critical shear stress at yield is $S_y/2$. The maximum shear stress for general loading we deduced in Section 4.5:

$$\tau_{max} = \frac{\sigma_1 - \sigma_3}{2}$$

(Recall the order: $\sigma_1 > \sigma_2 > \sigma_3$.) Thus, for safe design

$$\tau_{max} = \frac{\sigma_1 - \sigma_3}{2} \leq \frac{S_y}{2\text{FS}}$$

or

$$\sigma_1 - \sigma_3 \leq \frac{S_y}{\text{FS}} \tag{7.1}$$

where FS is the factor of safety, discussed in Chapter 5.

Relation (7.1) can be shown on a plot of principal stresses σ_1, σ_2, σ_3, as in Figure 7.1. The stress in the z direction, $\sigma_z = 0$, is one of these principal stresses. It will be σ_1, σ_2, or σ_3, depending on the signs of the other two. In the first quadrant, two of the principal stresses are positive, and the three Mohr's circles appear as sketched. The two positive principal stresses then become σ_1 and σ_2, and $\sigma_z = 0 = \sigma_3$. Relation (7.1) reduces to $\sigma_1 \leq S_y/\text{FS}$, a vertical line. In the fourth quadrant, one principal stress is positive, and the other is negative. The relative positions of the Mohr's circles are sketched in the figure, and $\sigma_z = 0$ is in the middle position, σ_2. The positive principal stress is σ_1, and the negative stress becomes σ_3. Relation (7.1) is as written (neither σ_1 nor σ_3 is zero) and becomes the line of slope $= 1$, as shown. In the third quadrant, two principal stresses are negative, hence $\sigma_z = 0$ is the largest and becomes σ_1. Relation (7.1) then reduces to $-\sigma_3 \leq S_y/\text{FS}$, or $\sigma_3 \geq -S_y/\text{FS}$, the horizontal line in the figure.

---

[1] Charles Augustin de Coulomb (1736–1806) was a French engineer and physicist. In 1773, Coulomb published a work on the strength of materials, from which is derived the theory discussed here. He is also known to engineers for his study of frictional forces (1779). Dry friction (constant frictional force, independent of velocity) is frequently called *Coulomb friction.* He also developed a method of working under water, which underlies modern techniques.

Coulomb is best known for his research in electricity and magnetism. Coulomb's law of electrostatic attraction and repulsion states that the force between charges is proportional to the product of the charges and varies inversely as the square of the distance between them. The unit of electric charge is today called a *coulomb.*

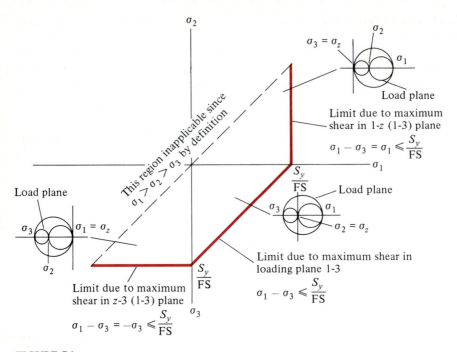

FIGURE 7.1
Safe stress levels per maximum-shear-stress theory for biaxial loading—ductile material. Maximum shear in all cases on planes at 45° to 1-3 directions. For axis labeling, see the Mohr's circles.

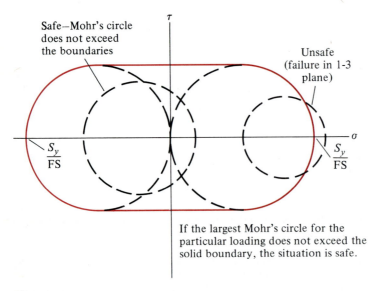

FIGURE 7.2
Maximum-shear-stress theory on τ-σ plot.

The theory can also be shown as an envelope within which the Mohr's circle(s) must fall for safety. This representation appears in Figure 7.2. If you sketch the Mohr's circles for the three possible arrangements depicted in Figure 7.1, you will see that it is sufficient to draw the circle for the loading plane only; if that circle exceeds the boundary, the largest Mohr's circle does also.

Example 7.1. A closed-end tube of material with a yield strength of 60 ksi has a 3-in outer diameter (OD) and a wall thickness of 0.050 in. It is subjected to an internal pressure of 1200 psi. Find the margin of safety in pounds per square inch (how many more pounds per square inch could we add before failure occurs?).

A stress element is shown in Figure 7.3a, and the stresses are computed by using the thin-wall equations as for Example 4.7:

$$\sigma_1 = \frac{pd}{2t} = \frac{1200 \times 3}{2 \times 0.050} = 36\,000$$

$$\sigma_2 = \frac{p\pi d^2/4}{\pi dt} = \frac{pd}{4t} = \frac{\sigma_1}{2} = 18\,000$$

$$\sigma_3 = 0$$

(In this thin-wall pipe problem, the diameter is taken as the OD, 3 in, for convenience. You could argue that the average diameter 2.975 should be used. The difference is insignificant in this case.)

The σ_1, σ_2, and σ_3 above are the principal stresses. They are labeled according to the convention established: $\sigma_1 > \sigma_2 > \sigma_3$. Failure will occur in the 1-3 plane, a shear failure at 45° to the 1 and 3 directions, as shown in Figure 7.3b. Since we want to find the margin of safety (MS) in terms of pressure, expression (7.1) is used as an equality with FS = 1. In other words, we will find the limiting value of stress:

$$\sigma_{1,\text{fail}} - 0 = S_y = 60 \text{ ksi}$$

Since the stresses are proportional to the pipe pressure,

$$MS = \frac{60 - 36}{36}(1200) = 800 \text{ psi}$$

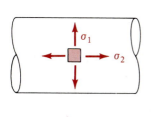

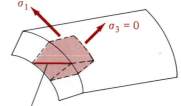

Plane of maximum shear stresses

(a)   (b)

FIGURE 7.3
Example 7.1. (a) Stress element. (b) Plane of maximum shear stress.

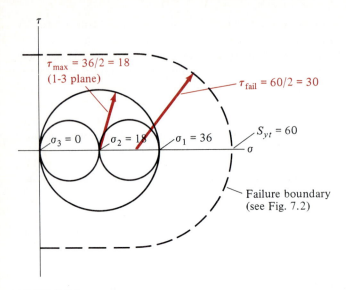

FIGURE 7.4
Figure 7.2 representation for failure for Example 7.1. Stresses in ksi.

One may also approach the problem by drawing Figure 7.2 and showing the element Mohr's circles on it. This is done in Figure 7.4, where the maximum shear stress is seen to be 18.

The failure shear stress is that causing yield in the tension test:

$$\tau_{yield} = \frac{S_{yt}}{2} = 30 \text{ ksi}$$

Since all the stresses are linear with the pipe pressure, the MS is

$$\frac{30 - 18}{18}(1200) = 800 \text{ psi of pressure}$$

Figure 7.5 is a sketch of a σ_1 vs. σ_2 plot (Figure 7.1) for this problem. The FS is, again, taken as 1. The point (36, 18 ksi) is plotted. Since the stresses are directly proportional to the pressure, we can construct a "load line" through this point. From the geometry the additional pressure necessary to reach the fail point is, as before,

$$\frac{bc}{ab}(p) = \frac{60 - 36}{36}(1200) = 800 \text{ psi}$$

Example 7.2. A $\frac{3}{4}$-in-drive socket wrench is shown in Figure 7.6. It is made of low-alloy steel with $S_{yt} = 90$ ksi. A 180-lb mechanic puts a prop under the extension and then stands on the end of the handle to loosen an obstinate bolt. Is the mechanic close to failing the handle or the drive?

We proceed by finding what weight W it takes to produce failure. Then we can compare that with the weight of the actual mechanic.

Figure 7.7a is a free-body diagram of the handle, which is a cantilever beam, and T and W are the reactions of the extension shaft. The bending moment increases from

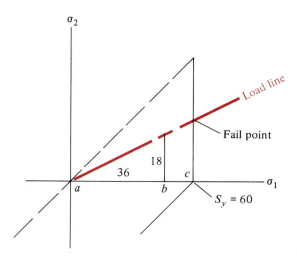

FIGURE 7.5
Plot of σ_1 vs. σ_2 (Figure 7.1) for
Example 7.1.

zero, where the mechanic stands, to $20W$ where the handle and extension meet. The
stress there is maximum on the top and bottom surfaces and is expressed by the classic
beam formula

$$\sigma_{\text{handle}} = \sigma_H = \frac{Mc}{I} = \frac{20Wd/2}{\pi d^4/64} = 483W$$

The stress element has only this tension stress, shown in Figure 7.7b. Thus $\sigma_1 = \sigma_H$ and
$\sigma_2 = \sigma_3 = 0$. By relation (7.1), the limit on σ_H is S_y. Making that exchange, we get

$$W_{\text{at fail}} = \frac{90\,000}{483} = 186$$

The handle barely survives the 180-lb load of the mechanic on its end.

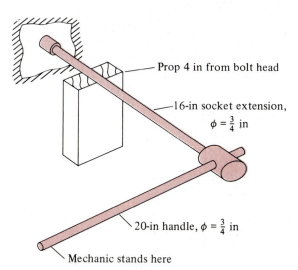

FIGURE 7.6
Socket-wrench arrangement for
Example 7.2.

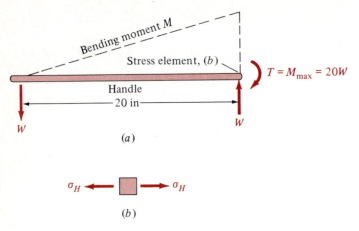

FIGURE 7.7
Handle analysis for Example 7.2. (a) Free-body diagram. (b) Stress element, top right of handle.

Turning now to the socket extension, we see the loading in Figure 7.8a. The extension is acting as a beam with a moment diagram as indicated in Figure 7.8b. The extension also resists twisting (that is its real purpose), which is constant along its length, as shown in Figure 7.8c. The bending stress Mc/I is maximum at the prop and is

$$\sigma_{ext} = \frac{12W(32)}{\pi d^3} = 290W$$

tension on the upper surface, compression on the lower.
The shear stress due to the twisting torque is

$$\tau_{ext} = \frac{20W(16)}{\pi d^3} = 241W$$

on the outer surface, everywhere on the bar.
Thus a stress element at the top or bottom of the extension at the prop (the worst place) has both a tensile and a shear stress, as in Figure 7.8d.
The Mohr's-circle representation of the stress situation and the failure envelope are sketched in Figure 7.8e. Since W is an unknown, the shear and bending stresses are simply shown in the proper proportions. At failure Mohr's circle would just touch the envelope, which is to say that the maximum shear stress would be equal to the maximum shear stress at failure in the tension test. Thus

$$W\sqrt{241^2 + 145^2} = 281.3W = \frac{S_y}{2}$$

hence $W = 160$ lb at failure. Since the mechanic weighs 180 lb, the mechanic fails the drive extension. That should not happen! What really does the drive in is the bending stress created by the use of the prop so close to the bolt.

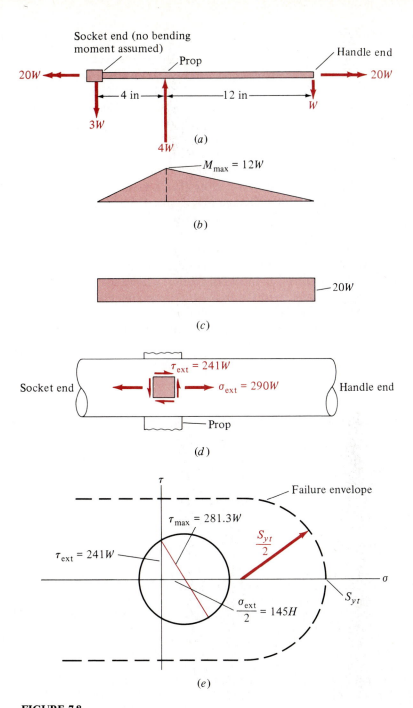

FIGURE 7.8
Socket-extension analysis for Example 7.2. (*a*) Loading. (*b*) Bending moment. (*c*) Twisting torque. (*d*) Stress element, top of bar, at prop. (*e*) Failure analysis.

Distortion Energy Theory

Although this theory was first proposed in 1856 by J. Clerk Maxwell,[1] it is commonly given the name the *Hueber–von Mises–Hencky theory*, of which the first name is frequently omitted. Hueber (Poland) proposed it in 1904; von Mises contributed to it in 1913 and Hencky in 1925. Von Mises and Hencky were German immigrants to the United States. The theory originated from the observation that failure did not correlate well with the total potential energy stored in a stress cube, called the *strain energy*. The strain energy can be divided into that associated with volume changes and that associated with shape changes, i.e., distortion. Material failure was found to correlate with the energy of distortion. Note that a uniform increase in all three dimensions is not distortion, but a change in volume. Other names for the theory are the *shear-energy theory* and the *octahedral shear-stress theory*. The names imply approaches leading to the same result.

The theory states that, for a given material, failure by yielding will occur at a critical level of distortion energy. Thus the task is to compute the distortion energy for a general loading and equate it to the distortion energy for a known situation, the tension test. The distortion energy is found by computing the total strain energy and then subtracting that due to volumetric change.

A normal stress, say σ_x, acting on a unit cube of material will cause an extension in its direction

$$\varepsilon_x = \frac{\sigma_x}{E}$$

and contractions per the Poisson effect in the y and z directions:

$$\varepsilon_y = \varepsilon_z = -v\,\frac{\sigma_x}{E}$$

These dimension changes are sketched in Figure 7.9.

---

[1] (James) Clerk (pronounced Klark) Maxwell, said to be the greatest theoretical physicist of the nineteenth century, was born in Scotland in 1831. Early in his career Maxwell reasoned that the rings of Saturn were comprised of loose particles. This was corroborated in 1983 by the flight of *Voyager*. Maxwell's investigations in color quantified Newton's theory that all colors are composed of combinations of the three primary ones: red, yellow, and blue.

Maxwell laid the groundwork for statistical mechanics with his work on the kinetics of gases. The maxwellian law of distribution (generalized by the Austrian physicist Boltzmann to become the Maxwell-Boltzmann distribution law) states that the velocities of gas molecules can be divided into groups, each representing a constant fraction of the molecules present. Maxwell also concluded that the average kinetic energy of a gas determines its temperature.

Maxwell's research in electricity and magnetism expanded on Faraday's work and laid out the interrelations among electric and magnetic fields, charge and current densities, and time.

Maxwell's prolific career was terminated by his death at age 48, in 1879.

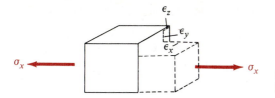

FIGURE 7.9
Dimension changes due to tension stress.

Considering now a unit cube with principal stresses σ_1, σ_2, and σ_3 acting, we see that the extensions become

$$\varepsilon_1 = \frac{\sigma_1}{E} - v\frac{\sigma_2}{E} - v\frac{\sigma_3}{E}$$

$$\varepsilon_2 = \frac{\sigma_2}{E} - v\frac{\sigma_3}{E} - v\frac{\sigma_1}{E}$$

$$\varepsilon_3 = \frac{\sigma_3}{E} - v\frac{\sigma_1}{E} - v\frac{\sigma_2}{E}$$

The energy in a unit cube of material subjected to these stresses will be the work done during the application of the stresses. For any one of them this work will be the shaded area $\sigma\varepsilon/2$ in Figure 7.10. Thus, the work done by σ_1 is

$$W_1 = \frac{\sigma_1}{2E}(\sigma_1 - v\sigma_2 - v\sigma_3)$$

and similarly for stresses σ_2 and σ_3. Adding the work of the three stresses gives

$$W = \frac{1}{2E}[\sigma_1^2 + \sigma_2^2 + \sigma_3^2 - 2v(\sigma_1\sigma_2 + \sigma_2\sigma_3 + \sigma_3\sigma_1)] \tag{7.2}$$

This expression includes work due to distortion and due to volumetric changes. It can be shown [e.g., 23, p. 138] that the latter is the work which an average stress $\sigma_{av} = (\sigma_1 + \sigma_2 + \sigma_3)/3$ acting in all three directions would do. Then $\sigma_1 - \sigma_{av}, \sigma_2 - \sigma_{av}$, and $\sigma_3 - \sigma_{av}$ are responsible for the distortion energy.

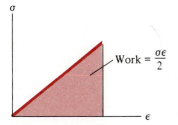

Work = $\frac{\sigma\varepsilon}{2}$

FIGURE 7.10
Energy in unit cube due to tension stress.

Expression (7.2) with all three stresses equal to the average stress becomes

$$W_{\Delta vol} = \frac{1 - 2v}{6E} (\sigma_1^2 + \sigma_2^2 + \sigma_3^2 + 2\sigma_1\sigma_2 + 2\sigma_2\sigma_3 + 2\sigma_3\sigma_1)$$

This must be subtracted from the total strain work W given by Equation (7.2), as indicated, and then we have

$$W_{distort} = \frac{1 + v}{6E} [(\sigma_1 - \sigma_2)^2 + (\sigma_2 - \sigma_3)^2 + (\sigma_3 - \sigma_1)^2]$$

This general expression for the distortion energy must be equated to that in a known situation, the tension test, at yield. At that point in the tension test, $\sigma_1 = S_y$ and $\sigma_2 = \sigma_3 = 0$, so

$$W_{tension\ test} = \frac{(1 + v)2S_y^2}{6E}$$

Setting the two works equal, we will get the condition for failure. The criterion for *avoiding* failure is

$$\sigma_{eff} = \frac{\sqrt{2}}{2} [(\sigma_1 - \sigma_2)^2 + (\sigma_2 - \sigma_3)^2 + (\sigma_3 - \sigma_1)^2]^{1/2} \leq \frac{S_y}{FS} \tag{7.3}$$

In writing this, the square root of each side was taken. The left side of expression (7.3) is called the *von Mises* or *effective stress*. A factor of safety has been included on the right.

Expression (7.3) is for triaxial loading. Most of our work will be in two dimensions, meaning that one principal stress is zero. It may be number 1, 2, or 3, depending on the size of the other stresses, as we saw earlier. If we name the principal stresses in the loading plane σ_p and σ_q, then the safety requirement is

$$\sigma_{eff} = \sqrt{\sigma_p^2 - \sigma_p\sigma_q + \sigma_q^2} \leq \frac{S_y}{FS} \tag{7.4a}$$

You will notice that it does not matter which of σ_p and σ_q is the larger. This expression can be plotted on principal-stress coordinates, as in Figure 7.1. The curve is an ellipse, as shown in Figure 7.11, where it is compared with the maximum-shear-stress theory.

For any given plane-stress situation, the principal stresses in Equation (7.4a) can be written in terms of σ_x, σ_y, and τ_{xy}. Then (7.4a) becomes

$$\sigma_{eff} = \sqrt{\sigma_x^2 + \sigma_y^2 - \sigma_x\sigma_y + 3\tau_{xy}^2} \leq \frac{S_y}{FS} \qquad \text{alternative} \tag{7.4b}$$

For pure torsion, $\sigma_x = \sigma_y = 0$, and we get

$$\tau_{xy} \leq 0.577 \left(\frac{S_y}{FS}\right) \qquad \text{torsion} \tag{7.4c}$$

This value is larger than that resulting from the maximum-shear-stress theory ($0.5S_y/FS$), as shown in Figure 7.11; it is closer to experimental observations.

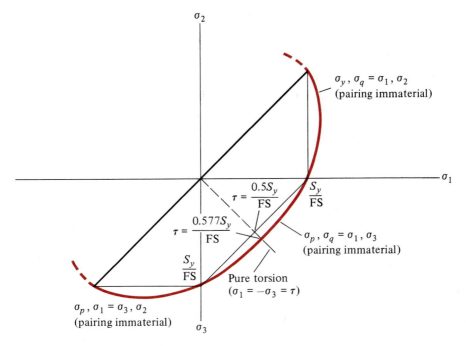

FIGURE 7.11
Comparison of distortion energy and maximum-shear-stress theories.

If there is only one tension stress, say σ_x, plus a shear τ_{xy}, as in bending with twisting, then (7.4*b*) becomes

$$\sigma_{\text{eff}} = \sqrt{\sigma_x^2 + 3\tau_{xy}^2} \le \frac{S_y}{\text{FS}} \qquad \text{single tension plus shear} \qquad (7.4d)$$

This situation is common where shafts are driven by gears.

Which theory fits the facts better? Test data show the distortion energy theory to be more accurate, and it has become the more popular. The maximum-shear-stress theory is more conservative, but not wastefully so, and it is somewhat easier to apply.

Example 7.3. Let us rework Example 7.1, using the distortion energy theory. The principal stresses are

$$\sigma_p = \frac{pd}{2t} = \frac{3p}{2(0.050)} = 30p$$

$$\sigma_q = \frac{dp}{4t} = \frac{\sigma_1}{2} = 15p$$

$$\sigma_3 = 0$$

Using (7.4*a*) for two-dimensional loading, we have for failure (FS = 1)

$$\sigma_{\text{eff}} = p\sqrt{30^2 - 30(15) + 15^2} = S_y = 60\,000$$

so $\qquad\qquad\qquad\qquad p = 2309 \qquad$ at yield

Hence the margin is $2309 - 1200 = 1109$ psi. The failure pressure of 2309 psi is 15 percent more than the 2000 psi predicted by the maximum-shear-stress theory.

Example 7.4. We will rework Example 7.2, using the distortion energy theory. We know already that the socket extension is the problem. The Mohr's circle at the prop is shown in Figure 7.8e. The principal stresses are then

$$\sigma_p = (145 + 281.3)W = 426.3W$$

$$\sigma_q = (145 - 281.3)W = -136.3W$$

The von Mises stress and the failure level for W are

$$\sqrt{426.3^2 + 426.3(136.3) + 136.3^2}(W) = S_y = 90E + 03$$

hence $W = 177$ lb, about 10 percent more than the failure load of 160 lb predicted by the maximum-shear-stress theory, as we would expect. Since the mechanic weighs 180 lb, either theory predicts that the mechanic fails the member.

Failure of Ordinarily Ductile Materials, but Not by Yielding

The theories just outlined for failure of materials exhibiting ductile behavior presume that the material yields, i.e., that the maximum shear stress reaches the point where yielding occurs. Situations can occur, even with uniaxial loading, in which the stress state is triaxial. Although the tensile stresses are high, the difference between them may be small, and thus the maximum shear stress is low. The Mohr's-circle representation of this combination is shown in Figure 7.12a.

Picture a rectangular plate with a reduced section, as in Figure 7.12b. The loading is in the x direction only. Section A is under higher stress than the sections at the sides B, because the area is smaller. Thus, due to the Poisson effect, section A would contract more in the y and z directions than sections B, except for the constraints to do so which B imposes. Section A is thus subjected to tensile stresses in the y and z directions, and a triaxial state of stress similar to that shown in the Mohr's-circle sketch is present. If either σ_y or σ_z were zero, the shear stress would become quite high. That is not the case, however; the shear stress is small, and the material cannot yield. Rather, it breaks in a brittle fashion at a critical stress approximating the ultimate stress.

Since a material will behave ductilely, i.e., yield, only when the maximum shear stress (or the von Mises stress) reaches the yield, care must be taken in using ductile failure theories where triaxiality of stresses limits the shear stress, creating the possibility of brittle rather than ductile failure. Where does this happen? Usually at notches. How can one avoid the problem? Make section transitions as gradual as possible, and use generous fillets. The problem *is* difficult. For some situations the stresses are calculable by finite-difference techniques. Then one can know if yielding, with subsequent stress redistribution, can occur. The state of knowledge in this area is not complete.

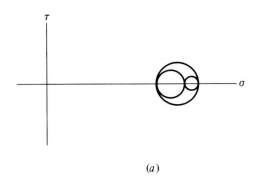

(a)

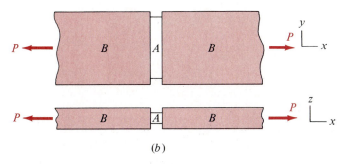

(b)

FIGURE 7.12
Brittle-failure situation for ductile material. (a) Mohr's circles—large tension, small shear. (b) Triaxial stresses can result from uniaxial loading.

7.2   BRITTLE BEHAVIOR

Brittle behavior occurs when the material shows no yielding; the stress-strain curve continues smoothly to fracture.

Mohr Theory

The Mohr theory[1] finds application with those brittle materials whose strength in compression is a great deal higher than that in tension.

Paralleling Figure 7.2 developed for ductile behavior, the theory proposes that the stress circles be drawn for the tension and compression tests, as seen in Figure

---

[1] The theory is also known as the *internal-friction theory*, because one arrives at the same result on that basis, and as the *Coulomb-Mohr theory*.

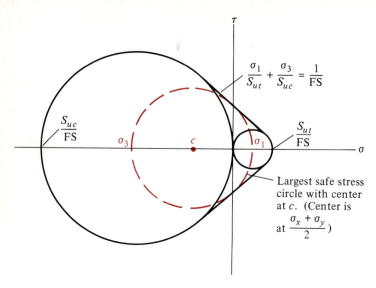

FIGURE 7.13
Construction of Mohr's theory of failure.

7.13, and that tangents be constructed to the circles.[1] Any loading producing a stress circle exceeding these boundaries is deemed to be unsafe.

For two-dimensional loadings, the boundaries for safety in Figure 7.13 are

$$\sigma_1 \leq \frac{S_{ut}}{FS} \qquad \begin{array}{l} \sigma_2 > 0, \text{ in-plane principal} \\ \text{stresses positive} \end{array} \tag{7.5a}$$

$$\frac{\sigma_1}{S_{ut}} + \frac{\sigma_3}{S_{uc}} \leq \frac{1}{FS} \qquad \begin{array}{l} \sigma_2 = 0, \text{ in-plane stresses of opposite sign} \\ (\textit{Note: } \sigma_3 \text{ and } S_{uc} \text{ are negative numbers.}) \end{array} \tag{7.5b}$$

$$\sigma_3 \geq \frac{S_{uc}}{FS} \qquad \begin{array}{l} \sigma_2 < 0, \text{ in-plane principal stresses negative} \\ (\textit{Note: } \sigma_3 \text{ and } S_{uc} \text{ are negative numbers.}) \end{array} \tag{7.5c}$$

The failure limits of the Mohr theory are also shown in Figure 7.14, which is a plot of Equations (7.5), paralleling Figure 7.1.

A modification of the Mohr theory, which extends the vertical boundary in the first quadrant down to the shear diagonal ($\sigma_1 = -\sigma_3$), as shown, generally fits experimental data better in the fourth quadrant. Equation (7.5b) above is replaced by

$$\sigma_1 \leq \frac{S_{ut}}{FS} \qquad (\sigma_2 = 0, \sigma_1 > -\sigma_3) \tag{7.5b'}$$

---

[1] Mohr actually proposed that the stress circles for all available test results, which would include torsion, for example, be drawn as on Figure 7.13 and a failure boundary of tangent curves unenclosed by the stress circles at the ends be established on that basis. One argument advanced for this arrangement is that it accommodates the observation that the hydrostatic stress state, whether tensile or compressive, does not produce yielding. (Mohr's circle becomes a point, hence the shear stress is zero.)

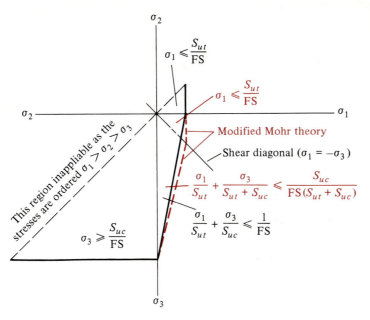

FIGURE 7.14
Safe stress limits per Mohr's theory of failure for biaxial loading. See notes on Figure 7.1 re axis labels.

$$\frac{\sigma_1}{S_{ut}} + \frac{\sigma_3}{S_{ut} + S_{uc}} \leq \frac{S_{uc}}{FS(S_{ut} + S_{uc})} \qquad (\sigma_2 = 0, \ \sigma_1 < -\sigma_3) \qquad (7.5b'')$$

The τ vs. σ representation of the modified theory is shown in the following example.

Example 7.5. Given the stress element of Figure 7.15a, how large may the compressive stress σ_x be if the material is brittle with $S_{ut} = 200$ MPa and $S_{uc} = -1000$ MPa? The factor of safety is FS = 2.

We will perform the analysis graphically first. The modified Mohr theory (Figure 7.13) is shown in Figure 7.15b. We do not know the greatest principal stress σ_1 yet, only that it is positive, as shown by the sketch of Mohr's circle. Also, clearly $\sigma_2 = \sigma_z = 0$ and $\sigma_3 < 0$. Point a on Mohr's circle for the stress element is located at $\sigma_y = 0, \tau = 80$, values which will not change (the sign of τ is immaterial). The center of the circle is then found by trial such that the circle passes through point a and is also tangent to bd as required. Then $\sigma_x \simeq 440$ MPa is read as shown.

The solution may also be arrived at analytically. The limiting expression is Equation (7.5b''), which represents the tangent line bd in the graphical solution. We see from the element's Mohr's circle there that

$$\sigma_1 = \frac{-\sigma_x}{2} + \sqrt{\frac{\sigma_x^2}{4} + \tau^2} \qquad \text{and} \qquad \sigma_3 = \frac{-\sigma_x}{2} - \sqrt{\frac{\sigma_x^2}{4} + \tau^2}$$

(Here σ_x is a positive number, the vector having been shown in the negative direction.) Substituting these in Equation (7.5b''), the only unknown is σ_x, which is found to be

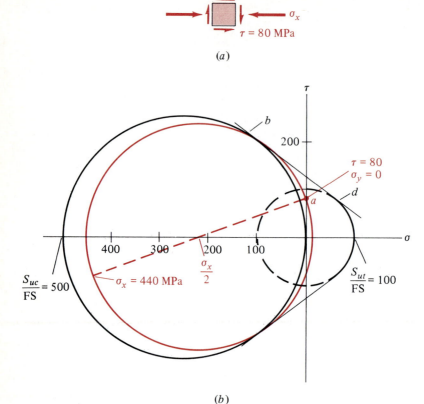

(a)

(b)

FIGURE 7.15
Analysis for Example 7.5. (a) Stress element. (b) Modified Mohr theory sketch.

$\sigma_x = 428$ MPa. (A second larger value of σ_x also results. The significance of this we leave as problem 7.15.)

The unmodified Mohr theory follows Figure 7.13 or Equation 7.5b, with the result $\sigma_x = 410$ MPa. This figure is somewhat less than that for the modified theory, as the unmodified theory is more restrictive. (See Figure 7.14.)

Maximum-Normal-Stress Theory

If the compressive strength S_{uc} of a brittle material is nearly the same as the tensile strength S_{ut}, you will see in Figure 7.13 that the Mohr theory reduces to the maximum-shear-stress theory (the circles on the ends become equal). By definition, a brittle material does not fail in shear. Failure happens when the largest principal stress reaches the limit found in the tension test S_{ut} or the smallest (largest compressive) exceeds the compressive limit ($S_{uc} = -S_{ut}$). The maximum-normal-stress theory thus requires for safety

$$\text{Max } \{\sigma_1, |\sigma_3|\} \leq \frac{S_{ut}}{\text{FS}} \tag{7.6}$$

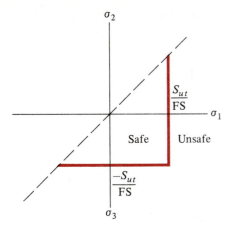

FIGURE 7.16
Maximum-normal-stress theory.

On a plot of principal stresses, the theory appears as shown in Figure 7.16. Comparing this plot with that of Figure 7.14, you see that the maximum-normal-stress theory is the same as the modified Mohr theory for the case where $S_{uc} = -S_{ut}$.

Example 7.6. We have the stress situation shown in Figure 7.17a in a brittle material, for which $S_{ut} = 250$ ksi. What is the FS?

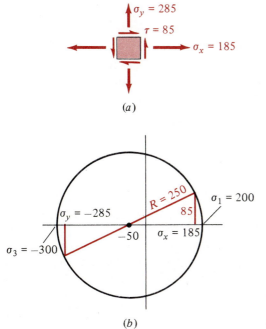

(a)

(b)

FIGURE 7.17
Element and analysis for Example 7.6. (a) Stress element. (b) Mohr's circle.

The Mohr's circle for this stress state is shown in Figure 7.17*b*. The critical stress is $\sigma_3 = -300$. Relation (7.5), with an equals sign to find FS, gives

$$350 = \frac{250}{\text{FS}} \qquad \text{so} \qquad \text{FS} = 0.71$$

Failure occurs.

7.3 STRESS CONCENTRATIONS

A simple illustration leads to the easiest understanding of the phenomenon of stress concentration. Figure 7.18 shows a plate of uniform thickness which carries a load longitudinally. Initially the plate is rectangular, and the load is evenly distributed across any section, i.e., the stress is uniform. If now the dotted areas are cut out, the load which had been carried by the removed pieces must be assumed by the reduced section. Most of the load will be borne by the material near the edges. Thus, we have a concentration of stress in the region of the fillet. Such concentrations appear whenever there is a change in geometry of a part.

In a few instances of simple geometry, the stresses can be computed exactly from elasticity theory. These are very few indeed, so that for most practical situations (steps in shafts, keyways, grooves, etc.) the severity of the concentration has been determined experimentally or estimated by computer programs. The experimental methods have usually been strain-gage or photoelastic techniques. The computer has made it possible to do the same work numerically through the finite-element method, and solutions for a number of difficult geometries have appeared in recent years in the literature. The result is expressed in the "theoretical" or "geometric" stress-concentration factor, which is

$$K_t = \frac{\sigma_{\text{at point, assuming material remains elastic}}}{\sigma_{\text{nom, at nearby reference position}}}$$

The stress in the denominator for the plate in Figure 7.18 would be F/A_2, that computable by elementary mechanics of materials in the lower section. The concentration factor is thus a measure of the increase in the larger nominal stress due to the de-

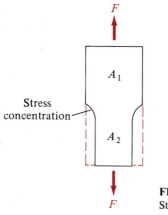

Stress concentration

FIGURE 7.18
Stress concentration.

viation in the geometry at the step. (Occasionally the reference is not the largest nominal stress. In any event it is always indicated.)

Stress-concentration factors have been determined for a large number of common situations. The largest single source is the compilation of R. E. Peterson [8], whose charts are reproduced in many texts, including this one (Appendix A). Those results were obtained by photoelastic and other methods, before the widespread use of computers.

The point of highest stress is usually a localized region of stress concentration. In the case of ductile behavior, if yielding in such an area cannot be tolerated, then the computed stress, incorporating K_t, will be required to fall within the failure limits discussed earlier. This, however, is a rare case; localized yielding in most static situations is not objectionable (less than about a thousand loading cycles). On the first load (or overload) application, yielding occurs; the material flows, and when the load is removed, a residual stress of opposite sign exists (e.g., compressive if the yielding was tensile). Then on subsequent loadings the stress level does not again reach the yield. The stress-concentration effect has lessened.

Thus, when ductile behavior and static loading are anticipated, the effect of stress concentration usually need not be taken into account in reckoning strength, unless yielding at the location of the concentration is prevented by triaxiality or is otherwise unacceptable.

The good engineer will seek to minimize stress concentrations through proper design. For example, in Figure 7.18 the transition from A_1 to A_2 would always be done with as large a radius as possible and never by straight lines, with a resulting sharp corner.

When the stress changes in magnitude and/or direction with time, which is termed *fatigue*, discussed in the following two chapters, stress concentration is the most important factor in failure. Over 80 percent of machine parts which fail do so by fatigue, and nearly all such failures initiate at a stress raiser (concentration).

Since brittle materials do not yield, the computation of stress must include the stress-concentration factor K_t. Steel becomes brittle when it is treated to high hardness, and we usually put cast irons in this category as well, though some are quite ductile (malleable iron, for example). Ceramics, which are coming into widespread engineering use, are naturally brittle.

Cast iron has many internal irregularities (graphite, slag, etc.) as well as a rough surface. These irregularities represent points of stress concentration, exactly as does a machined notch on the surface. Since published strengths are obtained on specimens with all these defects, the additional effect of notches, etc., on the surface will be small. The full value of K_t is thus not appropriate, and it is modified through a parameter q, known as the *notch sensitivity*. Notch sensitivity plays a large role in fatigue calculations, discussed in Chapter 9. The net stress-concentration factor is computed by

$$K = 1 + q(K_t - 1)$$

Note that q measures the amount of the stress due to the concentration ($K_t - 1$) to be added to the nominal stress (where K is 1). For cast irons q is quite low. A value of 0.2 is often suggested. Thus, for example, if K_t is 3, the factor to use would be $K = 1.4$.

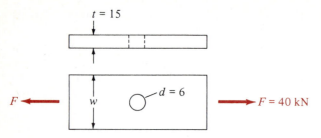

Dimensions in millimeters

FIGURE 7.19
Tension bar with stress concentration for Example 7.7.

Example 7.7. A bar of ASTM no. 30 cast iron (S_{ut} = 32 ksi = 221 MPa) with a transverse hole of 6-mm diameter is loaded in tension as shown in Figure 7.19. How wide must the bar be if a factor of safety of 1.5 is specified?

With FS = 1.5, we design for a force F = 40(1.5) = 60 kN. If we take the bar to be brittle, the stress concentration is included in the stress calculation, and the critical stress then is

$$\sigma_1 = 221E + 06 = K\sigma_{nom} = [1 + 0.2(K_t - 1)]\sigma_{nom}$$

where K_t can be taken from Appendix A-1. However, the width w is not known. The nominal stress used for this chart is, as we might expect, calculated on the basis of $A = (w - d)t$. We also note that the abscissa of the chart is d/w ($d = 0.006$). After these facts are taken into account, the above stress relation results in

$$K_t = \frac{1.66}{d/w} - 5.66$$

We determine the point on the K_t chart satisfying this expression by trial. After a few tries, we have $d/w = 0.20$ and $w = 30$ mm.

7.4 LINEAR ELASTIC FRACTURE MECHANICS

Over the years there has been constant pressure on the engineering community to utilize materials more efficiently. At the same time, metallurgists and metals producers have striven to improve the quality and reliability of their products. This general aggressiveness has pushed materials to the limit of their capacity to accommodate local plastic strain without ensuing fracture. The existence of flaws, manufacturing defects, and even cracks due to service loadings is now accepted as the starting point in design in many fields, most notably the aircraft industry. Fracture mechanics is the science of predicting under what circumstances an existing imperfection, usually a crack, will uncontrollably propagate and result in total fracture.

Stresses at Crack Tips

Figure 7.20 shows the three principal ways in which cracks are loaded. They are termed *modes* I, II, and III, as indicated in the figure. Most engineering situations involving cracks are of the mode I type. Mode III is clearly shear and would come

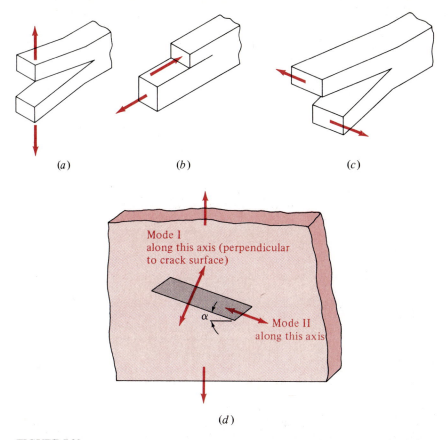

FIGURE 7.20
Modes of crack loading. (*a*) Mode I: tensile mode. (*b*) Mode II: sliding mode. (*c*) Mode III: tearing or shearing mode. (*d*) Example of mixed modes.

into play in cracked parts subjected to twisting torques. Mode II is seldom encountered. An occasional mix of modes occurs, such as when a crack is disposed at an angle to the loading direction, indicated in Figure 7.20*d*. Here, even with an inclination angle α as great as 60°, mode II makes little contribution to failure. In this introduction to fracture mechanics, we limit the discussion to the tensile loading of mode I.

A stress element in the vicinity of the tip of a mode I crack is shown in Figure 7.21. The stresses have been determined analytically:

$$\sigma_x = \frac{K_I}{\sqrt{2\pi r}} \cos \frac{\theta}{2} \left(1 - \sin \frac{\theta}{2} \sin \frac{3\theta}{2} \right) \tag{7.7a}$$

$$\sigma_y = \frac{K_I}{\sqrt{2\pi r}} \cos \frac{\theta}{2} \left(1 + \sin \frac{\theta}{2} \sin \frac{3\theta}{2} \right) \tag{7.7b}$$

$$\tau_{xy} = \frac{K_I}{\sqrt{2\pi r}} \cos \frac{\theta}{2} \sin \frac{\theta}{2} \cos \frac{3\theta}{2} \tag{7.7c}$$

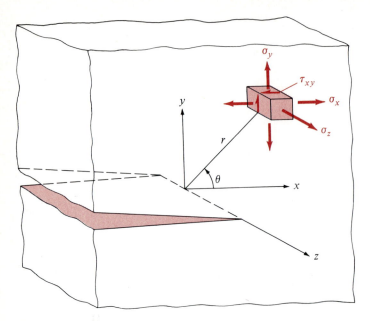

FIGURE 7.21
Stresses on an element in the vicinity of a mode I crack. The element is located by r and θ.

The expressions for mode II and mode III cracks are similar. K is a proportionality constant we will discuss presently; its subscript indicates the mode type.

The x and y stresses result in strains in the z direction due to the Poisson effect: $\varepsilon_z = -v(\sigma_x + \sigma_y)/E$. If the material is unrestrained in that direction, as in the case of a thin plate, then no stress σ_z results. This is known as *plane stress*, i.e., stresses in two dimensions only (x and y). On the other hand, if the plate is thick, then in its interior (no σ_z is possible on the surface) the material is restrained in the z direction, and $\sigma_z = v(\sigma_x + \sigma_y)$. This situation is known as *plane strain* (strains in two dimensions only). The z-type shear stresses (τ_{xz} or τ_{yz}) are zero in plane stress or plane strain. We indicate shortly what constitutes thick and thin plates.

Stress-Intensity Factor K_I

In Equations (7.7) the angle θ and the distance r are point coordinates. Then K_I determines the intensity of the stress field. Its value depends on the type and level of loading and the geometry of the crack. In (7.7) if r is put on the left side of the equation, you see that the units of K_I are stress $\times \sqrt{\text{distance}}$, i.e., MPa$\sqrt{\text{m}}$ or psi$\sqrt{\text{in}}$. Values of the stress-intensity factor have been worked out for a number of situations and are found in the literature [e.g., 50, 51]. A few charts for common configurations are shown in Figure 7.22.

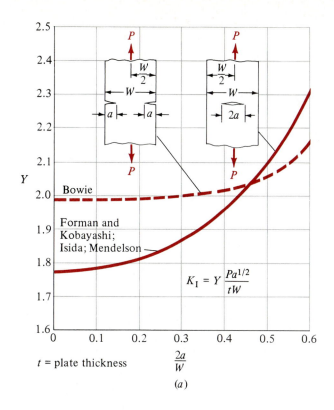

$$K_I = Y \frac{Pa^{1/2}}{tW}$$

Bowie

Forman and
Kobayashi;
Isida; Mendelson

t = plate thickness

$\frac{2a}{W}$

(a)

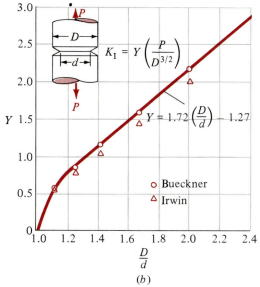

$$K_I = Y \left(\frac{P}{D^{3/2}} \right)$$

$$Y = 1.72 \left(\frac{D}{d} \right) - 1.27$$

o Bueckner
△ Irwin

$\frac{D}{d}$

(b)

FIGURE 7.22
Stress-intensity factors [73]. (*Copyright ASTM. Reprinted with permission.*)

(Continued)

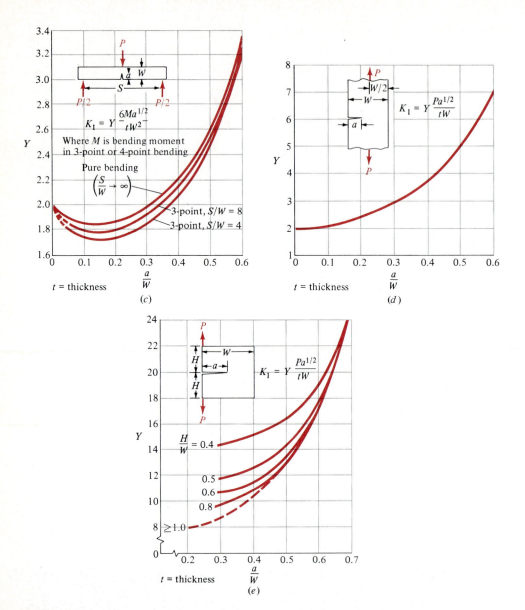

FIGURE 7.22
Continued.

Plastic Zone at Crack Tip

The stresses indicated by expressions (7.7) attain high levels as r becomes very small, which is what stress-concentration considerations would indicate. These stresses are based on the assumption of linear elastic behavior of the material, an assumption which reaches its limit at the yield point. Thus the material nearest the crack tip is in a plastic state. The extent of the plastic zone along the crack axis x is obtained by considering the stresses on elements along $\theta = 0$. With $\theta = 0$ in Equations (7.7a) and (7.7b), $\sigma_x = \sigma_y$, so that, with either the maximum-shear-stress theory or the distortion energy theory, at yield $\sigma_x = \sigma_y = S_y$. By making that substitution in (7.7a) or (7.7b), with $\theta = 0$, we get

$$r_y = \frac{1}{2\pi}\left(\frac{K_\mathrm{I}}{S_y}\right)^2 \qquad \text{plane stress} \qquad (7.8)$$

which is the distance along the crack axis that yielding extends. It is labeled *plane stress* because no σ_z was considered in the calculation. For plane strain [52]

$$r_y \simeq \frac{1}{6\pi}\left(\frac{K_\mathrm{I}}{S_y}\right)^2 \qquad \text{plane strain} \qquad (7.9)$$

which is smaller than that for plane stress, because of the additional restraint in the z direction.

The plastic zone makes the crack length effectively somewhat greater, and r_y and the physical length a of the crack are summed to produce an effective crack length:

$$a_\mathrm{eff} = a + r_y$$

If the plastic zone is large relative to the crack length, the background assumption of linear elastic material behavior comes into question. Elastic-plastic fracture mechanics is a growing area of this science for dealing with such situations. It is beyond the scope of this synopsis.

Critical Stress Intensity and Fracture Toughness

For any of the configurations represented in Figure 7.22, or others, it is important to note two things:

1. The stress (σ_x or σ_y) at any point around mode I cracks is given by Equations (7.7). Either stress can be expressed as $\sigma = K_\mathrm{I} f(\theta, r)$, where $f(\theta, r)$ is the same for any configuration involving a mode I crack.

2. As you see in Figure 7.22, K_I is directly proportional to the nominal loading. For example, for the edge cracks of Figure 7.22c, $K_\mathrm{I} = (Y\sqrt{a})\sigma$. Now $Y\sqrt{a}$ is a geometric parameter, so K_I varies directly with the stress σ. Thus K_I is a measure of the stress near the crack. As K_I increases, the extension of the crack reduces the overall stiffness of the piece and thus the potential energy stored in it. The crack extension forms a new surface, which requires energy. When a point is reached such that a small increase in crack size causes the release of more potential energy than is required to form the new surface, the situation has become unstable and the crack grows at an accelerating rate. Cataclysmic rupture occurs at this critical

value of K_I, which is given the symbol K_{Ic}. Thus with K_I being analogous to the applied stress, K_{Ic} is analogous to failure strength: yield strength for ductile behavior and ultimate strength for brittle behavior.

The critical stress intensity factor for a material is found in a standardized test: pulling a tension specimen with a crack on its edge. Figure 7.23 shows such a specimen mounted in a test machine. The crack is visible between the pins, through which the load is applied. The instrument measures the crack separation. The specimen is such a size that the value of K_{Ic} for plane strain results (thick parts). It is called the *plane-strain fracture toughness*. Generally, the minimum material thickness t and crack size a for plane strain to exist are [24]

$$a \text{ and } t \geq 2.5\left(\frac{K_{Ic}}{S_y}\right)^2 \tag{7.10}$$

We can find the approximate size of the plastic zone in plane strain in terms of a and t by utilizing expression (7.9). If it is solved for $K_I/S_y\,(=K_{Ic}/S_y$ at rupture) and substituted in (7.10), then

$$a \text{ and } t \geq 47r_y \qquad \text{or} \qquad r_y \leq 2\% \text{ of } a \text{ and } t$$

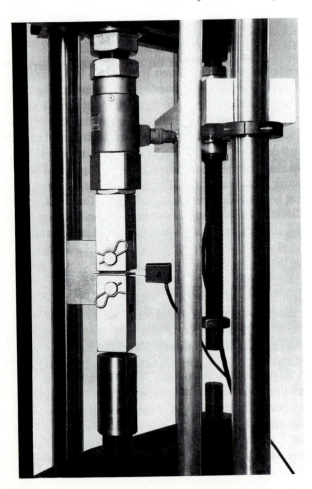

FIGURE 7.23
Fracture-toughness specimen in test machine. (*Courtesy, Fatigue Dynamics, Inc.*)

TABLE 7.1
Plane-strain fracture toughness of some engineering alloys

Material	K_{Ic} MPa$\sqrt{m}$	K_{Ic} ksi$\sqrt{in}$	S_{yt} MPa	S_{yt} ksi
2014-T651	24.2	22	455	66
2024-T3	~44.	~40	345	50
2024-T851	26.4	24	455	66
7075-T651	24.2	22	495	72
7178-T651	23.1	21	570	83
7178-T7651	33.	30	490	71
Ti-6Al-4V	115.4	105	910	132
Ti-6Al-4V	55.	50	1035	150
4340	98.9	90	860	125
4340	60.4	55	1515	220
4335 + V	72.5	66	1340	194
17-7PH	76.9	70	1435	208
15-7Mo	49.5	45	1415	205
H-11	38.5	35	1790	260
H-11	27.5	25	2070	300
350 Maraging	55.	50	1550	225
350 Maraging	38.5	35	2240	325
52 100	~14.3	~13	2070	300

(Reprinted by permission of John Wiley & Sons, Inc., from *Determination and Fracture Mechanics of Engineering Materials*, by Richard W. Hertzberg, copyright 1976 by John Wiley & Sons.)

For plane stress the K_{Ic} value may be significantly higher than that for plane strain, so design on the presumption of plane strain is conservative. While the plane-strain fracture toughness is a property of the material, the plane-stress fracture toughness is a property of the material and of the loading configuration and part size, especially the thickness t. This complicates the analysis.

Table 7.1 gives values of K_{Ic} (plane strain) for some metals. The yield strength is also indicated. Note that S_y and K_{Ic} can move in opposite directions—a stronger material does not generally provide greater fracture resistance. Some other values of K_{Ic} may be found in Appendix D.

The fracture criterion, as developed above, states that for a given crack size and crack/loading configuration, fracture will occur at a critical nominal-stress level (*nominal* as defined in the charts), or, for a given nominal-stress level and crack/loading configuration, rupture will occur at a critical crack size. The latter statement has bearing in fatigue situations where a crack may be growing slowly and the critical size for sudden rupture is important.

Example 7.8. A beam of 4340 steel heat-treated to $S_y = 220$ ksi is to be loaded as shown in Figure 7.24. It is estimated that a through-the-thickness edge crack $\frac{1}{8}$ in in length will be readily detectable. However, a factor of safety of 2 on that number will be applied, so that a critical crack size $a = 0.25$ in must be designed for. We need to establish the required thickness to avoid failure.

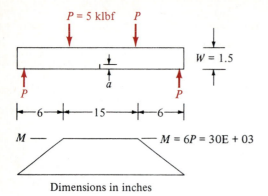

FIGURE 7.24
Beam configuration for Example 7.8.

Dimensions in inches

We first calculate the size of the plastic zone at the root of the crack, using expression (7.9), with $K_I = K_{Ic}$, at rupture.

$$r_y = \frac{1}{6\pi}\left(\frac{55}{220}\right)^2 = 0.003 \text{ in}$$

The effective crack length is thus 0.253 in.

We enter Figure 7.22c with $a/W = 0.253/1.5 = 0.17$ and find $Y = 1.86$. The chart, as indicated, is based on

$$K_I = Y\frac{6Ma^{1/2}}{tW^2}$$

At rupture $K_I = K_{Ic}$, hence

$$t = \frac{6YM\sqrt{a}}{K_{Ic}W^2} = \frac{1.86(6)\times 30\text{E}+03\times\sqrt{0.253}}{55\text{E}+03\times 2.25} = 1.35 \text{ in}$$

This calculation was based on the assumption of plane strain. That needs to be checked against (7.10):

$$a \text{ and } t \geq 2.5\left(\frac{K_{Ic}}{S_y}\right)^2 \geq 2.5\left(\frac{55}{220}\right)^2 = 0.16$$

Both a and t meet this requirement. If one or the other had not, the design would have been conservative, i.e., the thickness would be more than ample.

A further check must be made that gross yielding is avoided. We can either compute the thickness necessary to avoid it or see if the above design does in fact avoid it. We do the latter.

$$\sigma = \frac{Mc}{I} = \frac{6M}{tW^2} = \frac{6\times 30}{1.35\times 2.25} = 59 \text{ ksi}$$

Since S_y for this material is 220 ksi, a factor of safety of nearly 4 exists for yielding. This being the case, we might imagine that we could do the job better, and more cheaply, with a steel of lower S_y and greater K_{Ic}. The 4340 with $S_y = 125$ ksi and $K_{Ic} = 90$ ksi $\sqrt{\text{in}}$ results in a thickness requirement $t = 0.82$ in and a stress level $\sigma = 98$ ksi. The factor of safety against yielding is then

$$FS = \frac{125}{98} = 1.3$$

which might or might not be considered adequate.

The plane-strain requirement is

$$a \text{ and } t \geq 2.5 \left(\frac{90}{125} \right)^2 = 1.3$$

which neither a nor t meets. Thus the conditions are plane stress. Since K_{Ic} for plane stress is larger, the design is still satisfactory on the basis of crack propagation.

Elastic-Plastic Fracture Mechanics

The developments above presumed linear elastic behavior, which in turn required that the plastic zone at the crack tip be small in comparison to the crack size. In many practical cases, the plastic zone becomes too large for this assumption to be valid, and analysis techniques based on elastic-plastic behavior must be employed. This approach is still in the development stages.

7.5   CLOSURE

Limitations

This chapter has dealt with static failure theories, whose application therefore is limited to loading conditions approximating those of the tests from which the strengths are derived: ambient temperatures approximating "room" temperature, loadings which are not so rapid as to cause brittle behavior of normally ductile materials, and "few" (less than about a thousand) repetitions of the load.

Ductile Materials, Static Loading

For design under static loadings, the distortion energy theory is the most accurate. The maximum-shear-stress theory is conservative, but not wastefully so, and somewhat easier in its calculations.

Where stress concentrations occur, the stress concentration need not be taken into account if localized yielding can be tolerated. Caution must be used in situations where triaxiality of stress may happen, for then brittle fracture can occur.

Brittle Materials, Static Loading

When the ultimate tensile and compressive strengths differ, the Mohr theory applies. When those strengths are the same, the maximum-principal-stress theory should be used. In fact, it is automatic.

If there are stress concentrations, the nominal stress should be multiplied by a stress-concentration factor. In general, this is equal to the theoretical factor K_t. For cast iron a reduced value is used:

$$K = 1 + q(K_t - 1)$$

The factor q is usually taken as 0.2. The resulting stress is then used in failure calculations.

Fracture Mechanics

The presumption of cracks in structural members is becoming common practice. Designing on the basis of fracture mechanics to avoid critical crack sizes and consequent failure is then appropriate.

PROBLEMS

Sections 7.1 and 7.2

7.1. Assuming FS = 1.6, determine if the situations of Problems 4.7 and 4.8 are safe. No stress concentration exists.

(a) Material is ductile with S_{yt} = 50 ksi. Use the maximum-shear-stress theory for problems marked a, c, e, and g; otherwise, use the distortion energy theory.

Answer: 4.7a: Unsafe   4.8b: Unsafe
4.7b: Unsafe   4.8d: Unsafe
4.7d: Safe   4.8f: Unsafe
4.8a: Unsafe   4.8h: Safe

(b) Material is brittle with S_{ut} = 30 ksi and S_{uc} = −60 ksi. Use the modified Mohr theory.

Answer: 4.7c: Unsafe   4.8e: Safe
4.7e: Unsafe   4.8g: Unsafe
4.8c: Unsafe

7.2. A stress element has τ = −7 ksi. A compressive stress also exists in the x direction. How large may that stress be with FS = 2?

(a) Material is ductile with S_{yt} = 50 ksi. Use the distortion energy theory and the maximum-shear-stress theory.

(b) Material is brittle with S_{ut} = 30 ksi and S_{uc} = −60 ksi.

7.3. Repeat Problem 7.2, but τ = 8 ksi, σ_y = 2.5 ksi, and σ_x is tensile.

Answer: (a) Distortion-energy: σ_x = 21.9 ksi; maximum-shear-stress: $\sigma_x \simeq$ 21.7 ksi
(b) $\sigma_x \simeq$ 7.35 ksi

7.4. A container is to be made of ductile aluminum alloy having S_{yt} = 120 MPa. It has a wall thickness of 1.6 mm and a diameter of 70 mm. With these proportions, it may be analyzed as a thin-walled vessel. What pressure will fail the container? (Example 4.7 may be helpful for calculating stresses.) Use the maximum-shear-stress theory.

7.5. A bar of ductile steel (S_{yt} = 50 000 psi) is loaded by the arrangement shown in Figure P7.5. Calculate the factor of safety. (Note: The bar is twisted, bent, and pulled simultaneously. Example 7.2 may be helpful for calculating stresses.) Use the distortion energy theory.

Answer: FS = 4.3

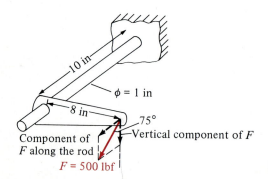

FIGURE P7.5

7.6. Repeat Problem 7.5, except that the angle of the force from vertical is in the opposite direction, so that the axial component of **F** applied to the bar is compressive.

7.7. Repeat Problem 7.5, except that the material is brittle with $S_{ut} = 180$ MPa and $S_{uc} = -670$ MPa.

   Answer: FS = 2.61

7.8. A stress element has $\sigma_x = \sigma_y = -30$ ksi and $\tau = -15$ ksi. Which plane is critical?
   (*a*) Material is ductile with $S_{yt} = 900$ MPa.
   (*b*) Material is brittle with $S_{ut} = 400$ MPa and $S_{uc} = -1000$ MPa.

7.9. The sketch in Figure P7.9 shows a chain-drive idler. The input is on the left (800 lb) and the output on the right. If the output chain drives a sprocket larger than 5 in, the object is speed reduction and torque increase. (The omission of the dimension of the input sprocket is not an error.) Assuming no stress concentrations, FS = 2, and a steel with $S_y = 44.5$ ksi, determine if the shaft size is satisfactory. The bearings are self-aligning; i.e., they transmit no bending moment to the shaft. Use the maximum-shear-stress theory.

   Answer: $\sigma_1 = 6113$, $\sigma_3 = -702$; size is OK

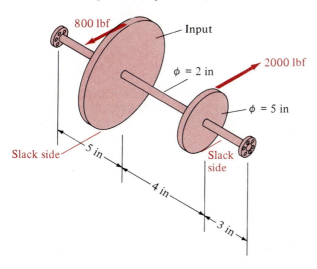

800 lbf
Input
$\phi = 2$ in
2000 lbf
$\phi = 5$ in
Slack side
5 in
Slack side
4 in
3 in

FIGURE P7.9

7.10. Repeat Problem 7.9, except that the chain on the left sprocket is vertical, not horizontal.

7.11. The lug wrench supplied with an automobile is shown in Figure P7.11. It is made of AISI 1040 steel, cold-drawn 20 percent (Appendix D) and is intended to torque lug nuts on or off at 120 pound-inches (lb·in). The two arms of the wrench are $\frac{7}{16}$ in in diameter. What margin of safety exists in terms of the torque transmitted, assuming the wrench is used in the worst way possible?

   Answer: MS = 481 lb·in

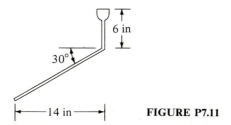

6 in
30°
14 in
FIGURE P7.11

Section 7.3

7.12. For the shaft shown in Figure P7.12, if FS = 1.3, how much more torque can be put on?
   (a) Material is ductile with $S_{yt} = 900$ MPa. A $K_t = 2$ exists. Use the maximum-shear-stress theory and the distortion energy theory.
   (b) Material is brittle with $S_{ut} = 400$ MPa and $S_{uc} = -1000$ MPa. A $K_t = 1.5$ exists (assume it is the same in tension and torsion). Take $q = 0.2$ (cast iron), and use the unmodified Mohr theory.

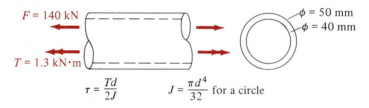

$F = 140$ kN

$T = 1.3$ kN·m

$\phi = 50$ mm
$\phi = 40$ mm

$$\tau = \frac{Td}{2J} \qquad J = \frac{\pi d^4}{32} \text{ for a circle}$$

FIGURE P7.12

7.13. A hollow shaft is supported at its ends on frictionless bars and is loaded by forces acting down on two levers, as shown in Figure P7.13. The stress-concentration effect (torsion or bending) at the points of attachment of the arms is estimated to be $K_t = 2.5$ $(q = 0.2)$. Find the margin of safety in terms of the vertical forces.
   (*Hint:* As the first step, the location of the worst stress situation must be found.)
   (a) The material is steel, with $S_y = 50$ ksi. Use the maximum-shear-stress theory.
      Answer: MS = 11 978 N
   (b) The material is cast iron, with a break strength of 30 ksi in tension and 150 ksi in compression.
      Answer: MS = 6908 N

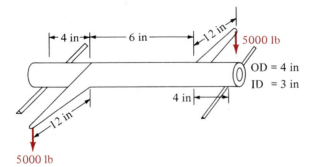

4 in
6 in
12 in
5000 lb
OD = 4 in
ID = 3 in
4 in
12 in
5000 lb

FIGURE P7.13

7.14. A stress element has $\tau = 90$ MPa, $\sigma_x = 3\sigma_y$, $K_t = 1.5$, and $q = 0.2$. How large may σ_x be for the materials of Problem 7.13? Use FS = 1.2 and the distortion energy theory.

7.15. In Problem 7.14, suppose σ_y is tensile. How large may σ_x be for the materials of Problem 7.13? Use FS = 1.2 and the maximum-shear-stress theory. For the brittle part, two values of σ_y result if the solution is performed analytically, rather than graphically. One value is obviously the solution, while the other appears not to make sense. Explain what this second value is associated with.
      Answer: Ductile: $\sigma_x = 337$ MPa; brittle: $\sigma_x = 92.4$ MPa

Section 7.4

7.16. A round bar, shown in Figure P7.16, has a circumferential crack.
  (a) Find the size of the plastic zone at the crack tip.
  (b) Find the rupture load, assuming crack propagation.
  (c) Find the load at yield and compare it with (b).

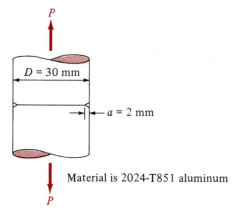

Material is 2024-T851 aluminum

FIGURE P7.16

7.17. A plate of 7075 T651 aluminum with an edge crack is shown in Figure P7.17.
  (a) How large may the crack be without rupture occurring?
      Answer: $a \simeq 105$ mm
  (b) Is the condition one of plane strain or plane stress?
      Answer: Plane strain
  (c) Is the plastic zone significant in the calculation of (a)?
      Answer: No, $r_y \simeq 0.1$ mm
  (d) In view of your response to (b), you may wish to reconsider how you answered (a).

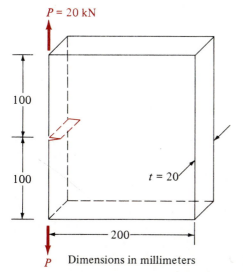

Dimensions in millimeters   **FIGURE P7.17**

7.18. Figure P7.18 is a schematic of a beam, which is to have a maximum deflection of 0.2 in. If a crack (anywhere) across the beam of 0.100 in is to be tolerated, what must the width of the beam be? The material is to be one of the 4340 steels.

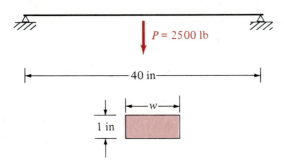

$P = 2500$ lb

40 in

w

1 in

FIGURE P7.18

7.19. A bar of Ti-6Al-4V titanium alloy (Table 7.1, $K_{Ic} = 55$ MPa) shown in Figure P7.19 is to support a load of 50 metric tons (t). A 2-mm edge crack has been discovered. What is the factor of safety on the load?

Answer: FS = 2.42

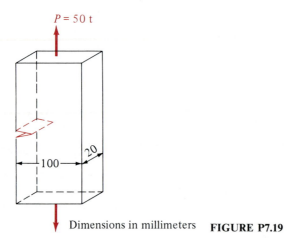

$P = 50$ t

20

100

Dimensions in millimeters   **FIGURE P7.19**

PHENOMENON OF FATIGUE IN ENGINEERING MATERIALS

In the tests described in Chapter 6, the loads were applied gradually and in one direction (monotonic). The material was subjected to static stresses, and this was the assumption underlying the failure theories in Chapter 7. In practice, loads are more often varying than they are constant, e.g., every moving part of any vehicle, the wing spars and skin of an aircraft, or the structural members of an offshore drilling platform. The nature of most of these elements demands a great many cycles of service life. For example, in 80 000 kilometers (km) an automobile valve spring goes through some 70 million compression-extensions, and the crankshaft is subjected to twice that number of load cycles. Failure under such conditions has been termed *fatigue*, and it always occurs at stress levels less, and often very much less, than what would be predicted by static failure theory.

The designer of devices which are to be subjected to repeated loadings must know the strength limitations of materials under such conditions. We start by relating a little background history and then look at the common tests performed and the material properties which derive from them.

8.1 HISTORICAL OVERVIEW

The earliest known studies of fatigue are some repeated loading tests on iron chain done by a German mining engineer in 1829. Later, in the mid-1800s, railway axles were found to be breaking at loads a great deal inferior to their static capacity, which had been the basis of their design. As this industry was then expanding rapidly, the problem attracted serious attention.

Railway wheels and axles were a unit, as many of them still are today, so that the axle rotated with the wheel. The axles were thus rotating beams; in the case of railroad cars, the bearings were outboard, as shown in Figure 8.1a, resulting in the moment diagram you see there.

The stress at any point in a beam is $\sigma = My/I$, where y is the distance from the neutral axis to the point of interest, as shown in Figure 8.1b. The greatest stress is, of course, at the top or the bottom, where y has the value of the radius. A stress element at the bottom of this beam is in compression. It will be at the top one-half

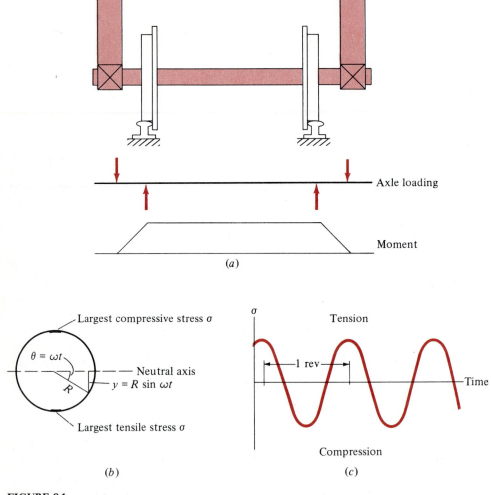

FIGURE 8.1
Railroad car axle. (a) Loading and moment diagrams. (b) Stress on axle. (c) Stress variation with time.

revolution later and subject then to the same stress, but tensile. The distance of the element from the neutral axis is given by $y = R \sin \omega t$ (Figure 8.1b), with ω being the rotational speed. The stress is then $\sigma = (MR/I) \sin \omega t$, which varies in time as shown in Figure 8.1c. With every 1000 miles (mi) something over a million cycles would accumulate, and unexpected failures would occur. The problem locations were near the bearing journals, where the stress was high due to the combination of the moment and the stress concentration which existed because of the section change.

William John M. Rankine (1820–1872), a Scottish engineer (the Rankine-cycle man), published a paper in 1843 entitled, "On the Causes of Unexpected Breakage of Journals of Railway Axles." A. Woehler, a German, studied the problem in great detail and published the results of 12 years of experimentation in 1870. The rotating-beam test scheme used by Woehler was until recently the basis of most material fatigue testing.

Of course, fatigue problems in engineering components and structures did not begin and end with railway axles. These problems have become, rather, one of the major concerns in engineering design. Efforts to overcome the fatigue difficulty have proceeded on two paths. One has been the study of the phenomenon at the material and theoretical level, which is discussed in this chapter. The other is the testing of components and assemblies, for it was early recognized that there is no completely satisfactory substitute for doing this, especially for critical parts. For some items that is not a great burden, home washing machines, for example. A good bit of simulated use can be accumulated by letting units run in a company laboratory and making corrections before a new model is released. Then if problems develop in customer use, faulty parts can be redesigned. The proving grounds of automobile manufacturers are a larger example. Here, of course, product safety is a main concern, which has become more and more acute with the development of product liability law in recent years. And the recall of cars to correct a design defect is very costly.

Over the years there have been a number of fatigue failures in aircraft. The British *Comet I* was the first jet passenger airplane. Two of these crashed in 1954, one of them 4 days following an inspection. The wreckage (recovered from the floor of the Mediterranean) was examined carefully and the conclusion reached that the pressurized cabin had failed, beginning with small cracks at a corner of a window. Each pressurization and return to atmospheric pressure at landing represented a cycle of stressing, of which there had been 1290 in the one plane and 900 in the other. All Comet aircraft of the type were grounded, and airframe fatigue design became the focus of increased study. One lesson deriving from these accidents was the necessity of testing complete major components, not just representative sections. Fuselage integrity seemed to be under control in recent years until a Boeing 737 lost one-third of the top of its cabin at 25 000 ft off the Hawaiian Islands in April 1988. The plane managed to land safely. Evidence pointed to fatigue failure at rivet holes in a frame stringer above the windows. This particular aircraft, part of an aging worldwide fleet of 737s, had logged 89 193 flights, hence that many pressurization cycles.

The loads on airplane parts and assemblies are far from a simple loading and unloading per flight; the loads change with every bump encountered and every

FIGURE 8.2
Boeing 757 wing/fuselage fatigue test setup. (*Courtesy, Boeing Commercial Airplane Co.*)

FIGURE 8.3
Boeing 757 main-landing-gear fatigue setup. (*Courtesy, Boeing Commercial Airplane Co.*)

maneuver made. Landing loads can be severe, and the loads imposed during taxiing are not insignificant.

Two of the many fatigue test arrangements for the Boeing 757 are displayed in Figures 8.2 and 8.3. The testing setup for the main rotor and blade control of a Sikorski helicopter is seen in Figure 8.4.

Aircraft pose a dilemma. On one hand, large safety factors, which one might employ to increase integrity, also increase weight, which implies decreased payload. This is poor economics, and in the extreme the machine might not even fly. On the other hand, small safety factors imply danger to human lives.

Two philosophies in aircraft design have evolved:

First, fail-safe design assumes that fatigue cracks will occur and parts may ultimately fail. Thus it is necessary to schedule inspections and backup in the form of multiple load paths and crack stoppers, so that cracks will be found and the parts

FIGURE 8.4
Fatigue-test arrangement for the main rotor and blade control of a Sikorski helicopter. (*Courtesy, Bell Helicopter Textron, Inc.*)

repaired before failure occurs. This philosophy is, however, not applied to all components of an airplane—the airframe and controls, yes, because they determine flight safety. It is generally not practical to do so with landing gear. An exception is very large aircraft, such as the Boeing 747 with four main gears; any three gears can bring the plane in safely. With multiengined aircraft, engines need not be fail-safe if the aircraft can fly with one or more dead. (This is generally not the case with small twin-engine planes.)

Second, damage-tolerant or safe-life design also assumes the existence of fatigue cracks. The aircraft is designed to resist failure due to such cracks (and other damage as well) for a specified period, termed the safe life, determined by fracture mechanics, before the cracks are detected and repaired. The growth rate of cracks is critical for the computation of the time to failure. A sizable factor of safety is applied to that time to arrive at the requisite inspection intervals. It is important that the critical parts be accessible for inspection.

Damage-tolerant design is required by the U.S. Air Force and Navy for obvious reasons. It is also used by builders of civil aircraft.

8.2   MECHANISM OF FATIGUE FAILURE

Theories attempting to explain fatigue failure have been through several generations of development. The term *fatigue* came into being because it was supposed that the material became tired in some way of the stress variations it was subjected to. People spoke of a gradual deterioration of the material under these conditions (which is, to some extent, true) and then, noting the nature of the broken surfaces, reasoned that it had "crystallized" from its original "fibrous" nature. They were, in fact, looking at the surface which broke at the very last, not in fatigue but in static failure, the brittle rupture associated with uncontrolled propagation of a crack.

There are three fairly distinct stages in fatigue failures: crack initiation, crack growth, and rupture.

CRACK INITIATION. Most metals consist of many grains or crystals. The crystals have a random orientation, unless the metal has been worked (rolled, forged, turned, etc.). In any case, some grains are oriented so that, under the loading conditions present, their planes of easy slip fall subject to higher shear stress than others, and slip occurs. The size of the bands of slip depends on the stress level and the number of times the stress is applied. Of course, the greatest slip activity in the part will be at the location of the greatest stress, a point of stress concentration, termed a *notch*. A notch can be a change in contour, a nick or scratch, or an internal imperfection, such as an inclusion or a porosity. The slip bands grow, and metal may extrude from the surface or intrude. You cannot see this without a microscope, however. The intrusions make matters worse, because they are excellent stress raisers. The slip bands finally become small ruptures, termed *microcracks*. They are indeed small, being only a few micrometers in size [1 micrometer (μm) = 0.001 mm]. Because of

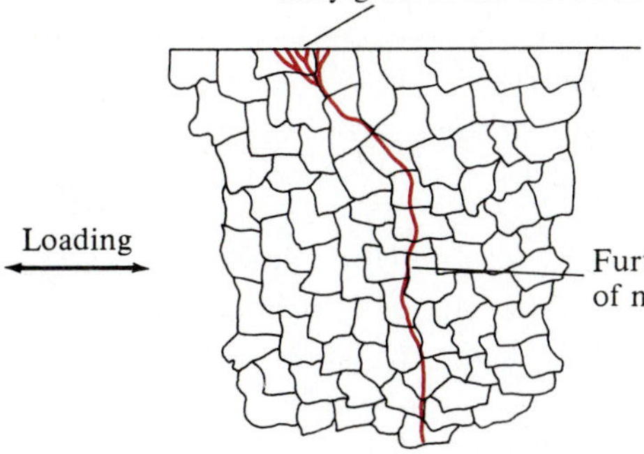

Several microcracks coalesce.
They grow in the direction of maximum shear.

Loading

Further growth occurs on planes
of maximum tensile stress.

FIGURE 8.5
Crack initiation and growth.

their association with shear stress and slip, microcracks are generally oriented with the maximum shear stress. They may grow across several grains, as shown in the sketch in Figure 8.5. There will usually be several cracks in a given locality, and as the process progresses, they tend to coalesce.

Materials which are brittle under the loading conditions, such as cast iron or high-strength steel at ordinary temperatures and loading rates, resist shear deformation, i.e., slip. In these cases the initiation of fatigue may occur at discontinuities in the structure or at some inclusion or void, rather than at the surface.

CRACK GROWTH. The tips of the microcracks created in the first phase are regions of intense stress concentration, and the material there is in a plastic state. The crack grows, but in a direction controlled by the maximum tensile stress rather than by the shear stress which initiated the process. This is seen in Figure 8.5. There may be several cracks, which will tend to come together. The advancing front of the crack is often evident on the fractured surface as a series of ripples, termed *beachmarks*, because they have the appearance of sand on a beach. Being concentric with the origin of the trouble, the beachmarks give notice of the crack's location. Figure 8.6 is a photograph of a typical fracture surface. This stage of the fatigue process is often termed *macrocrack growth*, or simply *crack growth*. The science of fracture mechanics has proved to be a valuable tool for analysis of this and the following phases of fatigue failure.

RUPTURE. The growth of the crack(s) as described above continues until a critical size is reached such that one more application of the load brings about instability and fracture.

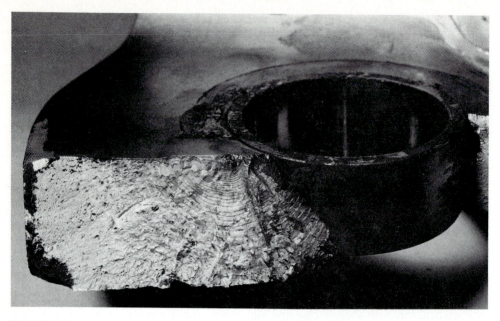

FIGURE 8.6
Structural failure resulting from a fatigue-initiated crack. The crack origin is at the center of the circular beachmarks. The static load fracture area is diagonally opposite. (*Courtesy, Boeing Commercial Airplane Co.*)

Factors Influencing Fatigue

Many factors influence the speed of the fatigue process, i.e., the length of time a part will resist the application of changing stresses:

1. The size of the applied stresses and their nature, i.e., whether completely reversed or consisting of an alternating stress superposed on a steady stress or the more general case of random stressing, as on a car axle; also whether the stresses are uni-axial or multiaxial
2. Stress concentration
3. Residual stresses, tensile or compressive, and stress gradients
4. Surface finish and/or coatings such as electroplating
5.. Temperature
6. Environment, especially corrosion
7. Size of part

8.3 STRESS-CONTROL FATIGUE TESTING

Traditional Tests

Any type of loading can be used for fatigue testing: axial, torsional, bending, or combinations of them. The simplest testing scheme for achieving cyclic stressing is the rotating beam, which produces reversed bending. The R. R. Moore machine, which operates on this principle and from which a great deal of the fatigue data on materials (note: *materials*, not components) in the United States have derived, is shown in Figure 8.7*a*. The loading and moment diagrams are displayed in Figure 8.7*b*. The moment is constant over the length of the specimen. The stress is greatest

(*a*)

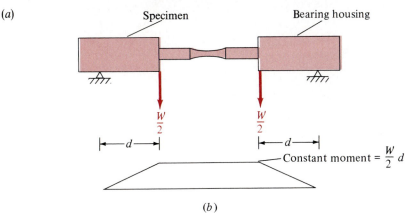

(*b*)

FIGURE 8.7
R. R. Moore fatigue-testing machine. (*Courtesy, SATEC Systems, Inc.*) (*a*) Testing machine. (*b*) Specimen loading and moment.

at the center, where the section is smallest. A point at the bottom is in tension (uniaxial—no shear), and one-half a revolution later, then at the top, it is in pure compression. The situation is the same as with the railroad axles described earlier.

A similar machine mounts the specimen as a rotating cantilever. Rotating-beam machines have the advantage of high testing speeds—they operate up to 10 000 rpm (1 r = 1 cycle of loading). A million cycles represent about 1 h 40 min of testing time at that speed.

In Figure 8.8 a machine for flexural testing of flat specimens is shown. A crank at the right applies a cyclic deflection; the stress must be computed or measured. The vise assembly can be adjusted vertically, which makes it possible to apply a constant mean stress, in which case the stressing is not one of complete reversal. We deal with that situation later.

Figure 8.9 shows the mechanism of a machine in which the loading is provided by the centrifugal force of a rotating imbalance. Depending on the specimen-mounting arrangement, the unit can be used for testing in the tension or bending mode. It can also apply a constant mean force.

Fatigue machines have counters to record the number of cycles and automatic shutoff when the specimens break—some tests may continue for a matter of days or weeks.

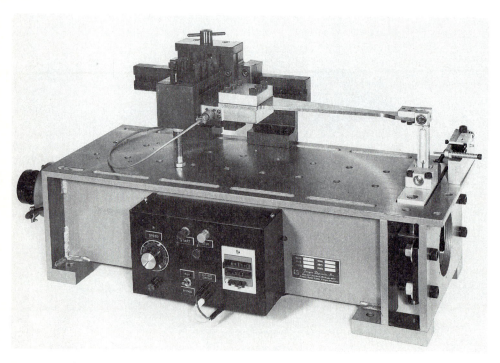

FIGURE 8.8
Machine for flexural testing of flat specimens. An eccentric at the right applies a cyclic deflection. (*Courtesy, Fatigue Dynamics, Inc.*)

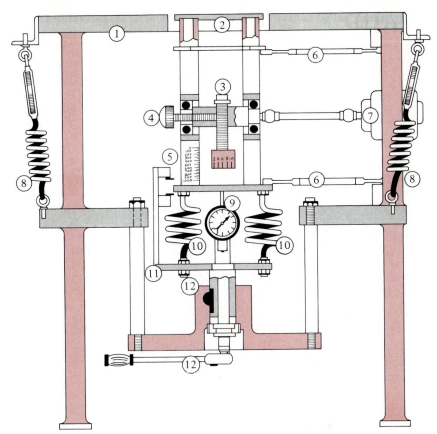

1. Stationary frame. Top provides workspace.

2. Reciprocating platen.

3. Rotating eccentric mass is source of dynamic force. Force is varied by screwing threaded rod in or out.

4. Thread screw locks threaded rod in position.

5. Scale reads pounds of vibratory force.

6. Flexure plates absorb horizontal centrifugal force so that only vertical force is transmitted to platen.

7. Synchronous motor drives eccentric mass at constant 1800 cycles per minute.

8. Springs provide seismic mounting so that no vibration is transmitted to, or from surroundings.

9. Dial indicates preload.

10. Compensator springs absorb inertia forces produced by reciprocating masses, preventing their transmission to the specimen.

11. Plate attaches one end of compensator springs to stationary frame.

12. Preload mechanism.

FIGURE 8.9
The load in this machine is provided by a rotating imbalance (3). (*Courtesy, SATEC Systems, Inc.*)

Modern Trends

In recent years there has been a movement away from the fatigue tests described above, especially the rotating-beam test. They are still used, however, with traditional materials. Moreover, rotating-beam tests are an efficient way to get large numbers of cycles. New materials, e.g., ceramics, and aerospace materials are no longer tested in these machines. The following quote from the handbook for the design of military aircraft (MIL-HDBK-5C, *Metallic Materials and Elements for Aerospace Vehicle Structures*, Sept. 15, 1976) is already more than a dozen years old.

> In the past, common methods of obtaining and reporting repeated load (fatigue) data included axial-loading tests, plate bending tests, and torsion tests. Rotating bending tests apply completely reversed (tension-compression) stressing to round specimens. Tests of this type are now seldom conducted for aerospace use and have therefore been dropped [from the handbook]. For similar reasons, flexure fatigue data also have been dropped. No significant amount of torsional fatigue data have ever been available. Axial loading tests, the only type retained herein, not only can consist of completely reversed loading (mean stress equals zero), but the mean stress can be varied.

The reason for these trends is the often questionable applicability of test results to loadings different from those of the tests. While rotating-beam and plate-bending tests are still routinely made and the results published and used for design, industries in which fatigue is critical have shifted toward axial-stress testing. Computer-controlled tensile testing machines have made this possible. Figure 8.10 shows one of these with its control equipment and an XY plotter, on which some load cycles can be discerned.

Test Specimens and Results

Fatigue specimens fall into two broad categories: those where the surface is made as smooth as possible and those where stress concentrations are deliberately created. The corresponding data are termed *smooth-specimen* or *unnotched* and *notched* data.

Each fatigue test yields one data point—the number of cycles to failure at a given level of cyclic stress. At sufficiently low stress, some specimens do not fail in a reasonable number of cycles, generally 10^6 to 10^8. To continue the test is considered a waste of time, for little useful information is gained. (A test to 10^8 cycles from 10^7 cycles takes 10 times as long!) Such data are called *run-outs* and are usually plotted with a small arrow pointing rightward.

A plot is made of these results, called an *S-N* curve, with S for stress and N for the number of cycles to failure. Since the reversal of stress is thought to have more significance than the completion of a cycle, the number of stress reversals is often plotted and labeled $2N$. The number of cycles or reversals tends to become very large, so the scale is usually logarithmic. If the stress scale is also logarithmic, the data for ferrous materials and titanium frequently fall on straight-line segments, as seen in Figure 8.11, with a break to horizontal near a million cycles. A stress

FIGURE 8.10
Axial fatigue test in a computer-controlled tension machine. (*Courtesy, MTS Systems, Inc.*)

level less than that of the horizontal portion implies indefinite life. That stress has been termed the *endurance limit*, but the term *fatigue limit* is now preferred. The strength for a finite life is called the *fatigue strength at x number of cycles*. You will see *fatigue limit* sometimes called *fatigue strength*; technically this is not correct.

Figure 8.12 shows the typical *S-N* curve shape for nonferrous metals and other nonmetallic materials. These plots usually show no clear break to horizontal, hence no fatigue limit. Rather, they continue slowly downward, though sometimes they may become nearly horizontal. The term *fatigue strength* is used for the strength at some selected number of cycles, frequently 500 million, which is greater than the life of most machine elements.

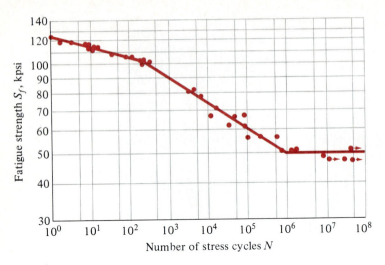

FIGURE 8.11
An *S-N* diagram for completely reversed axial fatigue tests. Material is UNS G41300 steel normalized; $S_{ut} = 116$ ksi; maximum $S_{ut} = 125$ ksi [2]. (*Data from NACA Technical Note 3866, December 1966.*)

You may wonder how DC-3s have managed to be operable after 50 years; also Air Force B-52s and KC-135s have seen a lot of service. There are two principal reasons:

1. If the stresses are kept low enough, lives of this order are quite possible.
2. The inspections which must be performed on aircraft have resulted in the replacement of critical parts over the years. The B-52 and the KC-135 are periodically rewinged.

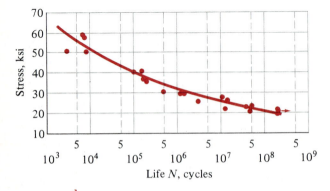

FIGURE 8.12
Completely reversed axial fatigue results for unnotched 7075-T6 aluminum alloy, various product forms, longitudinal direction. *Source: [78], 1985.*

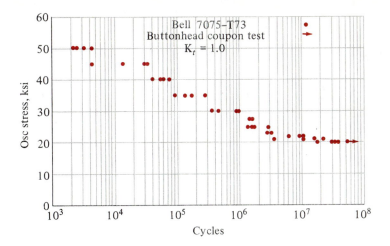

FIGURE 8.13
Axial fatigue-test results on shot-peened 7057-T73 aluminum buttonhead (type of grip which minimizes bending) flat specimens. (*Courtesy, Bell Helicopter Textron, Inc.*)

Stress Cycling about a Nonzero Mean Stress

The procedures here are the same as described above for reverse stressing. However, some laboratory test machines cannot accommodate the application of a mean stress, notably rotating-bending equipment.

Figure 8.13 shows results for cyclic axial stressing of 7075–T73 aluminum about a mean stress of 10 ksi. This type of test is of particular interest to the company that performed it, for most of their parts operate under conditions of steady plus fluctuating stress. The data suggest a leveling out of the strength and hence a fatigue limit for this material under this mean stress. Further testing at lower stresses would show the trend continuing downward, however slowly.

Survival Probability

In any testing, a batch of seemingly identical specimens tested under the same conditions does not yield identical results. There is always some scatter, as seen in Figures 8.11, 8.12, and 8.13. It is not uncommon for the life at a given stress level to vary by a factor of 10 or more, especially at high numbers of cycles, where the *S-N* curve has a small slope. There are three principal sources of this scatter:

1. Variations in specimen preparation (turning, grinding, etc.)
2. Variations in testing procedures
3. Variations in the property being tested

Some control can be exercised over the first two. The first has the greatest influence on scatter, especially when data from different laboratories (and/or different material

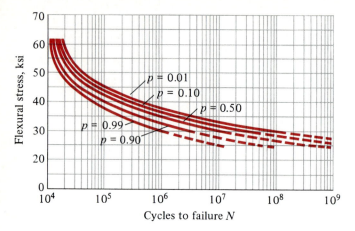

FIGURE 8.14
Survival probabilities for smooth specimens of 75S-T aluminum alloy in rotating-bending (*P-S-N* curve) [57]. (*Courtesy, ASME.*)

batches) are comingled. With care taken in the laboratory, the second source of scatter need not be significant.

Usually *S-N* curves are drawn through the center of the scatter, and hence they are an approximation to 50 percent survival probability. Quoted fatigue limits are generally on this basis, e.g., those listed in Appendix D.

At any particular stress level, we might ask what the probability is of survival of a specimen to a given number of cycles *x*. We could get an estimate by testing a number of specimens, say 20, at that stress and noting how many survived to *x* number of cycles. If 12 survived, then the estimate of the probability is 60 percent. Figure 8.14 shows the results of a statistical study of this kind. One hundred percent probability is approached at the bottom of the scatter and 0 percent at the top. Note that the scatter in the cycles to failure is very much greater at low stress levels than at higher ones.

The *S-N* data represent life to failure. They do not distinguish among the three stages of fatigue (crack initiation, crack growth, and failure).

Estimating the Fatigue Limit

The testing time to produce a reasonable amount of data for *S-N* curves is substantial, and the test specimens are expensive to make, so you do not always find published fatigue strengths. If the facilities, budget, or time is not available to perform the tests, a means of estimating the fatigue limit is needed. Figure 8.15 shows plots of fatigue strength vs. ultimate tensile strength for several classes of metals. The relationships implied provide some guidance in estimating the fatigue strength when it is

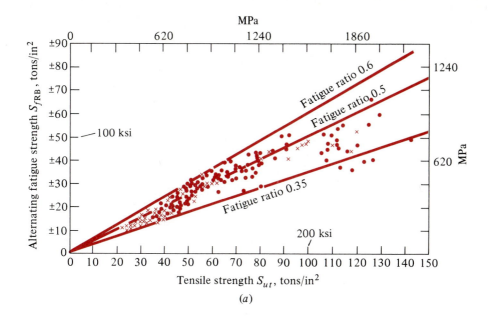

(a)

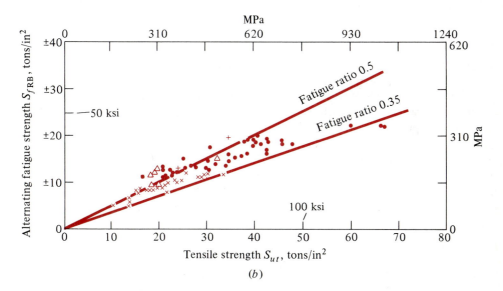

(b)

FIGURE 8.15

Relation between unnotched fatigue strength and ultimate tensile strength for rotating-bending [56]. (*Reprinted with permission of Pergamon Press from P. G. Forrest, Fatigue of Metals.*) (a) Carbon and alloy steels (10^7 to 10^8 cycles): ● = alloy steels, x = carbon steels. (b) Wrought and cast irons (10^7 cycles): x = flake-graphite cast iron, o = nodular cast iron, + = malleable cast iron, Δ = ingot iron.

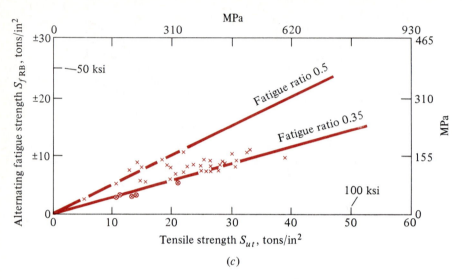

FIGURE 8.15 (Continued) (c) Aluminum alloys based on 10^8 cycles: x = wrought, x = cast.

not available from testing. In the case of steels,

$$S_{fRB} \simeq 0.5S_{ut} \qquad \text{with upper bound of } S_{fRB} = 100 \text{ ksi } (700 \text{ MPa}) \qquad (8.1)$$

is often used. Here S_{fRB} is the fatigue limit as measured in a rotating-beam (RB) test. As implied earlier, the fatigue limit of a real component is almost always less than that of a polished smooth rod.

Since the tensile strength of steels is about 500 times the Brinell hardness, the fatigue limit can be estimated from the Brinell hardness number:

$$S_{fRB} \simeq \begin{cases} 0.25H_B \text{ ksi} \\ 1.72H_B \text{ MPa} \end{cases}$$

with an upper bound again of 100 ksi (700 MPa). This is handy, since hardness tests are very easy to make and are nondestructive. The value of S_{fRB} is often the starting point for estimating a safe level of stress for design in a particular application.

8.4 STRAIN CONTROL

Section 8.3 described tests in which specimens were subjected to cyclic stressing. The analysis of fatigue on the basis of strain has gained many adherents in recent years, and it is the present state of the art. In strain tests, the maximum strain on the specimen is kept constant, rather than the maximum stress, thus the term *strain-controlled*. Usually the maximum stress changes during the test.

Strain-controlled testing has been made possible by computer-controlled testing machines actuated by a strain sensor on the specimen. The machines pictured in Figures 6.1 and 8.10 are capable of being so programmed.

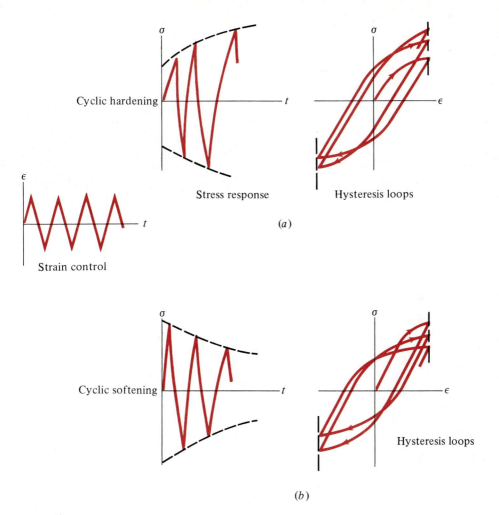

FIGURE 8.16
(a) Cyclic strain hardening and (b) cyclic strain softening under strain control [58]. (*Copyright ASTM. Reprinted with permission.*)

Figure 8.16 shows the typical response of materials to repeated reverse straining into the plastic region. Usually, materials which are initially soft become harder, as shown in Figure 8.16a. The stress corresponding to the imposed strain increases. Materials which are initially hard generally exhibit the reverse, cyclic softening, shown in Figure 8.16b. At any particular level of straining, an equilibrium stress-strain loop is reached, such as shown in Figure 8.17. The strain represented by such a loop consists of two parts: elastic and plastic. The latter, of course, is a greater proportion of the total as the amplitude of the applied strain is increased.

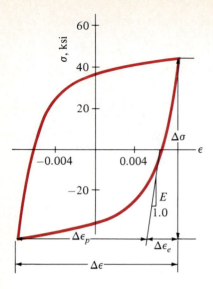

FIGURE 8.17
Stable stress-strain hysteresis loop [67]. (*Reprinted with permission of Society of Automotive Engineers, Inc.*)

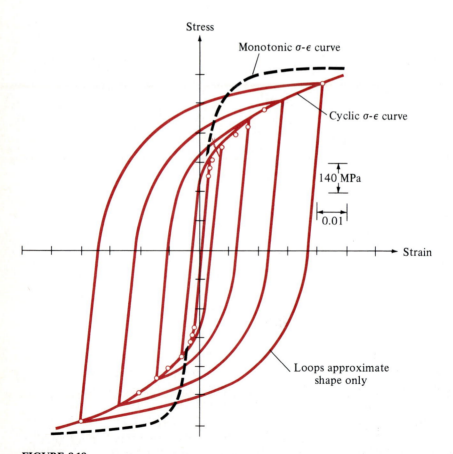

FIGURE 8.18
Cyclic and monotonic stress-strain curves for SAE 4340 steel [58]. (*Copyright ASTM. Reprinted with permission.*)

If a series of tests at various strain amplitudes is made and the tips of the equilibrium loops are plotted, the result is the cyclic stress-strain curve, of which Figure 8.18 is an example. It is analogous to the monotonic (static) true stress–true strain curve, which is also shown in the figure. Like the monotonic relation, the cyclic stress-strain relation on a log-log plot consists of two straight parts. The analytical relation is usually expressed in terms of the elastic and plastic strain components:

$$\varepsilon_{a,\text{true}} = \frac{\sigma_{a,\text{true}}}{E} + \left(\frac{\sigma_{a,\text{true}}}{K'}\right)^{1/n'} \tag{8.2}$$

You will see that the plastic part of Equation (8.2) is of the same form as Equation (6.2) (solved for the strain). Analogous to that expression, K' is known as the *cyclic strength coefficient* and n' as the *cyclic strain-hardening exponent*.

The cyclic yield strength σ'_y is determined as for monotonic testing by constructing a line parallel to the lower segment of the cyclic stress-strain curve through 0.2 percent strain and zero stress. The stress where the line intercepts the curve is taken as the 0.2 percent cyclic yield strength.

Values of cyclic strain constants are published for many materials; a few are given in Appendix E.

PROBLEMS

Sections 8.1 and 8.2

8.1. What is meant by fail-safe design?

8.2. What is meant by damage-tolerant design?

8.3. Name some factors of importance in the fatigue of metals.

8.4. Describe the several stages in fatigue failure.

8.5. Where do fatigue failures generally originate?

8.6. What does the term *beachmarks* refer to?

Sections 8.3 and 8.4

8.7. Describe the principles of operation of several fatigue-testing machines.

8.8. Define fatigue limit.

8.9. Define fatigue strength.

8.10. What materials have a fatigue limit?

8.11. Why do those materials for which only a fatigue strength is quoted not have a fatigue limit?

8.12. Suppose you wished to find the fatigue limit of a certain material in, say, pure tension. You would run some tension tests. What sorts of data would you take, and how would you plot them and find the fatigue limit?

8.13. Students in a laboratory session have taken the reversed-stress data shown in Figure P8.13. Plot an *S-N* curve and determine the fatigue strength or fatigue limit as appropriate.

Course _E.H. 293_ Term _71-1_ Section _2864_ Instructor _A.E. NEUMANN_

Material _7075-T6 ALUM._ Specimen _SEE SKETCH_ Machine _VSP-150 SER. 4986_

Machine Speed _1000 RPM_ Mean Stress _0_

Specimen Number	Stress Amplitude	Load	Start Date/Time	Student	Number of Cycles	Remarks
1	52000 PSI	89.1 LBS	1-11-71 10:43 AM	INSTR	12,400	
2	49,000 PSI	84.0 LBS	1-11-71 11:15 Am	JBR	21,600	
3	45000 PST	77.0 LBS	1-11-71 3:30 pm	PMW	65,300	
4	42000 PSI	72.0 LBS	1-12-71 8:45am	PMM	69,800	
5	40,000 PSI	68.5 lbs	1-12-71 2:20 Pm	JRB	211,200	
6	37,000 PSI	63.5 lbs	1-14-71 11:15 am	CPR	318,900	
7	34,000 psi	58.3 lbs	1-15-71 1:10 pm	BJP	2,674,300	No failure
8	33000 psi	56.6 lbs	1-18-71 3:40 pm	RJW	2,118,100	
9	35,000 psi	60.0 lbs	1-21-71 9:45 am	DRS	1,217,200	
10	38,000 PSI	65.1 LBS	1-25-71 4:10 pm	lqw	849,900	
11	40000 PSI	68.5 lbs	1-26-71 2:00 pm	JES	271,600	
12	44,000 psi	75.4 LBS	1-27-71 7:50 Am	JEP	60,800	
13	47000 PSI	80.5 lbs	1-29-71 3:20 pm	HST	55,500	
14	50,000 PSI	85.6 LBS	2-2-71 10:15 Am	JBR	20,500	
15	47,000 PSI	80.5 LBS	2-2-71 3:15 Pm	PMW	49,900	
16	43000 PSI	73.6 LBS	2-4-71 30 PM	PMM	100,100	
17	40,000 PSI	68.5 LBS	2-5-71 9:00am	JRB	145,700	
18	37,000 PSI	61.6 lbs	2-9-71 9:40am	CPR	1,119,900	
19	33,000 psi	56.6 lbs	2-12-71 8:50am	BJP	3,523,600	No failure
20	35,000 psi	60.6 lbs	2-16-71 1:10pm	RJW	2,610,100	No failure
21	38,000 psi	65.1 lbs	2-19-71 2:30pm	DRS	456,300	
22	42,000 PSI	72.0 LBS	2-23-71 11:10 am	lqw	151,000	
23	45000 PSI	77.0 lbs	2-24-71 9:15am	JES	74,700	
24	48,000 PSI	83.2 LBS	2-24-71 2:15 PM	JEP	35,150	
25	51000 PSI	57.3 lbs	2-25-71 2:14pm	HST	18,400	

JBR

FIGURE P8.13

8.14. What is the relation between the number of stress cycles and the number of stress reversals in fatigue testing?

8.15. Distinguish between stress control and strain control in fatigue testing.

CHAPTER
9

AVOIDING
FATIGUE
FAILURE

In Chapter 8 we described how material fatigue data are obtained and how the results are displayed in a stress or strain vs. cycles-to-failure curve. We also indicated that the fatigue limit for steels and the fatigue strength for other materials could be estimated, when necessary, from their ultimate tensile strength. Now we will see how to apply these results to the design of machine parts.

The early use of stress vs. cycles-to-failure (S-N) curves was to find the stress at which the curve leveled off, i.e., the stress more or less ensuring indefinite life. Parts were looked on as requiring a certain immortality. This is not always necessary and, when not necessary, uneconomic. Missiles are an example; their controls need to operate only a limited number of times during the expected life of the assembly.

Generally, static strength, which we discussed in Chapter 7, is predicated on less than about a thousand loading cycles and the assumption that the material of the part is nowhere strained plastically (more than once or twice). When these assumptions are violated, there may be a fatigue problem.

Fatigue analysis is generally divided into two zones. In the one, the straining is mainly within the elastic range, hence low stresses and long lives. This zone is thus known as *high-cycle fatigue*, corresponding to a fatigue life of about 10^3 or 10^4 cycles and greater. It includes, of course, indefinite life. Since the material action is mostly elastic, the fatigue strengths deriving from stress-life (S-N) curves are valid to use. This is termed the *stress-life model* and is described in the first part of this chapter.

The other zone, *low-cycle fatigue*, is associated with straining into the plastic range, thus higher stresses and shorter lives—less than about 10^3 or 10^4 cycles. Fatigue therefore can be a problem in the so-called static region. Low-cycle fatigue analysis is best done with strain-controlled fatigue data. Described in the second part of the chapter, it is called the *local strain model* and is the approach receiving most attention today. The local strain model may be used in high-cycle fatigue analysis as well.

PART A
STRESS-LIFE FATIGUE MODEL

This model is the classic approach. It is generally employed where there is no local cyclic plasticity and where a life greater than 10^3 or 10^4 cycles is needed. The material information may be limited to a fatigue strength or perhaps only an ultimate strength.

9.1   COMPLETELY REVERSED UNIAXIAL STRESSING, INDEFINITE LIFE

This section treats uniaxial, i.e., tensile or compressive, and torsional loading.

Correction Factors

It was early discovered that the fatigue strength of a given part was quite different from the results of smooth-specimen tests in the laboratory. Many factors affect fatigue performance, as indicated in Chapter 8. For initial design purposes, in the absence of test data on actual parts, smooth-specimen laboratory strength values must be translated to design strengths. This is done by modifying the laboratory fatigue limit or fatigue strength by empirically determined factors. We designate the resulting design fatigue strength S_f and compute it from

$$S_f = \frac{C_R C_F C_S C_I S_{f\text{RB}}}{K_f} \tag{9.1}$$

This approach is attributed to J. Marin [44]; it is used mostly with steels. The C's and $1/K_f$ are fatigue-strength reduction factors, described in the following.

RELIABILITY FACTOR C_R. As we noted and discussed in Section 8.3, S-N curves generally represent average lifetime vs. stress level. That is, the probability of surviving to the life indicated at a particular stress is 50 percent. When $S_{f\text{RB}}$ is estimated from the ultimate tensile strength, a 50 percent survival figure results. A survival rate considerably higher than 50 percent is required for most designs. Table 9.1 shows the correction factor C_R for reliability (probability of survival), based on the fatigue limit's having a standard deviation of 8 percent of its mean value. This is an upper limit generally for steels.

   The application of a reliability factor in this manner is standard practice in some design fields. Other engineers, however, prefer not to use it, relying rather on a safety factor derived from experience.

TABLE 9.1
**Reliability correction factor,
based on a standard deviation
equal to 8 percent of the mean
fatigue limit (approximate
maximum for steels)**

Reliability, %	C_R
50	1.000
90	0.897
95	0.868
99	0.814

FINISH FACTOR C_F. Fatigue failures commence at points of high local stress, often on the surface, where geometric discontinuities have been created by the part design (accounted for in K_f, discussed below) and where there are discontinuities on the smaller scale of the notches, grooves, etc., associated with machining or an as-cast finish. An unfinished rolled or forged element may have half the fatigue strength or less of the same part which has been ground and polished. The effect of the surface roughness is accounted for by the surface-finish factor C_F, values of which are shown for steel in Figure 9.1a. Some figures for 75S-T6 aluminum at 5×10^6 cycles are given in Figure 9.1b.

SIZE FACTOR C_S. Since it takes only one problem point on a microscopic level to start the process of fatigue failure, the greater the volume subjected to the higher stresses, the greater the probability of a failure's getting started. One author puts it eloquently by stating that "the larger the barrel, the greater the likelihood of finding a rotten apple." A number of approaches to the size factor have been proposed. Juvinall's [1, p. 214] recommendation is perhaps the easiest to apply:

For bending or torsion of round bars (ductile materials):

$$C_S = \begin{cases} 1 & D < 0.4 \text{ in (10 mm)} \\ 0.9 & 0.4 \text{ in} < D < 2 \text{ in (50 mm)} \\ 0.8 & 2 \text{ in} < D < 4 \text{ in (100 mm)} \\ 0.7 & 4 \text{ in} < D < 5 \text{ in (150 mm)} \end{cases}$$

(The R. R. Moore specimen is 0.3 in in diameter, for which C_S is seen to be 1.)

For shafts of diameter greater than 2 in, the experimental data for determining C_S are few. The ASME national standard "Design of Transmission Shafting" [69] suggests

$$C_S = \begin{cases} d^{-0.19} & 2 < d < 10 \text{ in} \\ 1.85d^{-0.19} & 50 < d < 250 \text{ mm} \end{cases}$$

These formulas are a compromise of several approaches from the literature. They give values agreeing with the figures in the above listing.

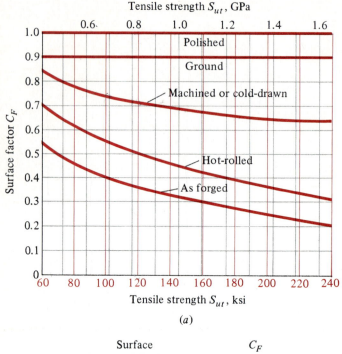

(a)

Surface	C_F
Longitudinal hand polish	1.00 (reference std)
Hand burnish	0.90
Smooth mill cut	0.87
Rough mill cut	0.79

(b)

FIGURE 9.1
Surface-finish factor C_F. (a) Values of C_F for steel. (*From* [2], *with permission.*) (b) Values of C_F for 75S-T6 spar-cap aluminum at 5×10^6 cycles, deduced from ref. 23, p. 196.

What should you do when the section is not circular? One approach is to use the dimension giving the worst C_S. As you might imagine, that is somewhat more severe than the reality of the situation. Kuguel [53] observed that the material stressed to 95 percent or more of the maximum stress in a part determines the fatigue resistance of the section. Based on that proposition, an equivalent diameter can be calculated for a round bar subjected to rotating-bending. For a rectangular section in bending,

$$D = 0.81\sqrt{A}$$

with A being the area of the section. This is good for bending about either major axis of the cross section. So, e.g., if we have a section $1 \times 2\frac{1}{2}$ in, the D to use is $D = 0.81\sqrt{2.5} = 1.28$ in, and then $C_S = 0.9$. If we had used the dimension (the largest) giving the worst (the smallest) C_S, then $C_S = 0.8$.

TABLE 9.2
Impact factor C_I

Type of load	C_I
Light (rotational machinery, such as motors, centrifugal pumps)	0.9–1.0
Moderate (slider-crank mechanisms, such as compressors)	0.7–0.8
Heavy (stamping presses, shears)	0.5–0.6
Severe (Forging hammers, rolling mills, stone crushers)	0.3–0.5

Source: Hindhede et al. [54].

Appendix B gives the details of the calculation of C_S by Kuguel's method and the results for some other cross-sectional shapes.

When the application is axial, the stress is uniform across the section, hence the barrel of possible problem sites is larger than in the rotating-beam (RB) test, and the strength is accordingly reduced. Shigley and Mitchell [2] recommend that $C_S = 0.71$ when S_{fRB} has been obtained by actual testing and $C_S = 0.60$ when S_{fRB} is taken from general curves or estimated from S_{ut}, as discussed earlier, i.e., without actual tests.

The fatigue strength of parts subjected to axial stressing is thus independent of part size. Some people link fatigue strength with the stress gradient in the part. In axial applications the stress is uniform across the part, and the stress gradient is zero, no matter what the size.

IMPACT C_I. A load applied suddenly creates greater stresses in any application than an equal one applied gradually. The factor C_I accounts for this effect. Hindhede et al. [54] suggest the values shown in Table 9.2 for this factor. The coefficient ranges are presented in somewhat general terms, because shock intensity is hard to quantify.

FATIGUE NOTCH FACTOR K_f. In Section 7.3 we stated that stress concentration is usually neglected in static loading. In fatigue, it must always be accounted for. However, the sensitivity of most materials to notches is less than the full value of the factor K_t would indicate. Notch sensitivity is measured by the parameter q.

$$q = \frac{K_f - 1}{K_t - 1}$$

in which K_f is the fatigue notch factor. From the above

$$K_f = 1 + q(K_t - 1) \qquad (9.2)$$

The -1 in the numerator and denominator of the definition of q results in values of q ranging from 0 when there is no notch sensitivity (K_f is then 1) to 1 when there is full notch sensitivity (K_f is then equal to K_t).

Notch sensitivity varies with the material and with the severity of the geometric deviation causing the stress concentration, as seen in Figure 9.2.

Rather than applying K_f to stress, we will use it as a factor decreasing strength, as for the other factors and as indicated in Equation (9.1). This is convenient when the alternating stress is superposed on a mean stress.

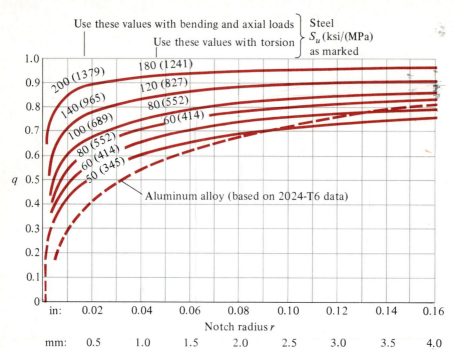

FIGURE 9.2
Notch sensitivity curves. (For *r* off the chart, use values at right edge.) [9], *with permission.*

OTHER EFFECTS. The reliability factor, finish factor, size factor, impact, and fatigue notch factor are the common effects in fatigue, and they suffice for this text. There are others: corrosive environments, fretting, higher temperatures. Some polishing and plating methods decrease fatigue strength; peening and some platings improve it.

The tests normally run in laboratories are done in an environment of air, which contains the corrosive elements oxygen and water vapor. Not many machine elements operate in vacuum, but fatigue behavior is known to improve slightly under such conditions. Many machines do operate in environments much worse than laboratory air—saltwater, salt air, and chemical plants are common examples. The fatigue strength may then be reduced by as much as 90 percent. The effect is due mainly to surface attack, promoting the formation of cracks through stress concentration, called *stress corrosion*. It is time- and stress-dependent. Data referring to specific conditions should be consulted for such applications.

Fretting is the wear that occurs between mating surfaces which have slight relative movement. The motion and clamping pressure result in cyclic shear stresses, plastic deformation, and sometimes high local temperatures. A certain amount of debris accumulates, and this results in stress-concentration sites and the ultimate initiation of fatigue failure. Of course, the action is much more severe in a corrosive atmosphere. It is time-dependent, hence more important in long-life situations. Fretting cannot be quantified, but its effect can be as great as those of stress concentrations and corrosion.

Generally for steels the fatigue strength decreases with increasing temperature. If specific information is not available, the following correction factor for wrought steel [62] can be used. T = degrees F.

$$C_T = \begin{cases} 1.0 & \text{for } T \leq 160°\text{F} \\ \dfrac{620}{460 + T} & \text{for } T > 160°\text{F} \end{cases}$$

The ASME shafting standard [69] notes that for most shafting steels (UNS G10350, G10600, G43400 are examples), the fatigue strength is essentially unchanged in the range $-70°\text{F}\,(-57°\text{C}) < T < 400°\text{F}\,(204°\text{C})$. Hence the temperature correction factor $C_T = 1$ for such steels within those temperatures.

In the case of aluminums and copper alloys, some testing needs to be done or other guidance sought for a particular alloy, for strength may be quite sensitive to temperature.

Example 9.1. A rotating stepped shaft is to operate under a steady bending load, as shown in Figure 9.3. The inner diameter (ID) of the bearings is expected to be about 1 in, and the fillet radius at the step will be about 0.1 in. The shaft will be machine-finished. A reliability of 99 percent is required. If a steel with an ultimate strength of 100 ksi is used, what fatigue limit may be expected of the material in the vicinity of the step?

We estimate the fatigue limit of rotating-beam specimens of this steel to be

$$S_{f\text{RB}} = 0.5S_{ut} = 0.5(100) = 50 \text{ ksi}$$

We must now determine the various correction factors:

Reliability: from Table 9.1, $C_R = 0.814$
Finish: from Figure 9.1, $C_F = 0.74$
Size: from paragraph following "Size Factor," $C_S = 0.9$
Impact: from Table 9.2, $C_I = 0.95$

Stress concentration: The theoretical stress-concentration factor is found in chart 9 of Appendix A: $D/d = 2$, $r/d = 0.1$, so $K_t = 1.7$. Notch sensitivity for a notch radius of 0.1 in is, from Figure 9.2, $q = 0.84$. Then

$$K_f = 1 + q(K_t - 1) = 1 + 0.84(0.7) = 1.59$$

The corrected fatigue limit, per Equation (9.1), is

$$S_f = \frac{0.814 \times 0.74 \times 0.9 \times 0.95 \times 50}{1.59} = 16.2 \text{ ksi}$$

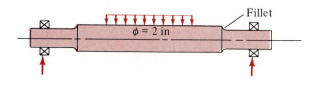

$\phi = 2$ in

Fillet

FIGURE 9.3
Shaft for Example 9.1.

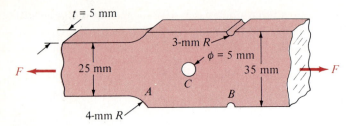

FIGURE 9.4
Bar for Example 9.2.

This figure, for a practical setting, is only about a third of the 50 ksi measured with smooth laboratory specimens. The S_f computed applies, of course, at the step, which would be the critical stress location in the design. The stress concentration having been accounted for in the material strength, it would not be included in the stress computations.

Example 9.2. A flat bar of steel has been machined to the shape shown in Figure 9.4. The strength S_{ut} is 700 MPa. Rotating-beam tests have been run on this material, and the bottom of the data scatter indicates a fatigue strength of 280 MPa. Determine the material fatigue strength at each of the three critical places A, B, and C and the critical stress location. The loading is axial.

The fact that the quoted fatigue strength is at the bottom of the scatter band (see Figure 8.14) means that the reliability factor is essentially equal to 1, for all locations: $C_R = 1.0$. In like manner, the finish factor, from Figure 9.1, is the same at all locations: $C_F = 0.73$. Since the fatigue strength was obtained from actual tests and the loading is axial, $C_S = 0.71$ everywhere. Recalling that

$$S_f = \frac{C_R C_F C_S S_{f\text{RB}}}{K_f}$$

it appears that the only item which will be different is K_f. So we can write

$$S_f = \frac{1.0 \times 0.73 \times 0.71 \times 280}{K_f} = \frac{145.1}{K_f}$$

For location A, K_t is found from Figure A.5 in Appendix A: $r/h = 4/25 = 0.16$, $H/h = 35/25 = 1.4$, so $K_t = 1.8$. From Figure 9.2, the notch sensitivity is $q = 0.85$, so $K_f = 1 + 0.85(1.8 - 1) = 1.68$. (The pertinent stress will be based on the 25-mm section.) We then get

$$S_f = \frac{145.1}{1.68} = 86.4 \text{ MPa} \qquad \text{at } A$$

At the notches, location B, Figure A.3 applies: $H/h = 35/29 = 1.21$, $r/h = 3/29 = 0.10$, so $K_t = 2.35$. From Figure 9.2, $q = 0.85$, so $K_f = 1 + 0.85(1.35) = 2.15$, and

$$S_f = \frac{145.1}{2.15} = 67.5 \text{ MPa} \qquad \text{at } B$$

(The reference section for stress is $35 - 2 \times 3 = 29$ mm.)

At the hole, location C, we refer to Figure A.1: $d/w = 5/35 = 0.14$, so $K_t = 2.6$. From Figure 9.2, $q = 0.86$, so $K_f = 1 + 0.86(1.6) = 2.38$. Then

$$S_f = \frac{45.1}{2.38} = 61.0 \text{ MPa} \qquad \text{at } C$$

(The reference for stress is $35 - 5 = 30$ mm.)

The fatigue strength at each location is shown in the following table. The critical location will be that with the smallest strength/stress ratio, which is C.

Location	Strength S_f	Stress σ	S_f/σ
A	67.5	$F/29t$	$1958t/F$
B	86.4	$F/25t$	$2160t/F$
C	61.0	$F/30t$	$1830t/F$

Shear Fatigue Strength

The emphasis in the above paragraphs has been on uniaxial (tension or bending) loading. A tensile design fatigue limit was the objective. The shear fatigue limit $S_{f,\text{sh}}$ may be estimated by using the maximum-shear-stress or the distortion energy theory:

$$S_{f,\text{sh}} = \begin{cases} 0.5 S_f & \text{maximum shear stress} \\ 0.577 S_f & \text{distortion energy} \end{cases}$$

These relations may be used as well for shear fatigue strengths for finite life, discussed next.

9.2   COMPLETELY REVERSED UNIAXIAL STRESSING, FINITE HIGH-CYCLE LIFE

Estimating the S-N Curve ($N > 10^3$)

If the S-N curve is available, then the smooth-specimen fatigue strength for a life of interest can be taken from it and corrected, as we discuss below, to arrive at a design value. In the absence of the S-N curve, the fatigue strength must be approximated. The usual shape of S-N plots on log-log axes for steels is seen in Figure 8.11. Generally the relation in the region between about a thousand and a million cycles is straight. Shigley and Mitchell [2, p. 284] showed that the fatigue strength in that region can be approximated as indicated in Figure 9.5, for rotating-bending. At 10^3 cycles it is estimated at $0.8 S_{ut}$, and at 10^6 cycles it is the fatigue limit $S_{f\text{RB}}$. The slope of the line is.

$$\beta = -\frac{1}{3} \log \frac{0.8 S_{ut}}{S_{f\text{RB}}}$$

and the intercept (at $10^0 = 1$ cycle) is

$$\gamma = \log \frac{(0.8 S_{ut})^2}{S_{f\text{RB}}}$$

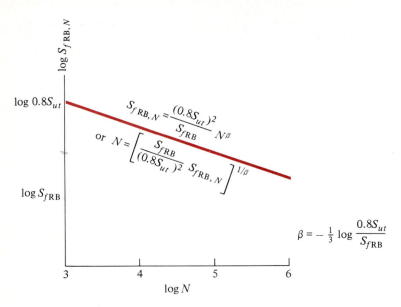

FIGURE 9.5
Approximating the high-cycle range of S-N curves for steel.

The fatigue strength is then

$$S_{f\text{RB},N} = 10^{\gamma} N^{\beta} = \frac{(0.8S_{ut})^2}{S_{f\text{RB}}} N^{\beta} \tag{9.3}$$

These expressions are, of course, approximations going through the center of the data scatter. They are for rotating-bending. If the fatigue strength is known for axial tests, then $S_{f\text{RB}}$ may be estimated by using the axial correction factor:

$$S_{f\text{RB}} = \frac{S_{f,\text{axial}}}{0.71}$$

The fatigue strength for a smooth specimen in rotating-bending $S_{f\text{RB}}$ must be corrected for size, finish, etc., and stress concentration to arrive at a design fatigue strength.

Correction Factors for Finite Life, Wrought[1] Steel

Little information is available on the determination of failure stress for finite life by applying strength-reduction factors to laboratory results. Clearly, the smaller the number of cycles, the closer the life approaches the static situation, and those factors become generally less applicable. We estimate the magnitude of each factor at 10^3 cycles and then consider that their product varies linearly on a log-log plot from 10^3 to 10^6 cycles.

---

[1] Wrought = worked: forged, rolled, etc.

RELIABILITY FACTOR $C_{R,1000}$. In the computation of fatigue strength for indefinite life, we included a reliability factor based on the fatigue limit for steels having a standard deviation of 8 percent of its mean value. Corresponding figures are not available for finite lives, so that in the approach below we will not use a reliability factor at either end of the life range, but will incorporate the uncertainties in a factor of safety (FS).

FINISH $C_{F,1000}$. At low lives, as static conditions are approached, surface finish has little importance, and this factor may be taken as unity.

SIZE FACTOR $C_{S,1000}$. Size also has little importance at short life, because only the material at the bottom of a notch matters. Hence this factor may be set to unity.

IMPACT $C_{I,1000}$. Impact will have about the same effect as for long lives (Table 9.2).

FATIGUE NOTCH FACTOR $K_{f,1000}$. The notch effect at 1000 cycles is computed by reference to the effect at long life K_f through a second sensitivity factor q_{1000}:

$$q_{1000} = \frac{K_{f,1000} - 1}{K_f - 1}$$

or
$$K_{f,1000} = 1 + q_{1000}(K_f - 1) \tag{9.4}$$

The value of q_{1000} correlates with the tensile strength, as shown in Figure 9.6.

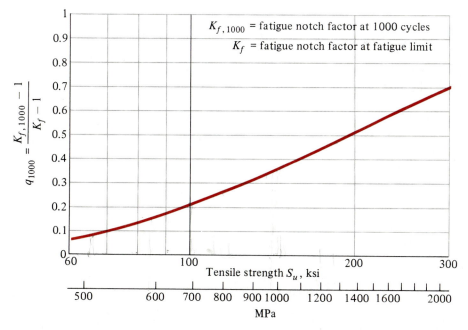

FIGURE 9.6
Sensitivity factor q_{1000} for steels [62]. (*Courtesy, Bethlehem Steel Corp.*)

OTHER EFFECTS. In the absence of specific data, one has to judge these effects according to the time of exposure at 10^3 cycles, for such effects as environment, corrosion, fretting, etc. Also consideration of the effects on static loading, which a life of 10^3 cycles more closely approximates than a life of 10^6 cycles, can be helpful. For example, shot peening, which has little effect on static tensile properties, could be expected to influence strength at 10^3 cycles only slightly.

Accounting for all the effects, at 1000 cycles we have

$$S_{f,1000} = \frac{C_I}{K_{f,1000}} S_{fRB} \tag{9.5}$$

If we designate by CK_N the effect of the correction factors (finish, size, impact, etc., and the fatigue notch factor) for some life of interest N, then

$$S_{fN} = CK_N S_{fRB}$$

As mentioned earlier, we assume that the decrease of CK_N from 10^3 to 10^6 cycles can be approximated by a straight line on log-log coordinates. That relation is shown in Figure 9.7. CK_N can also be expressed analytically:

$$CK_N = \frac{C_I K_f}{K_{f,1000}^2 C_S C_F} N^\lambda \tag{9.6}$$

where

$$\lambda = \frac{1}{3} \log \frac{C_F C_S K_{f,1000}}{K_f}$$

You should be able to derive Equation (9.6) (see Problem 9.8). As a quick check, find the boundary values. Does (9.6) yield the proper correction factors at $N = 10^6$ and at $N = 10^3$?

By combining (9.3) with (9.6), the design fatigue strength for N cycles, where $10^3 < N < 10^6$, can be written as

$$S_{fN} = \frac{(0.8 S_{ut})^2 C_I K_f}{S_{fRB} K_{f,1000}^2 C_F C_S} N^n \tag{9.7}$$

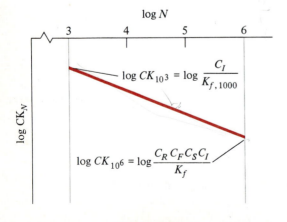

FIGURE 9.7
Variation of fatigue correction factors with life.

where
$$\eta = \frac{1}{3} \log \frac{S_{fRB} C_F C_S K_{f,1000}}{(0.8 S_{ut}) K_f}$$

If the life is required

$$N = \left(\frac{S_{fN} S_{fRB} K_{f,1000}^2 C_F C_S}{(0.8 S_{ut})^2 C_I K_f} \right)^{1/\eta} \qquad (9.8)$$

The calculations can also be accomplished graphically, as in Example 9.3.

Example 9.3. An estimate of the life of a grooved shaft subjected to rotating-bending is needed. Dimensions are shown in Figure 9.8; moment $M_a = 30$ N·m. The material is 4130 steel drawn at 400°F (data shown in Appendix F). The only applicable information given there is the ultimate strength $S_{ut} = 1034$ MPa. We estimate the fatigue limit at

$$S_{fRB} = 0.5 S_{ut} = 0.5(1034) = 517 \text{ MPa}$$

The alternating moment results in an alternating nominal stress, calculated on the basis of the minimum diameter:

$$\sigma_a = \frac{32M}{\pi d^3} = \frac{32 \times 30}{\pi (0.01)^3} = 306 \text{ MPa}$$

Our problem is to determine the life of the shaft under this stress, hence this value is S_{fN}. The correction factors are

$C_F = 0.89$ (Figure 9.1)

$C_S = 0.9$ (Paragraph in Section 9.1 following "Size Factor." The size $d = 10$ mm falls at the division between $C_S = 1$ and $C_S = 0.9$. We have chosen the conservative figure.)

$C_I = 1$ (no impact)

Recall C_R is not used (effectively $C_R = 1$).

$K_t \simeq 2.5$ (Figure A.12 in Appendix A)

$q = 0.80$ (Figure 9.2). Then

$$K_f = 1 + 0.80(1.5) = 2.20 \qquad (9.2)$$

$q_{1000} = 0.37$ (Figure 9.6). Then

$$K_{f,1000} = 1 + 0.37(1.20) = 1.44 \qquad (9.4)$$

The value of the exponent η in (9.7) can now be calculated:

$$\eta = \frac{1}{3} \log \frac{517 \times 0.89 \times 0.9 \times 1.44}{(0.8 \times 1034) \times 2.20} = -0.1615$$

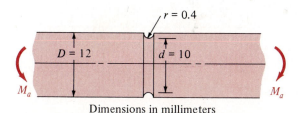

Dimensions in millimeters

FIGURE 9.8
Shaft for Example 9.3.

or

$$\frac{1}{\eta} = -6.191$$

With the above numbers, we can enter expression (9.8) for N:

$$N = \left(\frac{306 \times 517 \times 1.44^2 \times 0.89 \times 0.9}{(0.8 \times 1034)^2 \times 2.20}\right)^{-6.191} = 49\,344 \text{ cycles}$$

The graphical solution proceeds as follows. The fatigue strength for 1000 cycles is estimated at $S_{fRB,1000} = 0.8 S_{ut} = 827$ MPa (Figure 9.5). This is corrected per relation (9.5):

$$S_{f,1000} = \frac{827}{1.44} = 574 \text{ MPa}$$

The fatigue strength (limit) for 10^6 cycles is given by expression (9.1):

$$S_f = \frac{0.89 \times 0.9}{2.20} \, 517 = 188 \text{ MPa}$$

These points are shown on a log-log plot in Figure 9.9. At the value $S_{fN} = 306$ MPa, we measure $N \simeq 48\,500$ cycles. (Done more accurately, the graphic result would be identical to the analytical one.)

Example 9.4. An axially loaded bar, shown in Figure 9.10, must be machined. A factor of safety of 1.5 is to be applied to the load, and 50 000 cycles of loading are required. The project leader feels that 1040 steel, cold-drawn 20 percent, will be suitable. Is it?

The properties of this steel are listed in Appendix D. We will need

$$S_{fRB} = 370 \text{ MPa} \qquad S_{ut} = 805 \text{ MPa}$$

The alternating nominal stress on the part, including the FS on the load, is

$$\sigma_a = \frac{25\text{E} + 03 \times 1.5}{(\pi/4)(0.020)^2} = 119 \text{ MPa}$$

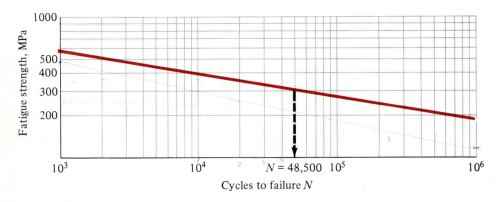

FIGURE 9.9
Fatigue curve for Example 9.3.

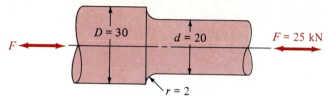

Dimensions in millimeters

FIGURE 9.10
Bar configuration for Example 9.4.

The material must have a corrected fatigue strength S_{fN} at $N = 50\,000$ cycles at least equal to this. To compute the fatigue strength, we need

$$C_F = 0.71 \text{ (Figure 9.1)}$$

$$C_S = 0.71 \text{ (axial application)}$$

$$C_I = 1 \text{ (assume no impact)}$$

$$K_t = 1.9 \text{ (Figure A.7 in Appendix A)}$$

$$q = 0.83 \text{ (Figure 9.2)}$$

Then

$$K_f = 1 + 0.83(0.9) = 1.75$$

$$q_{1000} = 0.26 \text{ (Figure 9.6)}$$

and

$$K_{f,1000} = 1 + 0.26 \times 0.75 = 1.20$$

The exponent η for expression (9.7) then becomes

$$\eta = \frac{1}{3} \log \frac{370 \times 0.71 \times 0.71 \times 1.20}{0.8 \times 805 \times 1.75} = -0.234$$

Then expression (9.7) gives

$$S_{fN} = \frac{(0.8 \times 805)^2(1.75)}{370 \times 1.20^2 \times 0.71 \times 0.71}\, 50\,000^{-0.234} = 215 \text{ MPa}$$

The strength is therefore adequate; in fact, it is a good bit more than needed.

Finite Life, Aluminum

The greater part of engineering design work in aluminum has been, as you might expect, done in the aerospace industry. Fatigue cracks in aircraft structures usually initiate at fastener holes, of which there are thousands. Fatigue data derive from specimens which typify the structure, so that smooth-specimen tests are rarely run. Test specimens are usually flat with holes, and the method of testing is axial. The prototype is based on lifetime data from those tests and on experience. The part is subsequently tested as part of its assembly, as discussed in Chapter 8.

In a problem where only smooth-specimen data are available, one approach would be to calculate a correction factor CK for long life, 10^6 cycles (or more), and apply it to the smooth-specimen fatigue strength over the entire range of life N. The

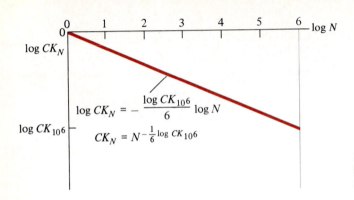

FIGURE 9.11
Straight-line approach to corrections for aluminum.

result would be conservative. Alternatively, on a log-log plot we can draw a straight line between $CK_{N=1} = 1$ (monotonic) and $CK_{N=10^x}$ (if the life used is 10^8, for example, $x = 8$) and use values drawn from the curve. This is illustrated in Figure 9.11 for $x = 6$. Analytically,

$$K_N = N^\lambda \qquad \text{where} \qquad \lambda = \frac{1}{x} \log CK_{10}x \qquad (9.9)$$

Use of this approximation is illustrated in Example 9.6.

9.3 EFFECT OF MEAN STRESS ON STRENGTH

Many materials applications involve fluctuating stresses about a nonzero mean. To define the effect for a particular material, a testing program involves constructing a curve of life vs. alternating stress, an S-N curve, for each of a series of mean stresses. Rotating-beam machines do not have the capability to apply both mean and alternating stresses. The machine of Figure 8.9, which produces an alternating force by means of a rotating imbalance, can apply a constant load through extension of a spring, as seen in the figure. Also, the programmed axial testing machines of Figures 6.1 and 8.10 can operate in this mode.

Ductile Materials

The results of the test program outlined above could be shown on a three-dimensional display such as in Figure 9.12. Representations like that do not lend themselves to reading off numbers, so the diagram commonly seen consists of cuts at constant life. Figure 9.13 shows typical results for long life; the scales have been made nondimensional. Various curves approximating the shape of such data have been proposed for design use when actual data are not available. Three are shown in the figure; they are described below. The inequalities in these relations indicate the condition for

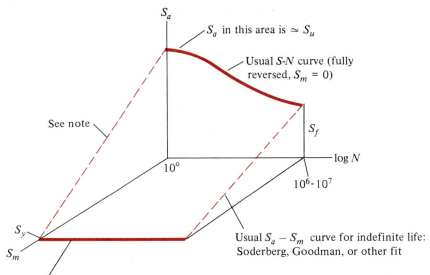

FIGURE 9.12
Spatial relation of S_a, S_m, and N.

safe design, with an FS on mean and alternating stresses included. The factor could be made different for each stress, if the problem conditions indicate it (this is discussed in Chapter 5). S_{fN} is the fully reversed design fatigue strength for N cycles.

The *Gerber* fit

$$\frac{\sigma_a}{S_{fN}} + \left(\frac{\sigma_m}{S_{ut}}\right)^2 \le \frac{1}{FS}$$

is used very little because it tends to be nonconservative—many of the data lie outside it.

Goodman's[1] proposal connects the fatigue strength on the alternating stress axis ($\sigma_a = S_{fN}$) with the ultimate strength on the mean stress axis ($\sigma_m = S_{ut}$) by a straight line:

$$\frac{\sigma_a}{S_{fN}} + \frac{\sigma_m}{S_{ut}} \le \frac{1}{FS} \qquad (9.10)$$

It represents a simple lower boundary to the experimental data.

---

[1] J. Goodman (1862–1935) was a professor of civil and mechanical engineering at the University of Leeds, England.

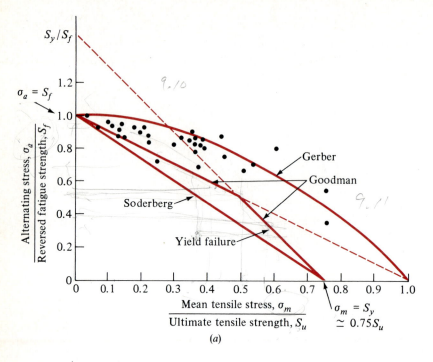

(a)

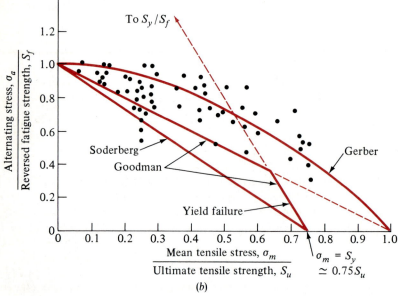

(b)

FIGURE 9.13
Influence of mean stress on long-life fatigue and showing several failure approximations [56]. (a) Steel ($\sim 10^7$ cycles). (b) Aluminum ($\sim S \times 10^7$ cycles). (*Reprinted with permission from P. G. Forrest Fatigue of Metals, Pergamon Books, Ltd.*)

Since the stress rises to a maximum of $\sigma_a + \sigma_m = \sigma_{max}$ each cycle, the yield strength would be exceeded on the first cycle of loading at the right-hand end of the Goodman line. Thus the further requirement is imposed

$$\sigma_{max} = \sigma_a + \sigma_m \leq \frac{S_y}{FS} \tag{9.11}$$

to prevent static failure. The result is a two-segment line, as shown in Figure 9.13. In using the Goodman relation in design, you can assume that one or the other of the lines governs and then check later. That is usually easiest; as a guideline, failure by yielding is clearly more likely if the alternating stresses are small. Alternatively, the design can be carried through on the basis of fatigue and then of static yielding and the governing dimension chosen.

The *Soderberg*[1] criterion is a straight line between the extreme of pure alternating stress, where $\sigma_a = S_{fN}$, and the other extreme of static stress, where $\sigma_m = S_y$:

$$\frac{\sigma_a}{S_{fN}} + \frac{\sigma_m}{S_y} \leq \frac{1}{FS}$$

The single-line Soderberg criterion is simpler than Goodman's relation. However, it is considered needlessly conservative today and has few adherents.

In this text we use the Goodman criterion.

Since $\sigma_a + \sigma_m = \sigma_{max}$ and $\sigma_m - \sigma_a = \sigma_{min}$, the above relations could be written in terms of σ_{max} and σ_{min}. You will also find the parameter $\Delta\sigma = 2\sigma_a$, the range of the alternating stress, used, as well as $R = \sigma_{min}/\sigma_{max}$ and $A = \sigma_a/\sigma_m$.

The above relations are approximations to the two-dimensional space of alternating vs. mean stress: σ_a vs. σ_m (Figure 9.12). The third dimension S_{fN} may be available from experimental data (*S-N* curves). When it is not, we can use the estimates of fatigue strength which we developed in Section 9.1.

Compressive mean stress results in an improvement in fatigue performance. This is because such stresses tend to close a crack, or, to put it another way, the net tensile stress across the crack is reduced. Advantage is frequently taken of this fact by shot-peening a critical surface, which is then left in a compression state with greater resistance to fatigue. The action is on the negative-mean-stress side of the σ_a-σ_m plot (Figure 9.13). Automobile and truck leaf springs are an example. They are simply flexible beams. The side which sees tension when the wheel hits a bump is commonly shot-peened.

Compressive mean stress has greater effect in raising the permitted alternating stress for short-life fatigue than for long-life fatigue, as is evident in the Goodman approximation (S_{fN} is raised, while S_{ut} remains fixed). When stress concentration is present (notches), the effect is more pronounced. In fact, the fatigue strength of a notched part can exceed that of a smooth one if the compressive stress is sufficiently high.

---

[1] C. Richard Soderberg (1895–1975) was a turbine engineer, professor and department head, mechanical engineering, and engineering dean at the Massachusetts Institute of Technology.

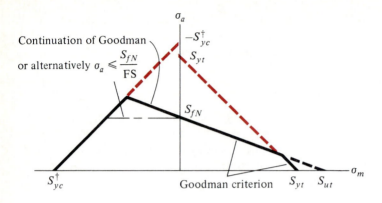

† In the absence of S_{yc}, use $-S_{yt}$.

FIGURE 9.14
Compressive mean stress and the Goodman criterion.

To estimate failure stress levels when the mean stress is compressive, the Goodman line can be extended into the negative region, or the alternating stress can be bounded by the fatigue limit S_{fN}, as shown in Figure 9.14. As seen in the figure, if the compressive mean stress is sufficiently large, static failure is the criterion; i.e., the sum of the alternating and mean stresses must not cause yielding:

$$\sigma_m - \sigma_a \geq \frac{S_{y,\text{comp}}}{\text{FS}} = \frac{-S_{yt}}{\text{FS}}$$

The negative of the tension yield strength $-S_{yt}$ is used in place of the compression yield strength $S_{y,\text{comp}}$ when the latter is not available. Note that σ_m and $S_{y,\text{comp}}$ are negative numbers.

Another representation often used for the Goodman criterion is that of Figure 9.15, in which $\sigma_{\max}$ and $\sigma_{\min}$ are the ordinate and σ_{mean} is the abscissa.

Pure Torsion

Data deriving from torsional tests (alternating torsional stress superposed on a mean torsional load) are shown in Figure 9.16, a diagram corresponding to Figures 9.13 and 9.14. As the diagram shows, the shear fatigue strength is affected little by the size of the mean stress until the mean stress is large enough to cause, when added to the cyclic stress, a static failure. The division between the one type of failure and the other can be expressed in terms of τ_{mean} (τ_m). If

$$\tau_m \leq \frac{1}{\text{FS}} (S_{sy} - S_{f,\text{sh}})$$

then a fatigue failure is indicated; otherwise, failure will occur by yielding in the first cycle. It is sometimes convenient to express the division between the two types in

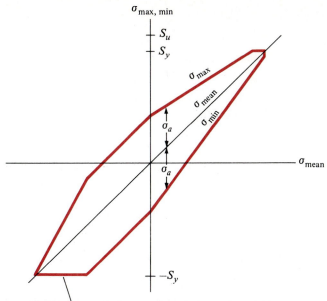

Smallest stress (largest compressive) equal to yield

FIGURE 9.15
An alternative representation of the Goodman criterion.

terms of the ratio of τ_a to τ_m. From the figure, this is

$$\tau_a = \frac{S_{f,\text{sh}}}{S_{ys} - S_{f,\text{sh}}} \tau_m \qquad (9.12)$$

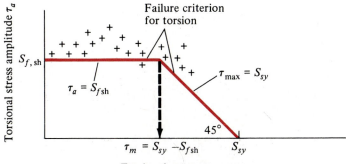

FIGURE 9.16
Effect of mean torsional stress on torsional alternating fatigue strength. (Since there is no distinction between + and − in shear, the negative left side of the diagram is absent.) (*From Shigley and Mitchell* [2], *with permission.*)

Thus, if

$$\frac{\tau_a}{\tau_m} \geq \frac{S_{f,\text{sh}}}{S_{ys} - S_{f,\text{sh}}} \tag{9.13}$$

then a fatigue failure will occur; otherwise, static failure will occur. To avoid fatigue failure,

$$\tau_a \leq \frac{S_{f,\text{sh}}}{\text{FS}} \tag{9.14a}$$

To avoid static failure,

$$\tau_{\text{max}} = \tau_a + \tau_m \leq \frac{S_{sy}}{\text{FS}} \tag{9.14b}$$

The shear-stress fatigue strength $S_{f,\text{sh}}$ may be estimated from the maximum-shear-stress theory or the distortion energy theory, as indicated in Section 9.1. The shear yield strength S_{sy} may be similarly estimated.

Some engineers apply a Goodman diagram to torsional loadings. This, of course, is a more conservative approach.

Brittle Materials

A σ_a-σ_m diagram paralleling Figure 9.13 for gray cast iron is shown in Figure 9.17. Since the compressive strength is typically several times greater than that in tension, the curve is skewed to the right, and you will note that a compressive mean stress raises the permissible alternating stress considerably. Until recently the use of brittle materials in fatigue environments has been quite limited, mostly gray cast irons in compression fatigue situations. Now, however, we see such applications as carbon fiber, a brittle material, in the wing of *Voyager*, which circumvented the globe in January 1987, and the use of ceramics in automobile turbo impellers. Of course, designs like these, involving high production or a need for extremely high reliability, are verified by exhaustive experimental programs.

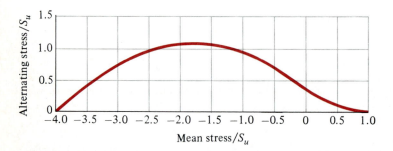

FIGURE 9.17
Alternating vs. mean stress for cast iron ($S_u = 20$ to 50 ksi); indefinite life, axial loads. [*13*], *with permission.*

9.4   FATIGUE UNDER VARYING STRESS LEVELS—CUMULATIVE DAMAGE

In the preceding paragraphs we have discussed fatigue situations in which specimens or parts were subjected to a fixed mean stress (which might be zero) and a fixed alternating stress. Here we ask what happens if those stress levels change in time. An early answer was suggested by Palmgren [15], who was interested in the service life of ball and roller bearings.[1]

The equivalent of *S-N* curves for bearings, bearing load vs. life, had been obtained by running them under various constant loads to failure. Since service loads are frequently not constant, the question arose how to estimate bearing life, given a load sequence and the load-*N* curve. In fatigue terms, we ask what the life will be if we know the stress sequence and the *S-N* curve. Palmgren's theory was restated by Miner in 1945 [16] and has since commonly been known as *Miner's rule*:

$$\frac{\text{Cycles at } \sigma_1}{\text{Life at } \sigma_1} + \frac{\text{cycles at } \sigma_2}{\text{life at } \sigma_2} + \cdots = D$$

or

$$\sum \frac{n_i}{N_i} = D \tag{9.15}$$

where D is the fractional damage, for example, $D = 0.3$ means three-tenths destroyed; $D = 1$ at failure.

The theory is frequently referred to as the *linear damage rule*, since it supposes that the damage at any stress level is directly proportional to the number of cycles, i.e., each cycle does the same amount of damage. The theory also supposes that it does not matter what the stress sequencing is, and the rate of damage accumulation at any stress level is presumed to be independent of past history. These propositions do not hold up in the laboratory, so that failure prediction on the basis of Miner's rule can be significantly in error. A number of theories have attempted to overcome these shortcomings, but none is completely satisfactory. Despite its problems, Miner's rule remains the most popular, largely because it is so simple. Also when the stress amplitudes are randomly applied, which corresponds more to real-life situations, stress sequencing and stress history are less consequential and Miner's rule generally is satisfactory.

Reversed Stressing (Zero Mean Stress)

If, say, 40 percent of the life of a part is used up at a certain stress level, Miner's rule states that at any other stress level 40 percent of the life is gone; 60 percent remains. Since the lifetime depends on the stress, the remaining life will be different from that at the first level of stress. Suppose we have a fully corrected *S-N* curve like the one sketched in Figure 9.18. The initial stress level is S_1, where a life of 2000 cycles is

---

[1] Arvid Palmgren joined Svenska Kullagerfabriken (SKF: Swedish Roller-Bearing Factory) in 1917 and spent some 30 years on research and engineering of ball and roller bearings, especially fatigue and life phenomena.

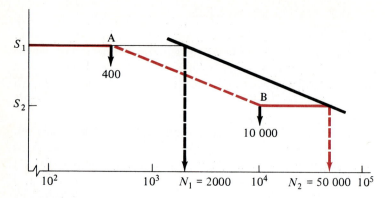

FIGURE 9.18
Miner's rule.

predicted. The part is run for 400 cycles (point A), using up 20 percent of its life. The stress is now lowered to S_2, where a life of 50 000 cycles is predicted. Although this is a different stress, we assume the amount of life used is still 20 percent. Since 20 percent of 50 000 is 10 000 cycles, the starting point at stress S_2 is $N = 10\,000$, point B, and the remaining life at this stress is 40 000 cycles. With this sequence, we predict failure in $400 + 40\,000 = 40\,400$ cycles—more than the life at S_1 and less than the life at S_2.

Example 9.5. A part is to be loaded axially. The correction factor at 10^3 cycles is $CK_{10^3} = 0.7$, and at 10^6 cycles $CK_{10^6} = 0.3$. Figure 9.19a shows the smooth-specimen S-N curve for the material in question, and we have drawn in the corrected curve from 10^3 to 10^6 cycles. Because some of the loading may be in the low-cycle region, we sketch that in as indicated, assuming that none of the correction factors are operative at 1 cycle (essentially monotonic loading). Rather than extending the curve horizontally above 10^6 cycles, which would signify no material damage at loads below about 15 ksi, the strength is shown slowly decreasing. The rate of decrease is indicated in Figure 9.19b.

Suppose the piece is to be loaded by the following sequence, or block:

$$25 \text{ cycles at } \sigma_1 = 30 \text{ ksi}$$

$$35 \text{ cycles at } \sigma_2 = 20 \text{ ksi}$$

$$196 \text{ cycles at } \sigma_3 = 13 \text{ ksi}$$

These stresses are completely reversed. How many such blocks will fail the piece? At the several stress levels, we note lives as follows from Figure 9.19a:

$$\sigma_1 = 30 \text{ ksi} \qquad n_1 = 4 \times 10^4 \text{ cycles}$$

$$\sigma_2 = 20 \text{ ksi} \qquad n_2 = 2 \times 10^5 \text{ cycles}$$

$$\sigma_3 = 13 \text{ ksi} \qquad n_3 = 5 \times 10^6 \text{ cycles}$$

The damage done in one block of loading is then

$$\frac{25}{4\text{E} + 04} + \frac{35}{2\text{E} + 05} + \frac{196}{5\text{E} + 06} = 8.39\text{E} - 04$$

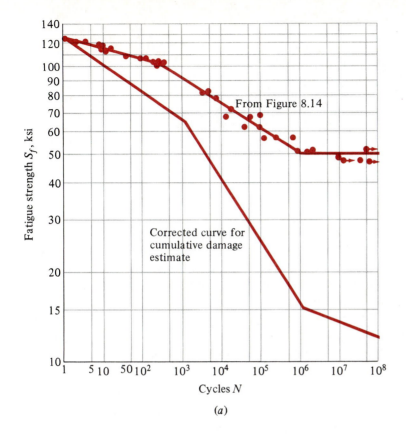

(a)

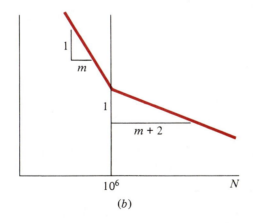

(b)

FIGURE 9.19
The S-N curve for Example 9.5. (a) Corrected S-N curve. (b) Determination of curve for $N > 10^6$. (*Suggested by H. S. Reemsnyder, Bethlehem Steel Corp.*)

Total blocks to failure are

$$\frac{1}{8.39\text{E} - 04} = 1192$$

In many cases the S-N curve will not be available, and the only information will be the ultimate strength S_{ut} and (not always) the fatigue limit S_{fss}. The corrected values of fatigue strength for 10^3 and 10^6 cycles are then estimated as in Sections 9.1 and 9.2. The fatigue strength at 1 cycle (actually $\frac{1}{2}$ cycle) can be approximated by the ultimate strength S_{ut}, corrected for impact C_I. Recall that our development supposes the loading to be in the high-cycle region. Thus, if some of it is low-cycle, the results may be somewhat questionable.

Alternating Stresses about a Mean Stress

Frequently the mean stress will not be zero, and it may change with each group of alternating stresses or even with each single cycle or half-cycle of stress. The problem is then a three-dimensional one, for we are dealing with combinations of the three variables S_a, S_m, and N. To compute cumulative damage, we have to know, for a

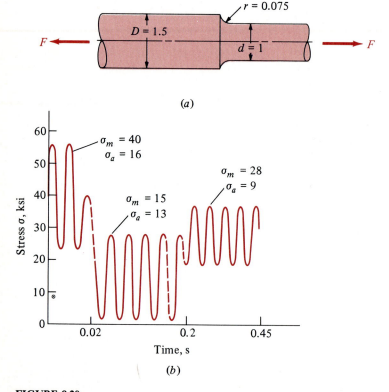

(a)

(b)

FIGURE 9.20
Data for Example 9.6. (a) Tension member. Dimensions are in inches. (b) Loading block diagram. Frequency is 50 Hz.

TABLE 9.3
Computation of damage in one block of loading, Example 9.6

S_m ksi	S_a ksi	S_{fN} (9.10)* ksi	log N_{fail} (Fig. 9.21)	N_{fail}	$N_{applied}$	Damage (N_{app}/N_{fail})
40	16	28.8	4.55	35 480	4	1.127E − 04
15	13	15.6	6.18	1 513 600	36	2.38E − 05
28	9	13.1	6.62	4 168 700	40	1.204E − 05
						$\Sigma = 1.485\text{E} - 04$

* $S_{ut} = 90.0$ ksi.

given S_a-S_m pair, what the life N is. The following example shows how this problem is handled. It is a simplified view of the real world, but we will get to that shortly.

Example 9.6. Figure 9.20a is a tension member made of 7075-T6 aluminum, for which the fatigue-life curve is given in Figure 8.12. The repeating stress sequence is sketched in Figure 9.20b. We need an estimate of the time to failure.

Figure 8.12 must be corrected for the stress concentration and other factors. From Figure A.7 in Appendix A, $K_t = 1.9$, and $q = 0.66$ in Figure 9.2. Hence $K_f = 1 + 0.66(0.9) = 1.59$. We will assume the C factors at $N = 10^8$ to be equal to 0.7. Then, applying Equation (9.9) to approximate the corrections for several lives along the curve, we find the corresponding fatigue strengths and draw the corrected S-N curve in Figure 9.21.

Table 9.3 shows the values of S_{fN} for the three segments of the loading and the corresponding lives taken from Figure 9.21. The number of cycles in each segment is obtained from the frequency (200 Hz) and the time length of the segment. The damage computed, $\Sigma = 1.485\text{E} - 04$, is for one block of 0.45-s duration. Failure occurs for damage = 1, so

$$\text{Time to failure} = \frac{0.45}{1.485\text{E} - 04} = 3030 \text{ s} = 50.5 \text{ min}$$

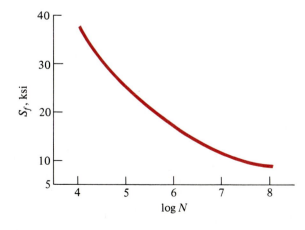

FIGURE 9.21
The S_f-N curve for Example 9.6 (from Figure 8.15).

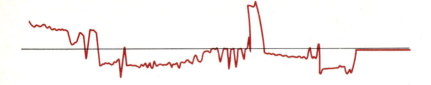

FIGURE 9.22
Loading history on vehicle part. (This is a clip from a much longer trace.) (*Courtesy, John Deere & Co.*)

Complex Stressing

With the exception of machinery running at constant speed and under constant load, most real loading histories are not at all like the sinusoids which we have used to characterize them. The loads on the axles of automobiles, for example, depend on the load in the car, the irregularities of the road surface, and the mood of the driver. Similarly, an airplane wing is subject to the random bumpiness of the air as well as the inertial effects endured in landing and taxiing. Figure 9.22 shows a typical history (load or strain, as the case may be). The problem is to relate patterns like these to *S-N* curves for the part or the material. Specifically, we must somehow find cycles of loading in such a chart, so as to estimate cumulative damage. A number of methods have been proposed to do this, the most popular of which is the one described below.

A sample (enlarged if necessary) from a history like that of Figure 9.22 might appear schematically as shown in Figure 9.23. The first step is to cut the sample so that it starts and ends at the same value of load, and then to arrange the block to start at the highest peak or the lowest valley. This is done by tacking the portion preceding the highest peak (or the lowest valley) onto the other end, as shown in the figure. The cycle count can now be made as follows.

From the starting point trace downward to the next peak, *A* to *B*, shown dashed in color in the figure. Run horizontally to the right to the next downward slope *C*, go down it to the next peak *D*, and continue in this manner until the lowest point is reached *E*. Continue upward in the same way to the endpoint. This step traces out the largest possible cycle. The rearrangement of the block to begin with the largest (or smallest) load was to ensure that this cycle gets counted, for it does the most damage. It has extreme values of 110 and −52.

Those parts of the trace unused in the above first cut are sketched in Figure 9.23c. The same procedure is now carried out, by going up or down as necessary. The results are shown in the figure.

The above scheme is continued until all parts of the history are accounted for, and the extreme values of the cycles can then be listed, as in the figure. These are then converted to mean and amplitude forces and used to compute the damage done by each cycle.

This method of load analysis is called *rain-flow counting*. The name derives from the similarity of the procedure to the flow of rain from a pagoda roof. To see the analogy, turn Figure 9.23 90°, so that time runs downward.

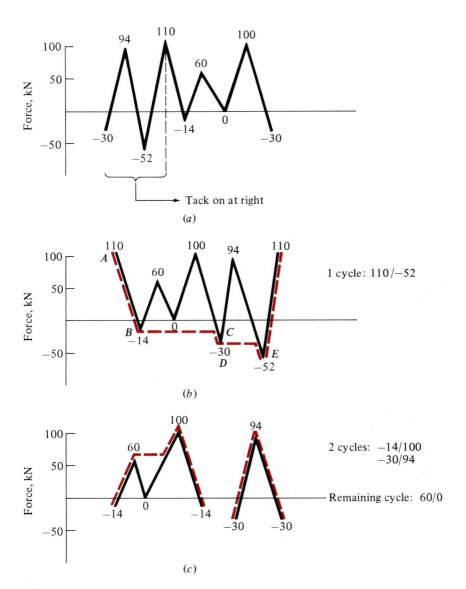

FIGURE 9.23
Rain-flow counting. (*a*) Sample starting and ending at same value. (*b*) Sample rearranged to start at highest load. First cut produces overall largest possible cycle. (*c*) Unused parts from (*b*) with second and third cuts.

Rain-flow (or other) counting is seldom done by hand. A short computer program will perform the above steps (see Problem 9.34).

Complex histories such as that of Figure 9.22 can be accurately described only by statistical models. We simulate the history by extracting a sample and assuming that it is representative. If the sample is at all sizable, then obtaining the expected life for each stress pair from an *S-N* curve is very time consuming. A computer is almost a necessity. The following example illustrates the method.

Example 9.7. The machine part of Figure 9.20 ($d = 25.4$ mm, $A = 506.7$ mm^2) is loaded with the sequence of Figure 9.23, which is of 0.63-s duration and can be considered typical of the loading. Impact is light. The material is 4340 steel, oil-quenched and tempered at 800°F (Appendix D). An estimate of the part life is required.

The maximum nominal stress the part will see is produced by the largest force, 300 kN: $\sigma_{max} = (300E + 03)/506.7 = 592$ MPa. We observe that this is lower than the yield when it is adjusted by the impact factor: $C_I S_{yt} = 0.95(1379) = 1310$ MPa. Hence there is no gross yielding; failure will be by fatigue.

We first convert the force cycles from Figure 9.23 to mean and alternating stresses:

Cycle peaks, kN	F_m, kN	F_a, kN	σ_m, MPa	σ_a, MPa
110/−52	29	81	57.2	160
−14/100	43	57	84.8	112
−30/94	32	62	63.2	122
60/0	30	30	59.2	59.2

Expression (9.8) could be used for finding the life N with S_{fN} resulting from Goodman's criterion (9.10). We perform the solution graphically.

We will need

$$S_{ut} = 1531 \text{ MPa}$$

$$S_{fRB} = 469 \text{ MPa}$$

$$C_F = 0.65 \text{ (machine finish)}$$

$$C_S = 0.71 \text{ (axial load)}$$

$$C_I = 0.95 \text{ (light impact)}$$

$$K_f = 1.59 \text{ (from Example 9.6)}$$

$$q_{1000} = 0.46 \text{ (Figure 9.6)}$$

$$K_{f,1000} = 1 + 0.46(0.59) = 1.27$$

We now have
$$S_{f,10^6} = \frac{0.65 \times 0.61 \times 0.95}{1.59} 469 = 129 \text{ MPa}$$

and
$$S_{f,10^3} = \frac{0.95 \times 469}{1.27} = 351 \text{ MPa}$$

Figure 9.24 shows the corrected *S-N* curve, which we have extended in the low region to $C_I S_{ut} = 0.95(1531) = 1454$ MPa at $N = 1$. In the high range beyond $N = 10^6$, the curve has been continued downward as in Figure 9.19. We can now fill in the following table:

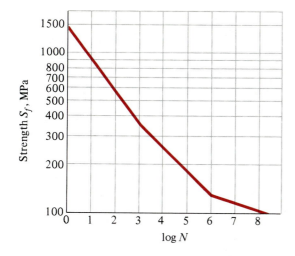

FIGURE 9.24
The S_f-N curve for Example 9.7.

Cycle	Equivalent S_{fN}, (9.10)	N (Figure 9.24)
1	166	$10^{5.2} = 1.58E + 05$
2	119	$10^{6.7} = 5.01E + 06$
3	127	$10^{6.0} = 1.0E + 06$
4	62	$10^{\sim 12} = 1E + 12$

Thus the damage in the sample interval is

$$D = \frac{1}{1.58E + 05} + \frac{1}{5.01E + 06} + \frac{1}{1.0E + 06} + \frac{1}{1E + 12} = 7.53E - 06$$

Since the sample is 0.63 s in duration,

$$\text{Life} = \frac{0.63}{7.53E - 06} = 8.3E + 04 \text{ s} = 23 \text{ h } 14.5 \text{ min}$$

Probabilistic Approach

The complex loadings, for which we developed the rain-flow counting method above, suggest the possibility of a probabilistic approach to cumulative-damage estimates. This is quite possible but seems not to have been developed practically. The reason may be that a rain-flow (or other) count would be necessary to obtain the loading probability curves, and thus the probabilistic approach would be no more efficient than the cycle-by-cycle calculation (by computer naturally) outlined at the end of the previous paragraph and illustrated in Example 9.7.

R. G. Lambert [47] has developed a closed-form solution for random loadings, both for the fully reversed case and for alternating stresses with an applied mean stress. He related the root-mean-square (rms) normally distributed stress to the number of cycles to failure and to the probability that a given part will last a given number of cycles. He also developed the probability of failure as a function of cycles run.

Vibratory Settings

In many applications, the loadings on parts can excite their resonant frequencies. Often the engineer can avoid the problem by causing the resonant frequency (frequencies) to be outside the frequency spectrum of the loading. Failing that, another dimension is added to the fatigue-life problem, that of inertial loadings. We discuss this in Chapter 14.

9.5   COMBINED STRESSES

The discussion to date has centered, except for pure torsion, on uniaxial stresses produced by a single type of loading, tension, or bending. We need now to examine how to combine the differing fatigue correction factors associated with the types of loading of a particular combination. We will then extend what we have learned to biaxial stressing.

Combination of Loading Types

When a part is subjected to a combination of different loadings, as in a combination of tension and torsion, the size effect C_S and the fatigue stress-concentration factor K_f will generally have different values for each loading. There is no totally satisfying way of combining the factors in the relations we have developed. A practical way to handle the problem is simply to weight the two effects according to the contribution of each stress to the alternating stress being used for the life computation. An example will help.

Example 9.8. The part shown in Figure 9.25 is subjected to reversed bending and tension by the force F applied at the end. The material is 4130 steel with $S_{ut} = 1034$ MPa, ground finish. Indefinite life is required with 50 percent reliability. How great may F be? The FS is to be 1.6.

In the usual manner, we find the following factors and stresses for the two types of loadings.

	Tension	Bending
C_R	1.00	1.00
C_F	0.89	0.89
K_t	2.68	2.17
q	0.91	0.91
K_f	2.53	2.06
C_S	0.60	0.90
σ_{nom}	$1366.60F$[†]	$14\,656F$[‡]

$$[†]\ \frac{F}{A} = \frac{0.966 \times F \times 4}{\pi(0.03)^2}$$

$$[‡]\ \frac{Mc}{I} = \frac{0.15 \times 0.259 \times F \times 32}{\pi(0.03)^2}$$

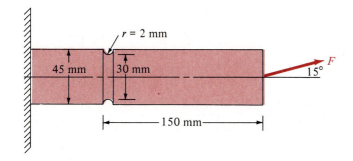

$$F_{\text{bend}} = F \sin 15° = 0.259F$$

$$F_{\text{ten}} = F \cos 15° = 0.966F$$

Tension		Bending		Sum
$\sigma_t = 1366.6F$	+	$\sigma_b = 14\,656F$	=	$\sigma = 16\,023F$

FIGURE 9.25
Combined-loading bar for Example 9.8.

The weighted value of K_f is

$$(K_f\sigma_{\text{nom}})_{\text{ten}} + (K_f\sigma_{\text{nom}})_{\text{bend}} = K_{f,\text{eff}}(\sigma_{\text{nom,ten}} + \sigma_{\text{nom,bend}})$$

$$2.53(1366.6F) + 2.06(14\,656F) = K_{f,\text{eff}}(1366.6F + 14\,656F)$$

$$K_{f,\text{eff}} = 2.10$$

The stresses are divided by the C_S values:

$$\frac{1366.6F}{0.6} + \frac{14\,656F}{0.9} = \frac{16\,023F}{C_{S,\text{eff}}}$$

to give

$$C_{S,\text{eff}} = 0.863$$

Both these figures are close to the values for bending because the bending stress is about 10 times the axial stress.

The sum of the nominal stresses, as shown in the figure, is $\sigma_a = 16\,023F$. It is a completely reversed stress. The remainder of the problem is developed as earlier. Then

$$S_f = \left(\frac{0.89 \times 0.863}{2.10}\right)\left(\frac{1034}{2}\right) = 189 \text{ MPa}$$

Equating this to the combined stress figure and applying the FS, we get

$$16\,023F = \frac{189\text{E} + 06}{1.6} \qquad \text{so} \qquad F = 7.4 \text{ kN}$$

The computation of effective K_f and C_s for two-dimensional cases follows the same procedure. Example 9.9 shows the detail.

Biaxial Stresses

As with static loadings, nearly all fatigue situations can be handled as loadings in a single plane, i.e., two dimensions, as shown on the general stress element in Figure 9.26. We also assume that the stress components are in phase, i.e., they increase and decrease simultaneously. Most design problems fall within these restrictions. A simple example is a cantilever loaded in bending and torsion by a fluctuating single force, as shown in Figure 9.27a. Since the tension and shear stresses are proportional to the force causing them, they are in phase. In the case where the stresses are out of phase, the designer can proceed on the basis that they are in phase, and the result will be on the conservative side.

Three theories are important. They parallel the static theories.

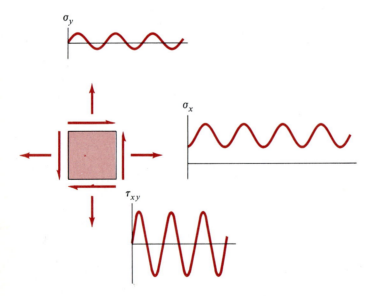

FIGURE 9.26
Two-dimensional element subjected to in-phase stresses.

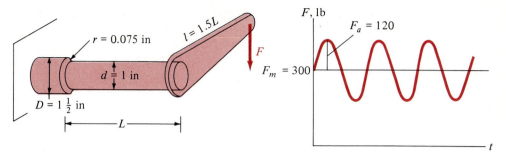

Material: 4340 steel, Q and T to S_u = 183 ksi
S_y = 170 ksi
S_{fRB} = 97 ksi

(a)

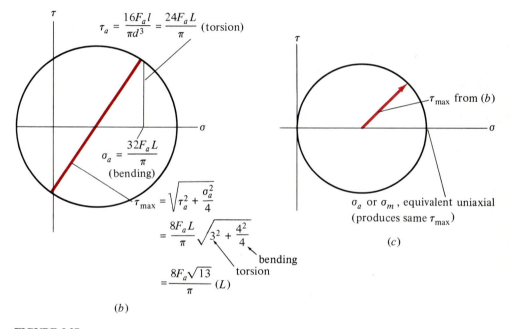

(b)

(c)

FIGURE 9.27
Cantilever beam analysis for Example 9.9. (a) Cantilever under fluctuating bending moment and torque. (b) Alternating-stress circle. The mean-stress circle is the same; change subscript a to m. (c) The equivalent uniaxial stress is simply twice τ_{max}.

MAXIMUM-SHEAR-STRESS COMBINED-STRESS FATIGUE FAILURE THEORY.
This theory holds that failure occurs at the same level of maximum shear stress in combined loading as in uniaxial loading. The most straightforward approach is to draw a Mohr's circle for the mean stresses and another for the alternating stresses. From each a maximum shear stress is computed, and uniaxial tension stresses producing the same mean and alternating stresses are written. These will be simply twice

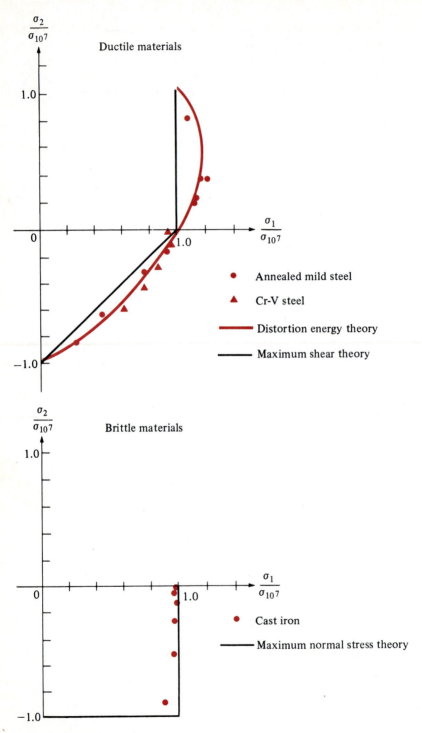

FIGURE 9.28
Comparison of biaxial fatigue failure theories with test data. (*From Sines & Naisman* [9], *with permission.*)

the size of the shear stresses. The failure criteria developed in Section 9.3 can then be applied as before. Figure 9.28 shows a comparison of the theory with a few data points, for reversed loading.

As with the static case, the maximum shear stress may occur in a plane other than the loading plane. The following example is an illustration of this.

Example 9.9. The cantilever of Figure 9.27 is to be designed with the dimensions and proportions shown. With an FS of 1.5 and reliability of 90 percent, how large can L be for indefinite life at the step in the shaft?

As before, we construct a table of factors necessary for the solution:

	Torsion	Bending
C_R	0.897	0.897
C_F	0.66	0.66
K_t	1.56	1.80
q	0.93	0.95
K_f	1.52	1.76
C_S	0.90	0.90

Figure 9.27b shows the mean-stress Mohr's circle with the shear and tension components and the maximum shear stress, which turns out to be in the loading plane. The diagram for the alternating stress is identical; the subscript is changed from m to a. We need now to calculate an effective K_f. (In this problem C_S is the same for both loadings.) Following the earlier procedure of Example 9.8, we have

$$\frac{8F_a L}{\pi}\sqrt{(1.52 \times 3)^2 + \frac{(1.76 \times 4)^2}{4}} = \frac{8F_a L}{\pi} K_{f,\text{eff}}\sqrt{13}$$

giving

$$K_{f,\text{eff}} = 1.60$$

The corrected fatigue strength then becomes

$$S_f = \frac{0.897 \times 0.66 \times 0.9 \times 97}{1.60} = 32.3 \text{ ksi}$$

The equivalent axial mean and alternating stresses are twice the maximum shear stresses, as shown in Figure 9.27c:

$$\sigma_{m,\text{eq}} = \frac{16F_m L\sqrt{13}}{\pi} = \frac{16(300\sqrt{13})}{\pi}L = 5509L$$

$$\sigma_{a,\text{eq}} = \frac{120}{300}\sigma_{m,\text{eq}} = 2204L$$

Goodman's criterion (9.10) results in

$$\frac{2204L}{32.3\text{E}+03} + \frac{5509L}{183\text{E}+03} \leq \frac{1}{1.5} \qquad \text{so} \qquad L \leq 6.78 \text{ in}$$

The yield must also be checked. At maximum load

$$\tau_{max} = \frac{8L\sqrt{13}}{\pi}(F_m + F_a) \leq \frac{S_y}{2FS}$$

$$\frac{8\sqrt{13} \times 420}{\pi}L \leq \frac{170E + 03}{3}$$

whence $L \leq 14.7$, hence yielding does not occur. Of course, we could have checked that the yield strength was not exceeded with a length of $L = 6.78$.

DISTORTION ENERGY COMBINED-STRESS FATIGUE FAILURE THEORY. This theory holds that failure occurs at equal levels of distortion energy in both combined-stress and uniaxial situations. As before, Mohr's circles for the alternating and mean stresses should be drawn, and the effective alternating and mean stresses written. The alternative forms of the theory for two dimensions are easiest to use:

$$\sigma_{\text{eff},a \text{ or } m} = \begin{cases} \sqrt{\sigma_x^2 + \sigma_y^2 - \sigma_x\sigma_y + 3\tau_{xy}^2} & \text{(general two dimensions in } x \text{ and } y) \\ \sqrt{\sigma_{a \text{ or } m}^2 + 3\tau_{a \text{ or } m}^2} & \text{(single tension plus shear)} \end{cases}$$

A comparison of the theory with data for reversed loadings is shown in Figure 9.28.

Example 9.10. We will solve Example 9.9 with this theory. The table of constants remains the same. Using the second of the above two relations and the stresses of Figure 9.27b, we have

$$\sigma_{a \text{ or } m} = \frac{8F_{a \text{ or } m}}{\pi}\sqrt{16 + 27}\,L$$
$$\underset{\text{bending}}{\uparrow} \quad \underset{\text{torsion}}{\uparrow}$$

The effective K_f is found from

$$\frac{8F_a L}{\pi}\sqrt{1.76^2(16) + 1.52^2(27)} = \frac{8F_a L}{\pi}K_{f,\text{eff}}\sqrt{43}$$

$$K_{f,\text{eff}} = 1.61$$

The corrected fatigue strength is then

$$S_f = \frac{0.897 \times 0.66 \times 0.9 \times 97}{1.61} = 32.1 \text{ ksi}$$

and the stresses are

$$\sigma_a = \frac{8 \times 120\sqrt{43}}{\pi}L = 2004L$$

$$\sigma_m = \frac{8 \times 300\sqrt{43}}{\pi}L = 5010L$$

The Goodman criterion then gives

$$\frac{2004L}{32.1E + 03} + \frac{5010L}{183E + 03} \leq \frac{1}{1.5} \quad \text{so} \quad L \leq 7.42 \text{ in}$$

This theory being less conservative than the maximum shear-stress-theory, a greater allowable overhang results. The possibility of yielding should also be checked (Problem 9.44).

MAXIMUM-NORMAL-STRESS COMBINED-STRESS FATIGUE FAILURE THEORY. This theory is used for brittle materials, as in the static case. We noted in Chapter 6 that materials showing less than 5 percent elongation in a 2-in tension specimen are generally regarded as brittle. The largest normal (principal) mean ($\sigma_{1,m}$) and alternating ($\sigma_{1,a}$) stresses are found and then used in conjunction with a diagram like Figure 9.17. The various correction factors require different treatment:

Alternating stress: Multiply by K_f, divide by various C's.
Mean stress: Multiply by K_t.

A comparison of the theory with data for completely reversed loading is shown in Figure 9.28.

PART B
LOCAL-STRAIN MODEL

The main body of a component is never intended to be stressed beyond the elastic limit, but a point of stress concentration, the root of a notch, may be subjected to alternating plasticity, and this location becomes the critical stress point. The fact that the majority of the material surrounding the site behaves elastically subjects the notch root to a strain-controlled condition, and life estimates are thus made on the basis of modeling (estimating) the local strain. This is the state of the art for the design and analysis of highly notched parts or of parts intended to have finite lives.

9.6 ELASTIC AND PLASTIC FATIGUE BEHAVIOR

With the local-strain model, alternating straining in what may be a quite small volume of material in a notch must be related to the results of strain-controlled tests, discussed in Section 8.4. Equilibrium strain loops were shown in Figures 8.17 and 8.18, and we noted that the strain consists of an elastic and a plastic portion. The data for those parts fall along fairly straight lines on a log-log plot of strain vs. reversals to failure, as sketched in Figure 9.29, and they are generally treated as straight lines. Sample equilibrium strain loops are also shown in the figure. The loop is, of course, larger at higher strains, and the plastic portion is then a greater fraction of the total.

In Figure 9.30 the elastic and plastic strains composing the strain vs. number-of-reversals (ε-$2N$) curve for axial tests on 4340 steel are shown. The specimens in these tests were unnotched, so the entire piece was subjected to the same strain. That strain and the stress to produce it were, of course, measured. The plastic component of the

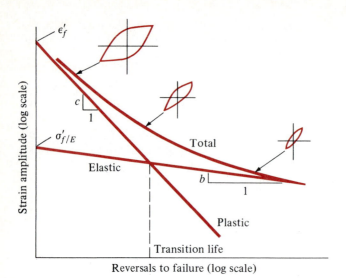

FIGURE 9.29
Schematic of strain amplitude vs. life with elastic and plastic components [58]. (*Courtesy, ASTM.*)

strain is one-half the width of the stable hysteresis loop in the strain-controlled test, as shown in Figure 8.17. The elastic part is then the difference between the total strain and the plastic portion, $\frac{1}{2}\Delta\varepsilon_e$ in Figure 8.17. Note in Figure 9.30 that, for low strains (long lives), life is governed by the elastic strain, which means that life is stress-controlled. But at high strains (short life), plastic strain is the dominant factor (remember that the strain scale is logarithmic), and the strain controls what happens. In the usual design case, this occurs where a stress concentration exists, and the situation there, rather than the average in the entire part, must be examined.

The intercept of the elastic contribution (Figure 9.30a) σ'_f is termed the *fatigue strength coefficient* (*coefficient* since that is how it appears in the equation of the line, shown in the figure). It can also be interpreted as the elastic contribution to the fatigue strength at $10^0 = 1$ reversal. As a single loading is a monotonic test, the fatigue-strength coefficient σ'_f is akin to the true monotonic fracture strength σ_f (Chapter 6) and may be taken equal to it as a first approximation, when σ'_f is not known. The two are not exactly equal, since the conditions of test are not the same (specimen size and shape, rate of loading, etc.). The slope of the line is b, as shown in the figure. It is called the *fatigue strength exponent*. The value of b, when not available, may be approximated by $-0.16 \log (\sigma_f/S_{ut})$.

The fatigue ductility coefficient ε'_f is the intercept of the plastic component of strain, as seen in Figure 9.30b. It may be approximated by the true monotonic fracture strain ε_f (Chapter 6). The slope of the plastic component is c, called the *fatigue ductility exponent*. When it is not available, the value of c may be approximated as -0.6.

If the elastic and plastic portions of the strain shown in Figure 9.30 are added to obtain the total true-strain amplitude,

$$\varepsilon_{at} = \varepsilon_{at,el} + \varepsilon_{at,pl} = \frac{\sigma'_f}{E} (2N)^b + \varepsilon'_f (2N)^c \tag{9.16}$$

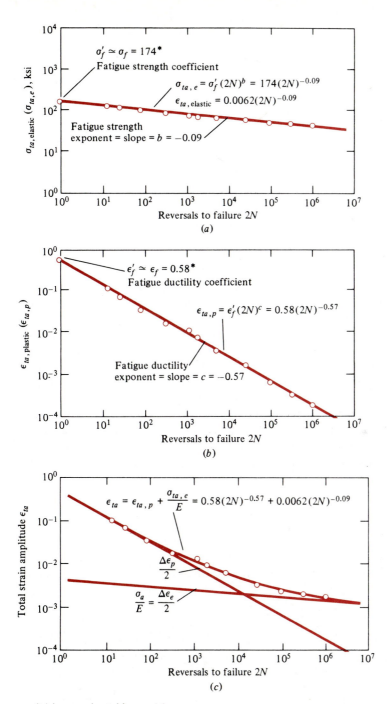

$^{*}\sigma'_f$ is approximated by σ_f, ϵ'_f by ϵ_f.

FIGURE 9.30

True-strain fatigue behavior of annealed 4340 steel [61] (axial tests). (*Courtesy, Society of Automotive Engineers.*) (*a*) Elastic contribution. (*b*) Plastic contribution. (*c*) Sum of (*a*) and (*b*).

where σ'_f/E has been substituted for the elastic strain coefficient. Values of the several constants in (9.16) are provided for polished axial specimens of some materials in Appendix E.

9.7   LOCAL STRAIN AND THE DESIGN PROBLEM, ZERO MEAN STRESS

For life prediction, the true strain amplitude at the critical point in the piece is needed. This will occur at a stress concentration. When the maximum stress does not exceed the elastic limit, the true and nominal strains are equal and follow Hooke's law: $\varepsilon_{\text{true}} = \varepsilon_{\text{nom}} = \sigma/E$. When the elastic limit is exceeded (low-cycle fatigue), plastic strain occurs and a relation between the stress or strain in neighboring material and the true strain at the critical point is required.

One method of estimating the strain at a location of stress concentration is the *linear rule*. It assumes that the *strain*-concentration factor when the material is in the plastic region is the same as the elastic *stress*-concentration factor. Since under elastic conditions strain is proportional to stress, the elastic stress-concentration factors are then also strain-concentration factors. The linear rule thus presumes that strain-concentration factors do not change when the material becomes plastic. For this reason the rule is sometimes called *strain-concentration invariance*. The linear rule is simple in its application, but predicts strains which are somewhat smaller than those actually existing. That is, the rule is nonconservative.

Neuber's rule [17] is much more widely used than the linear rule. It is accurate for plane stress, i.e., thin sheets under tension; for other forms and loadings it is conservative—it overestimates the strain. We limit our discussion to this approach.

Neuber's rule, adapted to fatigue [63], states that the fatigue notch factor is equal to the geometric mean of the true stress-concentration factor and the true strain-concentration factor:

$$K_f = \sqrt{K_\sigma K_\varepsilon} \tag{9.17}$$

The factors are defined as follows:

$$K_\sigma = \frac{\sigma_{\text{true, at notch root}}}{\sigma_{\text{true, away from but near notch root}}}$$

$$K_\varepsilon = \frac{\varepsilon_{\text{true, at notch root}}}{\varepsilon_{\text{true, away from but near notch root}}}$$

The values of σ and ε away from the root are normally calculated by the classic mechanics of materials expressions, which assume elastic material behavior. We term the resultant stresses and strains *nominal*. With elastic action in the bulk of the component, which is the usual case, the rule then becomes, for alternating stresses and strains,

$$K_f^2 = \frac{\sigma_{a,\text{true}}}{\sigma_{a,\text{nom}}} \frac{\varepsilon_{a,\text{true}}}{\varepsilon_{a,\text{nom}}}$$

or, by using $\sigma_{nom} = E\varepsilon_{nom}$

$$K_f^2 \frac{\sigma_{a,nom}^2}{E} = \sigma_{a,true}\varepsilon_{a,true} \qquad (9.18)$$

Neuber's rule is commonly applied to two types of problems, as follows.

Nominal Stresses Known, Estimate the Life

The presumption here is that a preliminary design has been made and a check on the life requirement is needed. Or an analysis is to be made of a failed part. The nominal stress $\sigma_{a,nom}$ and the fatigue notch factor K_f are known, i.e., the left side of (9.18). The objective is the true strain $\varepsilon_{a,true}$, with which the life can then be taken from a curve like Figure 9.30c or computed from the analytical equivalent, Equation (9.16). A further unknown in (9.18) is the true alternating stress $\sigma_{a,true}$, hence a second relation is needed. The parameters $\varepsilon_{a,true}$ and $\sigma_{a,true}$ are related by the cyclic stress-strain curve, which we discussed in Section 8.4, relation (8.2):

$$\varepsilon_{a,true} = \frac{\sigma_{a,true}}{E} + \left(\frac{\sigma_{a,true}}{K'}\right)^{1/n'} \qquad (8.2)$$

Thus we seek a simultaneous solution to (9.18) and (8.2). Solving (9.18) for σ_{at} (we abbreviate "true" to "t" and "nom" to "n") and inserting it in (8.2), we get

$$\varepsilon_{at} = \frac{K_f^2\sigma_{an}^2}{\varepsilon_{at}E^2} + \left(\frac{K_f^2\sigma_{an}^2}{\varepsilon_{at}K'E}\right)^{1/n'}$$

This, of course, cannot be solved explicitly for ε_{at}. It does suggest, however, the possibility of guessing a value of ε_{at}, calculating the right-hand side, and then using the result as a new, improved guess. This is known as an *iterative* or *recursive* solution, which we discuss more fully in Chapter 10. With the above form of the equation, the procedure does not converge. It will if we solve it for the ε_{at} in the second term on the right:

$$\varepsilon_{at} = \frac{K_f^2\sigma_{an}^2}{K'E}\left(\varepsilon_{at} - \frac{K_f^2\sigma_{an}^2}{\varepsilon_{at}E^2}\right)^{-n'} \qquad (9.19)$$

The first term in the parentheses of this relation derives from the total strain in (8.2), and the second term derives from the elastic part of it. Thus, if the plastic strain is small, the two terms are nearly equal, and the term in parentheses will approach zero; this becomes evident when a few trials are made. In such cases the solution is obtained by setting the term in parentheses equal to zero, which is tantamount to neglecting the plastic strain (the second term) in the expression immediately preceding.

> **Example 9.11.** The shaft of Figure 9.8, with $M_a = 50$ N·m, is to be made of SAE 1045 steel quenched and tempered to H_B 225 (Appendix E). We need an estimate of the life.
>   The cyclic stress-strain constants [for relation (8.2)] are $n' = 0.18$ and $K' = 1344$ MPa. The cyclic strain-life parameters [for relation (9.16)] are $\sigma'_f = 1227$ MPa, $b = -0.10$, $\varepsilon'_f = 1.00$, and $c = -0.66$.

The alternating nominal stress is

$$\sigma_{an} = \frac{32M}{\pi d^3} = \frac{32 \times 50}{\pi(0.01)^3} = 509 \text{ MPa}$$

The theoretical stress concentration is $K_t \simeq 2.5$, thus the stress on the notch surface, if elastic behavior were assumed, would be $\sigma_{a,\text{notch}} = K_t\sigma_a = 2.5(509) = 1273 \text{ MPa}$, twice the yield strength of 634. Clearly the action at the notch is plastic, and we must estimate the alternating strain. With the above numbers and $K_f = 2.02$, expression (9.19) becomes

$$\varepsilon_{at} = \frac{2.02^2(509E + 06)^2}{1344E + 06 \times 207E + 09}\left[\varepsilon_{at} - \frac{2.02^2(509E + 06)^2}{\varepsilon_{at}(207E + 09)^2}\right]^{-0.18}$$

Starting with a (pure) guess of 0.01 and recycling the result, we get

$$0.00942$$
$$0.00933$$
$$0.00936$$
$$0.00935$$
$$\varepsilon_{at} = 0.00935$$

With this value of notch-root alternating strain and the strain-life parameters listed above, we now must solve relation (9.16):

$$9.35E - 03 = \frac{1227E + 06}{207E + 09}(2N)^{-0.10} + 1.00(2N)^{-0.66}$$

We can arrange this for an iterative solution by solving it for one of the $2N$ terms, say the right-hand one:

$$2N = \left[9.35E - 03 - \frac{1227}{207E + 03}(2N)^{-0.10}\right]^{-1/0.66}$$

Starting with a guess of $2N = 2000$, we get, in succession,

$$2N = 2023$$
$$2021$$
$$2021 \text{ reversals}$$

Hence the expected life is $N = 1010$ cycles.

For a Desired Life, Size the Component

The objective in this type of problem is to find the size of the part to keep the strain within a determined safe value for a desired life. The procedure is the reverse of the routine discussed in the paragraph above. It is best illustrated by an example.

Example 9.12. A pinned link (tap fit), shown schematically in Figure 9.31, is to be made of the same steel as in Example 9.11. What minimum thickness of the link will ensure a life of 1000 cycles of reversed loading with light impact?

Because this problem is stated in USCS units, several parameters are different: $K' = 195$ ksi, $\sigma'_f = 178$ ksi, and $E = 30E + 06$. From Figure A.14 in Appendix A, we find $K_t = 4.4$. Then with $q = 0.86$ (Figure 9.2), $K_f = 1 + 0.86(3.4) = 3.92$.

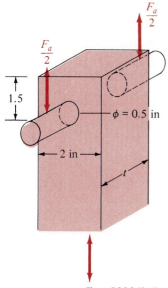

FIGURE 9.31

$F_a = 8000$ lb (including FS)   Pinned link for Example 9.12.

At the required life, the tolerable alternating strain is computed from (9.16):

$$\varepsilon_{at} = \frac{178E + 03}{30E + 06}(2E + 03)^{-0.10} + 1(2E + 03)^{-0.66}$$

$$= 2.77E - 03 + 6.63E - 03 = 9.40E - 03$$

The plastic part of the strain $(6.63E - 03)$ here is about $2\frac{1}{2}$ times greater than the elastic $(2.77E - 03)$.

The correction factor for light impact is $C_I = 0.9$. By analogy to the stress-life model, we will correct the tolerable alternating strain with this factor to get

$$\varepsilon_{at,\text{corr}} = 9.40E - 03 \times 0.9 = 8.46E - 03$$

The nominal stress on the link is

$$\sigma_{an} = \frac{F}{A} = \frac{8000}{1.5t}$$

Neuber's rule (9.18) can now be written to find the true alternating stress:

$$\sigma_{at} = \frac{3.92^2}{30E + 06}\left(\frac{8000}{1.5t}\right)^2 \frac{1}{8.46E - 03} = \frac{1722}{t^2}$$

The stress and the strain must satisfy the cyclic stress-strain relation (8.2), into which the above value of σ_{at} is substituted:

$$8.46E - 03 = \frac{1722}{t^2(30E + 06)} + \left[\frac{1722}{t^2(195E + 03)}\right]^{1/0.18} \qquad (8.2)$$

giving

$$t = \sqrt{\frac{1722}{30E + 06\left[8.46E - 03 - (3.88E - 12)/t^{11.111}\right]}}$$

To prevent the term in brackets in the denominator from being negative, t must be greater than 0.144. This will affect an iterative solution. A thickness of about 0.150 in satisfies the equation.

The above two steps—the application in succession of Neuber's rule (9.18) and the cyclic stress-strain relation (8.2)—could have been combined by using (9.19), with the same result.

9.8 LOCAL-STRAIN APPROACH TO THE MEAN-STRESS PROBLEM AND CUMULATIVE DAMAGE

Researchers in fatigue have found an analogy to the Goodman and Soderberg criteria for the relation between combinations of alternating true stress and mean true stress, as shown in Figure 9.32. The fatigue strength coefficient σ'_f, rather than S_y or S_{ut}, is the intercept on the abscissa axis, where mean stresses are plotted. The coefficient σ'_f was defined in Figure 9.30a and discussed in Section 9.6. In Figure 9.32 it is the mean stress which results in an alternating strength of zero; i.e., failure in such a test would occur in the first cycle (actually before the cycle begins), as in Figure 9.30a.

From the similar triangles of the figure,

$$\frac{S_{fN}}{\sigma'_f} = \frac{\sigma_a}{\sigma'_f - \sigma_{mt}}$$

giving

$$\sigma_a = (\sigma'_f - \sigma_{mt})\frac{S_{fN}}{\sigma'_f}$$

From Figure 9.30a, $S_{fN} = \sigma'_f(2N)^b$, which, when it is substituted above, results in

$$\sigma_a = (\sigma'_f - \sigma_{mt})(2N)^b$$

(b is defined in Figure 9.30a). If this, divided by E, is now substituted in (9.16) for the elastic alternating stress, then

$$\varepsilon_{at} = \frac{(\sigma'_f - \sigma_{mt})}{E}(2N)^b + \varepsilon'_f(2N)^c \tag{9.20}$$

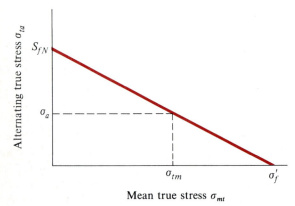

FIGURE 9.32
Local-strain approach to mean stress.

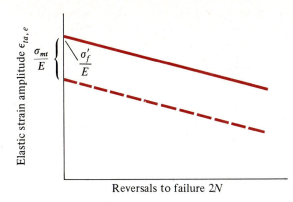

FIGURE 9.33
Application of a mean stress (strain) lowers the elastic contribution to alternating strain by σ_{mt}/E.

The plastic part of the expression is left unaltered, on the argument that a few cycles of plastic action relieve the mean strain.

Thus it is only necessary to reduce the fatigue strength coefficient σ'_f by the amount of the true mean stress to use expression (9.16) in cases where a mean stress exists. What this comes to is a lowering of the line in Figure 9.30a by σ_{mt}/E, as shown in Figure 9.33. An iterative solution for $2N$ in (9.20) is best set up by solving for the $2N$ on the extreme right:

$$2N = \left\{ \frac{1}{\varepsilon'_f} \left[\varepsilon_{at} - \frac{\sigma'_f - \sigma_{mt}}{E} (2N)^b \right] \right\}^{1/c} \tag{9.20a}$$

If the plastic part of the strain is negligible, the term in brackets becomes ε_{at}, which yields an explicit solution.

Expression (9.20) is often used to calculate damage in a cycle of loading, that is, $1/N$, per Miner's rule. If the variation of the load on the part with time is known, the calculation of σ_{mt} requires the simultaneous solution of Neuber's rule (9.18) and the cyclic stress-strain relation (8.2). The subscript a (alternating) is replaced here by the subscript m (mean). Inserting ε_{mt} from (8.2) in (9.18) and rearranging, we have

$$\sigma_{mt} = \frac{K_f^2 \sigma_{mn}^2}{E} \left[\frac{\sigma_{mt}}{E} + \left(\frac{\sigma_{mt}}{K'} \right)^{1/n'} \right]^{-1} \tag{9.21}$$

which can be solved iteratively for σ_{mt}. The strain amplitude ε_{at} is found as before using expression (9.19), and then the strain-life relation (9.20) can be solved for $2N$.

The more common case is that where strain gages have been used to obtain a strain history at the critical point. If $\varepsilon_{mt} = \sigma_{mt}/E$ is substituted in (9.20), we have

$$\varepsilon_{at} = \left(\frac{\sigma'_f}{E} - \varepsilon_{mt} \right) (2N)^b + \varepsilon'_f (2N)^c \tag{9.20b}$$

Rain-flow or other counting of the strain history yields pairs of ε_{at}-ε_{mt} values, which can then be entered to solve for $2N$ and the damage per cycle $1/N$.

Example 9.13. Strain gages were placed on the part of Example 9.11, and rain-flow counting has been done on the strain record. The range of strain for 1 cycle was ε: 8.293E − 03/4.975E − 03, from which

$$\varepsilon_{at} = 1.659\text{E} - 03 \qquad \varepsilon_{mt} = 6.634\text{E} - 03$$

Entering (9.20*b*) with these numbers, we have

$$1.659\text{E} - 03 = \left(\frac{828\text{E} + 06}{207\text{E} + 09} - 6.634\text{E} - 03\right)(2N)^{-0.11} + 0.95(2N)^{-0.64}$$

Solving this for the 2*N* on the far right gives

$$2N = \frac{1}{0.95}\left[1.659\text{E} - 03 - (2.634\text{E} - 03)(2N)^{-0.11}\right]^{-1/0.64}$$

which is solved iteratively to get 2*N* = 62 846, and the damage in this cycle is thus

$$\frac{1}{N} = \frac{2}{63\,846} = 3.182\text{E} - 05$$

In any practical setting there are a great many of the above computations to make, and they are always performed on a computer.

The life calculation using expression (9.20) or its equivalent (9.20*b*) is not the only method in use, but it is perhaps the most popular. For other approaches, see Reemsnyder [65]. One caveat: All these schemes are occasionally found to over-estimate the life, i.e., to be nonconservative.

9.9   A CASE STUDY[1]

The search for oil offshore is performed with drilling rigs which operate in water up to several hundred feet in depth. Bethlehem Steel Corporation has built a number of these rigs, most of which consist of a buoyant platform, a cellular mat, three cylindrical legs, and a hydraulic jacking system, as shown in Figure 9.34. The platform contains the derrick, drilling machinery, and living quarters. The mat rests on the ocean floor and serves as a foundation for the rig. The three tubular legs or columns are approximately 12 ft in diameter and 300 ft tall and are fabricated from steel plate 1.5 to 2.5 in thick. The legs are integral with the mat and pass through hydraulic jacking systems located on the upper deck of the platform. Rows of rectangular holes, called *pinholes*, seen in the figure along and around each column, receive large structural pins to raise, lower, and support the platform in conjunction with the hydraulic jacking system.

The pinholes in the legs, approximately 10 by 16 in, are cut by drilling a 1.5-in-diameter hole at each of the four corners and then cutting the sides with a gas torch. Gas cutting introduces tensile residual stresses of about one-half the tensile strength of the leg material at the locations of the drilled holes. Some legs were thermally

---

[1] We are grateful to Dr. Harold S. Reemsnyder, the Bethlehem Steel Corp., and the ASTM for permission to use this study. Details may be found in Reemsnyder [66].

FIGURE 9.34
Offshore drilling platform, showing pinholes. Note crack at upper left corner of inset. (*Courtesy, Bethlehem Steel Corp.*)

stress-relieved, while others were not. The stress-concentration factor K_t at the hole corner was 4.6 for bending.

During a "wet" tow to a drilling location, the platform serves as a buoyant hull from which the mat is suspended a short distance below the platform by the legs. The legs thus extend some 260 ft above the platform. The speed of this type of tow is about 4 knots; typical periods of pitch and roll are 17 and 23 s.

Another method of tow is the "dry" tow, where the rig is carried on a semi-submersible barge or ship. The towing resistance is lower than with the "wet" tow, so that towing speeds are about 6 to 8 knots. The rolling resistance is also less; thus the period of roll is closer to the wave period, 6 to 9 s, and the angle of roll is larger. For a dry tow on a barge, the upper portions of the legs are commonly removed and carried on the deck of the rig platform. Special-purpose semisubmersible self-propelled ships have recently come into use. These ships have a longer roll period and a smaller roll angle than a barge carrying a rig, and rigs can be transported with their legs intact.

Upon arrival at the work site, the mat is lowered to the sea floor, and the platform is then jacked up the legs to sufficient height for wave clearance and is locked to the legs.

A few years ago, cracks $\frac{1}{2}$ to 8 in long were detected at pinholes in several rigs after transit by wet tow and in one rig after a dry tow. The cracks were in the lower one-third of the legs just above the platform position during tow. The lengths of the cracks decreased with the distance above the platform.

For the dry tow, the problem was identified as the bending stresses induced by the inertia of the legs in rolling. The rig in question experienced very severe weather rounding the Cape of Good Hope. The length of upper legs removed and carried on the platform in this rig was only two-thirds of the length removed in those dry tows which did not result in pinhole cracking. This problem was corrected by carrying a greater part of the legs on the platform.

During a wet tow, the periods of pitch and roll are very large relative to the periods of waves likely to be encountered. Thus, the resulting angles of pitch and roll and the bending stresses due to inertial effects are small. The crews aboard the rigs observed significant cyclic deflections of the legs even in the absence of rolling. The motion was transverse to the relative wind and was due to vortex shedding. Vortices develop alternately on opposite sides of a cylindrical body moving in a fluid, resulting in an alternating force transverse to the direction of the fluid flow. The period of the alternating force is approximately 5 times the ratio of the diameter of the cylinder to the relative velocity of the fluid. Structurally the legs were vertical cantilever beams, and the period of vortex shedding equaled the legs' natural period (approximately 1 to 2 s, depending on the height of exposed leg) at a relative wind velocity of 24 miles per hour. Towing logs showed that such wind velocities were experienced the majority of the time at sea. The computed nominal bending stress range induced by the vortex shedding was about 15.4 ksi near the fixed end of the leg. Stress decreased, of course, with the distance from the support, i.e., the platform deck. A local-strain analysis estimated a life to crack initiation of 500 000 cycles for this stress range. This was the equivalent of 6 to 12 days of vortex shedding at the leg's natural frequency. Sea times for wet tows range from 1 to 3 months.

One obvious step to be taken was to lower the stresses in the critical areas by the elimination of vortex shedding. Spiral helical strakes on a cylindrical body act as spoilers and reduce the response of the body to vortex shedding. Strakes of rectangular cross section are called *Scruton spoilers*, after their inventor, and they are more effective than those of circular cross section. The optimum array is three strakes with a height 0.09 times the diameter of the body and a pitch 5 times the diameter. Also, Scruton found that the excitation of cylindrical bodies fitted with rectangular strakes was independent of the Reynolds number, thus increasing the reliability of model studies. When mounted on a flexible cantilever such as a smokestack, or jack-up leg, the spoilers need only to be applied to the top one-third of the member.

Removable Scruton spoilers are now bolted to the top 30 percent of each leg of rigs for transit and operation, as seen in Figure 9.35. They consist of three helices of thin plates 120° apart with a pitch of 5 to 6 times the leg diameter. The height of the spoilers is at least 0.07 times the column diameter. Vortex shedding has been eliminated, and no further cracking at pinholes has been observed.

Experiments were also designed to determine if thermal stress relief at the pinholes would be beneficial. Full-scale specimens and smooth, unnotched specimens were fabricated from plates of the steel used in constructing the rig legs. The full-scale

FIGURE 9.35
Offshore rig with spoilers attached to columns. (*Courtesy, Bethlehem Steel Corp.*)

specimens were cycled under load control, while the unnotched specimens were cycled under strain control. Both specimen configurations were tested in complete reversal. In general, thermal stress relief improved the fatigue resistance of the full-scale specimens. Also the behavior of the full-scale specimens was adequately modeled by the local-strain model using the results of strain-controlled fatigue tests on unnotched specimens.

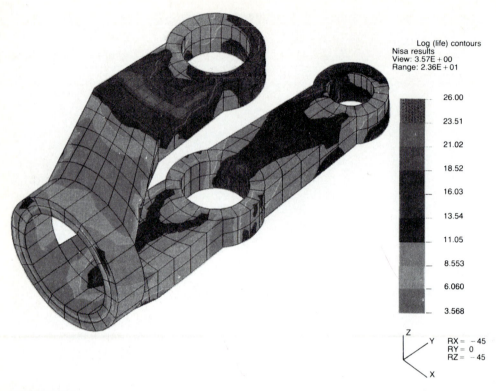

Log (life) contours
Nisa results
View: 3.57E + 00
Range: 2.36E + 01

— 26.00

— 23.51

— 21.02

— 18.52

— 16.03

— 13.54

— 11.05

— 8.553

— 6.060

— 3.568

RX = −45
RY = 0
RZ = −45

FIGURE 9.36
Yoke attached to the control arm and steering links of an automobile. The gridwork establishes the points where stresses and fatigue life are calculated by the program ENDURE. Log life is shown in various shades of gray, per the scale at the right, decreasing downward. Reprinted from *Mechanical Engineering*, ref. 79. *Courtesy, Engineering Mechanics Research Corp.*

9.10   COMPUTER-AIDED FATIGUE ANALYSIS

The expressions governing the stresses (and strains) within a body under load can be solved explicitly for only the simplest of structures. Finite element analysis (FEA) is a numerical solution based on a gridwork of a (finite) number of points on the surface and within the body under study. Such a gridwork is shown in Figure 9.36. The immense memory of modern computers and their computational speed and precision have enabled the computation of stresses in very complex shapes. A number of FEA software packages are commercially available, and their use is widespread.

FEA Combined with Fatigue Analysis

A logical extension of computerized FEA is its integration with fatigue and fracture analysis, and software to accomplish this is now obtainable. The organization of one

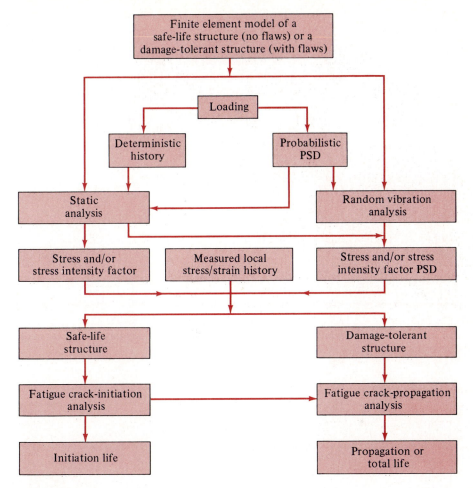

FIGURE 9.37
Schematic of the fatigue analysis program ENDURE. Reprinted from *Mechanical Engineering*, ref. 79. *Courtesy, Engineering Mechanics Research Corp.*

such program is shown in Figure 9.37. This particular package can perform either a stress-life or a strain-life analysis, depending on whether a high- or a low-cycle life regime is indicated. In the latter case the material's cyclic stress-strain curve and Neuber's rule are used, as described in Section 9.7. A load history is processed by rain-flow counting. Probabilistic (statistical) input is also provided for.

This program differentiates between crack initiation life and crack propagation life. Since the propagation phase may be used independently, damage-tolerant structural design analysis is possible, as shown in Figure 9.37.

Figure 9.36 shows typical output, with calculated lives in various regions of the object shown. Critical points are thus identified for further attention.

PROBLEMS

Section 9.1

9.1. Figure P9.1 shows a stepped bar, to be machine-finished, made of 4130 steel, which has $S_{ut} = 150$ ksi. Reliability is to be 99 percent, and impact will be mild. Estimate the material fatigue limit (indefinite life) for completely reversed axial loading.

   Answer: $S_f = 11.7$ ksi.

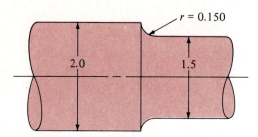

$r = 0.150$

2.0   1.5

Dimensions in inches   **FIGURE P9.1**

9.2. Repeat Problem 9.1, but the bar is loaded in torsion.

9.3. Figure P9.3 shows the loading on a ground steel bar and pin, which have $S_{ut} = 1500$ MPa. Also 95 percent reliability is required, and impact will be medium. How great may the reversing force be for indefinite life?

   Answer: $F = 18.3$ kN

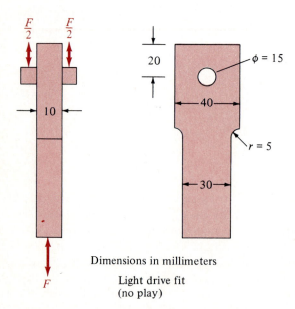

$\frac{F}{2}$ $\frac{F}{2}$

20

$\phi = 15$

10

40

$r = 5$

30

F

Dimensions in millimeters

Light drive fit
(no play)   **FIGURE P9.3**

9.4. The rotating steel ($S_{ut} = 1200$ MPa) shaft shown in Figure P9.4 must support a load $F = 2$ kN on an indefinite basis, with a reliability of 99 percent. Is the shaft sufficiently strong? Use an FS of 1.5 on F.

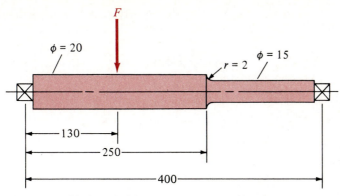

Machine finish. Dimensions in millimeters

FIGURE P9.4

9.5. The bar shown in Figure P9.5 is subjected to reversed bending, and $S_{ut} = 120$ ksi. Estimate the material fatigue limit at the critical location.

Answer: $S_f = 21.4$ ksi

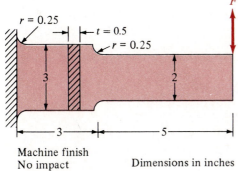

Machine finish
No impact   Dimensions in inches
Reliability = 90%   **FIGURE P9.5**

9.6. The plate shown in Figure P9.6 is loaded by forces which reverse several times a second with light impact. If the part is to be designed to have an indefinite life with 95 percent reliability, what is the largest value of the force which may be applied, if an FS of 1.5 is to be applied to *F*?

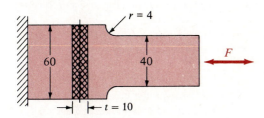

Dimensions in millimeters
Machined surface
Material: 1040 steel, CD 20%   **FIGURE P9.6**

9.7. The shaft shown in the figure rotates, so that a reversed-bending situation exists. Assuming FS = 2 and a reliability of 90 percent, what moment M may be applied, if the shaft is to last indefinitely?

   Answer: $M = 5533$ lb·in.

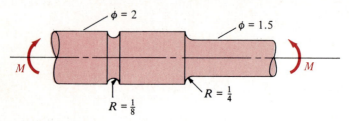

Dimensions in inches
Material: 4130 steel, water-quenched and drawn at 1200°F, ground finish

FIGURE P9.7

Section 9.2

9.8. Derive expressions (9.3) and (9.7).

9.9. Work Problem 9.1 for a life of 50 000 cycles.
   Answer: $S_{f,50\,000} = 29.6$

9.10. Work Problem 9.2 for a life of 3000 cycles.

9.11. Work Problem 9.3 for a life of 10 000 stress reversals. Use an FS of 1.5 on the force.
   Answer: $F = 45.8$ kN

9.12. Work Problem 9.4 for a life of 400 000 cycles.

9.13. Work Problem 9.5 for a life of 150 000 stress reversals.
   Answer: $S_{f,75\,000} = 39.3$

9.14. In Figure P9.1, the piece is to have a life of 1200 cycles. Material is 4340 steel, quenched and tempered at 1000°F to $S_{ut} = 183$ ksi (Appendix D). The surface is machined, and impact is mild. Estimate the material fatigue strength at the critical location for reversed axial loading.

9.15. Repeat Problem 9.14, but the loading is torsional.
   Answer: $S_{sf,1200} = 56.1$ ksi.

9.16. In Figure P9.3, the assembly is to have a life of 5000 cycles. Material of both parts is 1040 steel, cold-drawn 20 percent (Appendix D). Surfaces are machined. Impact is mild. How great may the reversing force be for a life of 5000 cycles? Use an FS of 1.5 on the force and 2.5 for the uncertainties in the computation of the fatigue strength of the pin itself.

9.17. In Figure P9.4, the piece is to have a life of 400 000 cycles. Material is 1040 steel, cold-drawn 20 percent (Appendix D). Estimate the material fatigue strength at the critical point. There is no impact.
   Answer: $S_{f,400\,000} = 188.1$ MPa

9.18. In Figure P9.5, $S_{ut} = 120$ ksi and $F = 3200$ lb. Estimate the fatigue life of the part.

9.19. The part of Figure P9.4 is made of 4340 steel, quenched and tempered to $S_{ut} = 1531$ MPa (Appendix D). There is no impact. The force $F = 1.5$ kN. Estimate the fatigue life of the part.
   Answer: $N = 256\,630$

Section 9.3

9.20. In Figure 9.13 prove that along the Goodman segment defining static failure, failure occurs in the first cycle of loading.

9.21. A machine-finished tension member is acted on by a force which varies continuously between 60 and 140 klb. Assuming no stress concentration, determine the limiting diameter if the material is 4130 steel.

   Answer: d = 1.64 in

9.22. Repeat Problem 9.21, but the forces are compressive.

9.23. The part of Figure P9.5 is made of 1030 steel (Appendix D). The force at the end consists of an alternating component $F_{alt} = 650$ lb plus a steady component. How great may the steady component be for indefinite life, if no margin of safety is allowed?

   Answer: $F_m = 3247$ lb

9.24. A machined tension member of 10-mm diameter sustains a force which varies continuously from -10 to 30 kN. A factor of safety of 1.5 is to be used. If the corrected fatigue limit $S_f = 400$ MPa, $S_{yt} = 850$ MPa, and $S_{ut} = 1200$ MPa, is the design okay for indefinite life?

9.25. A ground tension member of 10-mm diameter is loaded by an alternating force of 15 kN. The material is 4340 steel, oil-quenched and tempered at 800°F (Appendix D). How large a compressive force may be placed on the bar, in addition, if indefinite life is required? Assume no stress concentration and a reliability of 95 percent. Also FS = 2. Use the continuation of Goodman's criterion. (The question of buckling would be a separate issue.)

   Answer: $F_m = 44.6$ kN

9.26. The connecting rod shown in Figure P9.26 is forged from 1040 steel. (Use the properties for the 50 percent cold-drawn steel in Appendix D.) The force, transmitted through pins at the ends, varies from 0 to 4 kN, with no impact. Find the factor of safety based on indefinite life and a reliability of 99 percent.

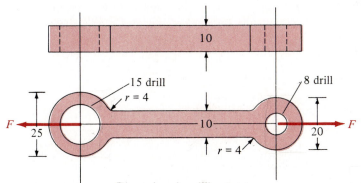

Dimensions in millimeters

FIGURE P9.26

9.27. The force on the connecting rod of Problem 9.26 varies from 3.5 to 9.5 kN. Estimate the life of the part, using a factor of safety of 2. Assume no impact.

   Answer: N = 791 000

9.28. The force on the connecting rod of Problem 9.26 consists of a steady compressive component of 2 kN plus an alternating component of 4 kN. Determine the factor of safety for a 50 000-cycle life.

9.29. Figure P9.29 shows a torsional spring. It is made of 4142 steel quenched and tempered to $S_{ut} = 205$ ksi ($S_{yt} = 200$ MPa). A steady torque of 33 klb·in is applied plus an alternating torque of 10 klb·in. Find the diameter D for 99 percent reliability, indefinite life, and FS = 2.5. Impact is mild.

Answer: $D \geq 1.95$ in

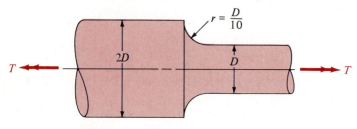

Machine finish

FIGURE P9.29

Section 9.4

9.30. The machined part shown in Figure P9.1 is made of 4130 steel (Appendix D). It is loaded by the following series of reversing torques, with moderate impact:

$$T = \begin{cases} 10 \text{ klb·in for } 50 \text{ cycles} \\ 25 \text{ klb·in for } 30 \text{ cycles} \\ 30 \text{ klb·in for } 3 \text{ cycles} \end{cases}$$

The time for this block of torques is 1 min 7 s. Assuming that it is representative of the loading and that it repeats, estimate the life of the part.

9.31. The shaft of Figure P9.4 is made of 4340 steel, oil-quenched and tempered at 800°F (Appendix D). The shaft turns at 2000 rpm; impact is light. The following is a representative block of loading.

$$F = \begin{cases} 526 \text{ N} & \text{for } 250 \text{ s} \\ 407 \text{ N} & \text{for } 185 \text{ s} \\ 3057 \text{ N} & \text{for } 1 \text{ s} \\ 1100 \text{ N} & \text{for } 65 \text{ s} \\ 1650 \text{ N} & \text{for } 7 \text{ s} \end{cases}$$

Estimate the life of the shaft.

Answer: $t = 14$ min 30 s

9.32. The connecting rod of Problem 9.26 is loaded by the following representive sequence of forces:

$$F_{alt} = \begin{cases} 3 \text{ kN} \\ 8 \text{ kN} \\ 9 \text{ kN} \\ 15 \text{ kN} \end{cases} \qquad F_{steady} = \begin{cases} 1.8 \text{ kN} & \text{for } 103 \text{ cycles} \\ 4.5 \text{ kN} & \text{for } 25 \text{ cycles} \\ 6.5 \text{ kN} & \text{for } 17 \text{ cycles} \\ 8.3 \text{ kN} & \text{for } 2 \text{ cycles} \end{cases}$$

Estimate the life of the part. Use a factor of safety of 2.

9.33. A loading history was taken on the connecting rod of Problem 9.26, and the following peaks (in kilonewtons) between points of equal value (-5) were read:

$$-5, \quad 6, \quad 1, \quad 7, \quad -4, \quad 6, \quad -6, \quad -3, \quad -7, \quad 4, \quad -6, \quad 0, \quad -5,$$

The elapsed time for the sequence was 300 ms. Assuming that this is typical of the loading and that the FS on the strength is 1.5, estimate the life of the part.

Answer: $t = 3$ h 53 min

9.34. Write a computer program to perform rain-flow counting. Do Problem 9.33, using the program.

9.35. Figure P9.35 shows a cantilever spring made of high-carbon steel. The minimum hardness is 452 H_B. The bend is gentle, so that no stress concentration need be considered. The surface is that of cold-drawn material. A history of the force at the end of such springs indicates a typical sequence (in pounds) as follows. Impact can be considered moderate.

$$0.5, \quad 2.5, \quad 1.8, \quad 2.5, \quad 0.0, \quad 2.0, \quad 0.7, \quad 3.2, \quad 2.5,$$
$$3.0, \quad 0.4, \quad 2.2, \quad 1.5, \quad 2.2, \quad 0.0, \quad 2.8, \quad 0.5$$

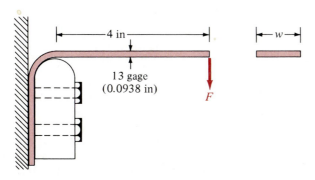

FIGURE P9.35

The elapsed time for the sequence is 4 s. Determine the width of the spring w, to the nearest $\frac{1}{8}$ in, necessary to ensure a life of 200 h with an FS of 1.5 on the strength.

Answer: $w = 1.5$ in

Section 9.5

9.36. We have a design for a thin-wall ($\frac{1}{4}$-in), 20-in-diameter vessel made of steel with a corrected $S_f = 50$ ksi and $S_y = 80$ ksi. The pressure varies continuously from 775 to 2250 psi. An FS of 3 is required. Is the design okay? For derivation of the stresses, see Example 4.7. Use the distortion energy theory.

9.37. At a point in a machine element, the stresses are $\sigma_{xm} = 60$, $\sigma_{xa} = 15$, $\sigma_{ym} = 20$, and $\sigma_{ya} = 10$ MPa. If the uncorrected material fatigue limit is $S_{fRB} = 270$ MPa, $S_{yt} = 400$ MPa, and $S_{ut} = 525$ MPa, determine the factor of safety. $C_F = 0.75$, $C_S = 0.8$, $K_t = 2$, and $q = 0.8$.

(a) Use the maximum-shear-stress theory.

Answer: FS = 1.44.

(b) Use the distortion energy theory.

Answer: FS = 1.68.

9.38. The stresses on an element are σ_{xm}, $\sigma_{xa} = \sigma_{xm}/3$, $\sigma_{ym} = \sigma_{xm}/2$, and $\sigma_{ya} = \sigma_{ym}/3$. Material and other data are the same as for Problem 9.37. Determine the safe level for σ_{xm} with an FS of 2.
(a) Use the maximum-shear-stress theory.
(b) Use the distortion energy theory.

9.39. For the part shown in Figure P9.39 select the cheapest of the first four listed steels in Appendix F such that the part will have a life of 400 000 cycles with FS = 1.5.

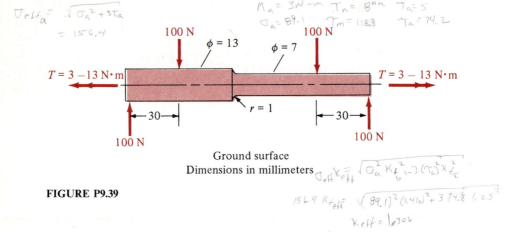

$$\sigma_{elta} = \sqrt{\sigma_a^2 + 3\tau_a^2}$$
$$= 156.4$$

$M_a = 3W - m \quad T_n = 8^{Nm} \quad T_a = 5$
$\sigma_a = 89.1 \quad T_m = 1188 \quad \tau_a = 74.2$

$T = 3 - 13\ \text{N·m}$ $T = 3 - 13\ \text{N·m}$

100 N $\phi = 13$ $\phi = 7$ 100 N

$r = 1$

$-30-$ $-30-$

100 N   100 N

Ground surface
Dimensions in millimeters

$\sigma_{eff} K_{eff} = \sqrt{\sigma_a^2 K_f^2, -3(\tau_a)^2 K_{fs}^2}$

FIGURE P9.39

$156.4\ K_{eff} = \sqrt{89.1)^2(1.416)^2 + 3\ 74.8\ 1.25^2}$

$K_{eff} = 1.306$

9.40. A stress element has $\sigma_{x,max} = 400$, $\sigma_{x,min} = 200$, $\sigma_{y,max} = -75$, and $\sigma_{y,min} = -225$ MPa. Evaluate the factor of safety. $S_f = 400$ MPa (corrected); $S_{ut} = 1000$ MPa. Use the maximum-shear-stress theory.

9.41. The L-shaped bar shown in Figure P9.41 is made of the 1060 cold-rolled steel shown in Appendix F, machine-finished. What force F_0 of the forms indicated below will the bar sustain on an indefinite basis with a reliability of 95 percent? Let FS = 1.7, and assume that the stress concentrations are insignificant. You must check two locations: at the wall and at the elbow. Use the maximum-shear-stress theory.
(a) The force varies from 0 to $+F_0$.
(b) The force varies from $+F_0$ to $-F_0$.
    Answer: $F_0 = 218$ N

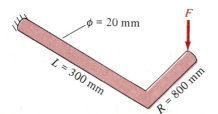

F

$\phi = 20$ mm

$L = 300$ mm

$R = 800$ mm

FIGURE P9.41

9.42. Rework Example 9.9, assuming a desired life of 250 000 cycles and a force which varies continuously between 0 and 500 lb. Since the reliability factor is set to 1 for finite life, increase the FS to 2.

9.43. The plate of Figure P9.43, made of 2024-T3 aluminum, is subjected to bending and tension. Find the width w necessary to ensure indefinite life with a reliability of 50 percent and an FS between 2 and 3. The chart for q, Figure 9.2, does not show this material; use the curve for 2024-T6. In the absence of information for the surface-finish factor, use a conservative $C_F = 0.5$. Similarly, take $C_S = 0.8$.

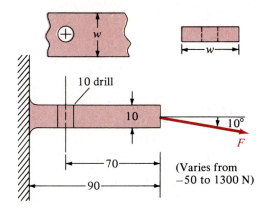

10 drill

10

$10°$

F

70

(Varies from
−50 to 1300 N)

90

Dimensions in millimeters   **FIGURE P9.43**

9.44. Check the possibility of yielding in the Goodman criterion solution in Example 9.10. (To be consistent, use the distortion energy theory.)

Sections 9.6 and 9.7

9.45. The part shown in Figure P9.1 is a proposed design to be loaded in rotating-bending by a moment $M = 20\,000$ lb·in. An FS of 1.5 is to be placed on the moment. A life of $3E + 05$ cycles is required. Material is SAE 1045 steel quenched and tempered to 225 H_B. Is the design okay?

Answer: No. The strain for the required life is $1.722E - 03$, but that existing is $4.969E - 03$.

9.46. The bar of Figure P9.3 is to be made of 1045 steel quenched and tempered to 410 H_B. A reversing force $F_a = 200$ kN is to be applied to the bar. A life of 150 000 cycles with an FS of 1.3 is required. Verify the design; if it is inadequate, suggest a redesign.

Section 9.8

9.47. The link shown in Figure P9.26 is to be subjected to a force which will vary continuously from 10 to 50 kN. A life of $2E + 05$ cycles is necessary. The material is 2024-T351 aluminum. Verify the design.

Answer: The design is inadequate. The existing strain is $1.405E - 03$, which is greater than the $0.97E - 03$ to get the required life.

9.48. Repeat Problem 9.46, except that the force varies continuously from -125 to 185 kN and the life requirement is 1000 cycles.

9.49. The part of Figure P9.1 is to be loaded in rotating bending by a moment $M = 25\,000$ lb·in. In addition, there is an axial force $F = 53\,000$ lb. Material is 1045 steel quenched and tempered to 410 H_B. Estimate the life of the part.

Answer: $N = 4636$

9.50. The flat part shown in Figure P9.50 is loaded by random forces. Rainflow counting on a sample produces the following cycles (kilonewtons): $300/-163$, $-128/282$, $-154/102$, $60/-120$. Material is 2024-T3 aluminum. Calculate the damage done in this segment of loading.

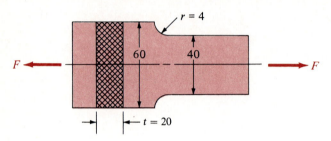

Dimensions in millimeters

FIGURE P9.50

9.51. A strain gage was placed on the side of the hole of the part shown in Figure P9.3 during service, and a strain history was taken. The first several cycles in a rain-flow count produced the following strain ranges:

$8.762E - 03/3.436E - 03$

$7.542E - 03/3.715E - 03$

$4.657E - 03/-2.631E - 03$

Material is 5160 steel quenched and tempered to H_B 430. Calculate the damage done by this series of strain cycles.

Answer: $D = 2.386E - 04$

CHAPTER
10

OPTIMALITY

A perceptive engineer once defined engineering as the "art of doing that well with one dollar, which any bungler can do with two after a fashion." We have put a glossier word to that idea: *optimality*. Much of engineering design is indeed the effort to accomplish some fairly ordinary task at minimum expense. We should not be sheepish about that. Less expense is another way to say less time. Less time spent on subsistence means more time for culture, in a word, civilization.

But engineering is not just an exercise in cutting costs. Our technological civilization is based on myriad products and techniques which would be impossible without optimal design. The turbines which power aircraft are a good example. As shown by the simplified sketch of Figure 10.1, air is brought to a high pressure in the compressor, energy is added in the burner, and power is extracted in the turbine. Even under ideal conditions, two-thirds of the power produced by the turbine is required by the compressor. If the design of either the turbine or the compressor is not optimum for the conditions, the power output of the turbine may not be sufficient even to drive the compressor. The idea of the gas turbine is centuries old, but only recently have we been able to optimize its design.

Engineering design is a striving toward optimality, then, to minimize cost and to maximize function. In any particular situation, these needs may seem contradictory, but that is often an illusion. The most functional design can be the simplest; the optimal solution is usually the least expensive.

Several techniques for optimization are reviewed here. Their use is demonstrated with models which are simple mathematical functions. This is not to say that optimization must always be done with a mathematical model, but it certainly helps.

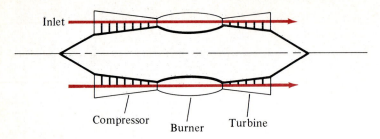

Inlet

Compressor

Burner

Turbine

FIGURE 10.1
Schematic of a gas turbine.

10.1 FUNCTIONS OF A SINGLE VARIABLE

The first step in any formal optimization is to write the thing to be optimized as a mathematical expression which we call the *criterion function*. Optimality is then defined as the value of this function being maximum, minimum, or zero.

When the criterion function is a simple expression in one independent variable, the problem is straightforward. For example, the structural engineer usually needs to know the location of the maximum bending moment in a beam, to ascertain that it is strong enough. If with x the distance along the beam, the bending moment is

$$M = 1200x - 140x^2 \quad \text{N·m}$$

then the maximum moment can be found by locating the point where the slope of this function is zero. That is, the derivative of the bending-moment expression with respect to x is set equal to zero:

$$\frac{dM}{dx} = 1200 - 280x = 0$$

$$x = 4.286 \text{ m}$$

The maximum bending moment is $M_{\text{max}} = 2571$ N·m. We have found the value of x which produced a maximum of an expression, the bending moment M, by setting its derivative equal to zero.

When we are dealing with functions of more than one variable, or when the power of the variable is greater than 3, it can be difficult to decide whether the optimum at hand is really the best available. Even in functions of a single variable, that variable may be imprisoned in a function whose derivative is not readily determined or solved. In such cases, instead of trying to solve for the one value of the independent variable which is the exact solution to the problem, we determine a value which is less than the solution and another which is greater. These can be called the *lower* and *upper bounds*, respectively. Then the distance between the bounds is reduced until the desired precision is attained.

Recursion

We have already seen the simplest of these, recursion, in Chapter 9. A single term containing the variable is isolated on the left, and a guess for the variable is substituted into the remaining terms on the right. The variable is then solved for and the result recycled until the difference between successive solutions is within the desired precision.

Example 10.1. A function is differentiated and set equal to zero, with this result:

$$3x - 2\cos x = 0$$

It is restructured:

$$x = \frac{2\cos x}{3}$$

Inserting a guess of $x = 1$ into the right side produces $x = 0.36$. Using this as a new guess produces 0.623. Successive calculations yield $x = 0.54, 0.57, 0.56, 0.56$. The correct value of x is 0.56, within two decimal places. The process can be tedious when it is done by hand, but quick and precise when it is produced, as these were, with a short computer program.

This example featured a very well-behaved function. The recursion process converges rapidly whether it started with a large guess or a small one, positive or negative. Such docility is not the general rule. Recursion has the virtue of simplicity, but some functions are not amenable to being thus separated. Others will converge only if the initial guess is close to the correct value, and some will not converge at all.

Interval Halving

Problems which are not readily solved with calculus or recursion may yield to a search routine. The simplest of these merely divides the interval in half and uses a test to decide which half contains the solution. Then the process is repeated with the half-interval chosen in the previous step.

Suppose we need to find the values of x (the roots) for which some function of x is equal to zero. The range of feasible values of x has been determined to be between $x = a$ and $x = b$. We calculate the value of the function at those points. If the sign of the function at $x = a$ is different from that at $x = b$, we know that an odd number of roots exists in the range. This situation, with a single root, is sketched in Figure 10.2.

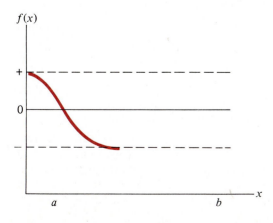

$f(x)$

FIGURE 10.2
A function with a root between a and b.

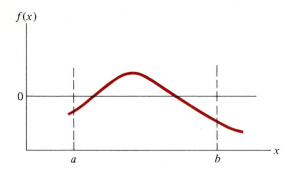

$f(x)$

0

x

FIGURE 10.3
A function with two roots between a and b.

If the sign of the function is the same at a as at b (an even number of roots), the situation is that shown in Figure 10.3. In this case, we reduce the interval until the sign of the function at one end is opposite to that at the other.

With an interval established like that of Figure 10.2, which contains an odd number (1) of roots, we place the first trial value of x at the midpoint:

$$x_m = \frac{a + b}{2}$$

and we calculate the value $f(x_m)$ of the function at that point. If that value has the same sign as at b, which is the situation shown in Figure 10.2, then we know that the root must be in the other half of the interval—between x_m and a. Therefore, the region from x_m to b is discarded by redefining the value x_m as b; that is, we set $b = x_m$.

If the sign of the function at x_m had been the same as that at a, then the left half of the interval would have been discarded by setting $a = x_m$.

The process is repeated until the width of the range—the difference between a and b—is less than the required precision.

A slightly more complex version of interval halving is used when the maximum or minimum value of a function is sought. In addition to the value of the function at the midpoint x_m, the values at the quarter points x_1 and x_2 are computed. These three points are shown in Figure 10.4a.

With three points in the interval evaluated, a new interval can be defined in three ways: the left half, the center half, or the right half. We choose the interval which has the most desirable value of the function at its center. Suppose a maximum is sought. Finding that the left-hand interval contains the highest value of the function, we choose it by setting $b = x_m$ as before. In this new interval, what was x_1 is now x_m, and the function is computed at two new points.

If the value of the function at x_m were the largest, as shown in Figure 10.4c, we would choose the center interval by redefining x_1 as a and x_2 as b.

Interval halving is a simple example of a gradient search, which moves always in the direction in which the criterion function improves the most rapidly. The rate of change of the criterion function is called the *gradient*. By evaluating a point at each end of each candidate interval, the interval-halving method estimates the mean gradient of that interval.

(a)

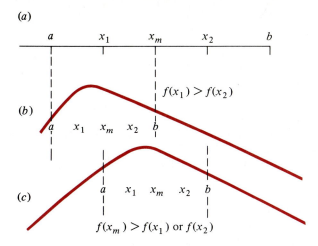

(b)

(c)

$$f(x_1) > f(x_2)$$

$$f(x_m) > f(x_1) \text{ or } f(x_2)$$

FIGURE 10.4
The function is calculated at two points x_1, and x_2 and compared to the previously calculated value at x_m, to determine which interval to split next.

Example 10.2. Consider the function

$$f(x) = 0.75x^2 + 8 \sin 0.85x$$

in the region $x = 0$ to $x = 4$. Figure 10.5 shows that it reaches a maximum value of about 11.5 at $x = 2.55$.

To find the maximum value of the function, we must convert it to an equation which will be satisfied at that point. The function will take an extreme value when its derivative is zero:

$$\frac{df(x)}{dx} = 1.5x + 6.8 \cos 0.85x = 0$$

or

$$x = -4.533 \cos 0.85x$$

A recursion solution of this relation diverges, whether the initial guess is greater than, less than, or nearly equal to the correct value.

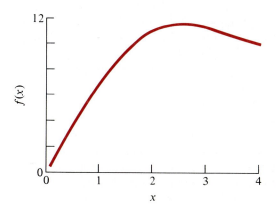

FIGURE 10.5
A function with an optimum (maximum).

The interval-halving method, however, works well here. The first step produces these values:

x	0	1	2	3	4
Label	a	x_1	x_m	x_2	b
$f(x)$		6.76	10.93	11.21	

The value of the function at x_2 is highest, so the right half of the space is kept by assigning the label a to the value of x that had been labeled x_m. That is, $a = 2$.

Two new points x_1 and x_2 are determined, and the corresponding values of the function are calculated:

x	2		3		4
Label	a	$x_1 = 2.5$	x_m	$x_2 = 3.5$	b
$f(x)$		11.49	11.21	10.51	

We already had the value of $f(x)$ at $x = 3$, which became x_m. Thus two new calculations are made at each step.

Since the value of the function at x_1 is the largest, we choose the left interval by the command $b = 3$. Two new quarter points x_1 and x_2 are then defined:

x	2		2.5		3
Label	a	$x_1 = 2.25$		$x_2 = 2.75$	b
$f(x)$		11.334		11.433	

and so on. Eight steps of this search reduced the interval to $2.539 < x < 2.5554$—about 0.6 percent. The uncertainty remaining at any step can be calculated from the fact that its size is halved at each step.

The Golden Section

Many centuries ago architects searching for a formula to determine the most aesthetically pleasing proportion for a room or window settled on a rectangle which would retain the same proportion—the same ratio between its sides—if a square were removed from it. Such a rectangle is shown in Figure 10.6 with a square of sides b marked off at one end. The retention of proportion requires that

$$\frac{b}{a+b} = \frac{a}{b} \qquad \text{thus} \qquad a = 0.618b$$

The golden section also has an important place in the technology of computer search routines. The golden section search evolved from the desire for a routine which would require the evaluation of only one new point at each stage. Maximum search efficiency requires that the trial points be symmetrically placed in the interval and that the interval be reduced by the same proportion each time.

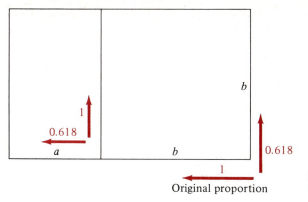

FIGURE 10.6
Original proportion   Golden-section rectangle.

Consider that the units of x have been adjusted so that the length of the starting interval is unity. Figure 10.7*a* shows a trial point x_1 placed a distance τ from the left end. To obtain the required symmetry, the other point x_2 is placed the same distance τ from the right end (Figure 10.7*b*). The function is evaluated at x_1 and x_2.

The next step is to shorten the interval by keeping either the segment defined in Figure 10.7*a* or that defined in Figure 10.7*b*. If $f(x_2)$ is more optimal than $f(x_1)$, then the segment which contains it is kept, as shown in Figure 10.7*c*. This new interval has one evaluated point, x_2, already in it. A new point is needed. To abide by the rules of the search, the new point must be located a distance from the right end equal

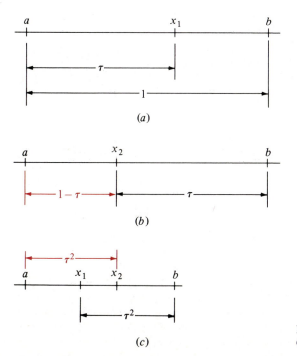

FIGURE 10.7
Golden-section search interval.

to τ times the width of the interval. The total width of the interval is τ, and so the distance from the right boundary to the new point is τ^2.

The second requirement is that the two points be spaced equal distances from the ends of the interval. Thus

$$1 - \tau = \tau^2$$

$$\tau = 0.618$$

the proportion of the golden section.

Example 10.3. Refer to the function of Example 10.2. The golden-section search, starting with the same interval, places two points instead of three:

x	0	1.528	2.472	4
Label	a	x_1	x_2	b
$f(x)$		9.457	11.484	

We discard the interval from a to x_1 and redefine. The position 1.528 was x_1; it becomes a. What was x_2 becomes x_1. A new x_2 is defined as the same distance from the left as x_1 is from the right.

x	1.528	2.472	3.056	4
Label	a	x_1	x_2	b
$f(x)$			11.145	

The value of the function at x_1 had been calculated in the previous step—it is 11.484. Thus only one new calculation needs to be made at each step. Since the value of the function is higher at x_1 than at x_2, the interval from x_2 to b is discarded. The location $x = 3.056$ is redefined as b, and a new x_1 is calculated.

x	1.528			2.472	3.056
Label	a	$x_1 = 2.112$			b
$f(x)$		11.145			

and so on. Eleven such steps reduce the range to $2.545 < x < 2.557$—a difference of 0.5 percent.

The golden section search requires more steps than the interval-halving search. But efficiency lies in the number of calculations rather than the number of steps. Eight steps of the interval-halving search required 16 calculations, while 11 steps of the golden-section search required only 12 calculations.

SEVERAL OPTIMA. Functions of higher order than the cubic can have more than one maximum. It is like the hikers who toil up to the peak they can see, only to find that it was shielding a higher one from view. Figure 10.8 shows a situation that any hiker will recognize. Suppose that a search has been started with a very small

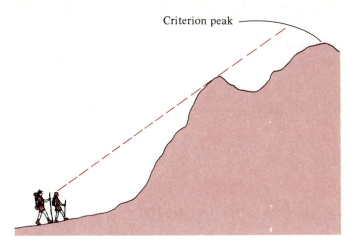

FIGURE 10.8
A local maximum can hide the final objective.

guess for the independent variable, and an optimum is found at a larger value. Other optima are suspected. The strategy is to repeat the process with a starting guess in the other direction; i.e., work down from a guess that is too high.

> **Example 10.4.** The simplest example of a function with two maxima is a linear fourth-order expression like
>
> $$f(x) = 12x - 20x^2 + 9x^3 - x^4$$
>
> which is shown graphically in Figure 10.9. The maxima can be located with an interval-halving computer code. If we start with a small value of x by making the limits $a = -2$ and $b = 2$, the computer produces an estimate of $x = 0.402$, at which the value of the function is 2.150.
>
> To set the starting range $a = -2$ and $b = 20$ is to start from the high side, with the result $x = 4.789$ and $f(x) = 61.286$.

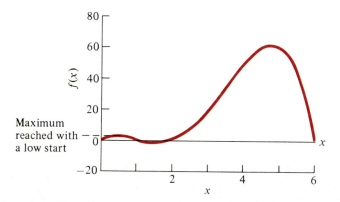

FIGURE 10.9
Sketch of a function with two maxima.

10.2 INEQUALITIES

Probably 99 percent of engineering analysis is performed through the manipulation of equations, that is, equalities. Frequently, in so doing, we are in fact dealing with an inequality. For example, the analysis of the strength of a shaft yields a minimum. That is, a diameter equal to or larger than the one calculated is required for strength. By designating the calculated diameter as d_{calc}, the conclusion can be stated as this inequality:

$$d \geq d_{calc}$$

Much of this text deals with inequalities.

Basic Rules

There are two simple principles.

RULE 1. The same quantities added to (or subtracted from) both sides of an inequality do not alter the inequality. This is analogous to the rule that an equation is not altered by adding or subtracting the same quantity from both sides. Thus

$$\text{If} \qquad 4 > 2$$

$$\text{Then} \qquad 4 + 3 > 2 + 3$$

$$\text{And} \qquad 4 - 5 > 2 - 5$$

"Greater than" is interpreted in the algebraic sense (-1 is greater than -3). The only exceptions are when it is understood that we are comparing only magnitudes (absolute values).

> **Example 10.5.** These rules apply as well for mixtures of numbers and variables. Thus, if $x > y$, then $x - 2 > y - 2$. Figure 10.10 shows a point (x_1, y_1), which satisfies the original inequality. The point $(x_1 - 2, y_1 - 2)$ also satisfies the inequality.

RULE 2. An inequality is not changed if both sides are multiplied (or divided) by a positive quantity. It is reversed if the quantity is negative. This is like the algebraic

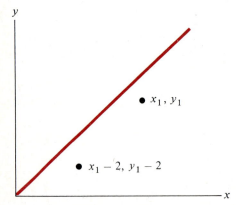

y

$\bullet \; x_1, y_1$

$\bullet \; x_1 - 2, y_1 - 2$

x

FIGURE 10.10
An inequality is not altered when equal quantities are added to (or subtracted from) both sides.

rule that the sign of a quantity is unchanged when multiplied by a positive quantity, but is reversed when the multiplier is negative. Thus

If	$r + 2 > 4d^2$	
Then	$(x)(r + 2) > (x)(4d^2)$	if $x > 0$
Or	$(x)(r + 2) < (x)(4d^2)$	if $x < 0$

It follows that if an inequality is multiplied by a quantity whose sign is unknown, the result depends on that unknown sign. Thus,

If	$r + 2 > 4d^2$	
Then	$(x)(r + 2) > (x)(4d^2)$	if $x > 0$
Or	$(x)(r + 2) < (x)(4d^2)$	if $x < 0$

Although the magnitude of the variable x may not be known, its sign often is. This is frequently the case in engineering problems where certain physical quantities are never negative: diameters, lengths, spring rates, and especially costs. It is not uncommon for a mathematical analysis to yield weird results because the analyst, without realizing it, assumed that a real physical quantity could have a negative magnitude.

The operations we will perform with inequalities in this book are all applications of the two principles given above. Sometimes extra care is required. Suppose that

$$x^2 > 16$$

If x can have only positive values, then $x > 4$. But suppose x is something like a stress, which can have a negative sense. Then the inequality tells us only that the *magnitude* of x is greater than 4. Either of these inequalities could be true:

$$x > 4 \quad \text{or} \quad x < -4$$

Example 10.6. Suppose

$$x^2 + 4x - 12 > 0$$

We can solve for x as though the inequality were an equation:

$$x \ (?) \ 2 \quad \text{and} \quad x \ (?) - 6$$

The original inequality is not satisfied by a value of x between these, such as $x = -2$. Therefore we conclude that the two inequalities which satisfy the original are:

$$x > 2 \quad \text{and} \quad x < -6$$

Combining Equalities and Inequalities

The force in a bar carrying only tension can be expressed as the cross-sectional area times the normal stress:

$$F = \frac{\pi d^2}{4} \sigma \tag{10.1}$$

If the objective is to maximize the force-carrying capacity of the bar, this is known as the *criterion function*. Suppose that limits exist on the diameter

$$d < d_{max} \tag{10.2}$$

$$d > d_{min} \tag{10.3}$$

and on the strength of the material

$$\sigma < \sigma_{lim} \tag{10.4}$$

By the second rule of inequalities, either (10.2) or (10.3) can be multiplied by the same positive quantity without changing the sense of the inequality. We use Equation (10.1), rearranged, as the multiplier:

$$\frac{1}{d} = \sqrt{\frac{\pi\sigma}{4F}} \tag{10.5}$$

Now the left side of (10.2) is multiplied by the left side of (10.5) and the right side by the right side, and the result is squared. The result is an upper limit on force:

$$F < d_{max}^2 \frac{\pi\sigma}{4}$$

A similar multiplication of (10.3) by (10.5) yields

$$F > d_{min}^2 \frac{\pi\sigma}{4}$$

a lower limit on force.

Frequently it is not obvious what the sense of an inequality should be. It may be possible to solve the problem through further analysis, such as taking derivatives. But it may be easier to sketch the graph of the function in question.

Example 10.7. A machine rests on a shock absorber which may be modeled as a perfect spring and viscous damper. As do all machines, it vibrates. This vibration is examined in some detail in Chapter 14; it can be modeled as a sinusoidal force with frequency equal to the machine speed. The force transmitted through the mount to the foundation can be expressed [28, p. 96] as a function of the ratio r of the machine speed to the natural frequency of the system and the ratio ξ of the damping to the critical damping:

$$F = F_R \frac{r^2\sqrt{1 + (2\xi r)^2}}{\sqrt{(1 - r^2)^2 + (2\xi r)^2}} \tag{10.6}$$

The term F_R is a constant, the disturbing force at resonance. The effect of the speed ratio on the transmitted force depends, of course, on the damping present. The relation F/F_R is shown in Figure 10.11 for a damping ratio of $\xi = 0.2$,

You can see that it would be essential to have a sketch of this function if you needed to know the effect of placing limits on the speed ratio. The points of zero slope *a* and *b* could be located by a single differentiation, but the plot makes it clear which is maximum and which is minimum.

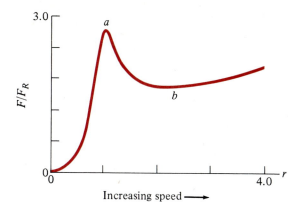

FIGURE 10.11
Sketch of the relative force transmitted by a vibration isolation mount as a function of frequency ratio.

10.3 FUNCTIONS OF MORE THAN ONE VARIABLE

Often engineering design requires that more than one variable be placed at its optimal value. If these variables are independent in their effect on the criterion function, then the optimum value of each can be found by holding all the others constant.

Exhaustive Searches

In general, the variables of a given situation are not independent. Then the most straightforward search simply calculates the value of the criterion function for every feasible combination of variables. The exhaustive search makes sense when only a few variables, each with a limited range of possible values, must be considered. For example, to evaluate a criterion function for 10 values each of 5 variables requires 50 calculations. This is no problem with a computer. However, the size of the search can quickly get out of hand as the number of variables or the range of possible values increases.

Programmed Searches

When an exhaustive search would use too much computer time, we must devise a method to arrive at an optimum by varying one variable at a time. The simplest method simply optimizes each variable in turn, as though the variables were independent. To be most efficient, the search can be started by ranking the variables according to their effect on the criterion function. The gradient, which is the change in the criterion per unit change in the variable, is calculated. The search starts with the variable of highest gradient and proceeds down the list.

Example 10.8. Consider a design criterion function

$$C = 2x + 5y \tag{10.7}$$

The other requirements of the design impose limits on the values of x and y:

1. $y \leq 3$

2. $x + 2y \leq 8$

3. $x + y \leq 6$

These are inequalities because they are upper limits on x and y.

Usually the variables in an optimization study represent real entities that cannot have negative values. Thus the lower limits on x and y are zero. The limits are shown in Figure 10.12.

We could search through this set by setting one variable constant and looking for the optimum value of the other. For example, we could set $y = 0$ and look for the optimal value of x. The criterion function would become $C = 2x$ and have a maximum value, at the limiting value $x = 6$, of 12. Then the value of y could be increased, say to 1, and the process repeated. The criterion, at the new limit of $x = 5$, would have the value 15. This is a change in the right direction for the criterion function, so the search would continue until an increase in y produced a decrease in the limit value of C. The one-dimensional search is a feasible procedure for this simple example, especially when the search can be guided by reference to a graph.

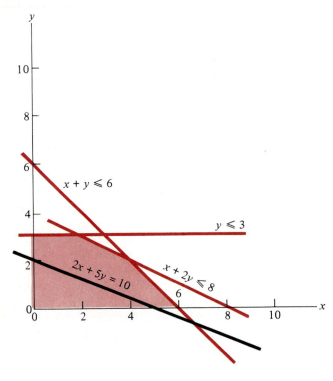

FIGURE 10.12
Graphical solution of a two-variable (two-dimensional) optimization problem.

The problem with the one-at-a-time search, however, is that the variables are seldom independent. That is, the optimum value of each variable will depend on the values of all the others. To minimize the error resulting from this fact, schemes are devised with the objective that the optimization of the last variable will find all the others also at optimal values. It isn't always possible to reach this ideal, but the result is usually much better than would be obtained without a formal scheme.

LINEAR PROGRAMMING. The more complex the criterion function, the more complex the search method. We illustrate only the simplest, in which the criterion is a linear function of the variables. This classic search method for finding the optimum value of a linear function of many variables is called *linear programming*. The algorithm is discussed here for a function of only two variables, because the resulting two-dimensional plot can be shown on a piece of paper, and the operation of linear programming is the same for 20 variables as for 2. The function is

$$C = Ax + By$$

which you will recognize as a general form of Equation (10.7). And x and y will be chosen to give C a maximum value. There are usually limitations on the values that the variables can have, and these are assumed also to be linear:

$$y < y_{max}$$
$$A_1 x + B_1 y < D_1$$
$$A_2 x + B_2 y < D_2$$
$$A_3 x + B_3 y < D_3$$

Since x and y are real physical quantities that can have only positive values,

$$x > 0 \quad \text{and} \quad y > 0$$

The limits shown in Figure 10.12 form the boundaries of a space within which a solution is possible. The space is called a *convex set* because a straight line from any corner to any other corner will not cross space outside the set. The set must be convex so that linear programming will find a true optimum.

A line is shown along which the criterion C has a constant value $C = 10$. This criterion has a value of zero when x and y are zero, as shown by Equation (10.7). Its magnitude increases as the line moves farther from the origin. The linear programming method moves this line outward from the origin until it touches only the farthest corner of the solution space.

Mathematically the linear programming method starts at a minimum point, such as zero, and expresses the criterion function and limits in terms of the variable of highest gradient. Then the other limits are examined and that limit chosen which will be first arrived at, i.e., which gives the smallest value of the chosen variable. Refer to Figure 10.12. If the highest gradient were in respect to x, the limits and the criterion would be expressed in terms of x. Then the limit first reached, for which the value of x would be least, would be $x + 0 = 6$. The next move would be along the limit $x + y \leq 6$. To stay on that limit line, x would be replaced in all expressions by $6 - y$, and a new limit and a new value of the criterion function would be sought. The process would be repeated until the value of the criterion function is less than at the end of the previous step.

Example 10.9. Linear programming can be demonstrated by applying it to the two-dimensional problem of Example 10.8. This is a maximization problem, and the variables cannot have negative values. Therefore, the search starts at the origin.

The gradient of the criterion function with respect to x is $dC/dx = 2$, and with respect to y it is 5. Therefore, the first move holds x at its limit of $x = 0$ and increases y until the first limit is reached. This is done by making the substitution $x = 0$ in all the expressions, which become:

1. $y \le 3$
2. $0 + 2y \le 8$
3. $0 + y \le 6$

and $$C = 0 + 5y$$

The first limit reached is $y = 3$, for which $C = 15$. To move along that limit, $y = 3$ is substituted in all the expressions:

2. $x + 2(3) \le 8$
3. $x + 3 \le 6$

$$C = 2x + 15$$

The first limit reached is $x = 2$, for which $C = 19$. This is greater than the last value of C, so another step is taken along the limit $x = 8 - 2y$. This is substituted into the remaining limit:

3. $(8 - 2y) + y \le 6$

$$C = 2(8 - 2y) + 5y$$

The limit is reached at $y = 2$, for which $C = 18$. This is less than the previous value. We conclude that the optimum value of C is 19, with $x = 2$ and $y = 3$.

MULTIDIMENSIONAL PROBLEMS. When the number of variables is greater than 2, a graphical representation becomes infeasible. The limits and the criterion function become "hyperplanes" of the form $Ax + By + Cz + \cdots$, and the solution space becomes a polyhedron. But linear programming can still be used, and its operation is identical to our two-dimensional example. The computer program for such multi-dimensional searches is known as the *simplex algorithm*. Details are given in specialized texts [80, p. 152]. The necessary computer software is available.

In the two-dimensional problem used here as an illustration, only one direction for motion was available at the end of each step. The only decision necessary was whether the completion of the next step would result in an increase in the criterion function. In a multidimensional problem, many directions are available at each step. The direction in which to move is chosen by examining the form of the criterion for each candidate direction and choosing the direction for which the criterion increases fastest. In other words, the algorithm proceeds along the steepest gradient.

NONLINEAR FUNCTIONS. Many engineering design decisions involve criterion functions and constraints that are nonlinear. In general, two courses of action are available. One applies when the approximate location of the optimum can be found. Then the constraints and criterion can be linearized. That is, they are replaced by linear approximations that are accurate for that limited region, and the simplex algorithm is used. Linear functions are so economical of computer time that it is often more efficient to repeat this process with ever-closer approximations than to use a more sophisticated search.

If a global search without linearization cannot be avoided, an algorithm similar to the simplex can be used. In addition to the complexity resulting from nonlinearity, the solution space is likely to be nonconvex. That is, there may be many "peaks." Searching along the steepest gradient will lead to a peak, but there is no way of knowing that it is the "highest" peak. A large body of literature is available on the subject of sophisticated search schemes, but the gist of all these strategies is that multiple searches must somehow be made, starting from different quadrants of the solution space.

PROBLEMS

Section 10.1

10.1. Using recursion, find the value of x which satisfies the equation

$$2x^2 - 3 \cos x = 0$$

Answer: $x = 0.9403$.

10.2. Find a positive root of the equation

$$0.75x^2 - 8 \sin 0.75x = 0$$

using either recursion or interval halving. Use a computer program.

10.3. Repeat Problem 10.2 for the equation

$$x^4 + x^3 - 19x^2 - 50x - 30 = 0$$

Answer: One root is 5.01.

10.4. The bending moment in a beam is expressed by

$$M = x^4 - 9x^3 + 9x^2 + 41x - 42$$

Find the location x of the maximum moment and its value. Use (*a*) an interval-halving program and (*b*) a golden-section search.

Section 10.2

10.5. Refer to Example 10.7. If the damping ratio is 0.15 and the speed ratio limited to $1 < r < 3$, does a maximum or minimum force result by placing the speed ratio at its largest value?

Answer: A minimum force results.

10.6. Suppose

$$z = \frac{y}{y^4 - y_0^4}$$

What results for z if y is set at y_{max} and

(a) $y > y_0 > 0$

(b) $y > y_0 < 0$

10.7. If $x = 4y^2 + 3y - 12$, what results for y if there is an upper limit on x?

10.8. In Problem 10.7, what results for x if y has an upper limit? Suppose the coefficient of y is -3. What happens?

Section 10.3

10.9. In an optimization problem, the criterion function is $C = 2y + 8x$. The limits on the variables x and y are

$$y \le 12$$

$$5x + y = 35$$

$$x \le 8$$

Find the maximum value of the criterion, using a graphical solution. Show the limits and two examples of straight lines along which the criterion has a constant value.

10.10. After doing Problem 10.9, solve it analytically by hand and then by using a computer program.

Answer: 60.8

CHAPTER
11

DESIGN OF TENSION MEMBERS

In this chapter and the next we deal with the design of what is perhaps the simplest of machine members, a bar loaded with a force along its axis. Most often the force is in the direction of stretching the bar, and we term it a *tension bar*. That is the subject of this chapter, with the exception of bolts, which are treated separately in Chapter 17. If the force tends to shorten the bar, it is called a *compression member* or a *column*, dealt with in Chapter 12.

11.1 INTRODUCTORY CASE— MINIMIZING DEFLECTION

In this problem a given static load P must be supported: $P = 1.5E + 05$ N, and the deflection of the bar under this load must be a minimum. There is a space problem, resulting in an upper limit on the diameter of the bar:

$$d \leq d_{max} = 30 \text{ mm} \qquad \text{or} \qquad A \leq A_{max} = 707 \text{ mm}^2$$

The bar's length must fall between fixed limits:

$$L \geq L_{min} = 200 \text{ mm} \qquad L \leq L_{max} = 300 \text{ mm}$$

Failure is defined as the yield of the material. We use FS = 2.
We first express the quantity to be optimized, the extension of the bar under load:

$$\delta = \frac{PL}{AE} \qquad E = \text{modulus of elasticity} \qquad (11.1)$$

This is called the *design criterion equation*; P is a fixed quantity, L and A have limits. If we place the area A at its upper limit, we obtain the inequality

$$\delta \geq \frac{PL}{A_{max}E} \tag{11.2}$$

With any given material, the modulus E is known, and this minimum deflection depends only on the length L. We can plot it against L and see then the region of possible design on the basis of the constraint $A \leq A_{max}$. The plots for two hypothetical materials A and B are shown in Figure 11.1. The area of possible solutions can be further defined by drawing the limits on L, as shown.

A further limitation, which is sometimes not explicitly stated since it seems so obvious, is that the bar must not fail. The stress in the bar is $\sigma = P/A$. If we substitute this in the criterion equation (11.1), we get

$$\delta = \frac{\sigma L}{E} \tag{11.3}$$

The term σ/E is the strain, millimeters of deflection per millimeter of length. When the strain is multiplied by the bar length L, the result is the stretch of the bar.

A limit is imposed on the stress σ. The maximum-shear-stress and distortion energy theories in this case give the same result: $\sigma \leq S_y/\text{FS}$. Making this substitution

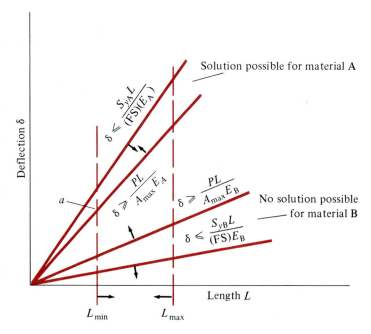

FIGURE 11.1
Minimum-deflection problem with limits on A, L, and stress.

in (11.3), we have

$$\delta \leq \frac{S_y L}{FS(E)} \tag{11.4}$$

and this is plotted in Figure 11.1 for the hypothetical materials A and B.

Now we examine the two limits on material B. The upper line, corresponding to the limit on area A, is a lower limit on the deflection δ. The lower line, corresponding to the maximum allowable stress, is an upper boundary. There is no solution which is simultaneously above the upper line and below the lower line. In other words, material B is not strong enough for this application.

For material A, however, solutions are possible between the stress-limited and size-limited boundaries. The lower corner of this space, point a, corresponds to the minimum deflection, and, of course, it is the shortest ($L = L_{min}$), thickest ($A = A_{max}$) bar we are allowed by the constraints. The stress for the design at point a will be less than the maximum permitted for the material.

Normally, you would have a greater choice than the two materials represented by A and B in Figure 11.1. The plots of that figure could be repeated for each candidate material, and the material resulting in the minimum deflection would be chosen. A study of the figure suggests, however, a somewhat more organized procedure for arriving at this result. Start with the observation that if the stress limit $\delta = S_y L/[FS(E)]$ lies above, or coincident with, the area limit $\delta = PL/(A_{max}E)$, then a solution for a given material is possible. This is the case of material A in the figure. We write

$$\frac{S_y L}{(FS)(E)} \geq \frac{PL}{A_{max}E}$$

giving

$$\frac{S_y}{FS} \geq \frac{P}{A_{max}} \tag{11.5}$$

This simply states that the material must have adequate strength. The minimum deflection will always be for $L = L_{min}$ and $A = A_{max}$:

$$\delta_{min} = \frac{PL_{min}}{A_{max}E}$$

So the least deflection δ can be determined by finding the material with the greatest stiffness E. Thus, the design procedure is to choose the stiffest material, then check condition (11.5) to see if the material is strong enough. If the test fails, the material with the next greatest E is tried, etc. The order of these steps could, of course, be reversed. That is, we could eliminate from the available materials all those failing (11.5) (not strong enough), then choose the greatest E from the remaining. For the problem specifications given at the beginning of the section, the strength test (11.5) is

$$\frac{S_y}{2} \geq \frac{1.5E + 05}{707E - 06} \qquad \text{so} \qquad S_y \geq 424 \text{ MPa}$$

Taking the list in Appendix F as candidate materials, we see that all the steels, except 1010 and 304 stainless, have yield strengths above the limit. Choosing the steels with the greatest E, we pick 1040, 2330, and 4130, which have essentially the same modulus (207 MPa). The choice among these surely would be on the basis of cost of the finished item. Since 1040 has the lowest base cost and would be easiest to machine, it is used. The final specifications are

$$d = d_{max} = 30 \text{ mm} \qquad L = L_{max} = 200 \text{ mm}$$

The stress is

$$\sigma = \frac{P}{A_{max}} = 212 \text{ MPa}$$

and

$$\delta = \frac{PL_{min}}{A_{max}} = 0.2 \text{ mm}$$

The principal purpose of the plots of Figure 11.1 in this problem was to provide a clear picture of how the analysis should proceed and the possibility of incompatibilities and how they may arise. The solution can be arrived at entirely analytically, but, as you will see in more complex problems to follow, a picture is much easier to look at than analytical expressions.

Although you might not be inclined to do so for this very simple introductory problem, the procedure outlined above could be performed on a computer. The flowchart is shown in simplified form in Figure 11.2.

Let us return to the deflection equation (11.1). Instead of plotting deflection vs. length, as we did in Figure 11.1, why not plot deflection vs. cross-sectional area A? That can be done, but instead of plotting against A, it is easier to use $1/A$ so as to have straight lines. The result is shown in Figure 11.3. The inequality (11.4) results in the limit $1/A \le S_y/[FS(P)]$ shown. For a solution to be possible, that limit must fall to the right of $1/A_{max}$:

$$\frac{S_y}{FS(P)} \ge \frac{1}{A_{max}} \qquad \text{or} \qquad \frac{S_y}{FS} \ge \frac{P}{A_{max}}$$

which is the strength requirement of relation (11.5).

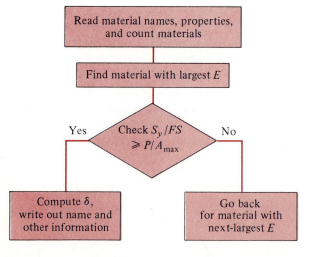

Read material names, properties, and count materials

Find material with largest E

Yes — Check $S_y/FS \ge P/A_{max}$ — No

Compute δ, write out name and other information

Go back for material with next-largest E

FIGURE 11.2
Simplified computer flowchart for minimum-deflection tension bar.

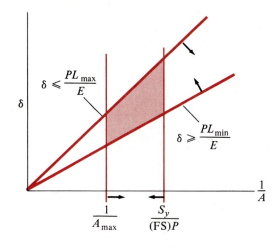

FIGURE 11.3
Deflection vs. 1/area for minimum-deflection problem.

11.2   MAXIMIZING THE FACTOR OF SAFETY

A tension bar must meet the restrictions $d \le d_{max}$, $L_{min} \le L \le L_{max}$, and the mass $M \le M_{max}$. A specified stiffness k is required. This is the stiffness of the bar to an axial load, the same as the spring constant of a spring. Since $\delta = FL/(AE)$,

$$k = \frac{AE}{L} = \text{specified number}$$

The force on the bar changes constantly in time from zero to a maximum, so this is a fatigue problem. An indefinite life is desired. The objective is to maximize the FS, which appears in the strength relation, Goodman's criterion (9.10). With the mean and alternating stresses equal, there will not be gross yielding. Expression (9.10) can be written in terms of either σ_a or σ_m. In terms of the alternating stress σ_a,

$$\frac{\sigma_a}{S_f} + \frac{\sigma_a}{S_{ut}} \le \frac{1}{\text{FS}} \tag{9.10}$$

Replacing σ_a with F_a/A, we solve for FS and write it as an equality:

$$\text{FS} = \frac{S_f A}{F_a(1 + S_f/S_{ut})} \tag{11.6}$$

This is the criterion equation.

Greater area will result in a larger FS (that is logical), but there is a limit on the area; it may not exceed A_{max}. This results in a limit on FS:

$$\text{FS} \le \frac{S_f A_{max}}{F_a(1 + S_f/S_{ut})} \tag{11.7}$$

Turning to the spring rate k, we have

$$k = \frac{AE}{L} \quad \text{or} \quad A = \frac{Lk}{E}$$

which we substitute in the criterion to get

$$\text{FS} = \frac{S_f k L}{E F_a (1 + S_f / S_{ut})} \tag{11.8}$$

This is an equality, because k is of a stipulated size, not a limit.

The length L is the only changeable quantity on the right side in expression (11.8). Through the algebraic substitutions made, we have thus defined the problem as that of finding the length L which results in a maximum FS. We can see the relation on a sketch of FS vs. L. Since the coefficient of L, for a given material, is a constant, expression (11.8) will be a straight line. Since it is an equality, our solution must lie *on* the line, not above or below it.

The problem specifications require that the mass be not greater than the limit $M_{\max}$. The mass of the part is given by

$$M = A L \rho \qquad \text{so} \qquad A = \frac{M}{L \rho} \qquad \rho = \text{density}$$

Inserting this in the criterion, we get an upper limit on the FS corresponding to the upper limit on the mass

$$\text{FS} \le \frac{S_f M_{\max}}{\rho F_a (1 + S_f / S_{ut})} \frac{1}{L} \tag{11.9}$$

a hyperbola.

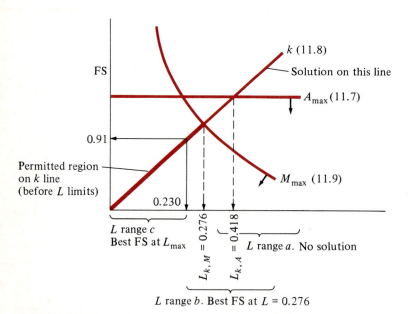

FIGURE 11.4
Maximum FS problem with limits on A, M, and L (material 1040 steel).

The three expressions [(11.7), (11.8), and (11.9)], which summarize the problem, are sketched in Figure 11.4. These curves are not drawn to scale, for the plots only indicate how to proceed to the solution, as we will see in the following. There are also limits on the length L—more on that presently.

First, observe that the relative positions of the curves in the sketch depend on the values of the properties of the material used in relations (11.7), (11.8), and (11.9). This means that the A_{max} limit may be above the intersection of the other two curves, as shown, or below it. There are only those two possibilities. (The situation will not always be so simple.) Which arrangement we have can be determined by finding the values of the length at the intersections of the k line with the other two curves, as shown in the figure, $L_{k,A}$ and $L_{k,M}$. If $L_{k,A}$ turns out to be less than $L_{k,M}$, Figure 11.4 is not correct; the A_{max} limit is below the intersection of the other two curves.

Setting (11.7) and (11.8) equal, we get

$$L_{k,A} = \frac{A_{max}E}{k}$$

and similarly with (11.7) and (11.9),

$$L_{k,M} = \left(\frac{M_{max}E}{k\rho}\right)^{1/2}$$

Example 11.1. Let us get specific with numbers and see how things work out for 1040 steel:

$$d_{max} = 30 \text{ mm} \qquad A_{max} = 7.06E - 04 \text{ m}^2$$

$$L_{max} = 230 \text{ mm} \qquad L_{min} = 130 \text{ mm}$$

$$M_{max} = 1 \text{ kg} \qquad F_{max} = 90 \text{ kN}$$

$$k = 3.5E + 08 \text{ N/m}$$

$$E = 207 \text{ GPa} \qquad \rho = 7.75E + 03 \text{ kg/m}^3$$

$$S_y = 490 \text{ MPa} \qquad S_{ut} = 586 \text{ MPa}$$

We estimate the endurance limit S_{fRB}, per Chapter 9, at $0.5S_{ut}$, and since this is an axial application, the size factor $C_S = 0.6$. The surface-finish factor from Figure 9.1 is $C_F = 0.76$. We will presume other factors at 1, including the stress concentration. Thus

$$S_f = 0.6 \times 0.76 \times 0.5 \times 586E + 06 = 133.6 \text{ MPa}$$

We compute

$$L_{k,A} = \frac{7.06E - 04 \times 207E + 09}{3.5E + 08} = 0.418$$

$$L_{k,M} = \left(\frac{1 \times 207E + 09}{3.5E + 08 \times 7.75E + 03}\right)^{1/2} = 0.276$$

The situation is thus that shown in Figure 11.4. For another material Figure 11.4 may not apply. Note that it is only the smaller of $L_{k,A}$ and $L_{k,M}$ which concerns us, since that fixes an upper limit and determines how far up the k line (constant stiffness) we can go. Thus, L cannot be greater than 0.276. However, other constraints on length

L were set out at the beginning of the problem. There are three possibilities:

a. The permitted L range is totally above the permitted region of the k line. In this case there is no solution. The smallest permitted length would require an impermissibly large mass to achieve the required stiffness k. The size limitation A_{max} might also be exceeded.

b. The L range brackets the upper limit of the permitted region of the k line, $L_{k,M}$. The largest FS is that corresponding to $L_{k,M} = 0.276$ mm. The solution is $L = L_{k,M}$ and $M = M_{max}$. Since $M_{max} = AL\rho$, then $A = M_{max}/(L\rho)$. And finally the value of FS can be found from any one of (11.6), (11.8), or (11.9).

c. The L range is entirely within the permitted region on the k curve, that is, $L_{max} < L_{k,M}$. In this case the largest FS is that corresponding to L_{max}. This happens to be the situation we have for the 1040 steel. Thus $L_{opt} = L_{max} = 0.23$ m (230 mm); k is as required, $3.5E + 08$. Since $k = AE/L$;

$$A = \frac{kL}{E} = \frac{3.5E + 08 \times 0.23}{207E + 09} = 3.89E - 04$$

or
$$d = 0.022 \text{ m (22 mm)}$$

and
$$M = AL\rho = (3.89E - 04)(0.23)(7.75E + 03) = 0.69 \text{ kg}$$

Note that d and M fall below their maximum permitted values—check that on Figure 11.4.

Finally, the optimum FS can be obtained from (11.6) by using the value of A calculated above or from (11.8) with $L = L_{max}$. We can also get it from expression (11.9), if we replace M_{max} with the actual mass, 0.69 kg. Relation (11.6) looks easiest:

$$FS = \frac{(133.6E + 06)(3.89E - 04)}{(45E + 03)(1 + 133.6/586)} = 0.94$$

And that is not too good; we would expect the part to fail. Since the FS was limited by the k curve at $L = L_{max}$, Equation (11.8) indicates that we should choose a material with a greater fatigue strength for another try.

You have perhaps observed that the above solution process consisted of a comparison of several critical values of variable L. Here the process is as a series of steps from which a computer routine could be written:

1. Calculate $L_{k,A}$ and $L_{k,M}$.
2. Compare the two values; call the smaller L_{small}.
3. If $L_{min} > L_{small}$, no solution is possible; otherwise, continue.
4. Compare L_{small} with L_{max}, and call the smaller L_{opt}. This is the solution value.
5. Calculate the corresponding FS_{opt} from (11.8); this is the maximum FS attainable.

Now, if we have a list of candidate materials, we repeat the above process for each and then pick the best FS. Possibly we could save some effort by selecting materials stronger than AISI 1040, i.e., with larger S_y and S_{ut}. If you look at Figure 11.4, you will see that the possible solutions occur on the stiffness line k at lengths equal to L_{max}, $L_{k,A}$, or $L_{k,M}$. The corresponding values of the FS are

$$FS_{L_{max}} = \frac{kL_{max}S_f}{F_aE(1 + S_f/S_{ut})}$$

$$FS_{k,A} = \frac{A_{max}S_f}{F_a(1 + S_f/S_{ut})}$$

$$FS_{k,M} = \frac{\sqrt{kM_{max}}S_f}{F_a\sqrt{E\rho}(1 + S_f/S_{ut})}$$

So parameters other than strength are involved, in particular the modulus of elasticity E and the density ρ.

If your computer has a plot routine, Figure 11.4 can be done on the screen or a plotter, obviating the necessity for some of the logic outlined above.

11.3 MANUFACTURED COST

In real-life problems the cost of a part is important. Cost depends on a number of factors; generally they can be broken into three groups:

$$\$ = \$_{fixed} + \$_{material} + \$_{machining}$$

The fixed cost is that having to do with overhead in the plant and office: administrative costs, insurance, taxes, janitorial costs, sales, etc. The name *fixed* comes from the fact that these costs are not sensitive to the marginal volume of production; i.e., they are the same whether the production rate is, say, 1000 items per day or 1001. Material cost depends on the market price for the material in question in the size and quantities required. Machining costs depend on the number and types of operations required to make a part. For a particular operation, the factors determining machining cost are:

Labor rate
Power rate
Machine-tool rate (depreciation, upkeep, etc.)
Machineability of the material
Tooling costs

Clearly, the faster you can cut off the material (other things being equal), the lower the cost per unit volume of material removed, hence, the shape of the cutting cost curve in Figure 11.5. Tool cost, however, rises as speed increases, because the edge dulls faster. So the process must stop periodically for sharpening or changing the tool. Those costs are not insignificant, and, of course, production is interrupted. The sum of these costs results in the parabola for total cost, with a minimum at some optimum cutting speed. Whether the operation is run at the optimum or not, the cost of removing a unit volume of material results, and we term that C_{mach}. For an operation like turning, then, the machining cost is

$$\$_{mach} = C_{mach}(A_0 - A)L$$

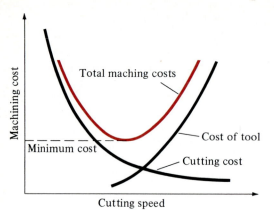

FIGURE 11.5
Components of machining cost.

where A_0 = cross-sectional area of purchased stock, so that $(A_0 - A)L$ represents the volume of material to be removed.

To get a dollar figure for the cost of machining a part, a company must be able to put numbers on the factors listed above, and the calculation will vary from firm to firm and job to job. It is not practical or instructive to assign numbers to all these details in a text like this. However, to demonstrate methods and to make it possible to arrive at solutions, we include in Appendix E the market cost C_{mat} (at time of writing) per pound for the materials listed and a reasonable value for the factor C_{mach} for the computation of machining costs. Thus we have

$$\$ = \$_{fix} + C_{mat}\rho L A_0 + C_{mach}(A_0 - A)L \tag{11.10}$$

where ρ = weight density. The reference is in general to round parts; other shapes would be handled in a similar manner.[1]

Example 11.2. We must turn a series of blanks of 50-mm-diameter cartridge brass to a finished size of 42.5 mm. The length of the parts is to be 200 mm. The fixed cost is figured to be \$1.80 per piece. What is the total manufactured cost?

From Appendix F we find

Material cost: $C_{mat} = \$4.95/kg$

Machining cost: $C_{mach} = \$3.10E + 04/m^3$

Density: $\rho = 8.54E + 03/m^3$

Applying, then, (11.10), we get

$$\$ = 1.80 + 4.95(8.54E + 03)(0.2)\frac{\pi}{4}(0.05^2) + (3.1E + 04)\frac{\pi}{4}(0.05^2 - 0.0425^2)(0.2)$$

$$= 21.78 \text{ per piece}$$

---

[1] Where the cost of machining is dominated by the complexity of the part and tight tolerances, expression (11.10) is admittedly an oversimplification.

11.4 MINIMIZING COST

Let us imagine the same tension bar as in Section 11.2 with the same sort of fatigue loading F varying from zero to F_{max}. There are similar restrictions on diameter and length, but this time, instead of a specified stiffness, the deflection is limited (lower bound on the stiffness):

$$\delta = \text{deflection under } F_{max} \leq \delta_{max}$$

We want to minimize the cost of making the bar. The design criterion will be the cost relation (11.10), rearranged slightly:

$$\$ = \$_{fix} + [(C_{mat}\rho + C_{mach})A_0 - C_{mach}A]L$$

We can insert the limit on A immediately, and we have

$$\$ \geq \$_{fix} + [(C_{mat}\rho + C_{mach})A_0 - C_{mach}A_{max}]L \tag{11.11}$$

A plot of $\$$ vs. L looks convenient. We could have inserted the limits on L instead, and a plot vs. A would develop (Problem 11.13).

Turning to the deflection, we have

$$\delta = \frac{F_{max}L}{AE} \qquad \text{from which} \qquad A = \frac{F_{max}L}{\delta E}$$

We insert this in the criterion:

$$\$ = \$_{fix} + (C_{mat}\rho + C_{mach})A_0L - \frac{C_{mach}F_{max}}{\delta E}L^2$$

By observing the upper limit on the deflection δ, inequality (11.12) results:

$$\$ \leq \$_{fix} + (C_{mat}\rho + C_{mach})A_0L - \frac{C_{mach}F_{max}}{\delta_{max}E}L^2 \tag{11.12}$$

The reasoning behind the inequality is this: If the extension δ gets smaller (which is the only change permitted), the coefficient of L^2 gets larger; and that term is negative, so the cost goes down (less machining).

There remains the stress to deal with. Goodman's criterion (9.10) gives

$$\sigma_a \leq \frac{S_f}{FS} - \sigma_a \frac{S_f}{S_{ut}}$$

where we have taken account, as before, of the fact that $\sigma_m = \sigma_a$ in this problem. We make the substitution $\sigma_a = F_a/A$ and sort out A to get the limit.

$$A \geq (FS)F_a\left(\frac{1}{S_f} + \frac{1}{S_{ut}}\right)$$

The stress-limited inequality results from inserting this in the criterion:

$$\$ \leq \$_{fix} + \left[(C_{mat}\rho + C_{mach})A_0 - C_{mach}(FS)F_a\left(\frac{1}{S_f} + \frac{1}{S_{ut}}\right)\right]L \tag{11.13}$$

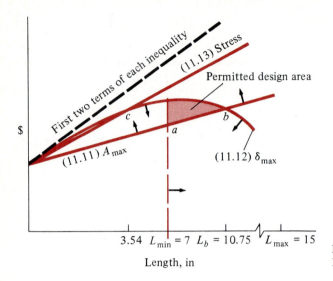

FIGURE 11.6
Minimizing cost.

The area A inserted was a minimum value, hence relation (11.13) sets an upper limit on cost.

In using relations (11.11), (11.12), and (11.13), the rod stock size A_0 (d_0) must be assumed. The bar diameter resulting from the analysis may require us to redo the calculations with a more appropriate stock size. The inequalities are shown in the sketch of Figure 11.6. The plots are easy to draw if you recognize that the first two terms of all the inequalities are the same and are straight lines, as shown.

Note that the stress-limit line (11.13) must be above the A_{max} line (11.11), or we do not have a solution. In other words, the slope of the former must be greater than that of the latter. Applying this condition, we find

$$A_{max} \geq (FS)F_a \left(\frac{1}{S_f} + \frac{1}{S_{ut}} \right) \tag{11.14}$$

which simply guarantees that the maximum size allowed is sufficient for the fatigue load on the part. Assuming we are safe on that score, it becomes clear that the minimum cost will lie at a in the lower left corner of the shaded area in the figure. This is the shortest, largest-diameter bar. It could happen, however, that the shortest length allowed L_{min} will be greater than the length corresponding to point b. In such a case, a solution is not possible—the maximum permitted area is not sufficient to keep the deflection within limits. Thus L_b is critical, and its value can be obtained by equating expressions (11.11) and (11.12):

$$L_b = \frac{A_{max}\delta_{max}E}{F_{max}} \tag{11.15}$$

Example 11.3. Following the above outline, we minimize the cost of a bar, using 1040 steel, available in $\frac{1}{4}$-in increments of diameter:

$$d_{max} = 1.85 \text{ in} \qquad A_{max} = 2.688 \text{ in}^2$$

$$A_0 = 3.14 \text{ in}^2 \qquad d_0 = 2 \text{ in}$$

$$L_{max} = 15 \text{ in} \qquad L_{min} = 7 \text{ in}$$

$$F_{max} = 15 \text{ klb} \qquad F_{cyc} = 7.5 \text{ klb}$$

$$\delta_{max} = 0.002 \text{ in}$$

$$S_y = 71 \text{ ksi} \qquad S_{ut} = 85 \text{ ksi}$$

Then $S_f = 0.6 \times 0.76 \times 0.5 \times 85 = 19.4 \text{ ksi}$

$$E = 30E + 06 \text{ psi} \qquad \rho = 0.28 \text{ lbm/in}^3$$

$$C_{mat} = 1.10 \text{ \$/lbm} \qquad C_{mach} = 1.00 \text{ \$/in}^3$$

$$\$_{fix} = 1.00 \qquad FS = 1.8$$

First, the strength test (11.14)

$$2.688 \overset{?}{>} \frac{1.8(7.5E + 03)(1 + 19.4/85)}{19.4E + 03} = 0.85 \qquad \text{Okay}$$

and the critical length (11.15)

$$L_b = \frac{2.688 \times 2E - 03 \times 30E + 06}{15E + 03} = 10.75$$

This point and the L range of the problem are shown in Figure 11.6. The minimum cost is computed from (11.11) with $L = L_{min} = 7$:

$$\$_{opt} = 1 + [1.10(0.28)(3.14) + 1(3.14 - 2.688)]7 = 10.93$$

The deflection

$$\delta = \frac{FL}{AE} = \frac{15E + 03 \times 7}{2.688 \times 30E + 06} = 0.0013 \text{ in}$$

which is within the limit of 0.002, as Figure 11.6 indicates.

The stock size A_0, corresponding to a diameter of 2 in, is not a problem, because the solution was restricted by A_{max}, corresponding to a diameter of 1.85 in. In other combinations of restrictions, the optimum diameter might go below the next stock size smaller, in which case some recomputations would be required.

Here again, the solution is facilitated if Figure 11.6 can be plotted on a computer. We show an example of such a solution in Chapter 12.

11.5 CLOSURE

We have shown in this chapter an optimization technique for members in tension. The cases of minimizing deflection, maximizing the FS, and minimizing manufactured costs were treated. Many other optimization criteria may be encountered, of course, and the limitations on the parameters will cause the solutions to vary. The following problems are illustrative.

PROBLEMS

Sections 11.1 and 11.2

11.1. Program the problem of Section 11.1 for which the flowchart appears in Figure 11.2. Use the following data:

$$P = 1.4E + 05 \text{ N} \qquad FS = 1.5$$

$$L_{min} = 150 \text{ mm} \qquad A_{max} = 350 \text{ mm}^2$$

Design the bar, giving the material and the dimensions. Consider the nonstainless steels of Appendix F.

Answer: $d = 21.2$ mm

11.2. In the problem of Section 11.1, minimizing deflection, assume these additional limits:

$$P = 10\,000 \text{ lb} \qquad FS = 1.3$$

$$d_{max} = 0.5 \text{ in} \qquad 15 \leq L \leq 20 \text{ in} \qquad M \leq 0.30 \text{ lbm}$$

Make a sketch as in Figure 11.1 (δ vs. L seems best). You will see that the stress-limited slope must be greater than the area-limited slope—this would be a first test for a given material. The intersections of the M limit curve (a parabola) with these two lines produce critical values of L, which should then be compared with the restrictions on L. Consider the nonstainless steels of Appendix F. The final solution should include all design parameters.

11.3. Design a tension bar for minimum mass with the following specifications:

$$F = 40\,000 \text{ lb (static)} \qquad FS = 1.25$$

$$\delta_{max} = 0.2\% \text{ of bar length} \qquad \delta = \text{total deflection of bar}$$

$$d \geq 1.75 \text{ in} \qquad 8 \leq L \leq 20 \text{ in}$$

Limit the material choice to the aluminums of Appendix F. Sketch a computer flowchart for the solution.

Answer: $M = 1.35$, $d = 1.48$, $L = 8$

11.4. We wish to design a tension bar to absorb a maximum of potential energy (PE $= F\delta/2$). The bar will be machine-finished; assume no stress concentrations. The force on the bar varies continuously from zero to some unknown maximum. There are geometric limitations:

$$d \leq d_{max} = 20 \text{ mm}$$

$$150 \text{ mm} = L_{min} \leq L \leq L_{max} = 300 \text{ mm}$$

Use FS $= 2$. Limit your choices to the steels. Find the bar dimensions, F_{max}, and the optimum PE.

11.5. Design a hold-down stud for maximum PE absorption capability (PE $= F\delta/2$). The force on the stud is cyclic, varying from 0 to F_{max}. The geometry would allow the shank to consist of two portions of different diameters, as shown in Figure P11.5. K_f is the fatigue stress-concentration factor at the thread root, which is the critical location. It has the surface finish and notch sensitivity factors built in.

After writing the expression for the PE, you will conclude that you want $L_2 = 0$ and $L_1 = L_{max}$, provided that d_1 passes what comes to a strength test. If the material fails that test, then the bar can be made larger ($d_{2max} > d_{1max}$), but the length will be less ($L_{2max} = 5$). However, d_2 must pass the strength test.

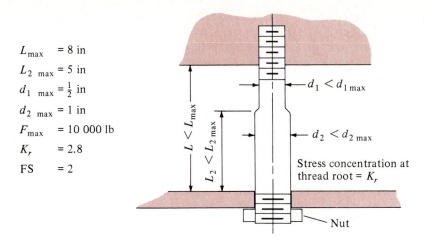

L_{max} = 8 in

$L_{2\,max}$ = 5 in

$d_{1\,max}$ = $\frac{1}{2}$ in

$d_{2\,max}$ = 1 in

F_{max} = 10 000 lb

K_r = 2.8

FS = 2

$d_1 < d_{1\,max}$

$d_2 < d_{2\,max}$

Stress concentration at thread root = K_r

$L < L_{max}$

$L_2 < L_{2\,max}$

Nut

FIGURE P11.5

11.6. A threaded tension member is to be designed to absorb a certain amount of potential energy (PE $= F\delta/2$). The force in the bar varies continuously from zero to a maximum, which is not specified. We have various requirements (USCS units):

$$PE = 75 \qquad d_{max} = 1 \qquad 6 \le L \le 20 \qquad FS \ge 1.5 \qquad K_r = 3.0$$

Note that FS is not fixed, but a minimum. And K_r is the fatigue stress-concentration factor at the root of the thread, the worst place. It includes the finish factor and notch sensitivity. Take the other fatigue correction factors as unity.

Occasionally the PE may be higher than 75 for short times. Yielding must be avoided during these periods. Hence we wish to maximize the ratio S_{yt}/σ_{max}, where σ_{max} is the stress when PE is 75 (normal operation). Solve the problem by computer, considering the steels of Appendix F as candidate materials. Give as output the material, bar dimensions, FS, and F_{max}.

11.7. A threaded tensile bar is to be subjected to a steady force plus a cyclic force. The amplitude of the cyclic force is one-half the steady force. For each cycle the tensile bar must absorb and release a specified amount of energy. The objective is to minimize the force transmitted to the machine frame; i.e., we wish to minimize F_{max}. The specifications are (USCS units)

$$PE = 100 \qquad FS = 1.2 \qquad L_{min} = 6 \qquad d = L/10 \qquad d_{max} = 0.75 \qquad K_r = 3.0$$

The thread root stress-concentration factor includes surface and notch sensitivity factors. Assume the other fatigue correction factors are unity.

Consider only the aluminums of Appendix F. Give the material, bar dimensions, and value of the steady force.

11.8. We have a tension bar like that in Section 11.2. The restrictions are

$$d_{max} = 30 \text{ mm} \qquad 130 \le L \le 230 \text{ mm}$$

$$\delta_{max} = 0.1 \text{ mm} \qquad (\text{under } F_{max})$$

$$F_{max} = 85 \text{ kN} \qquad F_{min} = 0$$

$$\$_{max} = 13.00 \qquad \$_{fix} = 1.50 \qquad M_{max} = 1.0 \text{ kg}$$

Use 1040 CR steel (Appendix F), which is available in 5-mm increments. The objective is to maximize the FS. Use the same fatigue limit as calculated in Example 11.1. In working this problem, you can follow a good bit of the outline in Section 11.2.
Answer: FS = 1.82; for cheapest bar $L = 173$ and \$ = 3.90

Section 11.3

11.9. The load on a ground tension bar varies constantly in time from 20 to 50 klb. We wish the bar to have a stiffness $k = 6E + 05$ lb/in. Other limitations are

$$0.875 \le d \le 1.5 \text{ in} \qquad 15 \le L \le 40 \text{ in}$$

$$\text{Cost} \le \$20 \qquad \$_{fix} = 2.00 \qquad \text{FS} = 1.3$$

Assume there are no stress concentrations. Stock is available in $\frac{1}{4}$-in increments. Design the bar for minimum length, using nonstainless steels.
Answer: $d = 0.875$, $L = 30.08$, \$ = 18.98

11.10. A tension bar is to be subjected to a constant force of 300 kN. The deflection under this force must be less than 0.5 mm. There are other limitations:

$$\text{Cost} \le \$20 \qquad \$_{fix} = 1.20$$

$$150 \text{ mm} \le L \le 1 \text{ m} \qquad d \le 47 \text{ mm}$$

We must maximize the reliability, hence we want to produce the largest possible FS. Thus we write $\text{FS} = S_y A / F$ and use this as the criterion. Note we use the equals sign, not an inequality. Material is available in 10-mm increments. (*Hint:* It is best to work down the list of materials from strongest to weakest.)

11.11. A machined stud must sustain a hold-down force of 20 klb, upon which is superposed a varying force, as shown in Figure P11.11a. The potential energy absorbed due to the varying load (from $F = 20k$ to $23k$) is 5 in·lb, Figure P11.11b. Other limitations are

$$d \le 1.5 \text{ in} \qquad \text{FS} = 1.8$$

$$\text{Cost per part} \le \$5.50 \qquad \$_{fix} = 2.40$$

Minimize the stud length. Use steels only, available in $\frac{1}{4}$-in increments.
Answer: $d = 0.625$, $L = 10.23$, \$ = 5.13

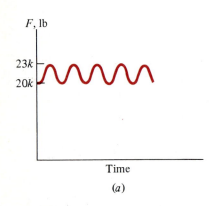

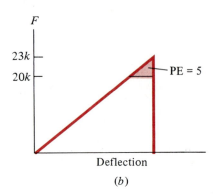

(a)   (b)

FIGURE P11.11

11.12. A machined tension member is to sustain a continuously varying force, which goes from 0 to a maximum of 20 klb. Use $K_t = 1.5$ $(r = 0.04)$ and FS $= 1.3$. There are these limitations:

$$\text{Cost} \leq \$13 \quad \$_{\text{fix}} = 1.10 \quad d \leq 1.5 \text{ in} \quad L \leq 26 \text{ in}$$

Design for a minimum spring rate $(k = AE/L)$, using nonstainless steels only, available in $\frac{1}{4}$-in increments. The specifications should indicate d, L, k, the material, and the cost. Construct a computer flowchart.

Section 11.4

11.13. Rework the problem of Section 11.4, aiming at a plot of $\$$ vs. A rather than $\$$ vs. L.

11.14. The force on a tension member varies continuously from 0 to 20 000 lb. Design for minimum manufactured cost with FS $= 1.3$ and (USCS units)

$$d \leq 1.625 \quad L \leq 15$$

$$\text{Stiffness } k = P/\delta \leq 3\text{E} + 06$$

Use the following materials:

	A	B
ρ	0.283	0.17
$\$_{\text{fix}}$	1.75	1.75
C_{mat}	1.70	5.00
C_{mach}	1.35	0.45
E	30E + 06	15E + 06
S_f (corrected)	25	15
S_{ut}	75	60

CHAPTER
12

DESIGN OF COMPRESSION MEMBERS

Frequently mechanical failures are due not to loads that exceed the yield or rupture strength of the materials in the part, but rather to a situation where the structure, once somehow disturbed, cannot restore itself to its original configuration. This is called *instability*.

A popular party trick is for a guest to balance himself or herself on an aluminum soft-drink can and then have someone touch the side of the can (best done *not* with the hand). The equilibrium is disturbed, and the can cannot return to its original straightness. The column strength decreases, and there is an immediate collapse.

Another classic instability occurs when the compressive load on a long rod is so high that, once the rod has been deflected from perfect straightness for any reason, the rod's stiffness is not sufficient to return it to its original shape, once the disturbance has been removed. When the rod deflects, the applied load results in a moment at the center (the rod becomes in essence a beam) equal to the load times the deflection. The moment causes more deflection, which causes more moment, etc., and a rapid instability failure, called *buckling* results. Everyone has done this with a long, thin stick, such as a yardstick. The instability of members under compression, or columns, is the subject of this chapter.

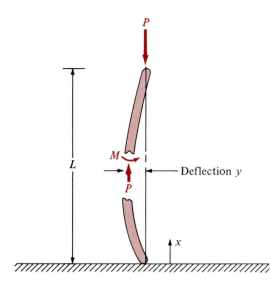

FIGURE 12.1
Free-body diagram of deflected pinned-end column.

12.1   THEORY OF BUCKLING

Basic Pinned-End Column

The load at which the buckling of a column occurs can be calculated by a procedure derived by the Swiss mathematician Euler[1] over two centuries ago. It is the smallest axial force for which a perfectly straight column collapses when it has been somehow deflected. Such a column, whose ends are kept in position but are free to rotate, is shown in Figure 12.1. From the theory of beams,

$$M = -EI \frac{d^2y}{dx^2}$$

---

[1] Leonhard Euler (1707–1783) was schooled in mathematics at the University of Basel, Switzerland. That institution was famous because of the teaching of Jean Bernoulli, father of Daniel Bernoulli, who is known to engineers for his work in fluid mechanics. Euler finished his master's degree at age 16, and by age 20 he had published his first paper.

In 1727, Euler moved to St. Petersburg (present-day Leningrad) to work at the new Russian Academy of Sciences. In a few years he became head of the department of mathematics, following Daniel Bernoulli. His two-volume text on mechanics was published in 1736.

In 1741 Euler was enticed to the Berlin Academy by the new King of Prussia, Frederick II. There Euler stayed for 25 years, where he wrote many papers and several volumes on calculus and, in 1759, became head of the academy.

Catherine II hired Euler to return to Russia and the Academy of Sciences in 1766. He had lost the sight in one eye some years earlier, and now the other failed him. Despite this handicap, with the help of his associates, he produced over 400 papers.

In engineering, Euler investigated the forms of elastic curves, i.e., of beams and columns with various loadings. He solved many of these—the one to which his name is most commonly attached is the buckling limit for columns, frequently known as the *Euler load*.

Euler also published analyses of the vibratory characteristics of beams and columns and was the first to write on a statically indeterminate structure problem.

Equilibrium of a section cut from the beam requires that $M = Py$. Substituting this, we get the differential equation

$$\frac{d^2y}{dx^2} + \frac{P}{EI} y = 0$$

whose solution is

$$y = A \sin \sqrt{\frac{P}{EI}} x + B \cos \sqrt{\frac{P}{EI}} x$$

Application of the boundary conditions $[y(0) = y(L) = 0]$ results in

$$B = 0 \quad \text{and} \quad \sqrt{\frac{P}{EI}} = \frac{n\pi}{L} \tag{12.1}$$

and thus

$$y = A \sin \frac{n\pi}{L} x$$

The column bends in a sine wave, and the maximum deflection A can have any value—that is the instability. From (12.1) we find the load when this happens:

$$P_{cr} = \frac{n^2 \pi^2 EI}{L^2}$$

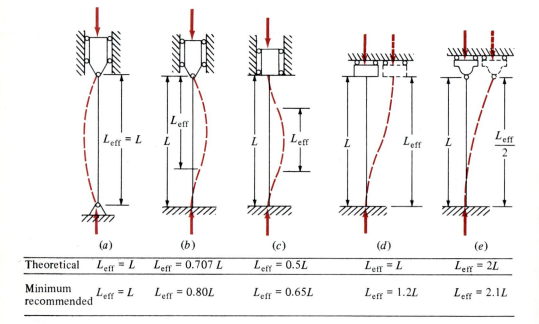

	(a)	(b)	(c)	(d)	(e)
Theoretical	$L_{eff} = L$	$L_{eff} = 0.707 L$	$L_{eff} = 0.5L$	$L_{eff} = L$	$L_{eff} = 2L$
Minimum recommended	$L_{eff} = L$	$L_{eff} = 0.80L$	$L_{eff} = 0.65L$	$L_{eff} = 1.2L$	$L_{eff} = 2.1L$

FIGURE 12.2
Column end conditions and equivalent lengths. (Buckled shapes are shown dashed.)

This critical load (subscript "cr") depends on the material stiffness E and the column dimensions. The failure is unrelated to material strength. It is due to elastic instability.

If $n = 1$, we see that

$$y = A \sin \frac{\pi}{L} x$$

which for beam length L is a half sine wave, as shown in Figure 12.2a. This is called the *fundamental mode*, exactly as in the vibration of a violin or piano string. (*Mode* means "manner," hence "manner of buckling.")

If $n = 2$, $$y = A \sin \frac{2\pi}{L} x$$

which is a full sine wave over the length, as seen in Figure 12.3. To produce this, we would need a support at the center to prevent the half sine wave at the lower load. We can continue the analysis in the same way for larger values of n, but the lowest is the one that counts, since the load reaches that value first, and the others require props:

$$P_{cr} = \frac{\pi^2 EI}{L^2} \tag{12.2}$$

Other End Conditions

The critical load above was worked out for a column with pinned ends, i.e., no moment applied at the ends. Since $y'' = M/(EI)$, the ends, where $M = 0$, are points where $y'' = 0$, or inflection points. These are positions where the slope y' stops decreasing and starts increasing, or vice versa. Figure 12.4 shows a curve with an inflection point on the usual xy plot.

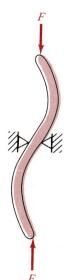

FIGURE 12.3
Buckling of a pinned-end column in a full sine wave would require a support at the center.

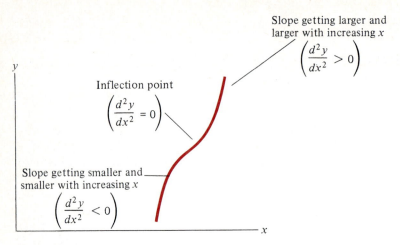

Slope getting larger and larger with increasing x

$$\left(\frac{d^2 y}{dx^2} > 0\right)$$

y

Inflection point

$$\left(\frac{d^2 y}{dx^2} = 0\right)$$

Slope getting smaller and smaller with increasing x

$$\left(\frac{d^2 y}{dx^2} < 0\right)$$

x

FIGURE 12.4
Inflection point.

If both ends of the column are fixed, i.e., not allowed to rotate or to move sideways with respect to one another (Figure 12.2c), the buckled shape is a full sine wave. The inflection points occur at the ends of the portion labeled L_{eff}, where "eff" is for *effective*. This column will have the same critical load as a pinned-end column half as long. Thus we can use expression (12.2) if we label the length L_{eff}. For the column with both ends fixed, then, we would insert one-half the actual length, and the buckling load would be 4 times as great as for a column with pinned ends. The effective lengths for other end conditions are shown in Figure 12.2.

Designers prefer to describe the geometry by the parameter L_{eff}/ρ, called the *slenderness ratio*, with ρ being the radius of gyration of the cross-sectional area ($I = \rho^2 A$). Of course, the radius of gyration to use is that about the neutral axis with the smallest I, that is, that corresponding to the weakest direction of bending. Making this change, the critical load becomes

$$P_{\text{cr}} = \frac{\pi^2 E A}{(L_{\text{eff}}/\rho)^2} \tag{12.3}$$

All the columns in Figure 12.2, except the basic one with pinned ends at the left, have at least one end which is not supposed to rotate, and two involve the presumption of no sideways relative motion. It is difficult in real structures to build in such perfect rigidity, especially rotational rigidity, and the effective lengths of these columns are therefore in practice somewhat greater than the theoretical values. The minimum effective lengths recommended by the American Institute of Steel Construction (AISC) [77] are indicated in the figure. Naturally, if you are uncertain of the rigidity, you can always assume a pinned end and err well on the safe side.

When columns buckle, they do not bend slightly at first and then gradually collapse as the load is increased. Rather, these failures are usually cataclysmic, that is, without warning the column suddenly collapses when the load reaches critical.

The Euler and Johnson Curves

The critical load, given by (12.3), divided by the cross-sectional area

$$\frac{P_{cr}}{A} = \frac{\pi^2 E}{(L_{eff}/\rho)^2} \tag{12.4}$$

is plotted against the slenderness ratio in Figure 12.5. This is called the *Euler curve*. It is for failure due to elastic instability, and the only material property on which it depends is the modulus of elasticity E, as we stated earlier. It has nothing to do with material strength. For this reason, P_{cr}/A is called the *critical unit-area load*, and the term *stress* is avoided. (The stresses will become excessive, however, as a result of the failure.)

   Notice that the unit-area load at which buckling is predicted by Euler's equation increases rapidly as the slenderness ratio decreases, because the slenderness ratio is squared. At some point, for what would be short, thick columns, the stress reaches the yield point before the column buckles, and we have material failure rather than buckling. Testing has shown that the transition from one type of failure to the other is a gradual one, so that, over a certain range of slenderness ratios, the unit-area load at failure is less than either the yield strength or the Euler value. A good approximation for the transition is a parabola fitted so as to start at the yield point on the y axis and to become tangent to the buckling curve at half the yield stress, as shown in Figure 12.5. This curve is known as the *parabolic formula*, or the *J. B. Johnson formula*, after its inventor:

$$\frac{P_{cr}}{A} = S_y - \frac{S_y^2}{4\pi^2 E}\left(\frac{L_{eff}}{\rho}\right)^2 \tag{12.5}$$

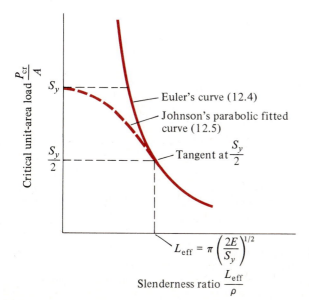

FIGURE 12.5
An Euler curve and Johnson's parabolic approximation for columns of low L_{eff}/ρ.

The tangent-point location on the L_{eff}/ρ axis can be obtained by setting $P_{cr}/A = S_y/2$. The result is

$$\frac{L_{eff}}{\rho} = \pi \sqrt{\frac{2E}{S_y}} \qquad \text{(divides Euler and Johnson)} \qquad (12.6)$$

Thus, a design is to be based on the Euler criterion [elastic instability per (12.4)] if the slenderness ratio falls above this critical value and on the J. B. Johnson fit between elastic instability and pure compression failure (12.5) if the slenderness ratio is below the critical value. In designing a column, with the slenderness ratio unknown at the start, it is necessary to assume failure per one criterion or the other to obtain a trial cross section and then check the assumption afterward.

Factor of Safety

Because failure due to elastic instability has nothing to do with stress and material strength, the factor of safety must be redefined as $FS = P_{cr}/P$. The P parameters would be subject to the same statistical treatment as were loads and strengths in Chapter 5, if it is intended to compute FS on a statistical basis. Usually, however, when failure is intolerable, designers use larger factors of safety, such as $FS = 2$ to 3, than would result from a statistical analysis.

Even when failure of a column is due to excessive material stresses [which is embodied, partly, in (12.5)], we have the same problem; we cannot express FS in terms of stress and material strength. We have to handle it in the same way as for elastic instability.

The Euler and Johnson Solutions

Example 12.1. We need a column 3 m long of square cross section with fixed ends to sustain a force of $1E + 05$ N. We use an FS of 3. The design critical load is then $P_{cr} = 3(1E + 05) = 3E + 05$ N. The material is to be steel with $S_y = 700$ MPa and $E = 203$ GPa.

For a square section, with w the section width, $I = w^4/12$ and $A = w^2$, so

$$\rho = \sqrt{\frac{w^4}{12w^2}} = \frac{w}{2\sqrt{3}}$$

A fixed-ends column has a recommended effective length

$$L_{eff} = 0.65L = 0.65(3) = 1.95 \text{ m}$$

hence

$$\frac{L_{eff}}{\rho} = 1.95 \frac{2\sqrt{3}}{w} = \frac{3.9\sqrt{3}}{w}$$

Now we have to guess whether the column corresponds to a Johnson type or an Euler type. The latter being an easier solution, let us try it first.

$$\frac{P_{cr}}{A} = \frac{\pi^2 E}{(L_{eff}/\rho)^2}$$

$$3E + 05 = \frac{\pi^2(203E + 09)w^2}{3.9^2 \times 3/w^2} \qquad \text{so} \qquad w = 51.1 \text{ mm}$$

Now we have to check to see if it really was an Euler column. From (12.6) the break point between the two types occurs at

$$\frac{L_{eff}}{\rho} = \pi \sqrt{\frac{2(203E + 09)}{700E + 06}} = 75.7$$

For our column

$$\frac{L_{eff}}{\rho} = \frac{3.9\sqrt{3}}{51.1E - 03} = 132$$

So we guessed right; it behaves as an Euler column.

Example 12.2. Suppose the load were a good bit more than in Example 12.1: $P_{cr} = 20E + 05$ N. Also let the length be 0.5 m. Then

$$L_{eff} = 0.65(0.5) = 0.325$$

and

$$\frac{L_{eff}}{\rho} = \frac{0.325(2\sqrt{3})}{w} = \frac{0.65\sqrt{3}}{w}$$

The Euler solution now gives

$$w^4 = \frac{(20E + 05)(0.65^2)(3)}{\pi^2(203E + 09)} \qquad \text{so} \qquad w = 33.5 \text{ mm}$$

and

$$\frac{L_{eff}}{\rho} = \frac{0.65\sqrt{3}}{0.0335} = 33.6$$

This puts us in the Johnson region, and we have to redo the calculations:

$$\frac{P_{cr}}{A} = S_y - \frac{S_y^2}{4\pi^2 E}\left(\frac{L_{eff}}{\rho}\right)^2$$

$$\frac{20E + 05}{w^2} = 700E + 06 - \frac{(49E + 16)(0.65^2)(3)}{4\pi^2(203E + 09)w^2}$$

So

$$w = 54.5 \text{ mm}$$

Just to check:

$$\frac{L_{eff}}{\rho} = \frac{0.65\sqrt{3}}{0.0545} = 20.7 \qquad \text{Okay}$$

12.2   ECCENTRIC LOADING

In real columns the load will never be aligned exactly with the column (i.e., pass through the centroid of the section), and/or the column will not be exactly straight. When the eccentricity e can be specified or estimated, the stress in the column can be calculated as the sum of the direct compression stress and a bending stress due to the load being applied off center, as shown in Figure 12.6. The bending stress causes the column to bend, so that the actual eccentricity of the load, measured from the original (straight) axis of the column, becomes greater at equilibrium than e. The

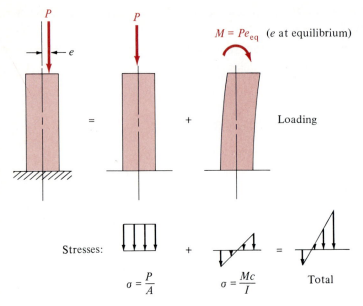

P

P

$M = Pe_{eq}$ (e at equilibrium)

e

$=$

$+$

Loading

Stresses:

$+$

$=$

$\sigma = \dfrac{P}{A}$

$\sigma = \dfrac{Mc}{I}$

Total

FIGURE 12.6
Eccentric loading of a column.

neutral axis ($\sigma = 0$) shifts toward the outside of the bend, as shown in the figure, and the largest stress, which is compressive, is at the inside. It is the sum of the bending and direct compressive stresses. This maximum stress can be expressed by what has become known as the *secant formula*:

$$\sigma = \frac{P}{A}\left(1 + \frac{ec}{\rho^2}\sec\frac{L_{\text{eff}}}{2\rho}\sqrt{\frac{P}{AE}}\right)$$

Here c is the distance from the neutral bending axis to the outer fibers, just as in beam bending; ρ, as before, is the radius of gyration about the bending axis. The quantity ec/ρ^2 is called the *eccentricity ratio*.

If the stress reaches the compression yield, the column fails. Of course, the failure is initially only local at the inside (right side in the figure). The deflection then increases, so the moment rises and the stress with it, which causes the failure region to grow inward, and so on. The column fails cataclysmically, due to excessive stress. Euler buckling, recall, was due to elastic instability and was not related to stress or strength.

If we set the stress in the secant formula equal to the compressive yield strength (which, since the material is ductile, is the same as the tension yield strength S_y), we have

$$S_y = \frac{P}{A}\left(1 + \frac{ec}{\rho^2}\sec\frac{L_{\text{eff}}}{2\rho}\sqrt{\frac{P}{AE}}\right) \tag{12.7}$$

If we wish to see how P/A varies with L_{eff}/ρ with a plot, we cannot solve Equation (12.7) explicitly for P/A in terms of L_{eff}/ρ. The reverse is, however, feasible:

$$\frac{L_{eff}}{\rho} = 2\sqrt{\frac{E}{P/A}} \arccos \frac{ec/\rho^2}{S_y/(P/A) - 1} \tag{12.8}$$

A plot of P/A vs. L_{eff}/ρ can be made with this expression. An example is shown in Figure 12.7 for a mild steel.

Both (12.7) and (12.8) are somewhat awkward for column design, because usually it is the cross section we need to find, and that is involved in area A as well as the radius of gyration ρ. Of course, a few guesses should bring us close with either expression or with a plot like Figure 12.7 for the particular material.

Recall that the tangent point separating the Euler and Johnson solutions is at $S_y/2$. For this particular steel, then, the chart indicates that for values of L_{eff}/ρ above about 90 (corresponding to a unit-area load of $S_y/2$, where the Johnson and Euler curves come tangent) and where there is no eccentricity, the critical unit-area load is determined by elastic instability, the Euler relation. If $L_{eff}/\rho < 90$ with no eccentricity, or if significant eccentricity is present at any slenderness ratio, buckling will be initiated by local yielding. The amount of eccentricity e to be used in the calculation is a decision based on the designer's knowledge of the precision expected in manufacture and assembly. Frequently a small eccentricity is assumed to be inevitable, and

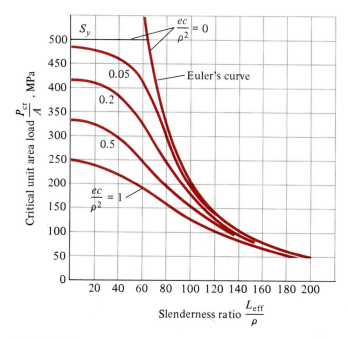

FIGURE 12.7
Secant formula for material with $E = 207$ GPa and $S_y = 500$ MPa (a mild steel).

the design is based on it. Some years ago a study of what one might term *unwanted* or *unavoidable* eccentricities was made, and a figure of 0.025 for the eccentricity ratio ec/ρ^2 was settled on. Of course, any off-centeredness deliberately incorporated to satisfy some other design need is treated with the numbers in question.

Expressions (12.7) and (12.8) describe failure in the plane of the bending moment, and e, c, and ρ pertain to that plane. The other plane, which may have a different effective column length L_{eff} and radius of gyration ρ, must also be checked, for the column may buckle by elastic instability in that plane at a lower load.

Example 12.3. For the columns of Examples 12.1 and 12.2, let us assume the inevitable unwanted eccentricity, $ec/\rho^2 = 0.025$, and calculate the column sizes with the secant relation. Neither form of the secant relation, (12.7) or (12.8), will give us what we want directly, and there seems little advantage in one over the other. We use (12.8).

For the first column ($P_{cr} = 3E + 05$ N, $L_{eff} = 1.95$ m), putting in the numbers, we have

$$\frac{3.9\sqrt{3}}{w} = 2w\sqrt{\frac{20E + 09}{3E + 05}} \arccos \frac{0.025}{(700E + 06)w^2/(3E + 05) - 1}$$

which simplifies to

$$4.106E - 03 = w^2 \arccos \frac{75}{(7E + 06)w^2 - 3000}$$

This expression can be handled with a hand calculator. One way is to try various w values until the sides balance. It works, but recursion is more sophisticated. We solve the expression for one of the values of w^2, of which the first one on the right of the equals sign is easier:

$$w^2 = \frac{4.106E - 03}{\arccos 75/[(7E + 06)w^2 - 3000]}$$

We now guess a value of w^2 and calculate the right side. This gives us a new trial value, and we go through the calculation again, etc., until the required precision is achieved. Let us guess $w = 0.070$, or $w^2 = 4.9E - 03$. We get in succession

$$w^2 = 4.9E - 03 \qquad \text{(initial guess)}$$

$$= 2.618E - 03$$

$$= 2.622E - 03$$

So $w = 0.0512$, that is, 51.2 mm (we would probably recommend 51.5 mm). This result is close to the Euler value of 51.1 obtained in Example 12.1. The reason is that the slenderness ratio $L/\rho = 132$ is well into the Euler range. (Do not forget $L_{eff}/\rho = 3.9\sqrt{3}/w$ in this problem—see Example 12.1.)

For the second column of the example, $P_{cr} = 20E + 05$ N and $L_{eff} = 0.325$ m. Using the same form of the secant formula (12.8), we get

$$1.767E - 03 = w^2 \arccos \frac{5}{(7E + 04)w^2 - 200}$$

which after several iterations results in $w = 33.1$ mm.

With the small eccentricity ratio we have used here and the size of the cross section that results (w) from the specified load and column length, we are very close to the corner where the material failure level ($P/A = S_y$) and the Euler curve intersect in Figure 12.5. The value of L_{eff}/ρ at that corner for our numbers is 33.1, and that from the secant formula results is

$$\frac{L_{eff}}{\rho} = \frac{0.65\sqrt{3}}{0.0331} = 34$$

The Johnson formula rounds out the corner with a parabola to fit empirical results, as we discussed earlier. Using it, we found $w = 56.9$ mm. This would be a better number to use here.

To summarize, if the slenderness ratio resulting from a secant formula solution places the column in the Johnson region, you should compute the column size on the basis of Johnson's parabola and then use the larger result.

The recursion routine we used to solve these problems converged very rapidly. That will not always be the case. Also, the system can be frustrated if you guess or land on a number such that the argument of the arccosine is greater than 1 (or less than −1). In the second part of the example, a w value of 54 mm produces arccos 1.21, which cannot be done. (In fact, a value of w anywhere between 52.78 and 54.12 mm will not work.) When this happens, change your guess.

12.3 A THREE-BAR TRUSS

In Figure 12.8 we show three axially loaded bars pinned together to form a rigid frame. A force F is applied parallel to the wall. To keep things simple, bars _1_ and _2_ are assumed to be round in cross section, though not necessarily of the same size. The assignment is to determine the angle θ such that the total material mass of the bars _1_ and _2_ is minimized. There may be limits on θ.

The criterion equation is the total volume of material:

$$V = h\left(A_1 + \frac{A_2}{\sin\theta}\right) \tag{12.9}$$

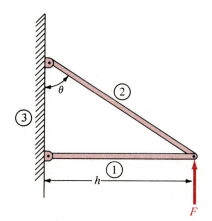

FIGURE 12.8
Three-bar truss.

The forces in the bars are obtained from the equations of static equilibrium:

$$F_1 = F \tan \theta \tag{12.10}$$

$$F_2 = \frac{F}{\cos \theta} \tag{12.11}$$

Bar 1 is loaded in tension. The stress is $\sigma_1 = F_1/A_1$, or, with (12.10),

$$A_1 = \frac{F \tan \theta}{\sigma_1} \tag{12.12}$$

The limit on σ_1 is

$$\sigma_1 \leq \frac{S_y}{FS_S} \tag{12.13}$$

Bar 2 is loaded as a column. Assuming that Euler buckling is the controlling mechanism, we get

$$(FS_F)F_2 \leq \frac{\pi^2 E I_2}{L_2^2} \tag{12.14}$$

For a round bar $I = A^2/(4\pi)$. Substituting this and (12.11) in (12.14), we obtain

$$(FS_F)F \leq \frac{\pi E A_2^2 \cos \theta \sin^2 \theta}{4 h^2}$$

or
$$A_2 \geq \frac{2h}{\sin \theta} \left[\frac{(FS_F)F}{\pi E \cos \theta} \right]^{1/2} \tag{12.15}$$

Substituting (12.12) first in the criterion relation (12.9), we get

$$V = h \left(\frac{F \tan \theta}{\sigma_1} + \frac{A_2}{\sin \theta} \right)$$

This clearly will be a minimum, no matter what the values of A_2 and θ, if σ_1 is placed at its limit, given by (12.13):

$$V \geq h \left[\frac{F(FS_S) \tan \theta}{S_y} + \frac{A_2}{\sin \theta} \right] \tag{12.16}$$

Relation (12.16) will be further minimized if A_2 has its smallest value, given by (12.15) (the bars are going to fail simultaneously: bar 1 by yielding, bar 2 by buckling):

$$V \geq h \left[\frac{F(FS_S) \tan \theta}{S_y} + \frac{2h}{\sin^2 \theta} \sqrt{\frac{(FS_F)F}{\pi E \cos \theta}} \right] \tag{12.17}$$

It remains now to determine the angle θ which will make (12.17) a minimum. This one-dimensionality makes the problem a good illustration of a mathematical search. The algorithm is best understood by a sketch; Figure 12.9 is an illustrative variation of volume with angle. We start by dividing the region of permissible angles

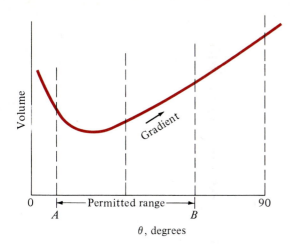

FIGURE 12.9
Volume as a function of angle θ.

in half. The volume is calculated. Then the angle is increased a small amount and the volume recalculated. In the situation of Figure 12.9, the volume increased, so we want to go in the other direction. Another way to say this is that the gradient or slope was positive. We are looking for a minimum, and so we proceed in the direction of the negative gradient. The space to the left is bisected and the gradient reevaluated. The process is repeated, always in the direction of the negative gradient, until the minimum is reached with acceptable precision. The flowchart is shown in Figure 12.10. We have used an increment for θ of 1 percent, and the gradient is checked each time before the interval is halved again.

Example 12.4. Let us suppose the truss supports a force $F = 8000$ lb, and $h = 27$ in. The round bars are made of steel with yield stress of 44 ksi and modulus of elasticity $E = 30E + 06$ psi. The factors of safety are $FS_S = FS_F = 1.5$. Relation (12.17) becomes

$$V \geq 7.36364 \tan \theta + \frac{16.4518}{(\sin^2 \theta)\sqrt{\cos \theta}}$$

The permissible range of angle is 0 to 90°. The output below shows that a precision of 0.1 percent was reached with six iterations.

θ, deg	45.00	67.5	56.25	50.63	53.44	54.84
V, in^3	46.49	48.93	42.95	43.54	42.97	42.89

The member dimensions are then

$L_1 = 27$ (value of h)

$A_1 = 0.387$ [(12.12) and 12.13)]   so $d_1 = 0.702$

$L_2 = 33.03(h/\sin \theta)$

$A_2 = 0.982$ from (12.15)   so $d_2 = 1.118$

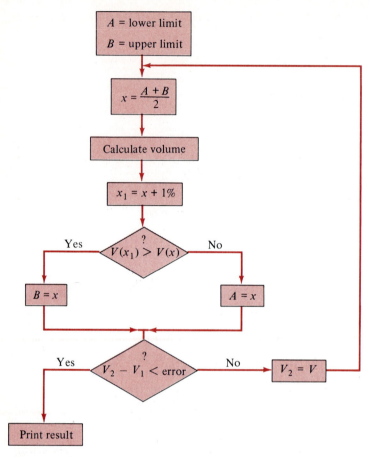

FIGURE 12.10
Algorithm for three-bar truss problem.

Our assumption that member 2 acts as an Euler column needs to be checked now. For a round section, the radius of gyration $\rho = d/4 = 1.118/4 = 0.280$. Since the column is pinned at the ends, $L_{\text{eff}} = L = 33.03$, and $L_{\text{eff}}/\rho = 118$. According to relation (12.6), this must be greater than $\pi\sqrt{2E/S_y}$ for an Euler solution. The latter quantity is 116, and we just make it.

In the event the compression bar (bar 2) of the truss requires the Johnson solution, the value of A_2 corresponding to (12.15) must be found on that basis [expression (12.15)] and used in (12.16).

As we noted earlier, the above development was based on failure of one bar of the truss by yielding *and* of the other by buckling. This was possible because the bars were permitted to have different diameters. If the bars are the same size, then failure will occur through yielding of bar 1 *or* buckling of bar 2. Two expressions for the volume result, the larger of which governs the design (Problem 12.19). (A lower limit of $\theta = 0$ in this problem leads to the undefined quantity $V = 0 \times \infty$. Such a lower limit is, in any event impractical.)

12.4   THE BEAM COLUMN

A beam under the combination of axial and lateral loads, as shown in Figure 12.11a, will exhibit greater deflections and be subjected to greater bending moment than if either load were acting alone. An expression for the resulting deflection can be obtained by solving the differential equation for the loaded beam. For a prismatic (constant-cross-section) beam under a single central lateral load Q and an axial load P, the resulting expression for maximum bending moment [11] is:

$$M_{\max} = \frac{QL}{4} \frac{\tan u}{u}$$

This is recognized as the bending moment $QL/4$ for a simple beam multiplied by a "magnification factor," $(\tan u)/u$, in which

$$u = \frac{\pi}{2} \sqrt{\frac{P}{P_{\mathrm{cr}}}}$$

The maximum deflection is

$$\delta_{\max} = \frac{QL^3}{48EI} \frac{3(\tan u - u)}{u^3}$$

Again, this is the formula for the deflection of a simple beam under lateral load Q, amplified by a function of P/P_{cr}. Both bending moment and deflection increase toward impossible values as u approaches $\pi/2$, that is, as P approaches the critical buckling load P_{cr}.

For a constant axial load P, deflections and bending moments (and therefore stresses) are proportional to the lateral loads. When several lateral loads (concentrated or distributed) are present, the principle of superposition can therefore be used, provided the axial load is included in each system. This is shown in Figure 12.11b.

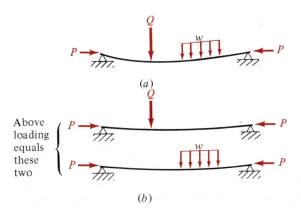

(a)

Above loading equals these two

(b)

FIGURE 12.11
Beam-column analysis.

The combined effects of more than one axial loading system cannot be treated by superposition, however, because the bending moments are not proportional to those loads. Such a problem might be a static plus alternating load P.

12.5 AN OPTIMUM COLUMN— MINIMUM MASS

In the derivation of buckling loads we assumed that the column had a constant cross section, i.e., it was prismatic. For a long column, that results in a lot of wasted material. That is, we would expect a properly tapered column to support as great an axial load as a prismatic one. Indeed, we see many examples of such columns.

The design of a column of minimum mass is not straightforward. We have to design against elastic instability for a column of varying cross section, and, assuming that the column deflects, we will want to avoid failure by excess stress all along the column. This involves us in calculating the moment at points along the length, and those depend on the deflections at each point and at the top of the column. But the deflections are, in turn, a function of the moments. It looks like an iterative solution.

Fortunately, the critical elastic-instability buckling load for a column can be estimated quite accurately by applying a work-energy relation to the deflection curve. If the true deflection shape is available, the calculation is exact. If the true deflection shape is not known (as in the case to follow), the buckling load can still be estimated quite accurately by using an assumed deflection curve.

Picture a column which supports a load P. Now suppose the column is bent into some "reasonable" shape. It shortens an amount λ, and the load P does work $W = P\lambda$. The curvature of the column results in bending moments, and energy is stored in the column equal to

$$U = \int \frac{M^2}{2EI} \, dx \tag{12.18}$$

If the work of P is less than the resulting stored energy, P is smaller than the critical load. If we find the size of P so that its work equals the stored energy (set $W = U$), we have the critical load P_{cr}.

What do we do for a "reasonable" deflection curve? Well, as an example, let the column be built in and the shape be that which results from the application of a lateral load Q at the top, as in Figure 12.12. The column is simulated by a cantilever beam. All we want is the deflection curve; we only imagine the load Q in order to get the shape. The boundary conditions of zero slope at the base and zero curvature ($M = 0$) at the top are met. From Appendix F the shape is

$$y = \frac{\delta x^2}{2L^3} (3L - x)$$

where δ is the maximum deflection, which occurs at the top. Its actual value in terms of Q does not concern us. This is the shape we assume for the column when it deflects under the real load P. (You say, "Why not use a sine curve, since that is the shape found at the beginning of the chapter for buckled columns?" Well, we could, but the

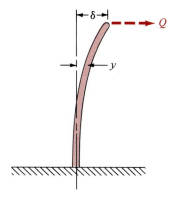

FIGURE 12.12
The deflection curve used will be that resulting from a fictional load Q.

final shape of this column will not be a sine curve, since the moment of inertia I will not be constant along the column.)

The energy stored in the column (called the *strain energy*), as given by expression (12.18), is

$$U = \int dU = \frac{1}{2EI} \int M^2 \, dx$$

Here $M = P(\delta - y)$, which, with the assumed shape y, is

$$M = P\left[\delta - \frac{\delta x^2}{2L^3}(3L - x)\right] = \frac{P\delta}{2L^3}(2L^3 - 3Lx^2 + x^3)$$

Thus, using (12.18), we get

$$U = \frac{P^2\delta^2}{8EIL^6} \int_0^L (2L^3 - 3Lx^2 + x^3)^2 \, dx = \frac{17P^2\delta^2 L}{70 \, EI}$$

To find the work done by P, we need the vertical distance it moves λ. If we look at any small length dx of the column, as shown in Figure 12.13, length a is

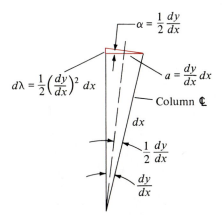

FIGURE 12.13
Computation of the vertical distance moved by load P.

found from the slope and the length dx. Then since angle α is $\frac{1}{2}dy/dx$, the side opposite a is

$$d\lambda = \frac{1}{2}\left(\frac{dy}{dx}\right)^2 dx$$

The work done by P is

$$P\lambda = \frac{P}{2}\int\left(\frac{dy}{dx}\right)^2 dx = \frac{3\delta^2 P}{5L} \tag{12.19}$$

We set $P\lambda = U$ to get

$$P = P_{cr} = \frac{42EI}{17L^2}$$

which is 0.13 percent high. The critical load predicted by the energy method will always be greater than the correct value when a fictitious deflection shape is used, because the column has been "constrained" into an "unnatural" shape.

We go back now to the minimum-mass column. Let the column have a fixed base and unconstrained top (Figure 12.2e), and it is to sustain an axial load of 40 klb, including the factor of safety. The length is to be 80 in, and the cross section will be rectangular with a constant thickness of 2 in. The material is steel with $E = 30E + 06$, and the width must be varied to keep the stress below a design figure of 24 ksi at all points.

The radius of gyration ρ of a rectangular cross section about the axis in the width direction is a function of only the thickness, as shown in Figure 12.14. The thickness is $t = 2$, so ρ works out to have the constant value $\rho = 1/\sqrt{3}$.

We assume that we have confidence in the rigidity of the foundation and use the theoretical effective length, which is twice the actual, so

$$\frac{L_{eff}}{\rho} = 160\sqrt{3} = 277$$

$$\rho = \sqrt{\frac{I}{A}} = \sqrt{\frac{\frac{1}{12}wt^3}{wt}} = \frac{t}{\sqrt{12}} = \frac{2}{\sqrt{12}} = \frac{1}{\sqrt{3}}$$

$$(w = \text{width}, \ t = \text{thickness})$$

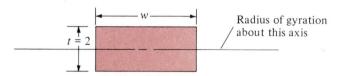

FIGURE 12.14
Finding the radius of gyration about the w axis.

The break point between Euler and Johnson columns is 157 with these material properties, so we are in the Euler category, with a required base width given by Equation (12.3) of

$$w = \frac{277^2(4E + 04)}{\pi^2(30E + 06)} = 5.19 \text{ in}$$

for elastic stability. This is for a prismatic column and gives us a starting point for the following process. Also we can calculate the prismatic column volume with which we will compare the results to come:

$$\text{Vol}_{\text{pris}} = w(2.0)(80) = 829.3 \text{ in}^3$$

We now compute a lateral deflection at the top of the column sufficient to cause the maximum stress at the base (P/A stress + bending stress) to be equal to the design stress. We decide on a number of segments to break the column into. Then, starting at the base and working our way up, we calculate the width at the top of each segment so as to keep the stress constant at the design level. Since the bending stress decreases as we approach the top, the size of the cross section will also decrease. During the process, we have to find the deflection at each stage (to get the moment). The deflections constitute our assumed deflection curve.

With a deflection shape to work with, we can calculate the strain energy in the column and the axial shortening due to the bending. The critical load can then be computed as described earlier in this section. This will turn out to be different from the desired critical load. We then adjust the deflection curve by multiplying all deflections by a constant $R = P_{\text{desired}}/P_{\text{computed}}$, compute new column widths on the basis of strength, and find a new critical load. We repeat the process until the load is as close as we require, and the column is designed. This column will fail simultaneously by elastic instability and by excessive stress.

Let us run through the process for five segments, a rather rough shot.

$$w_{\text{base}} = w_1 = 5.19 \text{ in} \qquad \text{(found above)}$$

$$\Delta L = \frac{80}{5} = 16 \text{ in} \qquad \text{(segment length)}$$

$$\sigma_{\text{design}} = \pm 24E + 03 \qquad \text{(given)}$$

At the base,
$$\sigma_{\text{design}} = \frac{-P}{A} - \sigma_{\text{base,bend}}$$

We can neglect the negative signs, since the critical stresses are compressive in this problem, and σ_{design} will also be negative. So

$$\sigma_{\text{base,bend}} = 24E + 03 - \frac{40E + 03}{5.19^2} = 20.15E + 03$$

Now
$$M_{\text{base}} = y_{\text{top}}(40E + 03)$$

and
$$\sigma_{\text{base,bend}} = \frac{Mc}{I} = \frac{3M}{2w} \qquad \left(\frac{I}{c} = \frac{2w}{3}\right)$$

giving $y_{top} = 1.74$

$$y''_{base} = y''_1 = \frac{M}{EI} = \frac{\sigma_{bend}}{cE} = 6.72E - 04$$

$$y'_1 = y''_{base}\frac{\Delta L}{2} = 5.37E - 03 \qquad \text{(average in 1st segment)}$$

$$y_2 = y_1 + y'_1(\Delta L)$$
$$= 0 + (5.38E - 03)(16) = 0.086 \qquad \text{(bottom of 2d segment)}$$

$$M_2 = P(y_{top} - y_2) = 66.16E + 03$$

$$\sigma_{2,bend} = \frac{M_2 c}{I} = \sigma_{design} - \frac{P}{A_2}$$

so that $w_2 = 4.97$

$$y''_2 = \frac{\sigma_{2,bend}}{E} = 6.66E - 04$$

$$y'_2 = y'_1 + y''_2(\Delta L) = 1.60E - 02$$

$$y_3 = y_2 + y'_2(\Delta L) = 0.342$$

and similarly on up, ending with a top deflection $y_6 = 2.025$. This figure differs from the y_{top} we used in the calculation (1.74, from the prismatic Euler load). Given the availability of the computer, we can refine the shape of the column by repeating the process, using this updated value for y_{top} and continuing until the two numbers converge. The results for this five-segment approximation are shown in Table 12.1.

The buckling load is now computed by the energy method outlined earlier in the section. The strain energy per expression (12.18) is

$$U = \sum M_n^2 \frac{\Delta L}{2EI_n} = P^2 \sum \left[(y_{top} - y_n)^2 \frac{\Delta L}{2EI_n} \right]$$

TABLE 12.1
Shape of five-segment column

Position	First pass		Refined, ready for computation of P_{cr} by energy method	
	Width w	Deflection y	Width w	Deflection y
Base	5.19	0	6.17	0
16	4.97	0.086	5.95	0.089
32	4.33	0.342	5.29	0.353
48	3.27	0.763	4.19	0.790
64	1.84	1.338	2.69	1.392
Top	1.55	2.025	0.84	2.134

TABLE 12.2
**First pass on critical load calculation,
five-segment column**

Segment	$U(E + 06)$	Work$(E + 03)$
1	$0.288P^2$	$0.245P$
2	$0.261P^2$	$2.189P$
3	$0.206P^2$	$5.973P$
4	$0.127P^2$	$11.298P$
5	$0.031P^2$	$17.234P$
Total	$U = (0.913E - 06)P^2$	$W = (36.938E - 03)P$

The work done by the force is, from (12.19),

$$W = \frac{P}{2} \left[\sum (y_n')^2 \right] \Delta L$$

In these expressions, M (hence y) and y' would be the averages in each segment. The strain energy U and the work W are shown in Table 12.2. Dividing one sum by the other yields the buckling load $P = 40\,459$ lb. The desired buckling load, however, is 40 000 lb. To get this, we multiply the deflections and slopes by a factor

$$R = \frac{P_{\text{desired}}}{P_{\text{calc}}} = \frac{40\,000}{40\,459}$$

(The shape thus remains the same.) The widths are now recomputed to keep the stress at the design value, and we find a new buckling load. The process is continued until the result is within the error we may set. For this five-segment approximation, the progression goes as shown in Table 12.3.

We had set a computational accuracy of 1 part in 10 000, which here means 4 lb in 40 000. Trial 4 hit within this number, with a column volume resulting of 684, a considerable improvement over the volume of the prismatic column (829). The five-segment grid is, however, somewhat coarse; with 300 segments the volume is 692.5, representing a savings in material of $16\frac{1}{2}$ percent. Computation time is a few seconds. The column shape is shown in Figure 12.15.

The procedure given here is an example of a method of optimization which uses a computer to "whittle" at a structural shape until an optimal shape is obtained.

TABLE 12.3
Progression of calculations of P, five-segment column

Trial	P, lb	Column volume (prismatic = 829.3), in^3
1	40 459	692
2	40 081	685
3	40 014	684.3
4	40 003	684.1

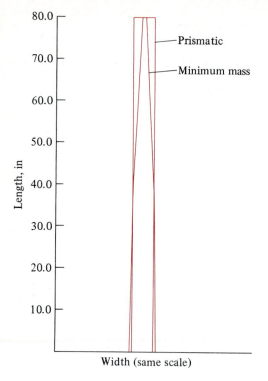

FIGURE 12.15
Shape of minimum-mass column.

Width (same scale)

Its use is not confined to simple columns; indeed, explicit methods have been developed to optimize such columns. See, for example, Haug and Arora [12].

12.6   CLOSURE

In this chapter we have dealt with the buckling of columns. Other buckling topics include buckling of axial members by torsion, the beam column, torque tubes, plates, arches, frames, etc. These problems are treated in specialized texts, which you should consult for guidance in their solution.

PROBLEMS

Section 12.1

12.1. Write a computer program for calculating column buckling loads, given all the necessary input.

12.2. Write a computer program for calculating the necessary design parameters to avoid buckling, given a load, column length, and material data.

12.3. Compute the buckling loads for the columns shown in Table P12.3. Use Appendix F for material properties. Material A is 1040 CR steel, material B is 4130 steel drawn at 400°F, and material C is 2011 tempered T8 aluminum.

Answer: (*a*) 2.37 MN,   (*c*) 9.92 kN,   (*d*) 345 klb,   (*f*) 87 klb

TABLE P12.3

Problem part	Column type in Fig. 12.2	L	Material	Section
a	b*	2.5 m	A	Solid round, dia. = 100 mm
b	b*	1 m	B	Solid round, dia. = 20 mm
c	d*	5 m	C	Tube, OD = 60 mm, ID = 40 mm
d	c*	50 in	B	Solid square, side = 2 in
e	e*	30 in	A	4 × 6 in rectangle
f	e*	40 in	B	Tube, OD = 3 in, ID = 2.5 in

* Use AISC recommended effective lengths.

12.4. Compute the length or section size as indicated for the columns in Table P12.4 so as to support a load of $P = 40$ kN with an FS of 2.5. Materials are as in Problem 12.3.
    Answer: (a) 40 mm;   (c) 13.3 × 26.6 mm;   (e) 183 mm

TABLE P12.4

Problem part	Column type in Fig. 12.3	L	Material	Section
a	b	2 m*	A	? Solid round
b	d	4 m*	C	? Solid square
c	c	½ m*	B	? Rectangle, width = 2 (thickness)
d	e	?*	C	Solid round, dia. = 40 mm
e	a	?*	B	10 × 20 mm rectangle
f	b	?*	A	10-mm square

* Use AISC recommended effective lengths.

12.5. We have some angle-section steel available ($E = 30E + 06$, $S_y = 80E + 03$ in USCS units) with dimensions and properties as shown in Table P12.5. A platform is to be built on four columns, which will be embedded in concrete on both ends. The platform will be 12 ft above ground and must support 3 tons (including its own weight and an FS). Choose the lightest angle for this purpose. There are two logical configurations for the columns, as shown in Figure P12.5. In the arrangement at the right, buckling will occur in the weak plane, shown dashed, for all columns, the mode being that of Figure 12.2d (fixed base and nonrotating top, but with a possibility of sideward motion). On the left, the situation is more complex. One mode of failure is that of all four columns buckling about the weak plane. That requires the top to remain stationary (Figure 12.2c). The other possibility is motion of the top at 45° to its axes, type d of Figure 12.2. This implies buckling about the weak plane for two of the legs (upper L, lower R, for example) and about the strong plane for the other two. How to find the strong I value is outlined in Problem 12.6.

TABLE P12.5

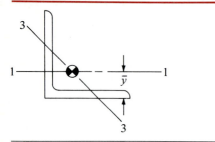

Size	A	$I_{1\text{-}1}$	$\rho_{1\text{-}1}$	$\bar{y}$	$I_{3\text{-}3}$	$\rho_{3\text{-}3}$
$1 \times 1 \times \frac{1}{8}$	0.23	0.02	0.30	0.30	0.008	0.19
$2 \times 2 \times \frac{1}{8}$	0.49	0.18	0.61	0.53	0.08	0.40
$3 \times 3 \times \frac{1}{4}$	1.43	1.18	0.91	0.82	0.49	0.58
$3 \times 3 \times \frac{3}{8}$	2.10	1.70	0.90	0.87	0.70	0.58
$4 \times 4 \times \frac{1}{2}$	3.75	5.46	1.21	1.17	2.26	0.78
$6 \times 6 \times \frac{1}{2}$	5.74	19.38	1.84	1.66	7.92	1.17

Note: $I_{3\text{-}3}$ is the smaller principal moment of inertia through the center of gravity, which means that 3-3 is the weakest axis in bending. Because of the symmetry 3-3 is at 45° to 1-1.

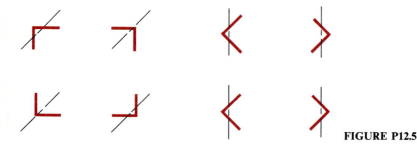

FIGURE P12.5

12.6. The platform of Problem 12.5 is to be built with two, rather than four, 12-ft lengths of angles. Choose the lightest angle for the job. Indicate the optimal orientation of the angles.

It will be necessary to find I in the strong plane for the angles. This involves finding an I on an axis inclined from one where it is known. The expressions turn out to be exactly the same as for stresses on rotated elements, and Mohr's circle can be used. In Table P12.5, I_3 is a principal moment of inertia, being the smallest; I_1 is for axes at 45°, and the I_{xy} on those axes would be a maximum (as with shear stress at 45° from the principal axes). The large principal I, which is associated with the axis at 90° to 3-3, that is, at the other end of the Mohr's-circle diameter, is found from the diagram in Figure P12.6.

Answer: $3 \times 3 \times \frac{1}{4}$ angle required

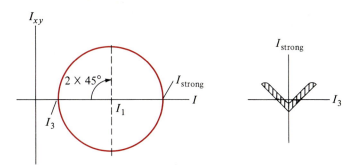

FIGURE P12.6

12.7. Work Problem 12.6, assuming that guy wires are run from the platform to the ground in the weak direction to prevent lateral motion.

12.8. We have a column made by welding four 20-mm steel rods together as shown in Figure P12.8. The ends correspond to Figure 12.2b. $E = 200E + 09$ Pa, and $S_y = 600E + 06$ Pa. Find the load supported by a length of
(a) 0.5 m   (b) 5 m   (c) 10 m
       Answer: (b) 11.6 kN

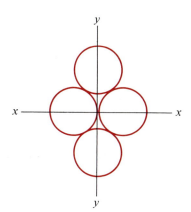

FIGURE P12.8

12.9. A column is made up per the sketch in Figure P12.8. The ends are fastened with pins which lie in the xx plane. They are very rigid, so that in the xx plane the column is restrained as in Figure 12.2c. Find the plane of buckling and the load we could place on a length of 3 m.

Section 12.2

12.10. Repeat Problem 12.1 for eccentric loading.

12.11. Do Problem 12.2 for eccentric loading.

12.12. Prove the statement in parentheses at the end of the last paragraph of Section 12.2.

12.13. Find the failure load for a steel column 50 in long made of the 2×2 angle of Table P12.5, with end conditions corresponding to Figure 12.2c. The load is applied as shown in Figure P12.13, and $S_y = 80$ ksi.

      Answer: P = 10 870 lb

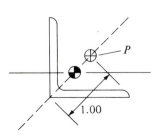

1.00

⬤ = centroid   **FIGURE P12.13**

12.14. Two 20-ft lengths of the 4×4 angles in Table P12.5 are welded together to form a column as shown in Figure P12.14. Determine the maximum load with an FS of 3 which can safely be applied. The ends are as in Figure 12.2e, and $S_y = 80$ ksi.

12.15. Select the lightest pair of angles from Table P12.5, arranged as shown in Figure P12.14, to support a load of 15 000 lb, assuming an inevitable unwanted eccentricity. The end conditions are per Figure 12.2c; length = 12 ft, and $S_y = 80$ ksi.

      Answer: $3 \times 3 \times \frac{1}{4}$

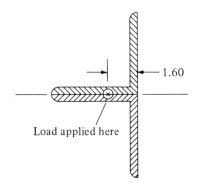

1.60

Load applied here

            FIGURE P12.14

Section 12.3

12.16. Rework Example 12.4, but use $h = 20$ in.

12.17. Assuming the bars are square in cross section, develop the equivalent of relation (12.17) and design the truss with the data of Example 12.4. Length $h = 23$.

      Answer: $t_1 = 0.129$, $L_2 = 28.88$, $t_2 = 0.907$

12.18. Repeat Problem 12.17, but $h = 30$ in.

12.19. Redo the development of Section 12.3 on the basis of bars of equal diameters. Find the optimum truss size for a load of 5000 lb, $h = 27$, $20° \leq \theta \leq 70°$, $FS_F = FS_S = 1.5$, $E = 30E + 06$, and $S_y = 40$ ksi.

Answer: $d = 0.994$, $L_2 = 33.21$

Section 12.5

12.20. Redesign the example column of Section 12.5 on the basis of
(a) A square cross section
(b) A square hollow cross section with wall thickness 15 percent of the side
(c) A round cross section
(d) A round hollow cross section with wall thickness $= \frac{3}{8}$ in

These problems lead to the necessity of solving a cubic equation to get the dimensions at each section. Newton's method or the exact solution can be used. The problem is that the exact solution in many handbooks is wrong. The correct one is (15.25) in Chapter 15. It may be programmed for computer solution as a subroutine of a dozen lines.

During the process of refining the shape of the column prior to the critical-load computation, it can happen that the deflection(s) (start to) exceed y_{top}, causing the moment to change sign. Since the moment determines the second derivative of the curve $[y'' = M/(EI)]$, the shape becomes unnatural—the column starts to curve back in the other direction. One way to handle this difficulty is to test the moment for sign and set it to zero if it is negative. This simply extends the deflection curve to the top as a straight line ($M = 0$ implies $y'' = 0$, which implies $y' =$ constant), so a new y_{top} is thus found, and the procedure is repeated.

12.21. Minimize the mass of a column, using the specifications provided by your instructor. It will be a fixed-bottom, free-top column as in the text example. However, assume a parabolic curve for the deflected shape. Start your algorithm by calculating the cross section needed to keep the compressive stress at the top within specifications. Assume a lateral deflection at the top. Use the cross section at the top as the first guess at the next station down, keeping the total stress constant. Use recursion to refine your value for the width. Sum the strain energy/P, and when you arrive at the bottom, calculate the critical load by the energy method. Devise an estimator to adjust the assumed top deflection, and repeat until the calculated critical load is close enough to specifications.

CHAPTER
13

DESIGN OF MEMBERS IN TWISTING

This chapter concerns machine members subjected to twisting by torsional loads. First, we discuss straight members of circular section. Some well-known examples are automobile drive shafts, torsion-bar suspension springs, and hatchback-door counterbalance springs. The latter have the attributes of being compact, economical, and very efficient—almost all stored energy is recovered. Later in the chapter we treat other shapes, including hollow members, whose cross sections may or may not be closed.

Shafts loaded in torsion and bending, which is very common, are discussed in Chapter 15.

The requirements of real machines lead to considerable complexity both in the shape and in the loading of torsion members, and their optimization can become correspondingly complex. Our objective here is to present methods which, while applied to simple cases in this chapter, will be useful in handling the complicated situations which can arise in practice.

13.1 BACKGROUND ANALYSIS

Stresses

Figure 13.1 shows a round bar under torsion. In Section 6.2, the stresses in such a bar were discussed. The stress (pure shear on the element shown in the figure) is maximum at the surface:

$$\tau_{surf} = \frac{Td}{2J}$$

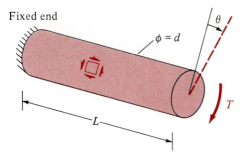

Fixed end

$\phi = d$

θ

L

T

Figure 13.1
Prismatic bar under torsional load.

This relation assumes elastic action of the material; T is the torque on the bar, d its diameter, and J the area polar moment of inertia. For a solid round bar, $J = \pi d^4/32$, and the above expression becomes

$$\tau_{\text{surf}} = \frac{16T}{\pi d^3} \tag{13.1}$$

If the bar is hollow, J will be equal to that of a solid round bar with a bar the size of the hole deducted:

$$J = \frac{\pi}{32}(d_o^4 - d_i^4) \tag{13.2}$$

where subscripts o and i stand for "outside" and "inside." If the wall of the tube is thin, this can be approximated by

$$J = \frac{\pi t d_m^3}{4} \tag{13.2'}$$

where t is the wall thickness and d_m the mean diameter. If the wall is 5 percent of the diameter, the approximate formula results in an error of only 0.2 percent.

Strength

We saw in Chapter 9 (Figure 9.16) that, in pure torsion, a mean torsional stress has almost no effect on fatigue life. And we wrote the following design criteria, both of which must be satisfied:

For fatigue safety:

$$\tau_a \leq \frac{S_{f,\text{sh}}}{\text{FS}} \tag{9.14a}$$

For static safety:

$$\tau_{\text{max}} = \tau_a + \tau_m \leq \frac{S_{sy}}{\text{FS}} \tag{9.14b}$$

In the above relations $S_{f,\text{sh}}$ is the shear fatigue strength and S_{sy} the shear yield strength. Expressions (9.14a) and (9.14b) apply to the two segments of the curve of Figure 9.16. Substituting them in the shear-stress equation (13.1), we get the following:

For fatigue safety:

$$\tau_a = \frac{16T_a}{\pi d^3} \le \frac{S_{f,\text{sh}}}{FS} \qquad \tau_m < \frac{S_{sy} - S_{f,\text{sh}}}{FS}$$

For static safety:

$$\tau_{\max} = \frac{16T_{\max}}{\pi d^3} \le \frac{S_{sy}}{FS} \qquad \tau_m \ge \frac{S_{sy} - S_{f,\text{sh}}}{FS}$$

Deflection and Stiffness

When a round shaft is subjected to a torsional load, a line originally parallel to the centerline will be rotated by an angle equal to the shear strain $\gamma = \tau/G$, as shown in Figure 13.2. (Recall that G is the shear equivalent of E: $\sigma = E\varepsilon$, $\tau = G\gamma$.) In a section of length dx, the cross section is rotated

$$d\alpha = \frac{\gamma\,dx}{d/2} = \frac{2\tau\,dx}{Gd}$$

Over the length L of the bar, the total twist is obtained by integrating this and inserting the expression for the shear stress τ for a round bar:

$$\theta = \int_{x=0}^{L} \frac{2\tau\,dx}{Gd} = \frac{TL}{GJ} = \frac{32TL}{\pi d^4 G} \tag{13.3}$$

The modulus of elasticity in shear G can be expressed in terms of Young's modulus for normal strain E:

$$G = \frac{E}{2(1+v)} \qquad v = \text{Poisson's ratio} = 0.3 \text{ for steel, aluminum, brass}$$

This relation assumes that the material is isotropic and elastic, i.e., that strains in all directions are proportional to the corresponding stresses. Wood, e.g., is not isotropic; elastic properties along the grain are quite different from those across the grain.

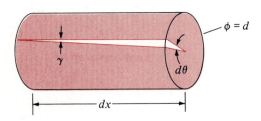

FIGURE 13.2
Deformation of shaft under torsional load.

The "spring rate" or "stiffness" of a torsion bar is

$$k = \frac{T}{\theta} = \frac{\pi d^4 G}{32L} \tag{13.4}$$

for a solid round bar. The units are newton-meters or pound-inches, both per radian of twist. The latter, being unitless, is not expressed.

13.2   TORSION-BAR MASS OPTIMIZATION

The mass of a torsion-bar spring can be expressed as its volume multiplied by the density of the material used. This becomes the criterion equation in this exercise

$$M = \rho \frac{\pi d^2 L}{4} \tag{13.5}$$

for a solid round bar.

There are usually constraints on the size:

$$d_{min} < d < d_{max} \qquad L_{min} < L < L_{max}$$

and we will want to specify limits on the stiffness:

$$k_{min} < k < k_{max}$$

The surface shear stress will also be limited by one or the other of the strength relations shown above [(9.14a) or (9.14b)]:

$$\tau_{surf} \leq \tau_{lim}$$

To find the optimum mass, we sketch a family of curves of mass vs. length, with the other variables as parameters. For the curves with stiffness k as the parameter, we need $M = f(k, L)$, and so the diameter d is replaced in the criterion equation with its equivalent from (13.4), $d^4 = 32Lk/(\pi G)$. This results in

$$M = \rho \sqrt{\frac{32Lk}{\pi G}} \frac{\pi L}{4} = \rho \sqrt{\frac{2\pi k}{G}} L^{3/2} \tag{13.6}$$

To express the criterion in terms of the surface shear stress, we find d in terms of the shear stress by using Equation (13.1): $d^3 = 16T/(\pi \tau_{surf})$. Solving this for d^2 and substituting in the criterion equation (13.5), we get

$$M = \rho \frac{\pi L}{4} \left(\frac{16T}{\pi \tau_{surf}} \right)^{2/3}$$

$$= \rho (4\pi)^{1/3} \left(\frac{T}{\tau_{surf}} \right)^{2/3} L \tag{13.7}$$

Expressions (13.5), (13.6), and (13.7) become inequalities when limits are imposed on d, k, and τ_{surf}. As we mentioned earlier, the length L is to be the independent variable in a plot of M vs. L; the limits on L will appear as vertical boundaries.

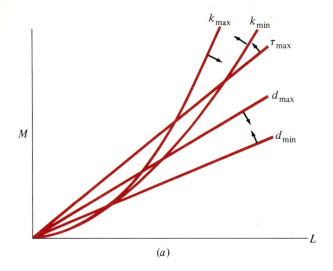

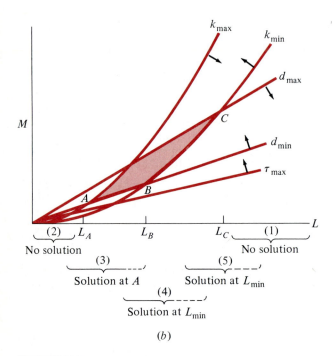

FIGURE 13.3
Minimizing torsion-bar mass. (*a*) No solution possible. (Material is not strong enough.) (*b*) Types of solution for various ranges of *L*.

Inserting the limits on the diameter in (13.5), we get

$$M \leq \frac{\rho \pi d_{max}^2}{4} L \tag{13.8a}$$

$$M \geq \frac{\rho \pi d_{min}^2}{4} L \tag{13.8b}$$

The stiffness limits in (13.6) give

$$M \leq \rho \sqrt{\frac{2\pi k_{max}}{G}} L^{3/2} \tag{13.9a}$$

$$M \geq \rho \sqrt{\frac{2\pi k_{min}}{G}} L^{3/2} \tag{13.9b}$$

and the stress limit in (13.7)

$$M \geq \rho (4\pi)^{1/3} \left(\frac{T}{\tau_{lim}} \right)^{2/3} L \tag{13.10}$$

A plot of the above inequalities is sketched in Figure 13.3. Note first that if the τ_{lim} (stress-limited) line falls above that for d_{max}, no solution is possible—the fattest bar we could make could not sustain the load. That situation appears in Figure 13.3a. If the material passes this first test, then the permitted design area, before the limits on L are considered, is that shown shaded in Figure 13.3b. The optimum locus is indicated by the heavy line ABC. The AB section will be on the d_{min} or τ_{lim} line, depending on which is higher, i.e., which governs. That may be determined by a comparison of the slopes of the two lines.

The limits on length L, that is, the L range, determine where along segment ABC the optimum is located. The five possibilities are shown in the figure.

1. If the shortest permissible L is greater than that corresponding to point C, or L_C, no solution is possible—the shortest, largest bar we can make falls short of being stiff enough, i.e., of k_{min}.
2. If the entire range of L is less than the length corresponding to point A, that is, $L_{max} < L_A$, there is also no solution. The longest, thinnest bar we could make is still too stiff, i.e., its $k > k_{max}$. If the τ_{max} line governs, i.e., lies above the d_{min} line, this statement still holds—the bar diameter is then limited by the τ_{max} restraint, rather than by the geometric restriction d_{min}.
3. If the range of L includes L_A, then the optimum is at A.
4. If L_{min} falls between L_A and L_B, then the solution is on segment AB at L_{min}.
5. Similarly, if L_{min} is between L_B and L_C, then the solution is on BC at L_{min}.

A flowchart to solve the problem on this basis is shown in Figure 13.4.

Example 13.1. A torsion-bar spring for a vehicle must have a length between 24 and 48 in, and stiffness between $2.4E + 05$ and $3.1E + 05$ lb·in and must carry an average torque of $1.5E + 03$ lb·in plus an alternating torque of $20E + 03$ lb·in. A steel is available

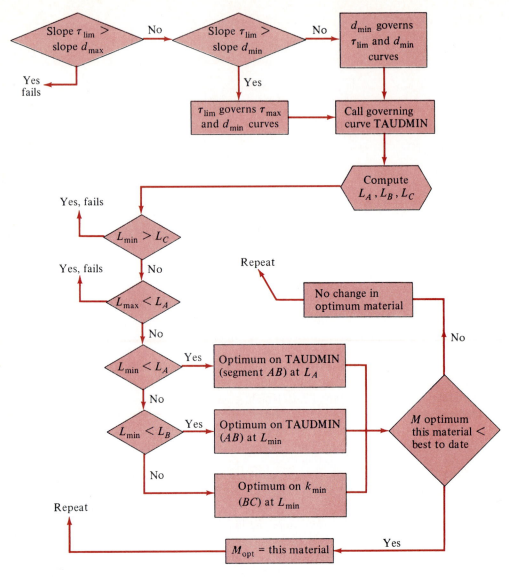

FIGURE 13.4
Rough flowchart for torsion problem.

($\rho = 0.28$ lb/in^3) for which we can use a design fatigue stress (includes fatigue corrections) of $S_f/\text{FS} = 30$ ksi and a design yield stress $S_y/\text{FS} = 65$ ksi. The part mass is to be minimized. Note that no limits are set on the diameter of the bar.

We need to determine first whether failure would be in fatigue or gross yielding. We use the distortion energy theory to estimate shear strengths. Since the alternating and mean stresses are proportional to the alternating and mean torques, respectively, we can determine whether a fatigue or a static failure is to be reckoned with by writing

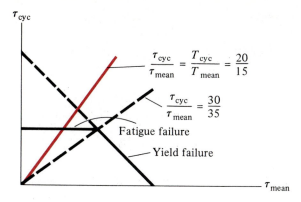

FIGURE 13.5
Load line for Example 13.1.

(9.14*b*). For a fatigue failure,

$$\frac{\tau_a}{\tau_m} \geq \frac{0.577 S_f}{0.577(S_y - S_f)} = \frac{30}{35}$$

Since for the present case

$$\frac{\tau_a}{\tau_{mean}} = \frac{20}{15}$$

we are dealing with fatigue failure. Our "load line" appears as in Figure 13.5 (corresponding to Figure 9.16). The limit on the alternating shear stress is thus

$$\tau_{a,lim} = \frac{0.577 S_f}{FS} = 0.577(30) = 17.31 \text{ ksi}$$

The plot of M vs. *L* corresponding to Figure 13.3 is shown in Figure 13.6. The minimum mass will correspond to the lower left corner of the permitted space, at length L_1. As long as the maximum length (48 in) is greater than L_1, a solution is possible. Here L_1 represents the shortest bar which will meet the stiffness requirements and not be overstressed. Setting (13.9*a*) and (13.10) (as equalities) equal and recalling that the torque in

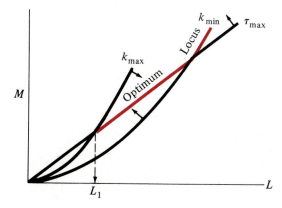

FIGURE 13.6
Torsion-bar mass problem, Example 13.1.

question is the alternating one, we have

$$\rho\sqrt{\frac{2\pi k_{max}}{G}}\, L^{3/2} = \rho(4\pi)^{1/3}\left(\frac{T}{17.31E+03}\right)^{2/3}L$$

from which $L_1 = 38.8$, which, being less than $L_{max} = 48$, is the optimum length for minimum mass.

Since the shear stress τ is at its limiting value, the diameter follows from

$$d = \left(\frac{16T}{\pi\tau_{lim}}\right)^{1/3} = \left[\frac{16(20E+03)}{\pi(17.31E+03)}\right]^{1/3} = 1.81 \text{ in}$$

The spring rate is k_{max}

$$k = 3.1E+05 \text{ lb·in}$$

and the mass is

$$M = \frac{\pi}{4}d^2 L\rho = \frac{\pi}{4}(1.81^2)(38.8)(0.280) = 28.0 \text{ lb}$$

13.3 MINIMIZING TRANSMITTED TORQUE

A torque bar, sketched in Figure 13.7, is to be designed to absorb repeatedly a certain fixed amount of potential energy, with the torque ranging from zero to a maximum value in one direction only. There are several geometric limitations, as indicated in Figure 13.7. We assume that there is a generous fillet at the attachment at the left, so that we need not worry about stress concentrations there. The bar is to have indefinite life. The design objective is to minimize the torque transmitted by the bar.

The potential energy in an elastic bar subjected to a torque is given by

$$\text{PE} = \frac{T\theta}{2}$$

where θ is the angle of twist of the bar. Since our assignment is to minimize the torque, we solve this relation for the torque to get the criterion equation. Since the torque is the same anywhere along the bar, it can be expressed in terms of the po-

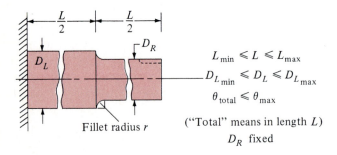

$$L_{min} \leqslant L \leqslant L_{max}$$
$$D_{L\,min} \leqslant D_L \leqslant D_{L\,max}$$
$$\theta_{total} \leqslant \theta_{max}$$

("Total" means in length L)

D_R fixed

FIGURE 13.7
Torque bar with constraints.

tential energy in the whole bar or that in the parts:

$$T = \frac{2\text{PE}_{\text{tot}}}{\theta_{\text{tot}}} = \frac{2\text{PE}_L}{\theta_L} = \frac{2\text{PE}_R}{\theta_R} \tag{13.11}$$

The relations among the limited quantities L, D_L, and θ will be needed:

$$\theta_{R \text{ or } L} = \frac{16TL}{\pi G D_{R \text{ or } L}^4} \tag{13.12}$$

(The 16 is not an error—the length in question is $L/2$.)

The expression for stress, (13.1), is

$$\tau_{\text{nom}} = \frac{16T}{\pi D_{R \text{ or } L}^3} \tag{13.13}$$

The stress is subject to concentrations, and this being a fatigue problem, we will require the fatigue strength reduction factors K_f for the critical points, which are the keyway and the shoulder at the center. Values for keyways are provided in Figure A.15 in Appendix A; $K_f = 1.6$ would be appropriate. Thus, the design fatigue strength, relation (9.1), is

$$S_{f,\text{key}} = \frac{C_D C_S S_{f\text{RB}}}{K_{f,\text{key}}} \qquad K_f = 1.6$$

Since no dimensions are given yet for the problem, $K_{f,\text{shoulder}}$ cannot be evaluated. However, we will see that the keyway is more critical. For now we write the design fatigue strength:

$$S_{f,\text{shoulder}} = \frac{C_D C_S S_{f\text{RB}}}{K_{f,\text{shoulder}}}$$

Looking first at the design criterion equation (13.11), the first version, we note that the only variable is the angle of twist θ_{tot}, upon which there is a limit θ_{max}. Placing θ_{tot} at its limit, we get the inequality

$$T \geq \frac{2\text{PE}_{\text{tot}}}{\theta_{\text{max}}} \tag{13.14}$$

To obtain a relation allowing an examination of the effect of D_L and L and their limits on the transmitted torque, we proceed as follows:

$$\text{PE}_{\text{tot}} = \text{PE}_L + \text{PE}_R = \frac{T\theta_L}{2} + \frac{T\theta_R}{2}$$

$$= \frac{T^2(8L)}{\pi G}\left(\frac{1}{D_L^4} + \frac{1}{D_R^4}\right)$$

whence
$$T^2 = \frac{\text{PE}_{\text{tot}}\pi G}{8L}\frac{1}{1/D_L^4 + 1/D_R^4} \tag{13.15}$$

To determine a minimum for the torque T, we see in this relation the prospect of plotting the torque against L, the total length, or against D_L, the diameter of the left segment. Choosing L, we see that the limits on D_L give two inequalities:

$$T^2 \geq \frac{PE_{\text{tot}}\pi G}{8(1/D_{L\min}^4 + 1/D_R^4)} \frac{1}{L} \tag{13.16a}$$

$$T^2 \leq \frac{PE_{\text{tot}}\pi G}{8(1/D_{L\max}^4 + 1/D_R^4)} \frac{1}{L} \tag{13.16b}$$

One other limited quantity remains, and that is the stress level, given by (13.13), and its limit for pure torsion, expressions (9.14a) and (9.14b). We will apply the maximum-shear-stress theory, which results in

$$\tau_{\text{cyc}} \leq \begin{cases} \dfrac{S_f}{2FS} & \text{if } \tau_{\text{mean}} \leq \dfrac{S_y - S_f}{2FS} \\[2ex] \dfrac{S_y}{2FS} - \tau_{\text{mean}} & \text{if } \tau_{\text{mean}} \geq \dfrac{S_y - S_f}{2FS} \end{cases}$$

Rearranging (13.13) and inserting the limit on τ, we get

$$T \leq \frac{\pi D_R^3 S_f}{32FS} \tag{13.17}$$

The fatigue strength S_f in this relation is that for the keyway or for the shoulder; the lower of the two must be used, i.e., the one with the greatest fatigue stress-concentration factor K_f. As indicated earlier, $K_{f,\text{keyway}}$ would be about 1.6. We see in Figure A.8 in Appendix A that we can get $K_{t,\text{shoulder}}$ values lower than this readily by making the fillet radius sufficiently large, and the value of $K_{t,\text{shoulder}}$ would be further reduced by the notch sensitivity relation to obtain $K_{f,\text{shoulder}}$. We thus conclude that the keyway stress governs, and so

$$T \leq \frac{\pi D_R^3 S_{f,\text{key}}}{32FS}$$

We have considered up until now that $D_L > D_R$. Suppose, though, that the diameter at the right D_R were larger than the minimum allowed diameter on the left: $D_R > D_{L\min}$. The possibility then exists of having the step in the other direction, as sketched in Figure 13.8. It is now no longer certain that the governing stress is that

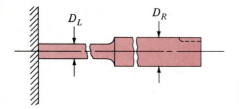

FIGURE 13.8
Possible situation where $D_R > D_{L\min}$.

at the keyway, and the question arises how this should be taken into account in arriving at the best design.

Consider first the variation of $K_{t,\text{shoulder}}$ with the geometry, as shown in Figure A.8 in Appendix A. For r/d (here r/D_L) greater than about 0.15, the variation in K is small. In order not to complicate the problem further, we assume that the stress-concentration factor K is constant, which is not unreasonable, since the fillet radius r may be adjusted. The nominal stress is now computed on the basis of the left diameter; it was previously based on D_R, a fixed dimension:

$$\tau_{\text{sh}} = \frac{16T}{\pi D_L^3} \tag{13.18}$$

Solving the above for $1/D_L^4$ gives

$$\frac{1}{D_L^4} = \left(\frac{\tau_{\text{sh}}\pi}{16T}\right)^{4/3}$$

and inserting this in (13.15) results in

$$T^2 = \frac{\text{PE}_{\text{tot}}\pi G}{8L} \frac{1}{[\tau_{\text{sh}}\pi/(16T)]^{4/3} + 1/D_R^4}$$

Substituting the limit on τ and rearranging give

$$\frac{8}{\text{PE}_{\text{tot}}\pi G}\left[\left(\frac{S_{f,\text{sh}}}{32FS}\right)^{4/3} T^{2/3} + \frac{T^2}{D_R^4}\right] > \frac{1}{L} \tag{13.19}$$

This cannot be solved for T without a struggle, if at all. But by looking at T as the independent variable and $1/L$ as the dependent one, it is clear that smaller T results in smaller $1/L$, and we could sketch this curve on our prospective plot of T vs. $1/L$. To get the shape of the curve, we would have to substitute numbers. And why not do just that?

With the computer accessibility and capability available today, all the expressions developed above can be plotted on a screen or on paper for a given material, and the optimum can be readily picked off. Then, by pressing a few keys, we can change candidate materials, and even geometric limitations, if we wish. Such a plot is shown in Figure 13.9. It is for 2330 steel, with the geometric constraints indicated in the figure. Note that the keyway stress governs for D_L greater than that corresponding to point a. For smaller D_L, the shoulder is critical. Also we could have made the plot against L rather than $1/L$.

The diameter corresponding to point a, where the two stresses become equally critical, may be found by equating expressions (13.17) and (13.18) for the torque (the latter with the limitation on τ_{sh} inserted), giving

$$D_L^3 = \frac{C_{D_R}K_{f,\text{sh}}}{C_{D_L}K_{f,\text{key}}} D_R^3$$

The solution to this problem may be obtained just as well by utilizing a plot of T vs. D_L (or D_L^3). This is left as an exercise in Problem 13.4.

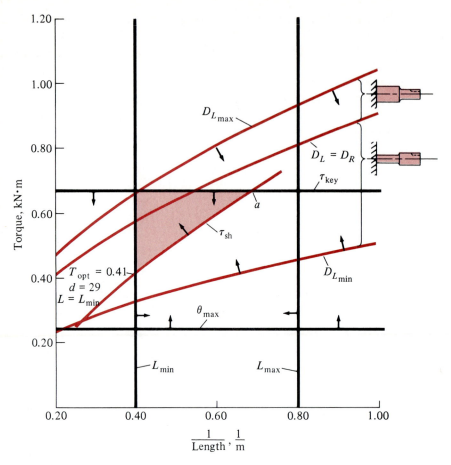

FIGURE 13.9
Graphical solution for minimizing transmitted torque.

13.4 OTHER DESIGN CRITERIA

In the preceding two sections, we discussed the problems of optimizing mass and transmitted torque in bars subjected to twisting moments. Many other criteria could govern the design: potential energy, cost, angle of twist, etc. Some are included in the problems at the end of the chapter.

13.5 STIFFNESS OF NONCIRCULAR FORMS

The discussion of the preceding sections has emphasized torsional members of circular cross section, i.e., bars or rods. Many machine and other mechanical members which are subjected, at least partly, to torsion do not fit that description.

The computation of the strength of forms not having the simplicity of the circle is a topic of some complexity, covered in advanced mechanics of materials texts. It is not treated here. We do discuss the stiffness of such members, which in many cases may be an issue greater than the strength. Machine bases, automobile and bus bodies, and arms for cranes and robots are examples.

Torsional Constant

The angle of twist for a bar was calculated in expression (13.3):

$$\theta = \frac{TL}{JG} \tag{13.3}$$

The constant J is, in general, known as the *torsional constant*, and in the case of round bars it was equal to the area polar moment of inertia. This will not be the case when the bar cross section is noncircular, and we will give it the symbol J_T. And $J_T G/L$ is the torsional stiffness defined earlier in the chapter for a round bar by (13.4). For a given length and material, J_T is thus a measure of the stiffness; the greater J_T is, the stiffer the structure.

If plane cross sections perpendicular to the axis of twist remain plane when torque is applied, the torsional constant J_T is equal to the polar moment of inertia of the cross-sectional area J. The polar moment J, in turn, is equal to the sum of the second area moments I_x and I_y. The only case where this obtains is that of a uniform circular rod or tube. Figure 13.10a shows elements on the wall of such a tube subjected to (torsional) shear stress. We recall from Chapter 4 that shear stresses always appear in equal pairs. Each element originally square becomes, under stress, a parallelogram, as seen in the figure. Because of symmetry, the longitudinal strain cannot vary around the tube, and two sides of the parallelogram remain perpendicular to the axis of the tube. Each plane cross section perpendicular to the axis of the tube thus remains plane, but rotated in the direction of the applied torque.

Members having unsymmetric (anything but round) and/or unclosed cross sections warp when they are twisted; i.e., cross sections perpendicular to the axis do not remain plane. Consider first the stresses in the circular tube discussed above if a cut is made in the tube wall parallel to the axis, as shown in Figure 13.10b. The restraint of circular symmetry is thus removed, and the tube will deform as shown in the figure. As always, less restraint means less stiffness; the torsional constant J_T is smaller.

Figure 13.11 shows a rectangular solid bar, whose cross section, of course, is closed. The longitudinal shear strain will not be uniform around the cross section because the shear stress is not uniform. The result is that cross sections originally plane become warped, as seen in the figure.

Calculating the Torsional Constant

It is a challenging problem in elasticity to calculate the torsional stiffness of noncircular cross sections. For solid bars of rectangular cross section, Wang [41, p. 89] developed an equation for the torsional constant

$$J_T = 16a^3b \left(\frac{1}{3} - 64 \frac{a}{\pi^5 b} \tanh \frac{\pi b}{2a} \right) \tag{13.20}$$

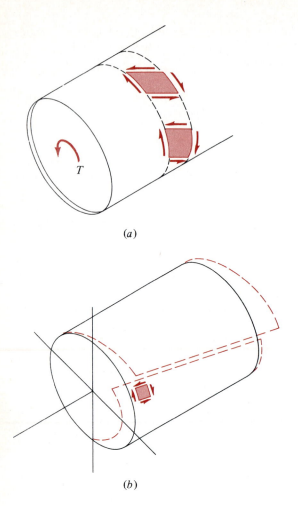

(a)

(b)

FIGURE 13.10
Stress elements in circular tubes under torsion. (a) Stress elements in closed circular tube. (b) Deformation of cut tube.

in which a and b are the half lengths of the short and long sides of the rectangle. The value of the terms in parentheses ranges from 2.25 for a square ($b/a = 1$) to 5.33 for a very thin plate ($b/a = \infty$). We can compare this expression for the torsional constant with the area polar moment of inertia for a square cross section ($b/a = 1$).

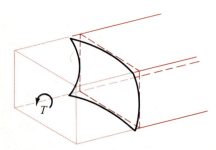

FIGURE 13.11
Cross sections originally plane become warped under torsion when the form is not circular.

The polar moment is

$$J = I_x + I_y = 2I_x = 2\frac{(2a)(2a)^3}{12} = 2.67a^4$$

while the torsional constant works out to be

$$J_T = 2.25a^3 a = 2.25a^4$$

a difference of about 18 percent.

The torsional constant of a thin plate can be expressed as that of a rectangle with $b/a = \infty$. Expressed in terms of length b and thickness t, the constant becomes

$$J_T = \tfrac{1}{3}bt^3 \tag{13.21}$$

An unclosed assemblage of thin plates will have a torsional constant equal to the sum of those of its members:

$$J_T = \tfrac{1}{3}\sum bt^3 \tag{13.22}$$

The shape of the assemblage contributes little to the stiffness.

The difference in torsional stiffness between a tube with a closed cross section and one with an open cross section can be startling. For example, a circular tube as in Figure 13.12, if not closed, will have a torsional constant, by Equation (13.21), of

$$J_T = \frac{1}{3}bt^3 = \frac{1}{3}(2\pi R)(0.1R)^3 = \frac{\pi R^4}{1500}$$

The closed tube, with $J_T = J = 0.540R^4$, computed per (13.2), is 258 times as stiff.

A useful expression for the torsional constant of closed thin-walled tubes [29, p. 164] is known as the *Bredt formula*:

$$J_T = 4t\frac{A_m^2}{L_m} \tag{13.23}$$

where t, A_m, and L_m are the thickness, area enclosed by the median line of the thickness, and length of the median line. For a circular tube, the Bredt formula reduces to Equation (13.2') if the difference between the mean line radius and the outside radius is ignored.

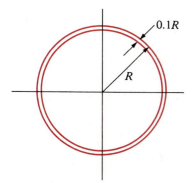

FIGURE 13.12
Circular tube with a wall thickness one-tenth of the radius.

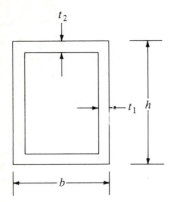

FIGURE 13.13
Rectangular tube of varying wall thickness.

To evaluate the torsional constant of tubes with varying thickness, we can use a more complete form of the Bredt equation:

$$J_T = \frac{4A_m^2}{\oint (ds/t)} \tag{13.24}$$

The integral $\oint ds/t$ is a summation of the reciprocal of the thickness of the wall per unit length. As an example of the use of this relation, consider the rectangular cross section of tube shown in Figure 13.13. We ignore the wall thickness in computing A_m. Thus, the area enclosed by the tube is $A_m = bh$, and

$$\oint \frac{ds}{t} = 2\frac{h}{t_1} + 2\frac{b}{t_2}$$

Then the torsional constant of the tube per relation (13.24) is

$$J_T = \frac{2b^2h^2t_1t_2}{bt_1 + ht_2} \tag{13.25}$$

Example 13.2. Consider a rectangular tube like that of Figure 13.13, made of steel, with $h = 20$, $b = 10$, $t_1 = 0.25$, and $t_2 = 0.5$, all in inches. From Equation (13.25), the torsional constant

$$J_T = \frac{(2)(10^2)(20^2)(0.25)(0.5)}{(10)(0.25) + (20)(0.5)} = 800 \text{ in}^4$$

Suppose the tube is 10 ft long and subjected to a torque of $T = 2.15\text{E} + 06$ lb·in. The total torsional deflection of the tube would then be

$$\theta = \frac{TL}{J_TG} = \frac{(2.15\text{E} + 06)(120)}{(800)(11.5\text{E} + 06)} = 0.0285 \text{ rad } (1.635°)$$

PROBLEMS

Section 13.2

13.1. Lay out the problem of Section 13.2, aiming at a plot of M vs. d (or d to some convenient exponent).

13.2. Lay out the problem of Section 13.2, including an upper limit on the cost of the part $\$_{max}$ (see Section 11.3). This results in more possible combinations of curves on a plot than it is worthwhile to sort out. Thus the strategy will be to sketch the plot by hand or by computer for a specific material. The optimum is then readily found.

Find the optimum mass, using the 1040 and 2330 steels of Appendix F. Assume that the fatigue limit is $0.3S_{ut}$ (includes all correction factors and FS).

$$1E + 05 < k < 1E + 06 \text{ in·lb/rad}$$

$$2 < d < 3 \text{ in}$$

$$50 < L < 100 \text{ in}$$

$$\$ < 350 \qquad \$_{fix} = 5 \qquad T_a = 55\,000 \text{ in·lb}$$

Give the specifications of the bar—its cost, mass, and stiffness.

13.3. A structural member is to be fabricated from hardened 5050 aluminum alloy. The member will be 2 m long and must support a load of 7.3 kN at a distance of 3 m from its centerline. The deflection at the load may not be more than 4 mm, and the thickness of the tube must be no less than 10 mm. Use the maximum-shear-stress theory of failure and an FS on the stress of 2. Design and proportion the tube for minimum mass (volume).

Answer: $t = 10$ mm, $d = 535$ mm

Section 13.3

13.4. Solve the problem of Section 13.3 by making a computer plot of T vs. D_L^3 for 2330 steel. Use the same geometric limitations as in Figure 13.9. You should, of course, reach the same design optimum. Figure P13.4 is a solution of this problem in the form of a graph of torque vs. D_L. (Since the computer does the work, we were not concerned with obtaining straight lines, which is more convenient when we make sketches of plots.)

Section 13.4

13.5. It is desired to design a torsion bar with a minimum spring rate. The torque cycles from 0 to an average expected maximum value of 30 000 in·lb. The FS is to be 1.6. Other pertinent dimensions, etc., are indicated in Figure P13.5.

The design criterion equation in this problem is that expressing the spring rate. Since the several sections of the bar act as springs in series,

$$\frac{1}{k_{eff}} = \sum \frac{1}{k_n} \qquad n \text{ covering all sections}$$

This expression indicates that it will be easier to maximize $1/k$, which is tantamount to maximizing the twist angle.

Obviously the full torque is not felt along the entire lengths of the keys. In computing the k values in L_1 and L_3, do not attempt to account for the material removed for the key—such material is effectively replaced to some extent by the key, and lengths L_1 and L_3 are stated to be the effective lengths which twist.

It will be determined at an early stage by examining the $1/k$ expression that L_{tot} should be assigned its greatest value L_{max}. Also d_R should be made as small as strength considerations permit. First, show that the stress at the bottom of the hole can be neglected. Then the easiest approach will be to plot τ/τ_{allow} vs. d_R rather than manipulating the criterion function.

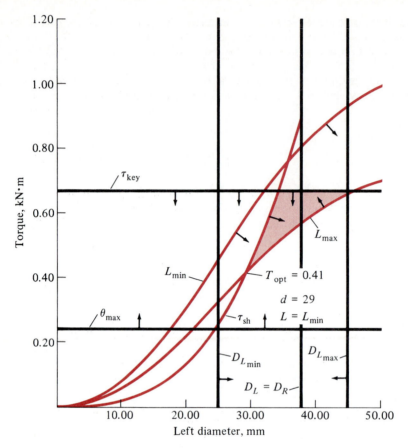

FIGURE P13.4
Graphical solution for Problem 13.4.

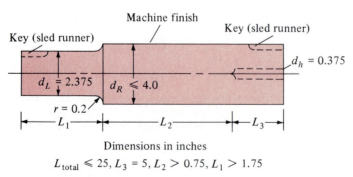

Dimensions in inches

$L_{total} \leqslant 25, L_3 = 5, L_2 > 0.75, L_1 > 1.75$

L_1 and L_3 are effective lengths which twist

FIGURE P13.5

It is not ensured that $d_R > d_L$, for d_L is fixed and may be grossly oversized for the load. Thus the shoulder may be in the other direction.

Candidate materials are the first three steels listed in Appendix F.

Answer: $d_R = 1.816$, $L_2 = 18.25$

13.6. A shaft is required on which a static torque of 12 kN·m will be placed. The deflection must not exceed 2.5°. The length must fall between 200 and 300 mm, and the diameter is limited to 55 mm. The objective is to minimize the cost, of which the fixed part is $3.10. Limit the materials to the steels of Appendix F, which are available in 10-mm increments. The FS is 1.3.

13.7. The torque on a torsion bar varies continuously between 5 and 15 kN·m. The specifications are

$$\$_{\text{fix}} = 4.00 \qquad 140 \le L \le 200 \text{ mm} \qquad 30 \le d \le 75 \text{ mm}$$

$$4\text{E} + 05 \le k \le 1.5\text{E} + 06 \text{ N·m/rad} \qquad M \le 5 \text{ kg}$$

The factor of safety is to be 1.2. Use 1040 and 2330 steels, available in 10-mm sizes. The objective is to minimize the cost.

There are too many restrictions to sort out for a computer solution, so make a plot of $ vs. L for each material. Do this by computer, if available.

Answer: $L = 140$ mm, $d = 72$ mm, $\$ = 25.86$

13.8. Work Problem 13.7, assuming the bar will be made of standard stock, without machining, available in 5-mm sizes.

13.9. A torsion bar is subjected to torque which varies from an unspecified static value T_{st} to $1.3T_{\text{st}}$. Also 200 in·lb of potential energy (static plus dynamic) is absorbed. Various geometric limitations exist as listed below. It is desired to minimize the transmitted torque. Draw the flowchart and solve by computer. Also FS = 1.5 (steels only).

$$2 \le d \le 4 \text{ in} \qquad L \le 20 \text{ in} \qquad \theta \le 1° \qquad \text{(total twist angle)}$$

Note: Torque may be plotted against d or L. Several materials will turn out to be equally good in this problem. It then becomes possible to satisfy a secondary design objective: cost, mass, spring constant, etc.

Answer: $T_{\text{max}} = 22\,918$ in·lb

13.10. We want to maximize the potential energy which a projected torsion bar can absorb. The specifications are the following:

$$1\text{E} + 04 \le k \le 2.5\text{E} + 04 \text{ N·m/rad} \qquad 15 \le d \le 40 \text{ mm}$$

$$M \le 3 \text{ kg} \qquad 150 \le L \le 800 \text{ mm} \qquad \text{FS} = 1.9$$

The torque is completely reversing; its size is not specified. Confine the solution to the nonstainless steels. Surface will be machine-finished. Lay out a computer flow diagram, and solve the problem by computer.

13.11. The torque on a machine-finished torsion bar consists of a static $T_{\text{st}} = 1.0\text{E} + 04$ lb·in plus an alternating $T_a = 1.3\text{E} + 04$ lb·in. Also 300 in·lb of energy must be absorbed at the peak torque. The FS is 1.5. The diameter must be between 1 and 1.7 in. Design the bar for minimum length. Consider nonstainless steels of Appendix F only.

Answer: $d = 1.55$, $L = 7.42$

13.12. The torque on a machined torsion bar varies from an unspecified maximum T_{max} to a minimum value $T_{\text{min}} = -2T_{\text{max}}/3$. Also 250 in·lb of energy must be absorbed at the peak torque with a total angle of twist less than 5°. The diameter must be greater than 2 and less than 4 in, and the length is greater than 25 in. Use FS = 2. The design objective is to minimize the peak torque. Consider the steels of Appendix F only.

Section 13.5

13.13. Find the torsional constant of a square cross section 100 mm on a side, with a thickness of 10 mm. Compare your answer with the polar second moment of area.

Answer: $J_T = 729$ cm^4, 26 percent smaller than J

13.14. Compare the torsional constants of the two cross sections shown in Figure P13.14. They have about the same cross-sectional areas.

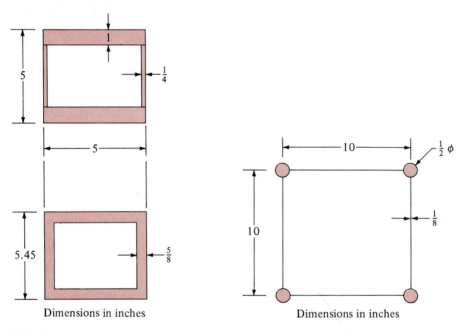

Dimensions in inches

FIGURE P13.14

Dimensions in inches

FIGURE P13.15

13.15. A structural member has been made by welding steel sheet between steel rods, as shown in Figure P13.15. Evaluate the torsional constant.

Answer: $J_T = 125$ in^4

13.16. Calculate the torsional stiffness per foot of length of the welded steel H section shown in Figure P13.16. Compare with the second-moment calculation.

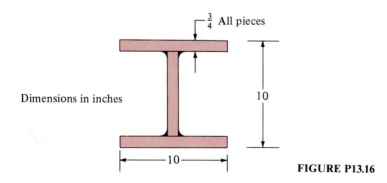

$-\frac{3}{4}$ All pieces

Dimensions in inches

10

10

FIGURE P13.16

13.17. Figure P13.17 shows the same section as in Problem 13.16, but with a plate welded to one side to increase the torsional rigidity. Evaluate the torsional constant.
        Answer: $J_T = 141.8$ in^4

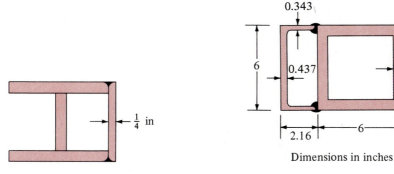

FIGURE P13.17   FIGURE P13.18

Dimensions in inches

FIGURE P13.17   **FIGURE P13.18**

13.18. A structural member is made by welding a C6 × 13 steel channel section to a $6 \times 6 \times \frac{1}{4}$ steel tube, as shown in Figure P13.18. Evaluate the torsional constant.

13.19. Two $3 \times 3 \times \frac{1}{4}$ steel angles from Table P12.5 are welded together as sketched in Figure P13.19. Compute the torque which would twist the pair $3°$ in a 10-ft length.
        Answer: 274 lb·in

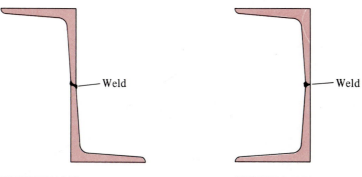

FIGURE P13.19   **FIGURE P13.20**

13.20. A channel is formed by welding two steel $6 \times 6 \times \frac{1}{2}$ angles (Table P12.5), as shown in Figure P13.20. Compute the torsional stiffness for 1 ft of the channel.

13.21. (a) A closed D tube of aluminum is sketched in Figure P13.21. Calculate its torsional stiffness per meter of length.
        Answer: $k = 5.36E + 04$

     (b) The tube is slit in the middle of the straight side. Compute its torsional stiffness per meter of length, and compare with the unslit tube.
        Answer: $k = 30.03$

13.22. An aluminum tube has OD = 50 mm and ID = 47 mm.

     (a) Calculate the stiffness of 1 m of the closed tube by the exact and by the approximate formulas.

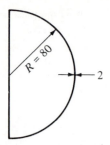

Dimensions in millimeters   **FIGURE P13.21**

(b) The tube is slit longitudinally. Find the torsional stiffness. Compare with the result of part (a).

13.23. An airplane wing spar is made of spruce top and bottom plates bonded to $\frac{1}{4}$-in plywood webs, as shown in Figure P13.23. Calculate the twisting deflection of an 8-ft length of spar due to a torque of 1800 lb·ft. Express this as the deflection at a point on a rib 9 in from the spar.

Answer: $\delta = 0.64$ in

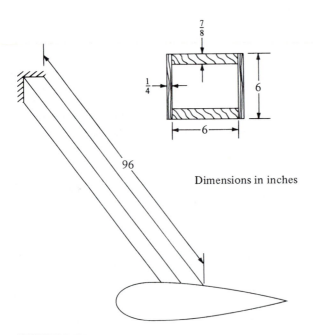

Dimensions in inches

FIGURE P13.23

BEAM
APPLICATIONS

This chapter presents, after a review of classic beam theory, engineering methods of evaluating beam stresses and deflections. Useful methods of evaluating the effects of damping are presented in the context of a beam used as a spring in a vibration isolator. This is not an unusual use; the leaf springs in automobiles are beams. In terms of energy stored per unit volume, a simple beam is one of the most efficient shapes a spring may take.

14.1   REVIEW OF BEAM THEORY

Beam Stresses

The well-known expression for normal stress (tension or compression) in a beam

$$\sigma = -\frac{My}{I} \tag{14.1}$$

is based on the assumption that stress is proportional to strain and that cross sections originally plane remain so. Results based on these assumptions are accurate when the deflection or depth of the beam is small compared to its span.

In Equation (14.1), M is the bending moment, I is the area moment of inertia of the beam cross section about the neutral axis, and y is the distance from the neutral axis to the point where the stress is to be computed. The minus sign is needed so that y and M can conform to the usual sign conventions, as shown in Figure 14.1. The neutral axis is the line across the beam cross section where the stress changes from tension to compression. That is, the normal stress at the neutral axis is zero.

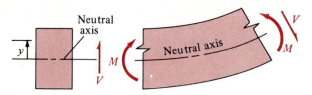

FIGURE 14.1
Conventional positive directions for beam computations.

When the beam is homogeneous, the neutral axis passes through the centroid of the cross section. Notice that the combination of a positive bending moment and a positive y produces a negative, or compressive, bending stress. The largest normal stresses are at top and bottom, where y has its largest value, usually denoted c:

$$\text{Maximum stress } \sigma_m = \frac{Mc}{I} \tag{14.2}$$

Bending stresses in beams are usually caused by loads which have a component perpendicular to the beam axis. This lateral component is called a *shear force*. If the shear force in the beam is V, then the shear stress at some point in the beam cross section is given by

$$\tau = \frac{VQ}{Ib} \tag{14.3}$$

where Q is the first moment about the neutral axis of the area to the outside of the point in question, and b is the width of the beam at that point. For a homogeneous beam with rectangular cross section of height h and width b, the relation is parabolic:

$$\tau = 6V \frac{h^2/4 - y^2}{bh^3} \tag{14.4}$$

The relation of Equation (14.4) is sketched in Figure 14.2. Notice that the shear stress is maximum at the neutral axis, decreasing to zero at the top and bottom surfaces. At the surface, there is no adjacent material to produce a stress, so this makes sense. If you compute the average stress from Equation (14.4), it will work out to be $\tau_{av} = V/A$, as it should.

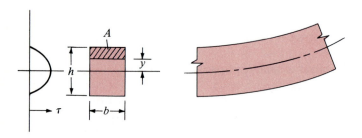

FIGURE 14.2
Shear stress in a beam. The term Q is the moment of area A about the neutral axis.

The preceding discussion concerned normal (tension and compression) and shear stresses resulting from transverse loads, where the beam carried a shear load V. The normal stresses are often called *flexural* or *bending stresses*. The maximum normal stress in a beam at failure is commonly referred to as the *modulus of rupture*.

"Pure" bending is bending without shear force, as with the central portion of the simply supported beam shown in Figure 14.3a. The shear force at any point along the beam is obtained from a free-body diagram such as that shown in Figure 14.3b or simply as the running sum of the applied forces. The forces and moment in the free-body diagram are shown in their conventional positive directions. The resulting shear diagram, in Figure 14.3c, shows that the shear force between B and C is zero.

If the loading is a distributed one, with the force per unit of span $w(x)$, then from a diagram like that of Figure 14.3b, it follows that

$$\frac{dV}{dx} = -w(x) \tag{14.5}$$

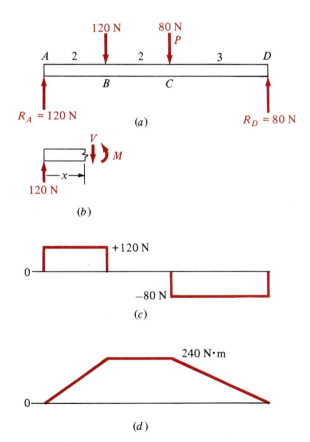

(a)

(b)

(c)

(d)

FIGURE 14.3
A simply supported beam. Section BC is in pure bending.

And the bending moment M is related to the shear:

$$\frac{dM}{dx} = V(x) \tag{14.6}$$

The moment diagram of Figure 14.3d can be constructed from the free-body diagrams or simply as the running sum of the area under the shear diagram, starting from the left.

Between the concentrated loads on the beam, the force per unit span $w(x)$ is zero, and as predicted by Equation (14.5), the shear is constant. Between B and C, the shear $V(x)$ is zero, and the bending moment is constant. This part of the beam is in pure bending.

Example 14.1. Figure 14.4a shows a beam rigidly supported at the left end, subjected to three concentrated loads. The free-body diagram shows that the support needs to supply only a bending moment; there is no net lateral force on the beam from the support to the 300-N force. Thus, the shear force in the AB section of the beam is zero, and the bending moment is constant. This portion of the beam is in a condition of pure bending.

When a beam is bent, plane sections can remain plane only if there is no deformation due to shear stress, which in turn means no shear force, as in section AB in Figure 14.4. A beam in any other condition will exhibit shear deformation, but

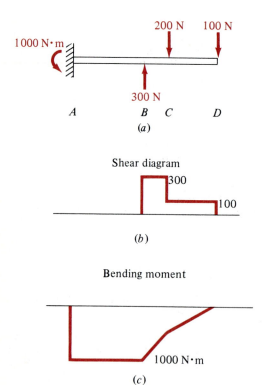

FIGURE 14.4
A cantilever beam. Section AB is in pure bending.

the effect of that deformation on normal stresses and on beam deflection is usually negligible. This is not true in very deep beams, where a sizable force (that is, V) is required to produce a significant flexural stress. Shear deformation may also be important in beams made up of sandwiches of materials with quite different shear stiffnesses, such as steel top and bottom plates bonded to a rubber center. In such cases, the elementary beam theory set forth above is questionable. Cross sections originally plane will not in general remain plane, and St. Venant's principle[1] may be violated.

In general, shear stress is maximum at the neutral axis, and normal stress is maximum at the top and bottom. The maximum stress at any point in the beam cross section can be obtained by using the methods of Chapter 4 to combine the shear and normal stresses. This maximum stress varies, of course, across the cross section. One might suspect it would attain its maximum value at some point other than the neutral axis or the outside, since at the neutral axis we find only shear stress and at the outside only normal stress. This does not, in fact, occur. The only two stresses that need be considered are those at the neutral axis and the outside surface; normally, the latter is greater.

Beam Deflections

Deflections of beams are calculated by twice integrating the basic relation

$$y'' = \frac{M}{EI} \tag{14.7}$$

The two constants of integration are determined by matching the resulting slope y' and deflection y with the restraints provided by the supports. For the beam in Example 14.1, with the moment a constant 1000 N·m in the first section from the support, the equation is $y'' = 1000/(EI)$, which integrates to $y' = [1000/(EI)]x + C$. The slope y' of this cantilever beam is zero at the mount, where $x = 0$, and therefore $C = 0$. Integrating the slope (y') equation, we get $y = [1000/(2EI)]x^2 + D$. Again, because $y = 0$ when $x = 0$, the constant D must be zero. (Reminder: y positive up.)

A catalog of frequently encountered beam loading situations can be found in Appendix G, along with formulas for shear, bending moment, and deflection obtained by integration of Equation (14.7).

Numerical Integration

When the loading on a simple beam is not symmetric, there is no simple way to determine what the slope is at either end or where the slope is zero. This is the case with the beam of Figure 14.3. Shigley and Mitchell [2] illustrate a way to obtain

---

[1] Elucidated by Barre de Saint Venant (1797–1886): "Two different distributions of forces, having the same resultant, acting on a body will have the same effect ... on parts of the body which are at a sufficient distance from the point of application." (From Stephen Timoshenko, "History of Strength of Materials," McGraw-Hill, 1953.)

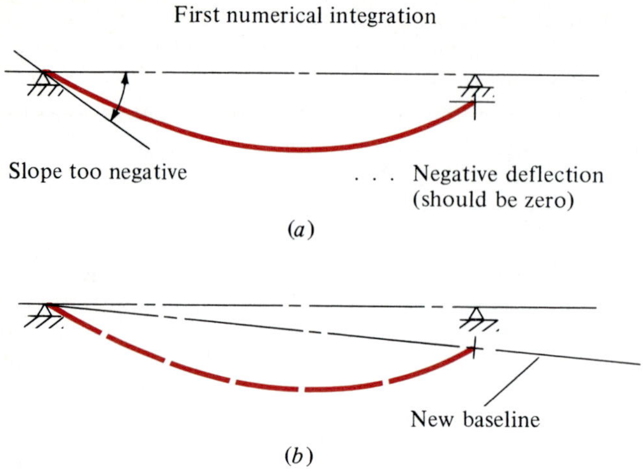

First numerical integration

Slope too negative   . . . Negative deflection
(should be zero)

(a)

New baseline

(b)

FIGURE 14.5
Sketch of a beam deflection obtained by numerical integration. The first slope estimate is too large.

the deflection of such a beam by a numerical integration of the fundamental beam equation (14.7). For the beam of Figure 14.3, the algorithm starts at the left support point, where the deflection is known to be zero. By starting with an estimate of the slope at the end, deflection is calculated at the end of each segment with a double numerical integration (summation). The calculated deflection at the other support is usually not correct, because the slope used to start the process was only an estimate. If the starting slope was too large, as shown in Figure 14.5a, the deflection calculated at the other end will be negative, and vice versa. The deflection calculated by this method is a linear function of the original starting slope. Therefore, the correct deflections can be obtained by simply adjusting the deflection at each point by an amount proportional to the error of the first run. This is accomplished by resetting the line of zero deflection, as shown in Figure 14.5b. The accuracy of the method is limited only by the number of increments used.

Example 14.2. We calculate the deflection of the beam of Figure 14.3 due to the 120-N load only. The reaction at the left support is 85.7 N, and thus the moment in that section of the beam is 85.7x N·m, where x is the distance from the left support. The slope at the support is estimated as that at the left end of a simple beam of 7-m span with 120-N load at its center: $PL^2/(16EI) = -367.5/(EI)$. (We could just let it be zero.)

We divide the beam into 35 increments, each 0.2 m long. The slope of each increment is estimated as the slope of the previous ("prev") increment plus the moment at the end of the increment times increment length over stiffness:

$$y' = y'_{prev} + \frac{M \, \Delta x}{EI}$$

(14.8)

The increase in deflection is estimated as the slope times the increment length:

$$y = y_{prev} + y' \, \Delta x$$

(14.9)

The results, for a few selected increments, are shown here.

x	Moment	Slope $\times EI$	Deflection $\times EI$
0	0	-367.5	0
0.2	17.1	-364.1	-72.8
0.4	34.3	-357.3	-144.3
2.0	171.4	-180.5	-585.4
7.0	0	-227.5	-73.5

The end of this list, $x = 7$, is the right support, and its deflection must be zero. So we simply add to each deflection the amount $73.5(x/7)$, which will bring the deflection at the right support to zero and that under the 120-N load to -564.4.

Superposition

Integration of the beam equation is the direct way to get an expression for the deflection of a beam at any chosen point. However, fitting the results to the boundary conditions can be a tedious job, especially if the beam supports are statically redundant. Redundancy means that there are more support forces and moments than are required for static equilibrium—and therefore more than can be solved for by using the equations $\Sigma F = 0$ and $\Sigma M = 0$. Real beams usually have redundant supports because the beams are thus made safer and more reliable. In a building structure the redundancies run into the hundreds, and in aircraft redundancy is the rule.

In the method of superposition, we calculate the deflection of the beam due to several loads by applying the loads one at a time and then summing the results. Superposition can be used when the behavior of the beam is linear, i.e., when twice the load results in twice the deflection.

Example 14.3. Using an expression from Appendix G, we can check the calculation of Example 14.2. The deflection of the beam of Figure 14.3 at the 120-N load, due to that load, is

$$\frac{(120)(5)(2)(7^2 - 5^2 - 2^2)}{(6)(7)(EI)} = \frac{571}{EI}$$

which agrees with the integration within 0.2 percent. By using the same expression, the deflection at the same point due to the 80-N load is

$$\frac{(80)(3)(2)(7^2 - 3^2 - 2^2)}{42EI} = \frac{411}{EI}$$

Summing these, we find that the total deflection at the 120-N load is $982/(EI)$.

Generally, beams subjected only to bending moments and lateral loads do, in fact, exhibit linear behavior, and superposition may be used. That is, the effect of two loads is the sum of the effects of each acting alone. However, when an axial load is included, the deflection due to that axial load will be out of proportion to the load.

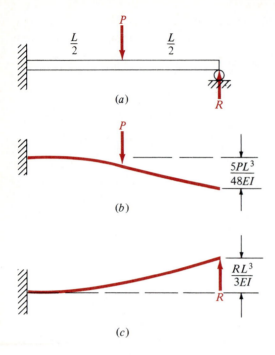

FIGURE 14.6
Force at a redundant support obtained by superposition.

This fact must be handled by including the axial load in calculating the effect of each lateral load.

> **Example 14.4.** The method of superposition can be used to find the reactions under a statically redundant beam. Consider the beam of Figure 14.6, which has a rigid support at the left end and a simple support at the right end. There are three unknown support reactions: a moment and force at the left and a force at the right. Only two equations of statics are available; we have more unknowns than equations. The beam is statically indeterminate.
>
> To find the reactions by superposition, first we consider the beam with the support at the right end removed, as shown in Figure 14.6b. Its deflection at the right end due to the load P is, from Appendix G, $y = -5PL^3/(48EI)$. Next we examine the same beam with only an upward force at the right end (the needed support reaction). The deflection corresponding to this force is $y = RL^3/(3EI)$.
>
> When we require the sum of these two deflections to be zero, we get the reaction $R = 5P/16$.

Moment-Area Method

The method of superposition worked easily for the examples we chose because the beam has a uniform stiffness, and formulas from Appendix G were applicable. Other methods, such as the moment-area method, may be better for beams of nonuniform stiffness.

The first moment-area theorem states that the change in slope between two points on a beam is equal to the area under a diagram of $M/(EI)$ between those points. The second theorem states that the vertical distance between point A on the beam and a tangent to the beam at point B is equal to the moment of the area referred to in the first theorem about point B.

The deflection at each point along the beam can be obtained with a computer program. Briefly, the algorithm is thus:

1. The $M/(EI)$ diagram is computed, point by point, starting at the end of the beam.
2. As $M/(EI)$ is computed at each point, the moment of the entire diagram about that point is computed. This represents the vertical distance from the point in question to a line tangent to the beam at the start.
3. When the second support is reached, the moment computed represents the vertical distance from the second support to a line which is tangent to the beam at the first support. But the actual deflection of this second support is zero. Therefore, the deflection of each point along the beam is the difference between a proportion of the end-to-end vertical distance and the distance computed in step 2.

Example 14.5. The deflection of one point on the beam of Figure 14.3 will be calculated. This is a uniform beam, so its $M/(EI)$ diagram shown in Figure 14.7a is identical in shape to the moment diagram. The deflected shape of the beam is sketched, with a tangent line placed at the left end A. The "vertical" distance from D to the tangent line (more precisely, the distance or deviation measured perpendicular to a line through the supports) is equal to the moment about D of the $M/(EI)$ diagram between D and A.

$$EIt_{DA} = 240\left(\frac{3}{2}\right)(2) + 240(2)(4) + 240(1)\left(5 + \frac{2}{3}\right)$$

$$t_{DA} = \frac{4000}{EI}$$

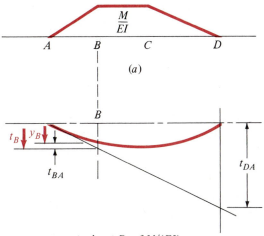

t_{DA} = moment, about D, of $M/(EI)$

t_{BA} = moment, about B, of $M/(EI)$ between B and A

(b)

FIGURE 14.7
Deflection of a simply supported beam obtained by the moment-area method.

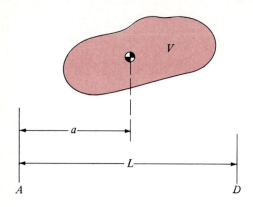

FIGURE 14.8
Conversion of a moment from one reference point to another.

The deviation at B from the original beam centerline to the tangent through A is calculated by ratio:

$$t_B = \tfrac{2}{7}t_{DA}$$

The deviation of B from the tangent to the beam at A is the moment of the area between them about B:

$$t_{BA} = \frac{240(1)(\tfrac{2}{3})}{EI} = \frac{160}{EI}$$

The deflection at B, from the original centerline, is obtained by difference:

$$y_B = t_B - t_{BA} = \frac{982}{EI}$$

as calculated earlier by superposition.

This example featured a simple beam of constant cross section with just two concentrated loads. The procedure works equally well with more complex load patterns and varying stiffnesses along the beam. That is, once the moment diagram is converted to an $M/(EI)$ diagram and a deflection curve is sketched, the procedure is exactly as above.

In constructing a computer program to calculate beam deflections by the moment-area method, it is convenient to get moments of the $M/(EI)$ diagram as a running sum. As the bending moment is computed and divided by EI for each segment of the beam, the area of this part of the $M/(EI)$ diagram and its moment about the starting point are added to the running sum. Before they can be used in the deflection algorithm, these moments about the left end of the diagram must be converted to moments about the right end with a simple transfer-of-axes formula. Consider an area V whose centroid is a distance a from point A, as shown in Figure 14.8. The moment of the area about A is Va. The moment of the area about point D is $V(L - a) = VL - Va = VL -$ moment about A.

Example 14.6. Figure 14.9 shows a steel shaft of 40-mm diameter which carries two gears of 60- and 30-kg mass. We have chosen a very slender beam for such heavy loads to get larger, more visible deflection. The table below shows static deflections calculated by a computer program written from the algorithm discussed in this section.

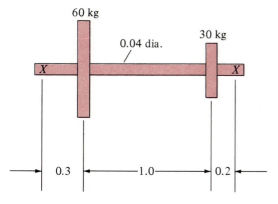

60 kg

30 kg

0.04 dia.

X

X

0.3   ←———1.0———→   0.2

FIGURE 14.9
A slender steel shaft carrying two concentrated masses.

x, mm	Static deflection, mm	x, mm	Static deflection, mm
10	0.35	80	1.37
20	0.69	90	1.29
30	0.95	100	1.17
40	1.16	110	1.0
50	1.3	120	0.8
60	1.38	130	0.5
70	1.40		

14.2 BEAM USED AS VIBRATION ISOLATOR

Vibration Isolation

Rotating machinery is everywhere, and it is never perfectly balanced. When such a machine is rigidly fastened to a floor, the force caused by the rotating unbalance can cause an objectionable vibration in the floor. Long ago it was found that a resilient material interposed between floor and machine could greatly reduce the vibration. Vibration isolation is measured by the reduction of the force transmitted to the floor by the vibration of the machine. It is summarized in this section in the context of a beam used as the resilient member of a vibration isolator.

The machine and its resilient support can be modeled as a mass supported by a spring of stiffness k and damping c, as shown in Figure 14.10. The disturbing force is modeled as a mass m at a radius e rotating with speed ω. The damping will be ignored for now.

When the disturbing frequency ω is very low, mass M moves with the unbalanced mass m. It is said to be in phase, and the transmitted force will be equal to the disturbing force $me\omega^2$. As the disturbing frequency is increased, the amplitude of vibration and the transmitted force increase. This increase reaches a destructive climax at the natural frequency of the spring-mass system. That natural frequency is given by [28, p. 29]

$$\omega_n = \sqrt{\frac{k}{M}} \qquad (14.10)$$

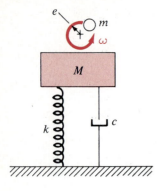

FIGURE 14.10
Schematic of a mass on a resilient mount subjected to a rotating imbalance.

A system is said to be *in resonance* when the disturbing force acts at the same frequency as the natural frequency of the system. Above the resonant frequency, the amplitude and transmitted force decrease.

The amplitude of vibration of an undamped system is

$$x = \left| \frac{P/k}{1 - r^2} \right| \tag{14.11}$$

in which $r = \omega/\omega_n$ is called the *frequency ratio* and $P = me\omega^2$ is the magnitude of the vibratory or disturbing force. The ratio of transmitted to disturbing force, called *transmissibility*, is found from Equation (14.11):

$$\text{Tr} = \left| \frac{F_{tr}}{P} \right| = \left| \frac{1}{1 - r^2} \right| \tag{14.12}$$

Equation (14.12) is shown graphically in Figure 14.11.

You can simulate these relations with a simple experiment. Suspend a heavy bolt from your hand with a rubber band. Pull the bolt down a short distance with your other hand and release it. The bolt will oscillate at the frequency ω_n given by Equation (14.10). Get a mental picture of this frequency. Now, starting with the bolt

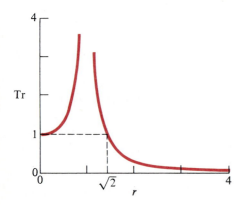

FIGURE 14.11
Transmitted force as a function of the frequency ratio for an undamped vibrating system.

at rest, move your hand up and down very slowly. The bolt will follow. Its motion is in phase with that of your hand. Increase the frequency and observe that the movement of the bolt (the amplitude of its vibration) is greater than that of your hand and increases beyond control as you approach ω_n. Finally, increase the frequency of your hand beyond ω_n. The bolt's motion is now displaced a half-cycle from that of your hand. As one moves up, the other moves down. As frequency increases, the motion of the bolt steadily decreases. If you can wiggle your hand fast enough, the bolt will become nearly stationary.

It seems clear that there is no point in trying to isolate a vibration unless the ratio $r = \omega/\omega_n$ is more than $\sqrt{2}$ and that the effectiveness of the isolation increases with r. This conclusion is valid, as we will see, for any amount of damping.

Example 14.7. A steel beam will be used as a vibration isolator. At its center, there will be a rotating machine with these characteristics:

Mass: $\quad\quad M = 50$ kg

Imbalance: $\quad m = 0.05$ kg at 0.01-m radius, 3600 rpm

Limits: $\quad$ Beam width (fixed), $b = 0.1$ m

$\quad\quad\quad\quad$ Maximum length $L = 1.5$ m

$\quad\quad\quad\quad\quad$ Deflection $X_{max} = 1.0$ mm

Steel: $\quad\quad\quad\quad$ Density $\rho = 7850$ kg/m^3

$\quad\quad\quad$ Young's modulus $E = 207$ GPa

$\quad\quad\quad$ Fatigue limit $S_f = 250$ MPa

$\quad\quad\quad$ Yield stress $S_y = 690$ MPa

$\quad\quad$ Factor of safety FS = 1.5

First, we determine the depth of steel beam required to support the mass of the machine with acceptable deflection and stress. We consider the beam to be simply supported, with a 50-kg (490-N) mass at the center of the longest permissible span. The mass of the beam itself will not be included at this time. The beam depth required so as not to exceed the yield stress is, from Equation (14.2) with $I = bh^3/12$,

$$h = \sqrt{\frac{3MgL}{2bS_y}} = 5 \text{ mm}$$

We will start with a beam sufficiently deep that its static deflection is one-half the total allowable. The required beam depth is then (Appendix G)

$$h = \left(\frac{MgL^3}{2EbX_{max}}\right)^{1/3} = 34 \text{ mm}$$

Next we estimate the natural frequency, in radians per second (rad/s), of the loaded beam with Equation (14.10):

$$\sqrt{\frac{k}{M}} = 140 \text{ rad/s}$$

With this estimated natural frequency, we can estimate the amplitude at running speed, by using Equation (14.11):

$$X = \left| \frac{71.06/(981E + 03)}{1 - (377/140)^2} \right| = 0.012 \text{ mm}$$

This is well within specifications. The transmissibility, from Equation (14.12), is also acceptable:

$$\text{Tr} = \left| \frac{1}{1 - 2.69^2} \right| = 0.183$$

This is a very satisfactory absorber at running speed, but we must find a way to bring the machine from rest to running speed and back without damage.

Vibration Isolation with Damping

A common strategy for bringing a machine safely through resonance is to provide damping via some device or construction which absorbs energy from the motion of vibration. An automotive shock absorber is a good example.

The usual assumption about damping is that it is *viscous*, i.e., that the damping force is proportional to velocity. This is true, more or less, when damping is provided by a mechanism such as the piston of a shock absorber moving through a fluid. The magnitude of damping is usually expressed as the damping ratio ξ. This is the ratio of the existing damping to that which would stop a free vibration during the first cycle. We use the assumption of viscous damping in this discussion and go into the nature of damping in greater detail later.

The natural frequency of a damped system is lower than that without damping:

$$\omega_{\text{damp}} = \sqrt{1 - \xi^2}\, \omega_{\text{undamp}} \tag{14.13}$$

And the expressions for amplitude X and transmissibility Tr become

$$X = \frac{P/k}{\sqrt{(1 - r^2)^2 + (2r\xi)^2}} \tag{14.14}$$

and

$$\text{Tr} = \frac{\sqrt{1 + (2\xi r)^2}}{\sqrt{(1 - r^2)^2 + (2r\xi)^2}} \tag{14.15}$$

The amplitude at resonance can be found from Equation (14.14), written for $r = 1$:

$$X = \frac{P}{2k\xi} = \frac{me\omega^2}{2k\xi} \tag{14.16}$$

Damping reduces both amplitude and transmissibility from start-up through resonance, as shown in Figure 14.12, but increases the transmissibility beyond the frequency ratio $r = \omega/\omega_n = \sqrt{2}$. As the frequency ratio increases beyond this point, however, the influence of damping decreases. The strategy for isolation, then, is to use damping to limit amplitude at resonance and to operate at high enough r to

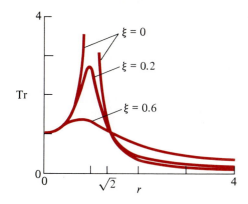

FIGURE 14.12
Transmissibility of a damped vibration absorber.

achieve the desired isolation. In other words, keep the natural frequency ω_n of the system low compared to the operating frequency of the machine.

Example 14.8. Find the amplitude of the machine of Example 14.7 at resonance. The damping ratio of steel, from Table 14.4, is $\xi = 1E - 04$. Inserting this in Equation (14.16) gives

$$X = \frac{(0.05)(0.01)(140)^2}{(2)(981E + 03)(1E - 04)} = 0.05 \text{ m} = 50 \text{ mm}$$

This is far too much deflection. The damping factor used is too low; that of a structure is always more than that of the materials used. But it is clear that this machine should not be allowed to run steadily at resonance.

Bringing a Vibration Absorber Up to Speed

We have shown that once the system is up to speed, the best isolation will be obtained with the minimum natural frequency. The mass of the machine is fixed, and so this means a beam of minimum stiffness. However, the beam must still support the weight of the load within limits of stress and deflection, and the beam's deflection must stay within limits during start-up and shutdown. In particular, the deflection may be excessive at resonance, a likely result for a unit with little damping, such as an all-metal beam. But naturally the beam will not be operated continuously at resonance. It will be brought through resonance on the way to operating speed. The amplitude at resonance does not, in any real system, reach its maximum value instantaneously. Instead, the amplitude near resonance grows at a rate which may be estimated as

$$\Delta X = \frac{me\omega_n \, \Delta T}{2M} \tag{14.17}$$

in which ΔT is the time spent at frequency ω_n. The maximum amplitude at resonance can be estimated now. We estimate the time spent within 1 percent of resonance as that same proportion, 1 percent, of the time required to come from rest to operating speed. The amplitude at resonance can then be estimated as the amplitude at 99 percent of resonant frequency plus the growth calculated with Equation (14.17).

Example 14.9. To continue Example 14.7, the amplitude is first calculated for 99 percent of resonant frequency, from Equation (14.14):

$$X = \frac{0.05(0.01)[(140)(0.99)]^2/(981\text{E} + 03)}{\{(1 - 0.99^2)^2 + [2(0.0001)(0.99)]^2\}^{1/2}} = 0.49 \text{ mm}$$

If the time required to come up to operating speed is 10 s, then the limit within 1 percent of resonance is estimated as 0.1 s. The growth in amplitude as the beam goes through resonance is, by Equation (14.17),

$$\Delta X = \frac{0.05(0.01)(140)(0.01)}{2(50)} = 0.07 \text{ mm}$$

It appears that this combination can pass through resonance without exceeding the stipulated amplitude.

In the examples in this section so far we have ignored the mass of the beam itself. To be accurate, that mass cannot be ignored in the calculation of natural frequency. The mass cannot be calculated ahead of time, since it depends on two of the variables in the problem, length L and depth h of the beam. Since the amplitude of vibration of the beam varies along its length, it is not correct simply to add the total beam mass to the machine mass. A solution to the problem of how to handle the mass of the beam is given in Section 14.5. It shows that if about one-half ($\frac{17}{35}$, to be exact) of the beam mass is added to the machine mass, the calculated natural frequency will be less than 1 percent in error.

Example 14.10. Consider a simply supported steel beam of cross section 10×20 mm with a span of 500 mm. The mass of the beam is 0.795 kg. This beam will have nearly the same natural frequency of vibration as a massless beam of the same stiffness with a single mass at its center of $(\frac{17}{35})(0.795) = 0.386$ kg.

Thus we simply replace mass M in the expressions for displacement and transmitted force by the equivalent mass of beam and load:

$$M_{\text{eq}} = M + \tfrac{17}{35}\rho bhL \tag{14.18}$$

Another factor ignored in the calculations so far is the effect of damping on natural frequency given by Equation (14.13). When more damping is used, the natural frequency is lowered. This effects an increase in the frequency ratio r and a reduction in transmitted force.

A complete design study for a vibration absorber should examine the effect of a range of damping coefficients on performance. These calculations can become tedious when several tries are needed to get satisfactory proportions. A short computer program makes it possible to compare many alternatives in a short time. Such a program merely does the calculations cited above, then checks the deflection of the beam (static plus dynamic) at each rotational speed against the specified limit. If the deflection exceeds the design limit, the beam thickness is increased and the calculations are repeated until the deflection is within limits.

TABLE 14.1
Required beam thickness and resulting transmissibility at 3600 rpm (377 rad/s)

Damping coefficient	Thickness, mm	Natural frequency	Frequency ratio	Transmissibility
0	49.8	246	1.53	0.74
0.0001	49.8	246	1.53	0.744
0.02	40	178	2.11	0.289
0.2	35	145	2.6	0.256
0.5	35	128	2.94	0.379

Example 14.11. We show in Table 14.1 the results of calculations done by computer for the beam of Example 14.7. The beam thicknesses shown are those required to keep the deflection within the stated limit.

Table 14.1 shows the effect of two influences on transmissibility. The natural frequency is decreased by damping, and the concomitant increase in frequency ratio reduces transmissibility. However, the disturbing frequency in all cases is more than 1.42 times the natural frequency. This is the region, as shown in Figure 14.12, where damping increases the force transmitted to the foundation.

Example 14.12. Once a computer program works properly, the effect of changing conditions can easily be evaluated. A lower operating speed will be closer to the natural frequency, and transmissibility will increase. Table 14.2 shows the result of cutting the operating speed in half.

With little or no damping, the natural frequency is actually greater than the forcing frequency, and the transmissibility is greater than 2. In other words, the force transmitted to the floor is more than twice the disturbing force. We'd be better off, in any situation where the transmissibility is more than 1, to bolt the machine to the floor. Short of a total redesign, there is really no alternative to heavy damping, and we see that the lowest transmissibility is achieved with a damping ratio essentially equal to 1, that is, critical.

TABLE 14.2
Operation at 1800 rpm (188.5 rad/s)

Damping ratio	Necessary depth, mm	Natural frequency	Frequency ratio	Transmissibility
0	50	251	0.75	2.295
0.0001	50	251	0.75	2.295
0.02	35	182	1.035	1.403
0.2	35.5	145	1.3	1.312
0.5	35.5	128	1.47	0.95
0.9	35.5	109	1.72	0.583
0.99	35.5	35	5.38	0.217

TABLE 14.3
Increased imbalance operation at 3600 rpm
(377 rad/s)

Damping ratio	Beam thickness	Transmissibility	Natural frequency
0		no solution	
0.02	45	0.47	213
0.2	35	0.247	145
0.5	35	0.379	128

The isolation problem is also tougher if the imbalance is increased by an increase in either unbalanced mass or eccentricity. Table 14.3 shows the result of a 60 percent increase in the mass, at the original forcing frequency of 3600 rpm. Acceptable operation is not possible without damping. As in the earlier calculation for this 3600-rpm condition, the optimum damping is about 20 percent of critical. While the transmissibility is slightly less than in the original case, the transmitted force is greater because the disturbing force is greater. The consequences of trying to run this device with both lower forcing frequency and greater imbalance are not shown here, but the conclusion is not hard to deduce: Redesign.

The design recommendations for the cases examined in these examples would be as follows:

rpm	Imbalance, kg	Thickness, mm	Damping ratio	Transmissibility
3600	0.05	35	0.2	0.256
3600	0.08	35	0.2	0.247
1800	0.05	35.5	0.99	0.217

In summary, the behavior of a vibrating beam is determined by the relationship between its stiffness, the mass carried, and the relative amount of damping. The most effective isolation results when the natural frequency is a fraction of the operating speed, and when just enough damping is provided to control amplitude while passing through the region of resonance.

14.3 DAMPING

Real beams, made of real materials, are not perfect springs. It is true that the relation between force and deflection is quite linear for small displacements, slowly applied. However, real beams will absorb more energy while being deflected than can be extracted from them as they return to their original shapes. The difference is called *hysteresis loss*. It is also referred to as *damping*, although the terms are not synonyms.

Textbooks, including this one, usually assume that the damping force is proportional to the velocity. This is called *viscous damping*. In reality, damping is a fairly complicated phenomenon. The force of damping can come from a variety of phenomena, singly or in combination. Some examples are fluid flowing through orifices, as in an automobile shock absorber, friction between imperfectly lubricated parts, or energy absorption within a solid body.

The ratio of the existing damping to the critical amount is known as the *damping ratio* or *damping factor*. It is commonly designated

$$\xi = \frac{C}{C_{cr}} \tag{14.19}$$

A system is critically damped when the mass, if displaced, moves back to its equilibrium position but not beyond. That is, the mass does not oscillate. Stand on the bumper of your car; then jump off. If the bumper returns to its rest position without oscillation, your suspension system is at least critically damped.

If we keep a record of the amplitude during each cycle of a freely vibrating (no external force) viscously damped system, we find that the logarithm of the ratio of successive amplitudes is constant. This dimensionless number δ, called the *logarithmic decrement*, is directly related to the damping ratio:

$$\xi = \frac{\delta}{\sqrt{\delta^2 + 4\pi^2}} \tag{14.20}$$

Damping ratios typically are in the range 0.05 to 0.10, in which case the δ^2 term in the denominator of (14.20) becomes very small compared to $4\pi^2$ and can be neglected. The relation becomes simpler:

$$\xi = \frac{\delta}{2\pi} \tag{14.21}$$

Example 14.13. Successive amplitudes are measured during free vibration. Their ratios, and the logarithms of those ratios, are calculated.

Amplitude	Ratio	$\log_n$ ratio
36		
	1.44	0.364
25		
	1.47	0.385
17		
	1.42	0.350
12		
	1.5	0.405
8		
	1.33	0.285
6		
Average logarithmic decrement $= \overline{0.357}$		

The damping ratio, calculated with Equation (14.20), is

$$\xi = \frac{0.357}{\sqrt{0.357^2 + 4\pi^2}} = 0.0567$$

Using the simpler relation of Equation (14.21), we get

$$\xi = \frac{0.357}{\sqrt{4\pi^2}} = 0.0569$$

The difference of 0.3 percent is negligible compared to the other uncertainties in a real situation. This damping ratio is therefore in the range where Equation (14.21) may be used.

The logarithmic decrement and damping ratio are often taken to be properties of the materials in question. Table 14.4 is a typical list [38]. Note that these values refer to damping within the material. Vibration texts often discuss only systems of idealized materials, with the damping provided by external devices.

The damping ratio can also be estimated from the difference between the natural undamped frequency calculated from stiffness and mass alone by Equation (14.10) and that measured by experiment. The calculated natural frequency, when the damping ratio is included, is given by Equation (14.13).

The damping ratio for steel is often cited as being 2 or 3 percent (0.02 to 0.03), much greater than the value cited in Table 14.4. The complexity of a real object and its environment can provide a substantial amount of damping. For example, experimental measurements of the natural frequency of a simple beam vibrating in air [37] indicated damping ratios from 0.370 to 0.484. This was a light beam vibrating with a large amplitude. In such a case, air resistance and the restraints in the real physical apparatus provide most of the damping.

While the logarithmic decrement is indeed related to the damping ratio of the material under test, the relationship of Equation (14.20) is true only for the configuration and conditions of the test.

When the amplitude is small, internal damping becomes more important. Admittedly, the viscous damping model is not rigorously accurate here, either. However, good results can be obtained by judicious use of viscous damping factors which have an equivalent effect in the particular application.

In sum, damping is usually modeled as a force proportional to velocity, called viscous damping. The ratio of existing damping to the critical amount which would prevent oscillation is usually measured in one of two ways: by the ratio of successive

TABLE 14.4
Internal damping

Material	Logarithmic decrement	Damping ratio
Steel	0.0006	1E − 04
Copper	0.0032	5.1E − 04
Quartz	0.0026	4.1E − 04
Glass	0.0064	10.1E − 04
Wood	0.027	43E − 04
Polystyrene	0.048	76E − 04
Hard rubber	0.085	135E − 04
Soft rubber	0.260	413E − 04

amplitudes of oscillation or by the ratio of damped to undamped natural frequency. As the complexity of an object increases, usually so does the effective damping.

14.4   SHEAR DEFORMATION

The last section dealt with the vital role played by damping in the control of mechanical vibrations. We now discuss how extra damping can be built into a beam by incorporating materials of high damping capacity. As shown in Table 14.4, softer materials generally have higher damping capacity. Rubber has several orders of magnitude more damping capacity than steel, and rubber is less stiff by many orders of magnitude.

The best place to put soft damping materials in a beam is near the neutral axis, where their low stiffness will least reduce the ability of the beam to withstand bending moments. This means that the damping materials will be located in the region of greatest shear stress. The combination of highest shear stress and lowest shear stiffness will result in the shear deformation being concentrated in the damping material. That is not a bad situation, but the shear deformation of the damping layer will affect the deflection of the beam as a whole, and we must take account of it. Also when a beam with a soft shear layer is caused to vibrate, a significant amount of energy will be absorbed in the shear layer because of its large deflection. The shear layer could get very hot.

The only beams that are entirely free of shear deformation are those in pure bending, with no net lateral loads and hence no shear stresses. The deflection due to shear deformation is usually ignored in beams which have a small depth in relation to their span (depth/span $< \frac{1}{10}$) and which are made of homogeneous material with a low ratio of tensile/compressive stiffness to shear stiffness (E/G). Steel, with $E/G = 2.54$, is a good example.

Beams in which the effect of shear deformation must be considered are usually called *Timoshenko beams*. The differential equation of such a beam is created by adding a term representing shear deflection to that for bending alone [11, p. 408]:

$$\frac{d^2y}{dx^2} = -\frac{M}{EI} - \frac{\alpha_s w}{GA} \tag{14.22}$$

The term w represents the lateral density of loading on the beam in terms of force per unit of span. The parameter α_s is called the *shear coefficient*. It is a function of the shape of the beam cross section.

In the case of a simply supported beam under a uniform load, the differential equation becomes

$$\frac{d^2y}{dx^2} = \frac{-w(xL - x^2)}{2EI} - \frac{\alpha_s w}{GA} \tag{14.23}$$

When this equation is integrated and boundary conditions are inserted, we get, for the deflection at the center of the span,

$$y_c = \frac{5wL^4}{384EI}\left(1 + \frac{48\alpha_s EI}{5GAL^2}\right) \tag{14.24}$$

TABLE 14.5
Fractional increase in deflection of a uniformly loaded simple beam due to shear deflection

L/h \ E/G	2.5	10	20	100	1000
10	0.03	0.12	0.24	1.2	12
20	0.0075	0.03	0.06	0.3	3
40	0.0019	0.0075	0.015	0.075	0.75

in which the second term in the parentheses represents the extra deflection due to shear deformation.

Example 14.14. A beam of rectangular cross section will have a shear coefficient α_s of 3/2 [11]. The expression for deflection at the center of the span becomes:

$$y_c = \frac{5wL^4}{384EI}\left[1 + \frac{6}{5}\left(\frac{h}{L}\right)^2\left(\frac{E}{G}\right)\right] \tag{14.25}$$

The conditions where shear deformation is important can be highlighted by displaying the results of some calculations in tabular form. The numbers shown in Table 14.5 are the value of the second term in the parentheses of Equation (14.25). They represent the increase in deflection due to shear deformation. The top left entry shows that for a steel beam ($E/G = 2.5$) of ordinary proportions, shear deformation increases deflection about 3 percent.

Vinson and Chou [36] studied the vibration of beams, including shear deformation, and developed an expression for their natural frequencies by variational methods. They found that the effect of shear deformation is to divide the undamped natural frequency $\sqrt{k/M}$ by a factor such that

$$\omega_n = \frac{\sqrt{k/M}}{1 + n^2\pi^2(E/G)(h/L)^2} \tag{14.26}$$

The denominator of Equation (14.26) is tabulated in Table 14.6 for several beam length-thickness ratios and ratios of normal to shear stiffness. The change in natural

TABLE 14.6
Reduction in natural frequency of simple beams due to shear deformation

L/h \ E/G	2	4	8	16	100
76	1.0003	1.0006	1.001	1.0026	1.03
15	1.009	1.017	1.035	1.07	1.9
10	1.019	1.039	1.078	1.16	2.9
5	1.079	1.15	1.31	1.63	8.9

frequency is negligible in a thin homogeneous beam (E/G is about 2 for most materials) but quite important in a thick composite beam, such as that listed in the lower right of Table 14.6.

The stiffness of a particular composite beam may be estimated by a simple procedure. A rubber layer inserted along the neutral axis of a steel beam has an extremely low resistance to shear deformation compared to the steel. When a lateral load is first applied to the beam, the beam deflects as though the top and bottom plates were acting independently, i.e., as though they were not fastened together. The effective moment of inertia of the beam is thus only the sum of the moments of inertia of the top and bottom plates, each evaluated about its own centroid. In the example we've been using, the replacement with rubber of the central 8 mm, of a beam of 20-mm total depth, would result in an effective moment of inertia of about one-eighth that of the original solid beam. As the beam deflects, the rubber layer deflects in shear, and the shear stress at the boundary between the rubber and the steel increases. When the magnitude of this shear stress reaches that which would be found at that spot in a homogeneous steel beam under the same loads, the beam will react to further loading as though it were a homogeneous beam in which the rubber layer had been replaced by a thin web capable only of carrying shear loads. In other words, the deflection of the beam is the deflection of a solid beam of the same cross section plus the deflection caused by shear deformation of the soft layer.

Example 14.15. Replace the beam of Example 14.7 by one fabricated by bonding steel plates to a rubber center. The top and bottom plates are 0.008 m thick. As before, we have assumed a static load of 490 N and a span of 1.5 m. The first row of figures in Table 14.7 is the deflection calculated by using the classic assumption of plane sections remaining plane. The second and third rows are the extra deflection due to shear deformation in two grades of rubber. Thus a layer of fairly stiff rubber whose thickness makes up one-fifth that of the beam will increase a deflection of 0.0025 by 0.0003, about 10 percent.

TABLE 14.7
Additional static deflection due to a central damping layer

	Thickness of damping layer, mm		
	4	**8**	**12**
Deflection with plane sections remaining plane	2.48	1.48	0.97
Deflection due to shear deformation			
$G = 2E + 08$	0.28	0.47	0.61
$G = 2E + 06$	28.	47.	61.

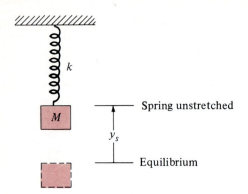

Spring unstretched

y_s

Equilibrium

FIGURE 14.13
An undamped spring-mass vibrating system.

14.5 SHAFTS AND BEAMS WITH DISTRIBUTED LOADING: THE RAYLEIGH METHOD

Earlier in this chapter we examined the vibration of a beam which carried a concentrated mass at its center. We asserted that the fundamental (lowest) natural frequency of vibration would be the same as if the beam itself had no mass and the mass at the center were equal to the concentrated mass plus about one-half the mass of the beam ($\frac{17}{35}$, to be exact). This can be derived by the application of an extremely useful technique for the analysis of vibrating beams and shafts, called the *Rayleigh*[1] *method*.

Rayleigh's method uses conservation of energy to calculate the natural or resonant frequency of vibration. The basis of the method can be illustrated by a simple example. A mass hung from a simple spring is shown in Figure 14.13. There is no damping. At equilibrium, the spring force must be equal to the weight, so the resultant force on the mass is zero. The spring will be extended an amount y_s equal to the weight divided by the stiffness of the spring:

$$y_s = \frac{Mg}{k} \tag{14.27}$$

If we now raise the mass that same distance y_s, to the point where the spring force becomes zero, and release it, then a vibration will ensue between the point of release and a lower position where the spring force is twice the weight. The motion will be simple harmonic, or sinusoidal:

$$y = y_s \sin \omega t \tag{14.28}$$

[1] John W. Strutt, Lord Rayleigh (1842–1919), was educated in physics at Cambridge, England. He was awarded the Nobel Prize for physics in 1904 for the discovery, with Sir William Ramsay, of the noble elementary gas argon. As successor to James Clerk Maxwell, he headed the Cavendish Laboratory at Cambridge. His research ranged over almost the entire field of physics, including sound, wave theory, optics, electrodynamics, flow of liquids, and elasticity. He was president of the Royal Society from 1905 to 1908.

The distance y is measured from the equilibrium point. The velocity

$$\frac{dy}{dt} = y_s \omega \cos \omega t$$

will be maximum at the equilibrium point:

$$V_{max} = \omega y_s$$

Thus the maximum kinetic energy is

$$\text{KE}_{max} = \tfrac{1}{2} M (y_s \omega)^2 \tag{14.29}$$

We are going to define potential energy as zero at the equilibrium point, where the kinetic energy is maximum. As the mass rises from equilibrium to the uppermost position, the gravitational potential energy increases by an amount Mgy_s. The potential energy (strain energy) in the spring, however, decreases by an amount

$$\frac{ky_s^2}{2} = \frac{Mgy_s}{2}$$

The net potential energy, then, at the top of the motion is

$$\text{PE}_{max} = \frac{Mgy_s}{2} \tag{14.30}$$

Since no energy is being added to or taken from the system, the total remains the same. Therefore, the maximum kinetic energy (at the midpoint, where potential energy $= 0$) must equal the maximum potential energy (at the extremes, where kinetic energy $= 0$). Setting them equal gives

$$\omega^2 = \frac{g}{y_s} \tag{14.31}$$

This is the form of the natural-frequency equation that we need in order to use Rayleigh's method. It can also be written by expressing Equation (14.10) in terms of the static deflection.

In the general case of a beam carrying several masses, the static deflection of the beam can be determined by superposition or the other techniques discussed earlier in this chapter. The mass of the beam itself can also be treated as a series of concentrated masses.

To generalize Equation (14.31), we use Equation (14.30) to express the maximum potential energy in a vibrating beam carrying many masses M_i:

$$\text{PE}_{max} = \tfrac{1}{2} g (M_1 y_1 + M_2 y_2 + M_3 y_3 + \cdots)$$

The maximum kinetic energy of this beam is, by Equation (14.29),

$$\text{KE}_{max} = \tfrac{1}{2} \omega_n^2 (M_1 y_1^2 + M_2 y_2^2 + M_3 y_3^2 + \cdots)$$

Equating, we find the square of the natural frequency is

$$\omega_n^2 = g \frac{M_1 y_1 + M_2 y_2 + \cdots}{M_1 y_1^2 + M_2 y_2^2 + \cdots} \tag{14.32}$$

Example 14.16. Rayleigh's method can be illustrated by solving for the natural frequency of the simply supported uniform beam shown in Figure 14.14. The only load carried by the beam is its own mass m per unit length. An exact solution is available for the natural frequency [43]:

$$\omega_n = 9.869 \sqrt{\frac{EI}{mL^4}} \qquad (14.33)$$

The deflection curve of a simple beam carrying a uniform load is obtained from Appendix G:

$$y = \frac{mgx}{24EI}(L^3 - 2Lx^2 + x^3) \qquad (14.34)$$

To start with a very crude approximation, we divide the mass of the beam into two parts. The centers of the parts are at the $\frac{1}{4}$ and $\frac{3}{4}$ span points; their deflections are equal:

$$y = \frac{mgL^4}{108EI}$$

Because the deflections and masses of the two segments are equal, the expression for ω^2 is simplified:

$$\omega^2 = g\frac{\sum my}{\sum my^2} = \frac{g}{y}$$

$$= 10.38\sqrt{\frac{EI}{mL^4}}$$

This result is 5 percent higher than the correct value. The error decreases as the segment size decreases:

No. increments	Natural frequency	Error, %
2	10.38	5.2
8	9.91	0.4
32	9.879	0.10
128	9.876	0.078

The natural frequency is thus estimated about as precisely as desired. The accuracy is impressive when you consider that the static deflections used in Rayleigh's method are not equal to the dynamic deflections which actually determine the energies in vibration.

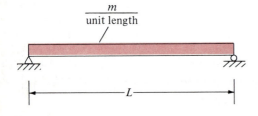

FIGURE 14.14
A simply supported uniform beam loaded only by its mass m per unit length.

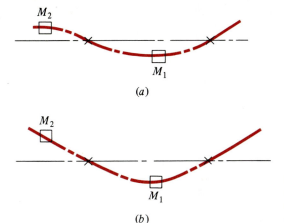

(a)

(b)

FIGURE 14.15
(a) Static deflection of a beam loaded with two concentrated masses. (b) Assumed static deflection for Rayleigh-method analysis.

The former are proportional to the mass at each station, but the latter are proportional to the product of mass and deflection.

It is a good idea to sketch the deflection curve of the beam after static deflections have been calculated, to make sure that the result thus obtained reasonably approximates the deflection shape during vibration. For example, the static deflection curve shown in Figure 14.15a is not a reasonable approximation of the shape taken by a vibrating beam at the lowest natural frequency. Instead, the shape shown in Figure 14.15b should be used, where the left end has been subjected to "reverse gravity." As we pointed out in Chapter 12, when an assumed deflection shape is used to approximate some property of a system, the fact that the assumed shape is not the same as the correct shape implies artificial constraints. In other words, the natural frequency calculated by the Rayleigh method will always be higher than the correct value, because the beam has been "constrained."

Example 14.17. We stated earlier in this chapter that, in computing the natural frequency, the mass of a uniform beam could be represented by an equivalent mass located at the center. Consider the simply supported beam with a load at the center shown in Figure 14.16.

From Appendix G, we get the static deflection curve of a simply supported beam with a concentrated load at its center:

$$y = y_{max} \frac{3xL^2 - 4x^3}{L^3} \tag{14.35}$$

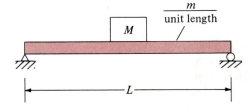

FIGURE 14.16
A simply supported beam loaded by its own uniform mass m per unit length and a concentrated mass M at the center.

The maximum velocity during vibration, at any point along the beam, will be

$$V_{max} = V_c \frac{3xL^2 - 4x^3}{L^3}$$

where V_c is the maximum velocity at the center. Integrating $mV_{max}^2/2$ over the length of the beam, with m representing the mass per unit length, we get the maximum kinetic energy of the entire beam:

$$\text{KE} = 2 \int_0^{L/2} \frac{m}{2} \left(V_c \frac{3xL^2 - 4x^3}{L^3} \right)^2 dx$$

$$= \frac{17}{35} mL \left(\frac{V_c^2}{2} \right) \tag{14.36}$$

The kinetic energy of the load alone is

$$\text{KE}_{load} = \tfrac{1}{2} M V_c^2$$

Summing, we get the total kinetic energy of the beam and the load:

$$\text{KE}_{tot} = \tfrac{1}{2}(M + \tfrac{17}{35}mL)V_c^2$$

Thus we see that mass distributed uniformly along a beam can be represented by about half as much mass concentrated at the center.

The maximum energy stored in bending of the beam can be expressed as $\tfrac{1}{2}ky_c^2$, where k is the stiffness of the beam to a lateral force at the center. If $k = g(M + \tfrac{17}{35}mL)/y_c$, then the maximum stored energy in the beam becomes

$$\text{PE}_{max} = \tfrac{1}{2}gy_c(M + \tfrac{17}{35}mL)$$

Equating the maximum potential and kinetic energies, we get Equation (14.31) again:

$$\omega_n^2 = \frac{g}{y_c} \tag{14.31}$$

In the extreme case $M = 0$, and we have a beam vibrating under the action of its own uniformly distributed mass. The estimated maximum static deflection becomes

$$y_c = \frac{\tfrac{17}{35}mL^4}{48EI}$$

and the natural frequency is

$$\omega_n = 9.94 \frac{EI}{mL^4}$$

which is about 1 percent in error. Again, this is good accuracy even though we did not use the correct expression for the deflection of a beam under a uniform load. Instead, we used the deflection of the beam with about one-half as much load concentrated at the center. This ought to have introduced error, because the static-deflection term in the Rayleigh equation is an estimate of the strain energy stored in the entire beam. That strain energy depends on the deflected shape of the entire beam, not simply the deflection of the center.

Example 14.18. The lowest natural frequency of the shaft shown in Figure 14.9 will be calculated. The two masses on the shaft are large compared to the mass of the shaft itself, and the latter will be ignored. We calculated the static deflection of the shaft due to the gears it carries as $9.58E - 04$ and $7.98E - 04$ at the locations of the 60- and

30-kg masses. The square of the natural frequency is then calculated, by using Equation (14.32):

$$\omega_n^2 = g \, \frac{60(9.58) + 30(7.98)}{60(9.58)^2 + 30(7.98)^2} \, (1E + 04)$$

$$\omega_n = 103.8 \text{ rad/s (991 rpm)}$$

14.6   POTENTIAL ENERGY STORAGE IN A BEAM

It is an interesting exercise in optimization to design a beam for maximum potential energy storage, subject to constraints. A cantilever beam of unvarying rectangular cross section and fixed length L subjected to a single load P at the end is shown in Figure 14.17. From Appendix G we find that the deflection under the load is

$$\delta = \frac{PL^3}{3EI}$$

As long as beam stresses are in the linear range, where deflection is a linear function of the load, the potential energy stored as a result of load P is

$$PE = \tfrac{1}{2}P \text{ (maximum deflection)}$$

$$= \frac{P^2L^3}{6EI} = \frac{2P^2L^3}{bh^3E}$$

The maximum stress is reached at the fixed end, where the moment is greatest. It is

$$\sigma_{\max} = \frac{6PL}{bh^2}$$

If we designate the design limit stress as $\sigma_{\lim}$, then the maximum allowed load is

$$P_{\lim} = \frac{\sigma_{\lim}bh^2}{6L}$$

which limits the maximum potential energy to

$$PE_{\lim} = \frac{\sigma_{\lim}^2 bhL}{18E} \tag{14.37}$$

This shows that, for a given beam length and allowable stress, the most important factor in potential energy storage is the strength of the material. Also the stiffness of the material E should be low and the cross section of the beam ab large.

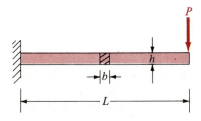

FIGURE 14.17
A uniform cantilever beam with a single concentrated load.

The expression for potential energy per unit volume is a simple one:

$$\frac{PE}{Volume} = \frac{\sigma_{lim}^2}{18E} \qquad (14.38)$$

which may help to explain the popularity of high-strength, low-stiffness fiberglass and Kevlar reinforced plastic for energy storage.

Instead of seeking a material of low stiffness, we might taper the beam to reduce its stiffness without increasing the maximum stress. The question is, How much taper? A tapered cantilever beam is shown in Figure 14.18. Note that the co-ordinate x is measured from the free (right) end of the beam. The expression for deflection at the end can be found by integration:

$$\Delta_{x=0} = \frac{12PL^3}{bE(h_l - h_r)^3} \left(\frac{4h_l h_r - h_r^2 - 3h_l^2}{2h_l^2} + \ln \frac{h_l}{h_r} \right) \qquad (14.39)$$

The potential energy stored as a result of the load P is $P^2 \Delta / 2$

$$\frac{6P^2 L^3}{bE(h_l - h_r)^3} \left(\frac{4h_l h_r - h_r^2 - 3h_l^2}{2h_l^2} + \ln \frac{h_l}{h_r} \right) \qquad (14.40)$$

The stress at any point along the beam is given by

$$\sigma = \frac{Mc}{I} = \frac{6Px}{b[(x/L)(h_l - h_r) + h_r]^2} \qquad (14.41)$$

This is differentiated to find the location of maximum stress:

$$x_{\text{max stress}} = \frac{h_r L}{h_l - h_r} \qquad (14.42)$$

The beam depth at this point is $2h_r$. The stress is

$$\sigma_{\text{max}} = \frac{3PL}{2h_r b(h_l - h_r)} \qquad (14.43)$$

If h_l is less than h_r, the point indicated by (14.42) will not be reached, and the maximum stress will occur at the wall: $\sigma_{\text{max}} = 6PL/(bh_l^2)$.

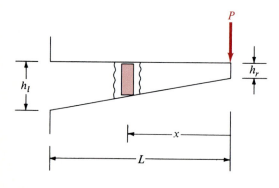

FIGURE 14.18
A tapered cantilever beam with concentrated load.

The procedure can be visualized with the help of the sample plot of Figure 14.19 of potential energy vs. the depth of the left end of the beam, as limited by stress and by upper and lower limits on the height of the right end of the beam and by a minimum allowable height at the left end. The maximum potential energy is obtained for the case shown with the heights of both the right and left ends set at minimum values.

The following example shows an iterative method of designing such a beam.

Example 14.19. Proportion a tapered cantilever beam such as that sketched in Figure 14.18 for maximum potential energy storage. Its specifications are as follows:

> Load: 5 kN
> Beam length: 0.8 m
> Width: 0.02 m
> Allowable stress: 350 MPa
> Modulus of elasticity: 207 GPa

The program starts by calculating the necessary depth, based on stress, for a uniform-depth beam and the potential energy for the uniform beam. The depth of the right end was then reduced and the potential energy calculated at each iteration and compared to the previous value. The h_l/h_r ratio was also checked, and when it passed 2, the beam depth at the wall h_l was recalculated such that the stress at the maximum stress point (the wall) was within limits. The maximum PE was reached at a ratio of h_l/h_r of 1.792.

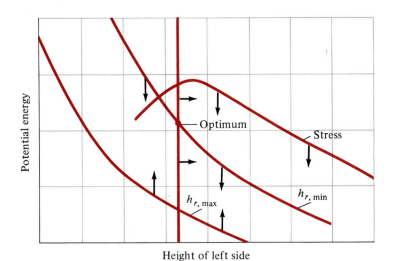

FIGURE 14.19
Computer-generated plot of potential energy in the beam of Figure 14.18, with limiting depths at the small and large ends.

PROBLEMS

Section 14.1

14.1–14.6. Determine the reactions at the supports, and construct the shear and bending moment diagrams of the beams shown in Figures P14.1 through P14.6.

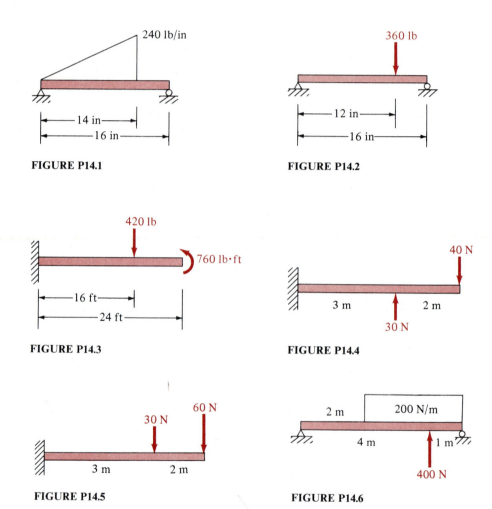

FIGURE P14.1

FIGURE P14.2

FIGURE P14.3

FIGURE P14.4

FIGURE P14.5

FIGURE P14.6

14.7–14.12. Do the tasks assigned in Problems 14.1 to 14.6, but as much as possible use the information about beams in Appendix G.

14.13. The beam of Figure P14.13 has a rectangular cross section of width 50 mm and thickness 60 mm.

    (a) Calculate the maximum shear stress in the cross section.

        Answer: 4.75 MPa.

    (b) Plot the variation in shear stress across the cross section, and demonstrate by numerical integration that the average shear stress is equal to V/A.

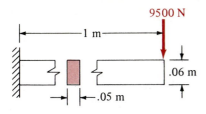

FIGURE P14.13
A uniform rectangular cross-sectional cantilever beam
with a concentrated load.

14.14–14.19. Follow the instructions of Problem 14.13 for the beams of Figures P14.1 to P14.6.

14.20. Prove or demonstrate that the maximum stress in a beam occurs either at the neutral axis or at the outside (top or bottom surface). Do this for a beam of solid rectangular cross section, with depth equal to twice the width.

14.21–14.23. In each of the beams shown in Problems 14.2 to 14.5, produce a state of pure bending somewhere along the beam by changing the magnitude of one load or adding one load.

14.24. It has been proposed to use a Douglas fir 2 × 10 in plank 10 ft long as a cantilever beam to support a load of 100 lb at the far end. Note that 2 × 10 lumber actually measures 1.5 × 9.5 in. Calculate the factor of safety for static loading, based on a modulus of rupture of 5000 psi.

14.25. The Uniform Building Code specifies that roof rafters must be 2 × 10 if their span is as much as 25 ft 6 in and if they are spaced 16 in on centers. Assume that the roof has a slope of 4:12 and that the total load is 23 lb/ft^2 projected on the horizontal, plus the weight of the rafter. For structural-grade Douglas fir, evaluate the resulting FS and maximum deflection.
    Answer: FS = 3.18.

14.26. Assuming that the plank of Problem 14.24 does not break, use direct integration of the beam equation to calculate its deflection.

14.27. The beam of Figure P14.13 has a constant width of $b = 100$ mm. Determine the depth h so that maximum stress will be constant, and equal to the flexural limit for Douglas fir, with an FS of 2.5.
    Answer: $h_{max} = 0.203$ m.

14.28. Use the depth-span relation determined in Problem 14.27, or one furnished by your professor. Check for optimality, i.e., whether a beam so proportioned will have minimum volume. Do this with "numerical variation" thus:
   (*a*) Calculate the volume of the given beam.
   (*b*) Change the relation by choosing, e.g., a different power of x. Adjust your parameters so that the maximum stress is unchanged. Calculate the volume.
   (*c*) Repeat part (*b*) with a change in the opposite direction. That is, if you had reduced the power of x, now increase it.
   (*d*) Try to vary the relation in some other way.

14.29. Consider the beam of Figure P14.3, with a single load $P = 4500$ N. The span is 12 m, and the cross-sectional proportion depth/width = 4. Use oak for the beam, and base your design on stress, with a factor of safety FS = 2.6 for a static load. Choose a moment M such that the total volume of the beam will be minimized.

14.30. Refer to Problem 14.29 for beam proportions. Assume the beam is made of aluminum 6061-T6. Use iterated numerical integration to calculate the deflection of the beam. Check your result with a formula from Appendix G.

14.31. Do Problem 14.30, using the moment-area method.

14.32. Use the method of superposition to find the support reactions for the beam of Figure P14.32.

Answer: $R_B = 1140$ N.

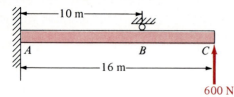

FIGURE P14.32
A uniform cantilever beam with a redundant support.

14.33. A Douglas fir beam 4 in wide, rigidly mounted to a wall, is shown in Figure P14.33. In an attempt to minimize both deflection and beam mass, the cross section is shaped such that $I = Bx^{0.6}$ (units are inches). Select B on the basis of stress, with FS = 2.5, and use integration to get the deflection curve of the beam.

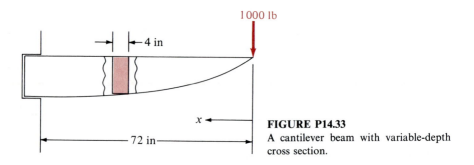

FIGURE P14.33
A cantilever beam with variable-depth cross section.

14.34. The steel shaft of Figure P14.34 is simply supported. Calculate the bearing reactions at A and B. To reduce deflection, another support is to be put at C. Calculate the bearing reactions at A, B, and C after the new support is added.

Answer: $R_c = 95.4$ N.

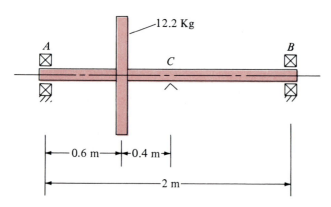

FIGURE P14.34
A shaft carrying a single mass. Another support is to be added at C.

14.35. Consider the shaft of Problem 14.34 without the center support. Write a computer program to calculate the deflection of the shaft, using the moment-area method.

14.36. Figure P14.36 shows the specifications for a beam with a hole at the center of the span, in which a pin is inserted, and a load P is hung from the pin. The design objective is to maximize the static load capacity P of the beam. The FS = 1.4, and yielding cannot be tolerated at any point.

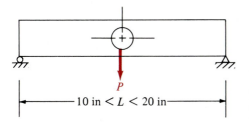

P

10 in $< L <$ 20 in

FIGURE P14.36
A beam with rectangular cross section. A pin is inserted into the hole, and a load is hung from the pin.

14.37. The beam of Figure P14.36 is to be as short as possible. Use AISI 1040, with FS = 1.4 and P = 12 000 lb. Typical properties of steel are found in Appendix C. Be sure to account for all stresses due to the applied load.

Section 14.2

14.38. Suppose the vibration isolator discussed in this chapter were to use a cantilever beam instead of a simply supported beam. Discuss the effect on the optimization procedure and results.

Section 14.4

14.39. Figure P14.39 shows a simply supported aluminum beam carrying a uniformly distributed load of 240 lb/ft of span.
(a) Calculate the maximum deflection of the beam, using a formula.
(b) Calculate the deflection of the same beam, but made of a composite material. The composite has the same modulus E as aluminum in the spanwise direction, but has an E/G ratio of 12 instead of 2.5.

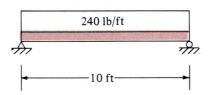

240 lb/ft

10 ft

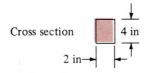

Cross section   4 in

2 in

FIGURE P14.39
Simply supported aluminum beam with uniform load.

Section 14.5

14.40. Figure P14.40 shows a steel shaft of 25-mm diameter, carrying two masses of 60 and 80 kg. Calculate the first critical frequency of vibration of the shaft.
Answer: $\omega_n = 110$ cycles/min.

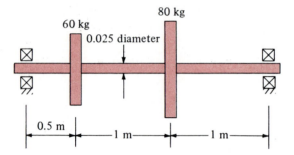

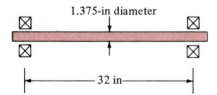

FIGURE P14.40
A uniform round rotating shaft with two masses.

14.41. The shaft of Figure P14.41 must carry two rotating masses of 20 and 40 lb each, and the masses must be no less than 4 in apart. You are to locate the masses on the shaft so that the fundamental (lowest) natural frequency will be as low as possible. Use the Rayleigh method, treating the shaft as a vibrating beam. Your instructor may require you to do this with a computer program.

FIGURE P14.41
The two masses must be located on this shaft to minimize the natural frequency of vibration.

14.42. Suppose that the masses of Problem 14.41 are centered on the shaft—i.e., are located 2 in on either side of the center. Assume that the shaft is made of AISI 1040 HR steel and that the damping ratio is 0.02.
(a) For the shaft size given, calculate the static deflection of each mass.
(b) Use Rayleigh's method to calculate the lowest natural frequency of vibration.
(c) Assume that the center of gravity of the 20-lb mass is 0.005 in from the center of the shaft. Find the deflection at the location of this mass when the shaft is rotating at its fundamental natural frequency.
(d) Find the deflection at the operating speed of 3600 rpm.
Answer: Deflection at the fundamental natural frequency of 2167 rpm is 0.042 in.

14.43. Use the Rayleigh method to show that, in the case of a single mass vibrating on a coil spring, the effect of the mass on the spring may be approximated by a concentrated mass equal to one-third the mass of the spring.

14.44. Solve Problem 14.43 by using a computer program which models the mass of the spring as a series of point masses strung on a massless spring. Start with a small number of point masses and increase until you have reduced the uncertainty to 1 percent.

14.45. Use a computer program, as in Problem 14.44, to verify the approximation developed in Section 14.5 for the mass of a vibrating beam (a concentrated mass at the center with magnitude equal to $\frac{17}{35}$ the mass of the beam).

Section 14.6

14.46. Choose the type of beam mounting (simple, cantilever, or a combination) to absorb the maximum amount of potential energy relative to any single concentrated load, per unit of steel beam weight.

14.47. Design the beam of Figure P14.47 for maximum potential energy storage. Notice that the beam is statically indeterminate. Consider two cases:
(*a*) The moment *M* is given as 10 000 in·lb.
(*b*) The magnitude of *M* is a variable in the design.

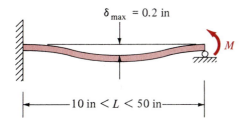

$\delta_{max} = 0.2$ in

M

10 in $< L <$ 50 in

FIGURE P14.47
A beam to be designed for maximum energy storage.

14.48. Keeping in mind that the objective, in using a beam as a vibration isolator, is to keep the natural frequency of vibration as low as possible, discuss the effects of:
(*a*) Applying an external moment anywhere
(*b*) An extra physical restraint, such as a third support
(*c*) Using a material of higher stiffness in the beam
(*d*) Changing the beam material from steel to aluminum, or to fiberglass/epoxy, or to graphite/epoxy.
(*e*) A different beam cross section:
    i. Hollow box
    ii. Wide-flange I beam
    iii. Rectangle, but with height $\gg$ width

CHAPTER
15

DESIGN FOR COMBINED LOADS

Real objects are three-dimensional, and if we are going to be precise, so are the stresses within them. As we pointed out in Chapter 4, sufficient accuracy can usually be attained by describing these stresses in two dimensions. In such cases, the stress analysis procedure of Chapter 4 and the failure criteria of Chapters 7 and 9 are appropriate.

However, there are occasions when a three-dimensional analysis is necessary. In this chapter, we discuss the analysis of complex loading in the context of a rotating shaft carrying overhung loads. The ASME transmission design standard is presented, followed by the optimization of a shaft subjected to gear loads and then of a hollow member containing internal pressure and subjected to bending and torsional loads.

15.1 TRANSMISSION SHAFTING DESIGN STANDARD—ASME

Transmission shafting is subjected to combined loads, since the forces on gear teeth or the pull of a belt or chain produces bending in addition to the torque, which is the main purpose of the shaft. Since the shaft rotates, the bending results in alternating stresses, shown on the surface stress element in Figure 15.1. In this ASME (American Society of Mechanical Engineers) standard [69], the torque on the shaft is assumed to be constant, and thus the shear stresses are also constant. Therefore, we have a combination of alternating normal and constant shear stress.

This problem could be handled with the methods developed in Section 9.5. If the distortion-energy combined-stress theory is chosen, the constant shear stress is

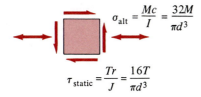

FIGURE 15.1
Stress element on the surface of a shaft
loaded in rotating-bending and torsion.

converted to an effective mean normal stress and the Goodman failure criterion is
applied.

The ASME standard evolves from work specifically directed to the problem of
cyclic tensile stresses due to rotating bending combined with static shear stresses due
to a constant torque, as indicated in Figure 15.1. Experimental measurements of
(cyclic) normal stress at failure are plotted against (mean) shear stress in Figure 15.2.

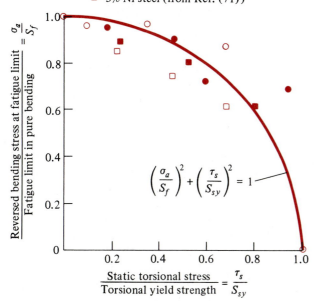

FIGURE 15.2
Fatigue limit for reversed bending in
combination with static torsion [72].

Both scales have been normalized to the appropriate material strength, as shown. The diagram is thus identical to Figure 9.13, from which the Gerber, Goodman, and Soderberg criteria derived. The ASME standard fits an ellipse to the data:

$$\left(\frac{\sigma_{alt}}{S_f}\right)^2 + \left(\frac{\tau_{static}}{S_{sy}}\right)^2 = 1 \tag{15.1}$$

Since the shear yield strength S_{sy} is frequently not found in published material data, it can be expressed in terms of the tension yield strength S_y by using expression (7.4) written at the yield point:

$$\sigma_{eff} = S_y = \sqrt{3}\tau = \sqrt{3}S_{sy}$$

or

$$S_{sy} = \frac{S_y}{\sqrt{3}}$$

Inserting this and the expressions for the stresses from Figure 15.1 in Equation (15.1), and attaching a factor of safety to the strengths, we get

$$\frac{32FS}{\pi d^3}\sqrt{\left(\frac{M}{S_f}\right)^2 + \frac{3}{4}\left(\frac{T}{S_{yt}}\right)^2} = 1$$

which gives for the diameter

$$d^3 = \frac{32FS}{\pi}\sqrt{\left(\frac{M}{S_f}\right)^2 + \frac{3}{4}\left(\frac{T}{S_{yt}}\right)^2} \tag{15.2}$$

The fatigue strength S_f is found through the process developed in Part A of Chapter 9. In this, the fatigue strength from smooth-specimen rotating-beam results is corrected for reliability, size, surface condition, stress concentration, etc. The ASME standard "does not recommend attempting to design shafts for finite fatigue life without obtaining prototype fatigue test data under simulated operating conditions."

> **Example 15.1.**[1] Determine the size for a drive shaft for a chain conveyer, sketched in Figure 15.3. The roller-chain sprocket between the bearings has a pitch diameter of 20 in and weighs 200 lb. The overhung roller-chain sprocket has a pitch diameter of 16 in and a weight of 275 lb. The shaft will be made of a 1040 steel, for which $S_{ut} = 120$ ksi and $S_{yt} = 99$ ksi. Operating temperatures are not expected to exceed 150°F (65.5°C), and the environment will be noncorrosive. The shaft must be designed for indefinite life with a survival probability of 90 percent. It will carry a steady driving torque of 138 880 lb·in and will rotate at 36 rpm. A sled-runner keyway will be used for the sprocket at the end of the shaft and a profile keyway for the center-mounted sprocket.
>
> With the stated driving torque and the 20-in pitch diameter of the center sprocket, the tension in the chain at that point is
>
> $$\text{Tension} = \frac{138\,880}{10} = 13\,888\text{ lb}$$

---

[1] We are grateful to the American Society of Mechanical Engineers for permission to use this example from the appendix to the shafting standard.

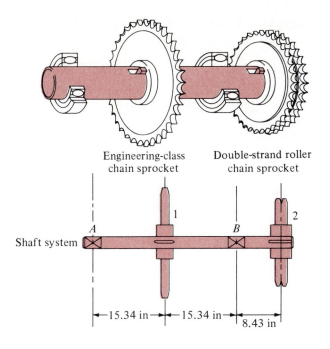

Engineering-class
chain sprocket

Double-strand roller
chain sprocket

Shaft system

←15.34 in→←15.34 in→

8.43 in

$P_1 = 14\,088$ $P_2 = 17\,635$

Loading, lb

$R_A = 2198$ $R_B = 29\,525$

$M_1 = 33\,717$

Moments, lb·in

$M_B = 148\,681$

FIGURE 15.3
Loading and moment diagrams for a chain conveyer shaft for Example 15.1.

The force P_1 on the shaft at the center is this figure plus the weight of the sprocket:

$$P_1 = 13\,888 + 200 = 14\,088 \text{ lb}$$

In like manner, the force at the end of the shaft is

$$P_2 = \frac{138\,880}{8} + 275 = 17\,635 \text{ lb}$$

These loads and the reactions at the bearings, determined from the condition of static equilibrium, are shown in the loading diagram in Figure 15.3. The moment diagram shows that the critical points are at the center sprocket and at bearing B. The shaft torque at both points is $T_1 = 138\,880$ lb·in.

We must now calculate the design fatigue strength for the shaft material at these two points. Lacking fatigue data on the actual batch of material to be used, we estimate the rotating-bending smooth-specimen strength at one-half the tensile strength:

$$S_{f\text{RB}} = \frac{120}{2} = 60 \text{ ksi}$$

This figure must be corrected per relation (9.1):

$C_R = 0.897$ for survival probability 90 percent (Table 9.1)
$C_F = 0.89$ for ground finish (Figure 9.1)
$C_S = 0.77$ for an estimated 4-in diameter, using $C_S = d^{-0.19}$
$C_I = 0.95$ assuming very light impact (Table 9.2)
$K_f = \begin{cases} 2.0 \\ 1.0 \end{cases}$ for the profiled keyway in quenched and drawn steel (Appendix I-15)
              at bearing B

Thus the design fatigue strengths become

$$S_{f,1} = \frac{(0.897)(0.89)(0.77)(0.95)(60)}{2.0} = 17.5 \text{ ksi}$$

$$S_{f,B} = \frac{(0.897)(0.89)(0.77)(0.95)(60)}{1} = 35.0 \text{ ksi}$$

With FS = 2, the shaft diameter is computed at the two critical points with (15.2):

$$d^3 = \frac{32(2)}{\pi} \sqrt{\left(\frac{33\,717}{17\,500}\right)^2 + \frac{3}{4}\left(\frac{138\,880}{99\,000}\right)^2}$$

$$= 3.59 \qquad \text{on basis of center keyway}$$

$$d^3 = \frac{32(2)}{\pi} \sqrt{\left(\frac{148\,681}{35\,000}\right)^2 + \frac{3}{4}\left(\frac{138\,880}{99\,000}\right)^2}$$

$$= 4.48 \qquad \text{based on situation at bearing } B$$

The nearest nominal shaft size is $4\frac{1}{2}$ in.

We had obtained a size correction factor by using an estimate of 4 in for the shaft diameter. For this larger shaft, the size factor becomes

$$C_S = 4.5^{-0.19} = 0.75$$

By using this more accurate figure, the design fatigue strength becomes

$$S_{f,B} = 34.1 \text{ ksi}$$

which leads via expression (15.2) to a diameter

$$d = 4.52 \text{ in}$$

which is slightly larger than the size chosen. Since this means only that the FS has been reduced slightly from the desired value of 2, we would probably not be justified in choosing a larger shaft.

15.2 MAXIMIZING THE TORQUE CAPABILITY OF A SHAFT

In this section, we discuss the particular problem of designing a shaft to transfer a maximum amount of torque. The technique, however, is suitable for optimizing nearly any facet of performance.

Figure 15.4 shows a shaft driven through a pair of gears by a motor. The loads on the shaft are like those in Figure 15.1, except for the addition of an axial load P.

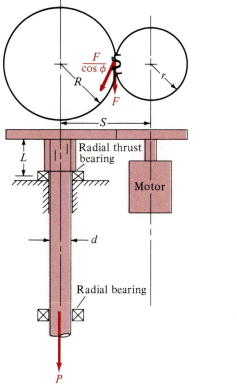

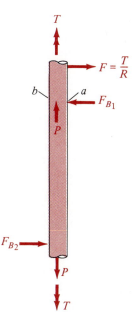

FIGURE 15.4
A gear-driven shaft with axial loading.

Power is transmitted between gear teeth by forces on the teeth, which cause torque and bending moment in the driven shaft. The bending occurs because it is not possible to put a support bearing directly in line with the driven end of the shaft. The tensile load is supported by a combination radial and thrust bearing at the upper end, next to the gear. We assume that the bearings act as simple supports, i.e., that they do not impose bending moments on the shaft.

The critical stress is likely to be where the shaft enters the thrust bearing, because that is where the overhung moment is greatest and where there is a considerable stress concentration due to the change in diameter. The force F transmitted between the gear teeth causes bending and results in the bearing reactions F_{B_1} and F_{B_2}. As shown in Figure 15.4, the force between the gear teeth is not perpendicular to the line of centers, but is at an angle ϕ, called the *pressure angle*. This means that the force between gears is larger, by a factor of $1/\cos\phi$, than would be calculated by dividing the torque by the radius. This extra force will be ignored here in the interest of simplicity.

The largest bending moment, at the thrust bearing, is

$$M = FL = \frac{TL}{R} \tag{15.3}$$

The greatest resulting stress is on the surface at points a and b, the former in compression, the latter tension:

$$\sigma_B = \frac{Mc}{I} = \frac{32TL}{\pi d^3 R} \tag{15.4}$$

As the shaft rotates, a point on its surface is one instant at a and half a revolution later at b. Thus, σ_B is an alternating stress.

The torque T and the tension P produce constant stresses in time and over the length of the shaft below the thrust bearing:

$$\tau_T = \frac{16T}{\pi d^3} \qquad \text{at surface} \tag{15.5}$$

$$\sigma_P = \frac{4P}{\pi d^2} \tag{15.6}$$

The stress situation for an element on the surface of the shaft is shown in Figure 15.5 along with the corresponding Mohr's-circle analysis. The combination of the constant torsional and tensile stresses τ_T and σ_P is equivalent to a constant maximum shear stress $\tau_{max,mean}$. The cyclic bending stress σ_B is equivalent to a cyclic maximum shear stress $\tau_{max,cyc} = \sigma_B/2$. Note that the direct shear stress in the shaft is ignored, since it approaches zero at points a and b.

We stated in Chapter 7 that the maximum-shear-stress theory of failure was more conservative than the distortion energy theory (but not greatly so) and simpler to apply. To focus attention on the optimization, we use it here. When the maximum-shear-stress theory of failure is combined with Goodman's criterion for a combination of an alternating and constant stress, we get from Equation (9.10):

$$\tau_{max,cyc} \leq \frac{S_f}{2FS} - \tau_{max,mean}\frac{S_f}{S_{ut}}$$

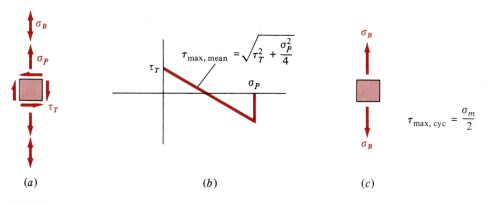

FIGURE 15.5
Stress analysis of a point on the surface of the shaft of Figure 15.4. (a) Stress element. (b) Mohr's-circle analysis for mean stress. (c) Cyclic stress.

As shown in Figure 15.5, the cyclic maximum shear stress is simply half the cyclic bending stress:

$$\tau_{\text{max,cyc}} = \frac{\sigma_B}{2} = \frac{16TL}{\pi d^3 R} \tag{15.7}$$

Inserting these maximum shear stresses in the failure criterion gives

$$\frac{16LT}{\pi d^3 R} \leq \frac{S_f}{2\text{FS}} - \frac{2S_f}{\pi d^2 S_{ut}} \sqrt{\frac{64T^2}{d^2} + P^2} \tag{15.8}$$

The objective is to obtain the greatest value of the torque T that can safely be carried by the shaft. That can be most readily accomplished by a different arrangement of inequality (15.8). First, all the terms containing T are moved to the left side:

$$\frac{16LT}{\pi d^3 R} + \frac{2S_f}{\pi d^2 S_{ut}} \sqrt{\frac{64T^2}{d^2} + P^2} \leq \frac{S_f}{2\text{FS}} \tag{15.9}$$

Now we have an expression on the left which increases monotonically with T and may not be larger than the term on the right, which is constant. It follows that the expression on the left will have its largest permissible value when the inequality is replaced by an equality:

$$\frac{16LT}{\pi d^3 R} + \frac{2S_f}{\pi d^2 S_{ut}} \sqrt{\frac{64T^2}{d^2} + P^2} = \frac{S_f}{2\text{FS}} \tag{15.10}$$

Example 15.2. Find the maximum torque that can be carried by the shaft set of Figure 15.4 with these constraints:

Shaft diameter	$d = 33$ mm (0.033 m)
Overhang	$L = 0.1$ m
Shaft spacing	$S = 0.5$ m
Gear radius	$R = 0.045$ m minimum, 0.35 m maximum

Motor power	$H = 75\ \text{kW}$
Motor speed	$n_{\text{mo}} = 1800\ \text{rpm}\ (30\ \text{r/s})$
Axial load on shaft	$P = 25\text{E} + 03\ \text{N}$
Factor of safety	$\text{FS} = 1.5$
Material	AISI 4130 steel tempered at 538°C

With corrections for size, finish, and stress concentration,

$$S_f = 138\ \text{MPa} \qquad \frac{S_{ut}}{S_f} = 6.65$$

The stress-limited torque equation (15.10) can be solved by recursion. For the first estimate, we set the axial load P equal to zero:

$$16\left(\frac{L}{R} + \frac{S_f}{S_{ut}}\right)\frac{T}{\pi d^3} = \frac{S_f}{2\text{FS}} \tag{15.11}$$

Equation (15.11), in which the axial load is ignored, produces a greater torque than permitted by the allowable stress, which is an upper limit. Inserting an intermediate value and the upper limit of gear radius gives

$$T = \begin{cases} 589\ \text{N·m} & \text{for } R = 0.25\ \text{m} \\ 744\ \text{N·m} & \text{for } R = 0.35\ \text{m} \end{cases}$$

The upper limit of gear radius R will be used.

The upper limit of T just obtained is inserted into the right side of a rearranged form of Equation (15.10), to produce a new estimate of the stress-limited torque:

$$\frac{16LT}{\pi d^3 R} = \frac{S_f}{2\text{FS}} - \frac{2S_f}{\pi d^2 S_{ut}}\sqrt{\frac{64T_{\text{est}}^2}{d^2} + P^2} \tag{15.12}$$

This is readily solved to yield $T = 740\ \text{N·m}$. This estimate of T is low because the value of T_{est} which was inserted into Equation (15.12) was too high. The difference between these first and second estimates, however, is less than 0.5 percent.

So far we have discussed only the limit placed on the torque by the strength of the shaft. There are a number of other constraints. The speed of the driven shaft must be kept between a lower and upper limit. The electric driving motor operates at constant speed, and we will assume that the power output from the motor is also constant. The available space requires that the shaft spacing S be fixed. The sizes of the gears are constrained because they must be larger than their shafts and yet must fit between the shafts. The power delivered by the motor is also a constraint.

Several design criterion equations can be developed. First we assume that the power transferred by the driven shaft is equal to that delivered by the motor

$$H = 2\pi n_{\text{sh}} T$$

giving

$$T = \frac{H}{2\pi n_{\text{sh}}} \qquad \text{N·m} \tag{15.13}$$

The limits on speed of the driven shaft result in limits on the torque:

$$T \leq \frac{H}{2\pi n_{\text{sh,min}}} \tag{15.14}$$

$$T \geq \frac{H}{2\pi n_{\text{sh,max}}} \tag{15.15}$$

Another relation results from the fact that the motor runs at constant speed, hence

$$n_{\text{sh}} = n_{\text{mo}}\left(\frac{r}{R}\right)$$

which, when it is substituted into Equation (15.13), given the relation between R and r, gives the equality

$$T = \frac{HR}{2\pi n_{\text{mo}}(S - R)} \tag{15.16}$$

This is a mathematical statement of what was said above: The output torque will increase with the gear radius R.

We can now summarize a procedure to solve this sort of problem to any desired precision. First calculate the stress-limited torque and that corresponding to the (constant) motor power at some arbitrary value of the gear radius R. A good place is halfway between the original limits. If the stress-limited torque is greater than that corresponding to motor power, increase R; if less, decrease R. The change in R should be proportional to the difference between the power-limited and stress-limited torques. Finally, the algorithm should recognize when the limits on gear radius prevent the attainment of an optimum.

Example 15.3. The problem posed in Example 15.2 is sketched in Figure 15.6, which shows that the shaft torque, for a given transmitted power and shaft diameter, is stress-limited at a gear radius of about 0.32 m. A computer program based on Equation (15.12) produced the same result:

Shaft diameter, mm	Optimum gear radius, mm		Torque, N·m	
20	45	(limit)	131	(stress limit)
33	318		699	
40	350	(limit)	928	(power limit)

For the smallest shaft diameter, the maximum torque is limited by stress in the shaft—the specified power may not be transmitted. With the intermediate diameter used earlier, an optimum is reached where the power and stress limit curves intersect. For the largest shaft diameter, the torque is limited by the available power.

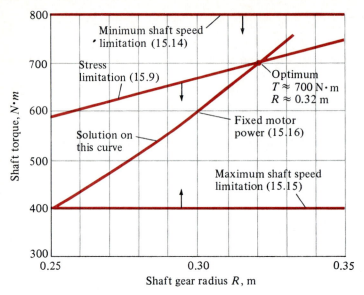

Material is 4130 steel.
Fatigue size factor $C_p = 0.75$, surface finish factor $C_S = 0.6$,
safety factor FS = 1.5, motor power $H = 75$ kW, motor speed $n_{mo} = 30$ rps,
shaft speed $n_{sh} = 15$ rps minimum, 30 rps maximum. Overhang $L = 0.1$ m,
shaft tension $P = 25\,000$ N, shaft spacing $S = 0.5$ m.

FIGURE 15.6
The torque carried by the shaft of Example 15.3 is limited, for a shaft size of 0.033 m, by stress in the shaft.

The equation which we have solved by recursion, Equation (15.10), can also be solved directly. To do this, isolate the term with a square root on one side, and square the equation. The result is a quadratic equation in T

$$\frac{16}{\pi d^3}\left(\frac{S_f^2}{S_{ut}^2} - \frac{L^2}{R^2}\right)T^2 + \frac{LS_f}{R(\text{FS})}\,T - \frac{S_f^2}{64}\left[\frac{\pi d^3}{(\text{FS})^2} - \frac{16P^2}{\pi dS_{ut}^2}\right] = 0 \qquad (15.17)$$

An upper limit for the limiting torque can be extracted with the quadratic formula:

$$T = \frac{-\dfrac{LS_f}{R(\text{FS})} \pm \dfrac{S_f}{S_{ut}}\sqrt{\dfrac{S_f^2}{(\text{FS})^2} - \dfrac{16P^2}{\pi^2 d^4}\left(\dfrac{S_f^2}{S_{ut}^2} - \dfrac{L^2}{R^2}\right)}}{\dfrac{32}{\pi d^3}\left(\dfrac{S_f^2}{S_{ut}^2} - \dfrac{L^2}{R^2}\right)} \qquad (15.18)$$

Equation (15.18) can yield several quite different values for T, depending on the relative values of the variables in it. These possibilities can be understood better with the aid of a graphical display of the failure criterion. For this purpose, we have plotted the left and right sides of inequality (15.8), each as a function of the torque T, in Figure 15.7.

The term on the right in inequality (15.8) is a constant minus a quadratic, whose plot resembles that of a parabola. The vertical position of the parabola depends on

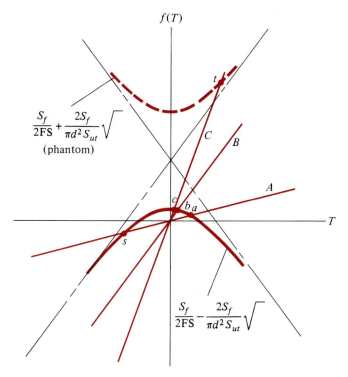

FIGURE 15.7
Graphical display of inequality (15.8).

the size of $S_f/(2\text{FS})$ relative to the size of the quadratic term when $T = 0$. The position shown in Figure 15.7 corresponds to

$$\frac{1}{2\text{FS}} > \frac{2P}{\pi d^2 S_{ut}}$$

The equations of the asymptotes to the parabola can be determined by considering very large values of T. The P^2 term then becomes insignificant, and the square root can be eliminated, leaving

$$\text{Ordinate} = \frac{S_f}{2\text{FS}} \pm \frac{16 S_f T}{\pi d^3 S_{ut}}$$

These are sketched in Figure 15.7.

The term on the left of inequality (15.8) is a linear function of T, whose plot is a straight line through the origin. The slope of the line is $16L/(\pi d^3 R)$. We can identify three possibilities for this line:

Case A Slope is less than the left-hand asymptote of the parabola.
Case B Slope is equal to the left-hand asymptote of the parabola.
Case C Slope is greater than the left-hand asymptote of the parabola.

Notice that the expressions for the slopes of both the parabola asymptote and the straight line contain the term $16/(\pi d^3)$. Eliminating this term from both expressions, we find that case A represents the situation wherein $L/R < S_f/S_{ut}$. The solution is at a in the figure and apparently at s. The solution indicated at s is a negative torque and is associated with the extension of Goodman's criterion into the negative τ_a region, which does not have physical meaning. The solution for case A, then, is given by Equation (15.18) using the plus sign.

Case B is the very unlikely situation wherein the ratios S_f/S_{ut} and L/R are exactly equal to each other. The first term of Equation (15.17) would vanish, leaving an easy solution for T:

$$T = \frac{R(\text{FS})}{L} \frac{S_f}{64} \left[\frac{\pi d^3}{(\text{FS})^2} - \frac{16P^2}{\pi d S_{ut}^2} \right] \tag{15.19}$$

which is point b in Figure 15.7.

Case C occurs when the straight line has a greater slope than the asymptote to the parabola. The solution is at point C, and there seems to be an extra solution at point t on the upper image of the parabola. It does not, in fact, exist, because the upper branch of the parabola, shown dashed, is a mathematical phantom created when Equation (15.10) was squared to produce Equation (15.17). The solution for case C is then given by Equation (15.18) using the minus sign.

15.3   THREE-DIMENSIONAL (NONPLANAR) STRESS

Sometimes it is easier to treat a three-dimensional stress situation as such rather than to go through some elaborate rigmarole to reduce the number of dimensions. What follows here is not a treatise on elasticity, but rather a few practical methods.

The stress at a location within a body is expressed as the force per unit area acting on the faces of an infinitesimal cube. There are three pairs of faces and three such forces. Each of these is resolved into three components: a normal stress acting perpendicular to the face and two shear stresses acting parallel to the face. These nine components of stress are shown in Figure 15.8. They are frequently shown in an array, called a *tensor*:

$$\begin{vmatrix} \sigma_{xx} & \tau_{xy} & \tau_{xz} \\ \tau_{yx} & \sigma_{yy} & \tau_{yz} \\ \tau_{zx} & \tau_{zy} & \sigma_{zz} \end{vmatrix} \tag{15.20}$$

The first subscript indicates the plane, and the second indicates the direction, as shown in Chapter 4. Thus the stress components on the top row act on the x plane, which is perpendicular to the x axis. And σ_{xx} is the normal component, perpendicular to the plane, acting in the x direction. Also τ_{xy} acts on the x plane, in the y direction. As its symbol indicates, it is a shear stress.

The definition and calculation of the maximum or principal stresses existing in a particular situation are discussed in Chapter 4. Some of that discussion is repeated here. The nine stress components of expression (15.20) are visualized in Figure 15.8

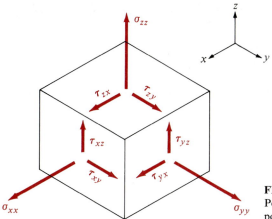

FIGURE 15.8
Positive directions for stress components at a point.

as forces acting on a cube of unit size. This cube is not a physical reality, but a set of boundaries, each having a unit area. It can be moved and rotated to suit the convenience of the analyst. As the cube is rotated, the stress components acting on its sides change.

There will always be an orientation of this cubical element for which all the shear components vanish. The three normal (tensile or compressive) stress components which remain on the faces for this orientation are termed *principal stresses*. One of these three will be the greatest normal stress existing at the point. It is designated σ_1. The least of the three is called σ_3.

It follows logically that there is another orientation of the cube for which the shear components are maximum. The largest of these, as pointed out in Chapters 4 and 7, is equal to one-half the difference between the two extreme (principal) normal stresses:

$$\tau_{\text{max}} = \frac{\sigma_1 - \sigma_3}{2} \tag{15.21}$$

The usual design tactic is to assume that the material in question is isotropic, with properties the same in all directions. With this assumption, our concern is only with how large the extreme stresses are—not their orientation. The extreme values of normal stress can be found by solving this determinant equation, in which S represents the principal stress:

$$\begin{bmatrix} S - \sigma_{xx} & \tau_{xy} & \tau_{xz} \\ \tau_{yx} & S - \sigma_{yy} & \tau_{yz} \\ \tau_{zx} & \tau_{zy} & S - \sigma_{zz} \end{bmatrix} = 0 \tag{15.22}$$

When expanded, (15.22) is a cubic equation in S:

$$S^3 + a_1 S^2 + a_2 S + a_3 = 0 \tag{15.23}$$

where
$$a_1 = -(\sigma_{xx} + \sigma_{yy} + \sigma_{zz})$$

$$a_2 = \sigma_{xx}\sigma_{yy} + \sigma_{xx}\sigma_{zz} + \sigma_{yy}\sigma_{zz} - \tau_{xy}^2 - \tau_{xz}^2 - \tau_{yz}^2 \qquad (15.24)$$

$$a_3 = -(\sigma_{xx}\sigma_{yy}\sigma_{zz} + 2\tau_{yz}\tau_{zx}\tau_{xy} - \sigma_{xx}\tau_{yz}^2 - \sigma_{yy}\tau_{xz}^2 - \sigma_{zz}\tau_{xy}^2)$$

The three values of S (the roots) which satisfy Equation (15.23) are the principal stresses.

A solution for cubic equations is available. The equations given here are the easiest to use when the roots are real, which stresses always are.

$$S_1 = 2\sqrt{-Q}\,\cos\frac{\theta}{3} - \frac{a_1}{3}$$

$$S_2 = 2\sqrt{-Q}\,\cos\left(\frac{\theta}{3} + 120°\right) - \frac{a_1}{3} \qquad (15.25)$$

$$S_3 = 2\sqrt{-Q}\,\cos\left(\frac{\theta}{3} + 240°\right) - \frac{a_1}{3}$$

in which

$$Q = \frac{3a_2 - a_1^2}{9} \qquad \theta = \cos^{-1}\frac{R}{\sqrt{-Q^3}}$$

and
$$R = \frac{9a_1a_2 - 27a_3 - 2a_1^3}{54}$$

Equations (15.25) can readily be programmed on a computer (Problem 15.14). In setting up such a computer program it may be necessary, if $\cos\theta$ is negative, to replace θ with $180° - \theta$.

Example 15.4. Consider a situation with the following stresses (all given in megapascals):

$$\sigma_{xx} = +50 \qquad \tau_{xy} = +15$$
$$\sigma_{yy} = +20 \qquad \tau_{xz} = -10$$
$$\sigma_{zz} = +10 \qquad \tau_{yz} = 0$$

Inserting these values in (15.23):

$$S^3 - 80S^2 + 1375S - 5750 = 0$$

The use of (15.25) produces the principal stresses

$$S_1 = +58 \text{ MPa} \qquad S_2 = +15.7 \text{ MPa} \qquad S_3 = +6.34 \text{ MPa}$$

15.4 MINIMIZING MASS UNDER TRIAXIAL STRESS

In the Section 15.2 we examined the case of an overhung shaft subjected to a combination of biaxial stresses. That is, the stresses at each critical point could be treated as vectors lying in a single plane. However, there are times when the significant stresses must be defined in three dimensions. An example is a tube containing a very high

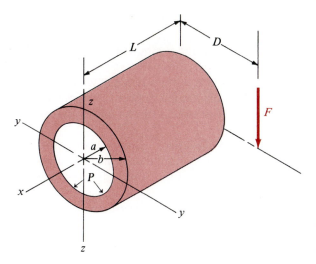

FIGURE 15.9
A thick tube loaded by internal pressure and an external force.

internal pressure which is also subjected to bending and twisting loads. Such a tube is shown in Figure 15.9.

The stresses in the tube due to the internal pressure P are given by expressions known as the *Lamé equations*. The radial ("rad") component of stress is

$$\sigma_{\text{rad}} = P\left(\frac{a^2}{b^2 - a^2}\right)\left(1 - \frac{b^2}{r^2}\right) \tag{15.26}$$

The parameters a and b are the inner and outer radii, respectively, and r is the radius to the point in question. At the inner surface $\sigma_{\text{rad}} = -P$; that is, the pressure constitutes a compressive stress. The radial component of stress is maximum at the inner surface and decreases to zero at the outer surface, since the equation is based on the assumption that the outer pressure is zero.

The tangential ("tang") component of stress is

$$\sigma_{\text{tang}} = P\left(\frac{a^2}{b^2 - a^2}\right)\left(1 + \frac{b^2}{r^2}\right) \tag{15.27}$$

The tangential component is also maximum at the inner surface. Substituting $b = na$ in Equation (15.27), we get

$$\sigma_{\text{tang}} = \frac{n^2 + 1}{n^2 - 1}P$$

at the inner surface and

$$\sigma_{\text{tang}} = \frac{2}{n^2 - 1}P$$

at the outer surface. The tangential stress is a function of only the internal pressure and the ratio n of the outer to the inner radius. When the thickness of the cylinder is small compared to the radius, Equation (15.27) reduces to the familiar hoop-stress equation $\sigma = PR/t$.

The load F, applied with both an overhang L and a moment arm D about the shaft centerline, causes bending, torsion, and direct shear stresses. These must be combined with the radial and tangential components of the stress due to the internal pressure. The nature of that combination will be different at different locations within the cylinder. Figure 15.10 shows stress elements at two locations where the largest combined stress might be expected.

The stress element shown at (a) is within the upper cylinder wall along the vertical centerline. Here the tensile bending stress σ_{xx} and the torsion shear stress τ_{xy} will reach maxima when the element is at the outer surface. The radial and tangential stresses σ_{zz} and σ_{yy}, on the other hand, are maximum at the inner surface. The transverse shear stress τ_{xz} falls to zero at both inner and outer surfaces. An element taken from the opposite side of the tube would show the same stresses, except that the bending stress would be compressive, or negative, as seen in the figure.

The stress element shown at (b) is located along the horizontal centerline. This is the neutral axis for the tube acting as a beam, so the bending stress is zero and the direct shear stress τ_{xz} is maximum. The torsional, radial, and tangential stresses have the same magnitudes as at (a), but they appear in different directions with regard to the xyz axes. The torsion stress is in the same direction as, and simply adds to, the transverse shear stress. The radial and tangential stresses are identified as σ_{yy} and σ_{zz}, respectively.

First we analyze the stresses along the vertical centerline (Figure 15.10), starting at the top. In this location, the bending stress is positive, or tensile:

$$\sigma_{xx} = \frac{4FLr}{\pi(b^4 - a^4)} \tag{15.28}$$

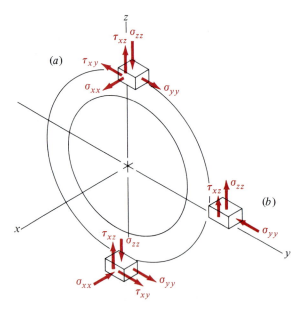

FIGURE 15.10
Stress situations at critical points in the tube of Figure 15.9.

The torsional shear stress has the same value at any radius around the tube

$$\tau_{xy} = \frac{2FDr}{\pi(b^4 - a^4)} \tag{15.29}$$

although in reference to the axis set we're using, it will be negative at the top of the tube and positive at the bottom, as shown in Figure 15.10.

The transverse shear stress is normally obtained from Equation (14.3):

$$\tau_{xz} = \frac{VQ}{Ib} \tag{14.3}$$

In Figure 15.10, it is maximum along the y axis and minimum along z. It decreases to zero, along z, on the inner and outer surfaces of the tube. The evaluation of the area moment Q in Equation (14.3) is, for this geometry, a tedious chore which would not change the conclusions of the example. Therefore, we adopt the conservative tactic of setting the direct shear equal to its average value at all points.

The bending stress is usually much greater than the direct shear stress in a situation like this, so our analysis will concentrate on the stress situation shown in Figure 15.10a. For simplicity, the conservative maximum shear theory of failure is used in this illustration. The individual stress components are calculated in the usual way and then inserted in Equation (15.23) to get the principal stresses. With the principal stresses known, the allowable stress S_y/FS is compared with the greatest difference between principal stresses, $S_1 - S_3$. If the test fails, the trial design is not strong enough, so we add material and repeat the calculation.

Of the several objectives that a design may have, we discuss one of the simplest: to achieve a satisfactory state of stress with minimum volume. We start with the applied loads, geometric limitations, and available material. Inside and outside diameters are set at minimum limits, and the failure criterion is checked. If the stress is too high, the outside diameter is increased. Otherwise, we increase only the inside diameter (which reduces volume and thus mass) until the calculated stress is at its limit.

The outside diameter is increased one increment, and the process is repeated. The process is stopped when an increase in outside diameter results in an increase, rather than a decrease, in volume.

Example 15.5. Proportion the tube of Figure 15.9, with these specifications:

Minimum inner radius	$a = 0.19$ m
Overhang	$L = 0.6$ m
Moment arm	$D = 0.4$ m
Internal pressure	$P = 10$ MPa
Force	$F = 0.4$ MN
Material, AISI 4130	$S_y = 918$ MPa
Factor of safety	$FS = 4$

The first calculations will assume the inner and outer radii to be $a = 0.19$ and $b = 0.20$ m. At the inner surface at the top, we find these stresses, in megapascals:

Radial	Tangential	Bending	Torsion
σ_{zz}	σ_{yy}	σ_{xx}	τ_{xy}
-10	195.1	195.6	-65.2

These are substituted into Equation (15.23), along with the average direct shear stress of 32.6 MPa, which was assumed to be the same at all points. The principal stresses, from Equation (15.23), are 262.6, 133.7, and -15.6 MPa. The largest difference, $262.6 - (-15.6) = 278.2$, divided by 2, is the maximum shear stress. It is larger than the allowable stress $S_y/(2\text{FS}) = 918/[(2)(4)] = 114.75$. Evidently, the tube will have to be made thicker. Before we go on to that, it is interesting to calculate the individual components, and the maximum shear stress, as the radius increases from inside to outside. This is shown in Table 15.1.

With the high internal pressure of 10 MPa, the tangential stress dominates, and the maximum shear stress is greatest at the inner surface. Bending and torsional stresses are greatest at the outer surface.

If the bending and torsional stresses are made greater by increasing the force F, a situation results wherein the maximum shear stress is greatest at the outer surface, as shown in Table 15.2.

Another possible location for the greatest maximum shear stress is the bottom of the tube. Here the bending stress is compressive, or negative. Also the torsional stress becomes positive. Table 15.3 shows that, for the original loading, the maximum shear stress is indeed highest at the bottom of the tube.

We have shown that the location of the critical maximum stress may be on the inside or outside surface, depending on the situation. The point to remember is that, in a combined load situation, the location of the critical maximum stress may not be obvious. Prudence dictates that the algorithm check stresses all the way through the

TABLE 15.1
Stresses in the tube wall at the top

Internal pressure $P = 10$ MPa, force $F = 0.4$ MN; dimensions in meters; stresses in megapascals

Radius	Radial	Tangential	Bending	Torsional	Maximum shear
0.190	-10	195	196	-65	139
0.192	-7.9	193	197	-66	138
0.194	-5.8	191	200	-67	138
0.196	-3.8	189	202	-67	137
0.198	-1.9	187	204	-68	137
0.200	0.0	185	206	-69	136

Greatest maximum shear stress = 139 MPa at $r = 0.19$

TABLE 15.2
Stresses in the tube wall at the top with larger force

Internal pressure = 10 MPa, external force = 1 MN, stresses in megapascals

Radius, m	Radial	Tangential	Bending	Torsional	Maximum shear
0.190	−10	195	489	−163	299
0.192	−8	193	494	−164	301
0.194	−6	191	499	−166	302
0.196	−4	189	505	−169	304
0.198	−2	187	510	−170	306
0.200	0	185	515	−172	307

Greatest maximum shear stress = 307 MPa at $r = 0.200$

cylinder wall each time that a significant change is made in proportions or loads. The algorithm used is simple; the inside and outside radii are set at minimum values, and the maximum shear stress is calculated. Then the outside radius is adjusted until the maximum shear stress is within a preset tolerance of the allowable stress $S_y/(2FS)$. The volume per unit length is calculated. The inside radius is increased a small amount, and the process is repeated. The new volume is compared with the old. If the increase in inside radius has produced a smaller volume, then this new inside radius and volume become the standard, and the inside radius is increased again. When the increase in radius has caused an increase in volume, we know that the previous radius and volume are optimal values.

Example 15.6. Assuming that the stress at the bottom outside surface remains critical, we now find inner and outer radii for the tube of Example 15.5 which minimize the volume per unit length.

The results of a computer run are shown in Table 15.4. We might conclude from Table 15.4 that an inside radius of 0.192 is more optimal than 0.190. However, the program contained a "glitch." A second computer run, with the tolerance on stress reduced from 1.0 to 0.1 percent, yields a different conclusion, as shown in Table 15.5.

TABLE 15.3
Stresses (in megapascals) at the bottom of the tube with loads as for Table 15.1

Radius, m	Radial	Tangential	Bending	Torsional	Maximum shear
0.190	−10	195	−195	65	209
0.192	−8	193	−197	66	209
0.194	−6	191	−200	67	209
0.196	−4	189	−202	67	209
0.198	−2	187	−204	68	209
0.200	0	185	−206	69	210

Greatest maximum shear stress = 209.7 MN at $r = 0.200$ m

TABLE 15.4
Optimization

Radius,* m			Volume
Inner	Outer	Maximum shear stress, MPa	Unit length
0.1919	0.208	114.7	0.0224
0.1919	0.210	115.6	0.0222
0.194	0.211	115.8	0.0223

Optimal proportion: $r_i = 0.1919$ $r_o = 0.210$
Volume/unit length = 0.0222 m³/m

* Minimum internal radius is 0.19 m.

TABLE 15.5
Optimization with a closer tolerance

Radius, m			Volume
Inner	Outer	Maximum shear stress, MPa	Unit length
0.190	0.208	114.7	0.0222
0.192	0.210	114.7	0.0224

TABLE 15.6
Optimization without inside-radius limit

Radius, m		Volume
Inner	Outer	Unit length
0.12	0.149	0.02406
0.16	0.181	0.02201
0.17	0.189	0.0220
0.18	0.199	0.0221

Table 15.5 shows that the minimum specified inner radius is the optimum; any increase results in an increase in volume. Would a smaller inside radius be better? We check this by starting the optimization at the much smaller inside radius of 0.12 m. Table 15.6 identifies an optimal inside radius of 0.17 m.

To calculate the stresses used in Example 15.5, Equation (15.23) was solved by a subroutine based on relations (15.25). The three roots are returned unranked.

The principal stresses have to be ranked from greatest to least to calculate the maximum shear stress. This ranking is accomplished with a sorting procedure called a *bubble sort*. These routines are available on most computers. However, for the many microcomputers that do not have this ability, we include the flowchart for a bubble sort as Figure 15.11.

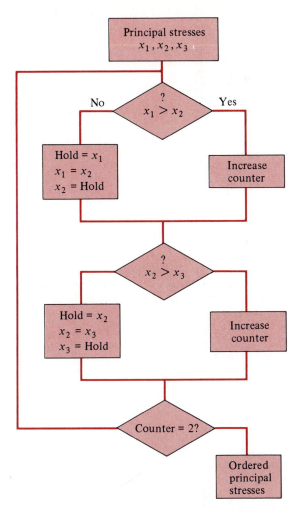

FIGURE 15.11
Flowchart to place principal stresses in order.

A complete optimization program would, after optimizing on the location thought to be critical, check the resulting geometry at other locations, where the tensile bending stress is maximum and where the shear stress is maximum.

PROBLEMS

Section 15.1

15.1. The sketch in Figure P15.1 shows a gear-driven shaft supported on two bearings. The gear has a pitch diameter of 160 mm. The shaft, to be made of 1040 steel cold-drawn 20 percent (Appendix D), is to transmit 900 horsepower (hp) at 800 rpm with 95 percent reliability. Size the shaft according to the ASME standard, with FS = 2.
   Answer: 87 mm.

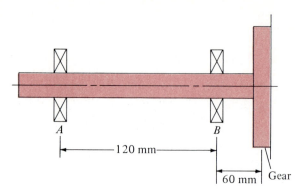

FIGURE P15.1
A gear-driven shaft supported on bearings at *A* and *B*.

15.2. Estimate the factor of safety of the shaft of Example 15.1 for fatigue failure by the ASME standard. Use the shaft size and material chosen in the example and the same power transferred, but double the shaft speed. That is, use a shaft speed of 72 rpm. Also assume a moderate impact.

15.3. In Figure P20.9 (page 626) the pitch diameter of the gear is 70 mm. The intertooth force $F = 700$ N, and the speed is constant at 1500 rpm. Bearing *A* is a no. 106 ball bearing (Table 20.2, page 610), and bearing *B* is a no. 304 ball bearing. Estimate the factor of safety for fatigue failure by using the ASME standard, with 98 percent reliability. The shaft material is steel with a tensile strength of 900 MPa.

Section 15.2

15.4. Use your microcomputer to do the calculations (and plotting, too, if you have such a program) for the curves sketched in Figure 15.7. Use shaft diameters of 20 and 40 mm. Do you agree with the remarks in the text concerning this figure?

15.5. Recalculate Example 15.2 for an overhang of 0.5 m. Use the recursive technique of Equation (15.12).

15.6. Do Problem 15.5, but use Equation (15.18) and show the results on a plot like Figure 15.7.

15.7. Repeat the calculations of Example 15.2, but include the effect of the gear's separating force shown in Figure 15.4.

15.8. Suppose, in Example 15.2, we were to double the motor power available. What shaft and gear sizes will permit maximum torque?

15.9. Do Problem 15.6 for the conditions of Problem 15.7.

15.10. (A class project) Program your computer to solve Equation (15.18) for any given set of inputs such as those of Example 15.2. Check the considerations of Figure 15.7, and plot the results as in Figure 15.6.
  Answer: 742, 2400 (a "phantom").

15.11. Figure P15.11 shows a gear mounted on a shaft between bearings. The shaft is turned by a swash plate, not shown, which delivers 8 kilowatts (kW) at 3600 rpm. Because the function of the swash plate is to convert force parallel to the shaft to torque, a force of 3200 N pulsating (not reversing) 4 times per revolution is carried by the shaft to a thrust bearing at *B*. Reliability and cost are paramount needs in the machine of which the

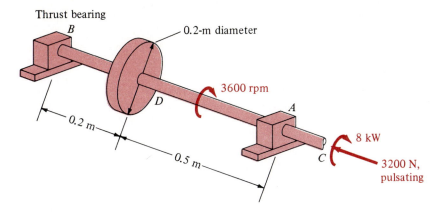

FIGURE P15.11
A shaft delivers power from a swash plate at C to a gear at D.

shaft is a part. With this in mind, choose a suitable steel, bearing locations, and shaft configuration. Provide drawings and specifications as required by your professor.

15.12. Use the problem specifications of Problem 15.11, but design for minimum weight and a 2-h lifetime.

Section 15.3

15.13. Use the routine of Equation (15.25) to solve the cubic

$$S^3 - 5.5S^2 - 6S + 18 = 0$$

15.14. Put a program on your microcomputer to produce the roots of a cubic equation, using Equations (15.25). Demonstrate your program with the equation

$$S^3 - 9.1S^2 - 0.1S + 44.2 = 0$$

Answer: -2, 8.5, 2.6.

15.15. Write a computer code to accept three numbers, such as the three roots produced by (15.25), and place them in order of greatest (largest positive magnitude) to smallest.

15.16. The stresses in a machine part have been calculated as $\sigma_{xx} = -24$, $\sigma_{yy} = +56$, $\sigma_{zz} = +10$, and $\tau_{xy} = +12$ kpsi. Calculate (*a*) the principal stresses and (*b*) the maximum shear stress. As directed by your instructor, do this by hand or modify the computer codes of Problems 15.14 and 15.15 to start with stresses.

Answer: 10, 57.7, -25.7.

15.17. Follow the instructions of Problem 15.16 for this stress state, expressed in megapascals:

$$\sigma_{xx} = 215 \qquad \sigma_{yy} = 45 \qquad \sigma_{zz} = -135 \qquad \tau_{xz} = +67$$

Section 15.4

15.18. For Example 15.5, calculate the maximum shear stress for outside radius $r_o = 0.20$ m and inside radius chosen to use the same minimum volume of material as given in Table 15.6.

15.19. (Should be assigned to follow Problem 15.18). Construct a computer program which will ask for an outside radius, and then choose an inside radius to keep the stress below the allowable limit. Output should include the inside radius and the total material volume.

15.20. (Follows Problem 15.19). Expand the program to search for the optimum inside and outside radius, using an interval-halving routine. Use material volume as a criterion of optimality.

15.21. A rocket booster tank is filled with liquid at a pressure of 10 atmospheres (atm). Lateral accelerations (wobbles, if you will) cause an alternating bending moment at the base of the tank which we assume to be constant at $8E + 08$ N·m. The material to be used is aluminum 7076-T6. The diameter of the tank must be between 8 and 20 m. Determine an appropriate factor of safety and fatigue life.

15.22. (To follow Problem 15.21). After you and your instructor have agreed on an answer to Problem 15.21, determine an optimum tank diameter and wall thickness (criterion: minimum weight).

15.23. A typical axle shaft on a rear-wheel-drive automobile carries not only the driving torque from the engine but also the other loads which result from its function of connecting the wheel to the car. These loads include the weight of the car and the side loads due to cornering and hitting curbstones. Assuming that the shaft will be made of AISI 1060 steel stress relieved at 900°F, that the maximum engine power is 120 hp, and that the vehicle's gross weight is 1180 kg (60 percent on driving wheels), determine the minimum permissible axle diameter. Assume that the car uses size R-13 tires and that maximum cornering forces correspond to the maximum friction coefficient between rubber and dry pavement.

   Answer: $d = 0.036$ m.

15.24. Refer to Example 15.5. Write a computer code to do the problem as done in the example, and check that you get the same results.

15.25. With the code of Problem 15.24, check the effect of ignoring the direct shear-stress component.

15.26. With the code of Problem 15.24, check against the possibility that the optimal solution given in the text results in unacceptable stresses at other locations in the tube.

15.27. Write a computer code to calculate the transverse shear stress at any point in the tube of Example 15.5 by a numerical solution of Equation (14.3). Use this to check the results of Example 15.5 at the critical points.

   Answer: Maximum shear stress = 163 MPa.

15.28. Correct the computer code of Problem 15.24 so that, in the first iteration, an external radius exactly equal to the minimum specified is used.

15.29. Use the computer code developed for one of the previous problems to produce a table of optimum radii for the problem of Example 15.5, for the following conditions:

   Maximum outer radius: 0.24 m
   Minimum inner radius: 0.12 m
   Overhang: 0.4 to 1.6 m
   Moment arm: 0.4 to 1.6 m
   Internal pressure: 4 to 40 MPa

CHAPTER 16

DESIGN WITH COMPOSITE MATERIALS

Engineering became a profession in the age of iron, and we think of ourselves as primarily artisans of the special iron alloy known as *steel*. We don't think of steel as a composite, but it is. Composites, both natural and artificial, have been used for as long as there have been historical records.

This chapter starts with a discussion of an old composite known as portland cement concrete. We do this because concrete is an engineered composite and because it may be easier for you to understand the science of composites when it is applied to a familiar material. There is nothing new, then, about composites. What is new is a vastly improved knowledge of how the constituents of composite materials work together and a profusion of new, synthesized materials to use in the making of composites.

16.1 FAMILIAR COMPOSITES

A Natural Composite

Wood, when available, has always been a favorite construction material for people. All woods are composites of cellulose fibers held in a matrix of lignin. As we all know, different woods have different properties. The differences are due in large part to the size and length of the fibers and how they are arranged. We will see that the properties of all composites are affected strongly by those factors.

An Artificial Composite

PORTLAND CEMENT CONCRETE. Ordinary clay mud, when it is reinforced with straw and dried, makes the serviceable building material called *adobe*. Sand makes the material better. When the right mixture of clay and sand is baked, it is transformed to the hard, waterproof material we call *brick*. Similarly, when a mixture of clays and limestone is ground, "fired" in a kiln, and then reground, the resulting powder can be used as cement. In use, portland cement[1] is mixed with sand, gravel, and water. The water reacts with the cement to form the artificial stone that we call *portland cement concrete*. Its strength comes largely from the sand and gravel; in the right proportions[2] [31], these angular particles fit closely together and are held in place by the cement.

Even the best portland cement concrete has little tensile strength. A high-quality mix having 4000-psi compressive strength may have only 650-psi tensile rupture strength, as measured in a bending test. Structures of marvelous complexity and grace have nevertheless been built by utilizing the properties of the arch, in which even the material on the bottom side is always loaded in compression.

Despite its weakness in tension, concrete is routinely used, with reinforcement, to make beams. Since concrete is weak in tension, we provide strength on the tensile side by incorporating stiff reinforcing rods. These rods are usually made of steel. This may seem contradictory, because steel, as heat-treated and fabricated for reinforcing rods, is very ductile. Concrete, on the other hand, is brittle. Yet steel is more than 12 times stiffer than concrete. At any given overall strain, the stress in the steel will be much higher than that in the concrete. Thus, we are able to fabricate a beam in which the ductile but stiff steel rods have been stretched enough to take up the load before the brittle, but not nearly as stiff, concrete has reached its fracture stress. The steel supports the tensile loads. The concrete holds the steel rods in position and provides compressive strength and protection from the elements.

To design a composite beam is more complicated than to design one made entirely of one material, because the moment of inertia of the cross section is not simply a function of the shape. The problem is usually handled by assigning to each material in the cross section an area proportional to its stiffness. This produces a new cross-sectional shape that can be analyzed as though it were composed of all the same material.

> **Example 16.1   Reinforced concrete beam.** Find the moment of inertia of a concrete beam 12 in wide and 24 in deep. The beam is reinforced with six no. 6 ($\frac{6}{8}$- or 0.75-in-diameter) steel reinforcing rods.
>
> When the beam is loaded, the strain in steel and the strain in concrete are equal, but the stress in the steel is greater than that in the concrete by the ratio of their

---

[1] It is so called because of its resemblance to stone quarried on the isle of Portland, England.

[2] For each part by volume of cement

High-strength: 1 part sand, 2 parts gravel

Low-strength: 3 parts sand, 6 parts gravel

and 5 to 7 gal water per sack of cement.

stiffnesses. Here that ratio is 15. Thus, the force exerted by the steel in this example is equal to that exerted by 15 times as much concrete. In calculating the moment of inertia on the tension side, the stiffness of the concrete is ignored. The actual area of steel is replaced by its equivalent, in terms of stiffness, of concrete.

Figure 16.1 shows the actual beam and its "all-concrete" equivalent. The neutral axis passes through the centroid of the all-concrete cross section. Referring to Figure. 16.1b, we write

$$15(6)\left(\frac{3}{4}\right)^2\left(\frac{\pi}{4}\right)(22 - h) = 12h\left(\frac{h}{2}\right)$$

$$h = 9.21 \text{ in}$$

The moment of inertia is

$$I = \frac{12(9.21)^3}{3} + 15\left[\frac{6\left(\frac{3}{4}\right)^2\pi}{4}\right](22 - 9.71)^2$$

$$= 9629 \text{ in}^4$$

A "balanced" reinforced-concrete beam is one in which the steel on the tension side and the concrete on the compression side reach their respective failure stresses at the same time. The classic beam formula $\sigma = My/I$, Equation (14.1), predicts that these maximum stresses will be in the ratio of their distances from the neutral axis: steel stress/ concrete stress $= (22 - 9.21)/9.21 = 1.388$. The ratio of the failure stress of steel to that of concrete is about 10, so the steel appears not to be fully utilized. However, the beam

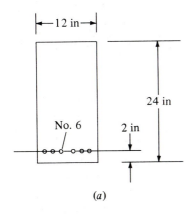

(a)

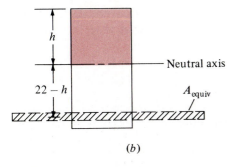

(b)

FIGURE 16.1
(a) Cross section of a reinforced-concrete beam. (b) The equivalent beam, showing a concrete area equivalent to the steel rods.

is not as out of balance as it appears in this calculation, because the compressive stress in the concrete is not maximum at the surface, as the classic beam theory predicts. Instead, it falls nearly to zero at the surface. The maximum compressive stress is found well inside the concrete, where the lateral restraint makes it possible for the material to reach its full compressive strength. When this behavior is taken into account, the beam is much closer to "balance."

Steel-reinforced concrete is not limited to roadways, bridges, or building frames. By imaginative use of steel mesh or pierced plate for reinforcement, architects such as Nervi[1] have produced buildings of great beauty and complexity. Concrete boats are sailing the seas, and reinforced concrete is a candidate material for parabolic solar reflectors. The combination of lightweight concretes (by incorporating a volatile foaming agent or hollow glass beads) and reinforcing fibers such as glass or carbon will make possible some surprising uses of concrete in the future.

HIGH POLYMERS. The popular term for high polymers is *plastic*. Plastics certainly do not seem like composites. They usually appear smooth, shiny, and homogeneous, and we would expect to find that uniformity existing down to the molecular level. But such is not the case. High polymers are composites, and we can understand their behavior only when we think of them as composites. Consider the ubiquitous clear plastic of throwaway cups and cheap sunglasses: polystyrene. While the polystyrene of today is synthesized from petroleum, it is chemically a cousin of the naturally occurring material called *amber*. The molecule of polystyrene is an extremely long (some 40 000 units) chain of carbon atoms with bulky six-carbon groups called *benzene rings* attached at regular intervals. The molecules may be branched, like a grapevine. Although the carbon units which are the backbone of the molecule are held at definite angles with respect to each other, these "links" can rotate freely. In other words, the chain is flexible.

The bulky side groups, branching, and the flexibility of the polystyrene molecules result in their becoming entangled with each other. This entanglement and a relatively weak electrostatic attraction called the *van der Waals force* account for the stiffness of polystyrene. When the temperature is raised, the resulting increased molecular motion effectively weakens the van der Waals bond, and the molecules can be made to move relative to one another, just as one can sort out a gob of fishing worms by soaking them in water. Thus, because polystyrene can be reshaped by heating, it is called a *thermoplastic*.

The properties of a high polymer are determined in large degree by how the "fibers" (molecules) are arranged. The polyethylene of squeeze bottles and electric insulation is also a branched chain of carbon atoms, but without the attached side groups of polystyrene. It is, therefore, much less stiff than polystyrene.

---

[1] Pier Luigi Nervi (1891–1979), an Italian engineer and architect, did most of his work in concrete. He developed "ferrocimento," consisting of layers of steel mesh sprayed with a cement mortar. His structures include a stadium in Florence, the Pirelli building in Milan (Italy's first skyscraper), two sports palaces for the 1960 Rome Olympic Games, and the UNESCO headquarters in Paris. He was professor at the University of Rome and held a named chair at Harvard.

Both polystyrene and "squeeze-bottle" polyethylene are regarded as *amorphous*, or noncrystalline. This is because the irregularity of the chains, and the bulky side groups of the polystyrene, prevent the molecules from forming the neat rows which are known to polymer scientists as *crystallinity*.

Polyethylene can be made with molecules which are longer and more regular in shape. In this form, it is possible for the molecules to arrange themselves in neat rows, like telephone poles stacked in a storage yard. With all units of adjoining molecules thus very close to one another, the material is stiffer and stronger and has a higher melting point. Recent chemical research has produced polymer molecules which are not only unbranched but also very stiff. Fibers made from such polymers have strengths which exceed those of the finest steel.

Table 16.1 lists the physical properties of some popular thermoplastics, as reported by manufacturers [33].

We have inferred that the bonds between the carbon units in a polymer molecule are very strong, which indeed they are. Diamond, for example, is a three-dimensional linkage of carbon atoms, one of the hardest materials known. These links are called *covalent bonds*. When high polymers are made or treated so that there is an extensive three-dimensional covalent linkage between molecules, a material is created which does not melt. The process is called *cross-linking*. The resulting polymer is called a *thermoset*, because the cross-linking reaction often requires elevated temperature. See Table 16.2.

The original thermoset plastics were materials which occur naturally. Rubber is an excellent example. Natural rubber is actually the sap of the rubber tree (*Hevea Brasiliensis*), a "latex" of particles suspended in water. When most of the water is driven off, it becomes a soft, springy material very useful as an eraser. With further processing, natural rubber becomes a soft, rubbery (sorry, but there is no better word) plastic known as a *plastisol*.

Charles Goodyear discovered accidentally in 1839 that if natural rubber containing 5 percent sulfur were heated, the rubber became the tough material we know so well. In this condition, rubber is a genuine thermoset. More sulfur produces more cross-links, so that rubber can be made into a material hard and heavy enough to be a bowling ball.

TABLE 16.1
Properties of a few thermoplastics

	Specific gravity	Tensile strength, ksi	E, Mpsi	Impact (notch), ft·lb	Distortion temperature, °F
ABS*	1.05	6	0.32	4.4	195
Acetal	1.42	8.8	0.4	1.3	230
Nylon 6 alloy	1.14	11.8	0.4	1	167
Polycarbonate	1.2	9	0.33	60	265
Polyethylene HD	0.95	2.6	0.2	0.4	120
Polystyrene	1.07	7	0.45	0.45	180
Polyamide-imide (Torlon 4203L)	1.39	27.8	0.7	2.7	532

* ABS is an alloy or copolymer of acrylonitrile, styrene, and butadiene rubber.

TABLE 16.2
Properties of a few thermosets

Natural rubber (*Hevea*)
    Elastic modulus at 300 percent elongation   1770 psi
    Tensile strength   4050 psi
    Elongation at fracture   490%
Synthetic rubber: Many, many types and compounds are available.

	Specific gravity	Tensile strength, kpsi	Elastic modulus, Mpsi	Impact, ft·lb	Heat distortion temperature, °F
Polyester	1.31	8.5	0.34	1.2	130
Polyester, 30 percent fiberglass	1.52	19.5	1.4	2.5	430
Epoxy	1.1–1.4	4–13	0.35	0.2–1.0	115–550
Epoxy glass	1.6–2	10–20	3.04	2–30	250–500

Rubber can also be cross-linked by oxygen; automobile tires are thus doubly attacked by the elements. Oxygen diffuses into the rubber and, by cross-linking it, reduces its flexibility. At the same time, ultraviolet radiation breaks the covalent bonds in the main polymer chains, reducing strength. Tire manufacturers defend against this attack by incorporating carbon black. That's why tires are black.

FIBER-REINFORCED COMPOSITES. So far, we have discussed the naturally occurring composite wood, the artificial composite of portland cement and steel "fibers" called reinforced concrete, and the high polymers. In this section we discuss deliberate compositions of strong, stiff fibers in a matrix of softer material.

The science of fiber reinforcement could be said to have begun at the end of the 19th century when strong, tough insulators were needed for the developing electrical industry. Phenolic plastic had the insulating properties, and the addition of cellulose or asbestos fiber provided the necessary toughness. Polyester and epoxy casting resins have since made possible a wide range of inexpensive articles. As fiberglass developed from "spun glass" to the long, consistent, and strong filaments available today, it has displaced the older fibers.

Polyester plastic is widely used as a matrix. It is inexpensive and easy to use. However, the family of thermosetting polymers known as *epoxies* has come to be the standard for engineered composites because of their superior properties. Another advantage of the epoxies is that they are "addition" polymers. Nothing is given off during the curing process. All the ingredients used become part of the finished object. *Condensation polymers*, such as polyester, release water during the cure, which must be vented somehow.

A composite of glass fibers in epoxy is similar to steel-reinforced concrete in that the glass fibers, being about 20 times stiffer than the epoxy, bear most of the applied loads.

Example 16.2. A laminate of 60 percent, by volume, straight "S" glass fibers and 40 percent epoxy is subjected to a tensile load in the direction of the fibers. Perfect bonding is assumed, which means that the strain in the fibers is exactly equal to the strain in the epoxy. In a cross section of unit area, the load carried by the glass at strain ε is

$$F_{\text{glass}} = 0.6(13E + 06)\varepsilon = 7.8E + 06\varepsilon \text{ lb}$$

and that carried by the epoxy is

$$F_{\text{epoxy}} = 0.4(0.35E + 06)\varepsilon = 0.14E + 06\varepsilon \text{ lb}$$

The epoxy, in addition to holding the fibers in place and protecting them from abrasion, greatly increases the strength of the composite by transferring loads from the occasional broken fiber to its unbroken neighbors.

Fiberglass is still the most widely used reinforcement fiber, due to its low cost and high tensile strength. The fibers are made of a glass much like that used for light bulbs. The fibers are soft and pliable to the touch, almost silken. The softness is an illusion, caused by the very small diameter of the fibers (usually about 3/10 000 in). The strength of glass filaments, like the illusion of softness, is due to their small diameter. The glass in a light bulb has a breaking strength of about 6 ksi (44 MPa), but its strength approaches 1000 ksi (6895 MPa) when it is spun into a filament.

The smaller the fiber, the smaller the flaw which can exist. Griffith [43] showed many years ago that the fracture strength of fine glass fibers could approach the theoretical limit of more than 1 million psi. The average breaking strength of the glass fibers now used in high-performance composites is, in fact, above 600 000 psi. The principal drawback of glass fiber as a reinforcement is its low stiffness—less than one-half that of steel.

Carbon and boron fibers have the advantage of high stiffness. The fibers themselves are as much as twice as stiff as steel, and they have nearly the same tensile strength as fiberglass. Fibers of tungsten, sapphire, iron, and other exotic materials have very attractive properties, but their cost has kept them from substantial use.

Nylon fibers, the inexpensive material of so many fishing lines (and stockings), are widely used to reinforce thermoplastics. A new family of "tailored-molecule" synthetic fibers is appearing, which combines the light weight and some of the resiliency of nylon with a strength and a stiffness comparable to that of steel. A well-known example is aramid fiber, first used as a reinforcement in tires. Table 16.3 lists the properties of some high-performance fibers.

TABLE 16.3
High-performance fibers, listed in order of elastic modulus

	Density, lb/in³	Elastic modulus, Mpsi	Tensile strength, kpsi
S glass	0.09	13	580
Aramid (kevlar 49)	0.052	18	400
Iron	0.283	29	1900
Steel	0.28	30	580
Nickel	0.324	31	560
Beryllium	0.066	35	128
Alumina	0.143	60	2800
Boron carbide	0.091	65	930
Carbon (PAN)	0.072	75	300
Silicon	0.115	122	1600
Graphite	0.06	142	3000

16.2 DESIGN OF ENGINEERED COMPOSITES

Design for Stiffness

ISOTROPIC REINFORCEMENT. The adjective *isotropic* denotes a material whose properties are the same regardless of the direction of measurement. This is the standard assumption of any beginning study of strength of materials, and it is a good assumption for unreinforced portland cement concrete, cast iron and steel, and the "plastic steel" sold by your local hardware store. In all these cases, the matrix is reinforced by hard particles about as long as they are wide. In the case of the concrete, the reinforcement is gravel and sand; in cast iron and steel, it is a mixture of graphite and crystals of the metal. Plastic steel is, in the better quality stuff, epoxy resin reinforced with particles of steel. Short fibers are also used to improve the stiffness and strength of ductile materials, such as glass fibers in nylon. The primary reason for using short fibers rather than long ones is that the fibers can be added to the matrix and the resulting mixture dispensed and formed as though it were, in fact, homogeneous. Short-fiber reinforcement is also a good deal less expensive. Many automobile parts are made of short-fiber composites, formed by injection molding.

The simplest design rule for a material composed of a homogeneous matrix reinforced by particles of random shape is the *rule of mixtures*, which states that the property of the composite is the sum of the contributions of matrix and reinforcement:

$$P_c = P_m v_m + P_r v_r \tag{16.1}$$

where v_m is the proportion by volume of the matrix and v_r is that of the reinforcement. The rule of mixtures is exact for the density of the composite, as one would expect. It is also quite good for the elastic modulus and tensile strength in the range from 20 to 80 percent reinforcement by volume.

> **Example 16.3.** We are to estimate the properties of a plastic steel composed of epoxy resin reinforced with 30 percent by volume steel flakes, well mixed in. Except for such porosity as may have been caused by the mixing operation, the density can be calculated exactly.
>
> $$\text{Density} = 0.043(0.7) + 0.28(0.3) = 0.115 \text{ lb/in}^3$$
>
> The stiffness and strength can be estimated:
>
> Stiffness (Young's modulus):
>
> $$E = 0.35(0.7) + 30(0.3) = 9.245E + 06 \text{ psi}$$
>
> Yield strength:
>
> $$S_y = 10(0.7) + 40(0.3) = 19E + 03 \text{ psi}$$

ORTHOTROPIC FIBER REINFORCEMENT. The highest possible stiffness and strength are obtained in composites in which the fibers are continuous throughout the finished article and placed parallel to each other.

The reinforcement may be supplied as a tape of parallel filaments held in place by a partially cured matrix or as woven or felted cloth. In general, engineered products

FIGURE 16.2
The Lear fan, prototype of an airplane made entirely of fiber-reinforced plastic.

must have strength and durability in more than one direction. For this reason, a composite made with tapes or unidirectional mats must be built up in layers placed at angles to one another. The properties of the resulting composite are not isotropic; their values depend on the direction of measurement. Those properties can, however, be described with reference to a set of mutually perpendicular axes. Such axes are called *orthogonal*, and such a laminate is called *orthotropic*. Orthotropic laminates are found in filament-wound pressure vessels, snow skis, and major components of aircraft, such as wings. Figure 16.2 shows an example.

ELASTIC PROPERTIES OF A SINGLE LAYER (LAMINA). We will assume that each layer of the composite is a sheet of parallel, endless fibers. Laminae that have been laid up by using unidirectional tape closely approximate that ideal. When woven cloth is used instead of tape, the approximation is not as good because the fibers in woven cloth must cross over one another. As shown in Figure 16.3, the direction of the fibers is called no. 1, and that perpendicular to the fibers is no. 2.

To estimate the tensile modulus E_1 in the fiber (no. 1) direction, we assume that the lamina has been subjected to a tensile load in that direction, as shown in Figure 16.3a. Fiber and matrix are assumed to be perfectly bonded to each other. Thus, they are subjected to the same strain. The rule of mixtures [Equation (16.1)] results.

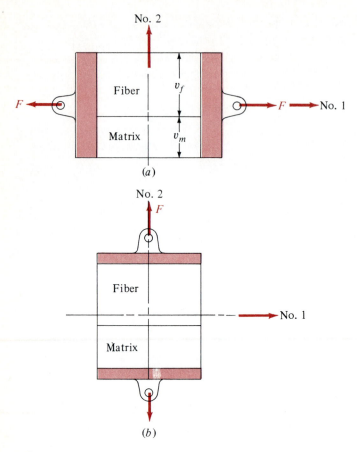

FIGURE 16.3
(a) The equal strain model for evaluation of composite modulus. (b) The equal stress model for composite modulus.

In the direction perpendicular to the fibers (no. 2), the tensile modulus and the in-plane shear modulus G_{12} can be estimated by assuming that fiber and matrix are subjected to the same stress, that they are acting in series. This is shown schematically in Figure 16.3b. Equations (16.2) and (16.3) result:

$$E_2 = \frac{E_f E_m}{v_m E_f + v_f E_m} \tag{16.2}$$

$$G_{12} = \frac{G_m G_f}{v_m G_f + v_f G_m} \tag{16.3}$$

The validity of this "equal stress" model is very sensitive to the shape and packing of the fibers and to what extent they touch each other. Solutions have been obtained by using the mathematics of elasticity, but they are very complex. For design purposes, the Halpin-Tsai equations [35] may be used:

$$\frac{M}{M_m} = \frac{1 + \xi n v_f}{1 - n v_f} \tag{16.4}$$

in which

$$n = \frac{M_f/M_m - 1}{M_f/M_m + \xi} \tag{16.5}$$

where M refers to the modulus in question (E_2 or G_{12}), and subscripts identify fiber (f) or matrix (m). The factor ξ refers to the effectiveness of the reinforcement. Good results have been obtained for E with $\xi = 2$ and for G with $\xi = 1$.

Example 16.4. Consider a lamina made of 60 percent by volume fiberglass in an epoxy matrix. We calculate the no. 1 elastic modulus as

$$E_1 = 0.6(12) + 0.4(0.5) = 7.4E + 06 \text{ psi}$$

From the equal-stress model, the modulus in the 2 direction is estimated as

$$E_2 = \frac{(12)(0.5)}{v_m E_f + v_f E_m} = 1.176E + 06 \text{ psi}$$

Using the Halpin-Tsai equations, however, gives

$$E_2 = \frac{(0.5)[1 + 2(1)]}{1 - 0.6} = 3.75E + 06 \text{ psi}$$

which is a much larger, and more realistic, value.

Reasoning from the same model as that used in developing Equation (16.1) leads to the same form of equation for the major Poisson's ratio:

$$v_{12} = v_m v_m + v_f v_f \tag{16.6}$$

The major or longitudinal value of Poisson's ratio obtained by Equation (16.6) gives the strain in the 2 direction, perpendicular to the fibers, caused by a unit strain in the 1, or fiber, direction.

The minor Poisson's ratio v_{21} gives the strain in the 1, or fiber, direction resulting from a unit strain in the direction perpendicular to the fibers. It is usually much less than the major ratio; i.e., a unit elongation perpendicular to the fibers will cause much less strain in the fiber direction than vice versa. Recall that we have assumed that the stress perpendicular to the fibers is the same in fiber and matrix. The matrix is much less stiff than the fiber. Therefore, the strain in the matrix will be more than the overall strain; that in the fiber much less. It follows that the strain described by the minor Poisson ratio, which is a strain in the fiber, will be much less than that described by the major ratio, which is a matrix strain. Leslie [45] gives

$$v_{12} = 0.27 \qquad v_{21} = 0.09$$

for a fiberglass-epoxy lamina, and Vinson and Chou [36] give

$$v_{12} = 0.36 \qquad v_{21} = 0.033$$

for a boron-epoxy lamina. Beyond noting that greater disparity in E leads to greater disparity in v, we will not try to generalize further from these data.

REINFORCEMENT AT AN ARBITRARY ANGLE. The preceding equations for estimating lamina properties in the 1 and 2 directions would be sufficient if all laminates were "cross-ply," like ordinary plywood. In such a laminate, each lamina is at right angles to its neighbor. However, in nearly all practical cases, optimal laminate properties can be obtained only if the individual plies or laminae are at acute angles to one another. Therefore, we need to be able to estimate lamina properties at any acute angle to the 1, or fiber, direction. The simplest way to do this is to ignore the stiffness of the matrix entirely. Consider a tape 1 unit wide, containing n fibers, as shown in Figure 16.4. The tape is at an angle θ to the x axis, and the tension in each fiber is T. The force per unit of tape width is nT. The component of that force in the x direction is

$$F_x = nT \cos \theta$$

This unit width of tape, projected on a plane perpendicular to the x axis, has a width of

$$W = \frac{1}{\cos \theta}$$

Dividing force by "area," we find that if the tension in each fiber is T, the tension per unit width in the x direction is

$$\frac{F}{W} = nT \cos^2 \theta \tag{16.7}$$

and that in the y direction is

$$\frac{F}{W} = nT \sin^2 \theta \tag{16.8}$$

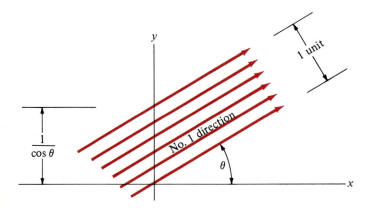

FIGURE 16.4
Model for "netting" analysis, in which the strength and stiffness of the matrix are ignored.

Expression (16.7) can be used to express the tensile modulus of a lamina in this form:

$$E_\theta = E_1 \cos^2 \theta \qquad (16.9)$$

Equation (16.9) provides a good first estimate of the modulus.

Better relations for the elastic constants of the lamina can be derived by transforming stress and strain from the 1 and 2 axes parallel and perpendicular to the fibers to the axes of the lamina [35]. Coupling effects (such as the Poisson effect) are incorporated. Equations (16.10) are the resulting expressions for the elastic constants of the lamina in terms of the angle θ between the x and y axes and the 1 and 2 axes aligned with the fibers:

$$\frac{1}{E_x} = \frac{1}{E_1} \cos^4 \theta + \left(\frac{1}{G_{12}} - \frac{2v_{12}}{E_1} \right) \sin^2 \theta \cos^2 \theta + \frac{1}{E_2} \sin^4 \theta \qquad (16.10a)$$

$$v_{xy} = E_x \left[\frac{v_{12}}{E_1} (\sin^4 \theta + \cos^4 \theta) - \left(\frac{1}{E_1} + \frac{1}{E_2} - \frac{1}{G_{12}} \right) \sin^2 \theta \cos^2 \theta \right] \qquad (16.10b)$$

$$\frac{1}{E_y} = \frac{1}{E_1} \sin^4 \theta + \left(\frac{1}{G_{12}} - \frac{2v_{12}}{E_1} \right) \sin^2 \theta \cos^2 \theta + \frac{1}{E_2} \cos^4 \theta \qquad (16.10c)$$

$$\frac{1}{G_{xy}} = 2 \left(\frac{2}{E_1} + \frac{2}{E_2} + \frac{4v_{12}}{E_1} - \frac{1}{G_{12}} \right) \sin^2 \theta \cos^2 \theta + \frac{1}{G_{12}} (\sin^4 \theta + \cos^4 \theta) \qquad (16.10d)$$

We have omitted from this list two coefficients which relate shear strain to normal stress. The result, for a composite of 60 percent fiberglass by volume in epoxy resin, is shown in Figure 16.5.

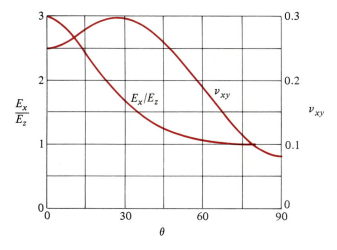

FIGURE 16.5
Calculated elastic moduli for a fiber-reinforced material.

To optimize a laminate is to choose the number of layers and the best orientation for each layer. When the relations of Equations (16.10) must be used for each iteration, the task is monumental. The designer needs a way to create a reasonable structure which will at least serve as a starting point for optimization. To this end, Tsai and Pagano [35] have suggested a design method based on calculations of properties for each lamina which will be invariant with respect to the angle between the lamina's 1 axis and that of the laminate.

First, we express the elastic properties of a lamina in terms of a stiffness matrix:

$$
\begin{bmatrix} \sigma_1 \\ \sigma_2 \\ \tau_{12} \end{bmatrix} = \begin{bmatrix} Q_{11} & Q_{12} & 0 \\ Q_{12} & Q_{22} & 0 \\ 0 & 0 & Q_{66} \end{bmatrix} \begin{Bmatrix} \varepsilon_1 \\ \varepsilon_2 \\ \gamma_{12} \end{Bmatrix}
\tag{16.11}
$$

If you are a bit rusty with matrix operations, Equation (16.11) is a compact way of writing several equations, the first of which is

$$
\sigma_1 = Q_{11}\varepsilon_1 + Q_{12}\varepsilon_2 + (0)\gamma_{12}
$$

This is for an orthotropic lamina in plane stress; there are no stresses perpendicular to the plane of the lamina. As before, 1 is the fiber direction. The Q's are defined as follows:

$$
Q_{11} = \frac{E_1}{1 - v_{12}v_{21}}
$$

$$
Q_{12} = v_{12}Q_{11}
$$

$$
Q_{22} = \frac{E_2}{1 - v_{12}v_{21}}
\tag{16.12}
$$

$$
Q_{66} = G_{12}
$$

in which E_1 and E_2 are the familiar Young's moduli and G_{12} is the modulus of elasticity in shear.

As we remarked earlier, the axes of the laminate are at some acute angle θ to those of each lamina. The stiffness matrix of the lamina, in reference to those axes, is more complex:

$$
\begin{bmatrix} \bar{Q}_{11} & \bar{Q}_{12} & \bar{Q}_{13} \\ \bar{Q}_{21} & \bar{Q}_{22} & \bar{Q}_{23} \\ \bar{Q}_{31} & \bar{Q}_{32} & \bar{Q}_{33} \end{bmatrix}
\tag{16.13}
$$

Each stiffness coefficient is a function of the angle θ. For example,

$$
\bar{Q}_{11} = Q_{11} \cos^4 \theta + 2(Q_{12} + Q_{66}) \sin^2 \theta \cos^2 \theta + Q_{22} \sin^4 \theta
$$

The suggestion of Tsai and Pagano was that the stiffnesses $\bar{Q}$ be expressed as relatively simple functions of several parameters which would not change as the orientation of the lamina was changed. With this system, for example,

$$
\bar{Q}_{11} = U_1 + U_2 \cos^2 \theta + U_3 \cos^4 \theta
\tag{16.14}
$$

where

$$U_1 = \tfrac{1}{8}(3Q_{11} + 3Q_{22} + 2Q_{12} + 4Q_{66}) \qquad (16.15a)$$

$$U_2 = \tfrac{1}{2}(Q_{11} - Q_{22}) \qquad (16.15b)$$

$$U_3 = \tfrac{1}{8}(Q_{11} + Q_{22} - 2Q_{12} - 4Q_{66}) \qquad (16.15c)$$

$$U_4 = \tfrac{1}{8}(Q_{11} + Q_{22} + 6Q_{12} - 4Q_{66}) \qquad (16.15d)$$

$$U_5 = \tfrac{1}{8}(Q_{11} + Q_{22} - 2Q_{12} - 4Q_{66}) \qquad (16.15e)$$

They then showed, by an analysis far too involved for this discussion, that U_1 and U_5 were the only invariants needed to describe the average stiffness of the finished laminate. They are called the *isotropic stiffness* and the *isotropic shear rigidity*. For an isotropic material,

$$U_1 = Q_{11} = \frac{E}{1 - v^2} \qquad (16.16a)$$

$$U_5 = Q_{66} = G_{12} = \frac{E}{2(1 + v)} \qquad (16.16b)$$

The greatest value of the stiffness invariants U_1 and U_5 is in the first stage of design, where they may be used to determine the approximate size and shape of a cross section. Then the shape is refined, by using the other stiffness constants and the required gross stiffnesses.

Example 16.5. Find the invariant stiffnesses of the lamina of Example 16.4. We had estimated the moduli parallel and perpendicular to the fibers as

$$E_1 = 7.4\text{E} + 06 \text{ psi} \qquad E_2 = 3.75\text{E} + 06 \text{ psi}$$

The typical shear modulus and Poisson's ratio for epoxy-fiberglass laminae are

$$G = 1.3\text{E} + 06 \qquad \text{and} \qquad v_{12} = 0.25$$

The stiffness coefficients are thus (all in pounds per square inch)

$$Q_{11} = \frac{7.4}{1 - 0.25^2} = 7.89\text{E} + 06$$

$$Q_{12} = \frac{0.25(7.4\text{E} + 06)}{1 - 0.25^2} = 1.973\text{E} + 06$$

$$Q_{22} = \frac{3.75\text{E} + 06}{1 - 0.25^2} = 4.0\text{E} + 06$$

$$Q_{66} = 1.3\text{E} + 06$$

From these, we calculate the invariants U, using Equations (16.15):

$$U_1 = 5.602\text{E} + 06 \qquad U_2 = 2.96\text{E} + 06$$

$$U_3 = 0.3343\text{E} + 06 \qquad U_4 = 2.316\text{E} + 06$$

$$U_5 = 1.643\text{E} + 06$$

ELASTIC PROPERTIES OF A LAMINATE. To be usable, a laminate must consist of at least two plies, or laminae. The simplest way to estimate the stiffness of a laminate in extension or compression is simply to add the stiffnesses of the individual laminae. A stiffness matrix results:

$$\begin{bmatrix} N_x \\ N_y \\ N_{xy} \end{bmatrix} = \begin{bmatrix} A_{11} & A_{12} & A_{16} \\ A_{12} & A_{22} & A_{26} \\ A_{16} & A_{26} & A_{66} \end{bmatrix} \begin{Bmatrix} \varepsilon_x^o \\ \varepsilon_y^o \\ \gamma_{xy}^o \end{Bmatrix} \tag{16.17}$$

where each N is the force per unit width (normal stress multiplied by thickness) and each A is the sum of the stiffnesses Q of the laminae, each multiplied by its thickness. The subscript xy refers to shearing force and strain, and the superscript o means that the strains are those of the middle surface of the laminate (or the central lamina, if the laminate is symmetric).

Example 16.6. The concept of a sailplane wing spar is sketched in Figure 16.6. It calls for a light wooden core with top and bottom reinforcing skins (called *spar caps*) of epoxy fiberglass. We will assume the same composite formulation as in Example 16.4. The design calls for a load in the spar cap of 3600 lb/in of width and a maximum strain of $\varepsilon = 8E - 03$. How many layers of fiberglass should be used?

In Example 16.5, we calculated the isotropic stiffness of this material: $U_1 = 5.6E + 06$ psi. A typical fiberglass tape will be about 0.010 in thick when in place, so the effective stiffness per layer will be

$$5.6E + 06(0.010) = 5.6E + 04 \text{ lb/in}$$

The necessary laminate stiffness is

$$\frac{3600}{8E - 03} = 45.0E + 04 \text{ lb/in}$$

which indicates that 45/5.6, or at least 8, layers should be used.

Classical plate theory assumes that a line segment which is originally straight and perpendicular to the middle surface will remain so. This can happen only if shear deformations between laminae are ignored. In addition the tensile or compressive

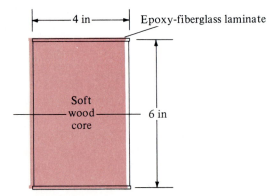

├─4 in─┤ Epoxy-fiberglass laminate

Soft
wood ── 6 in
core

FIGURE 16.6
Cross section of a sailplane wing spar.

strain (and therefore stress) perpendicular to the layers is assumed zero. This is often called the *Kirchhoff hypothesis for plates*. It is the same as the beam theory reviewed at the beginning of Chapter 14. With these assumptions, the bending stiffness of the laminate can be calculated as the sum of the stiffnesses of its layers, each calculated with respect to the neutral axis:

$$\text{Laminate stiffness} = \sum (UA)y^2 \tag{16.18}$$

in which UA represents the area-modulus product of each layer and y the distance from its centroid to that of the laminate.

Normally we also assume that the laminate is balanced, i.e., that layer properties are symmetric about the center of the laminate. A balanced laminate behaves under load as a sheet of homogeneous orthotropic material. It will stretch or compress, but it retains its shape. An unbalanced laminate, however, will change shape under load. In practice, balanced laminates have an odd total number of layers, such as a center layer with two layers symmetrically disposed on each side. Plywood is normally made as a balanced laminate.

Example 16.7. Calculate the moment of inertia of the spar whose cross section is shown in Figure 16.6. The calculation has two parts. First, we calculate the moment of inertia of the two laminate caps about its own centroid:

$$\frac{2bh^3}{12} = \frac{2(4)(0.080)^3}{12} = 1.7\text{E} - 04 \text{ in}^4$$

Second, we transfer that moment of inertia to the neutral axis with the transfer-of-axes formula:

$$I = 2[(0.080)(4)(3)^2 + 1.7\text{E} - 04] = 5.7603 \text{ in}^4$$

The bending stiffness of the spar cap about its own centroid is clearly insignificant. Thus, in the calculation of the moment of inertia, the thickness 0.080 can be ignored in comparison with the distance 3 to the neutral axis. The bending stiffness of the soft, light core is also ignored.

Strength of Laminates

Determination of the strength of a fiber-reinforced laminate, the stress at which it is expected to fail by yielding or fracture, can be a very involved matter. The reason is that the fibers are not uniform in strength or uniformly loaded.

We start with the strength of the simplest of laminates: a bundle of straight fibers fastened together at their ends. We know the average fiber breaking strength from tests on many individual fibers. Those tests have also shown that the breaking strength varies widely from fiber to fiber. While the largest number will break at loads near the average, a significant number will break at very low loads. When a load has been reached at which the stress in surviving fibers is equal to the average strength of the fibers, about half of them will have already broken. This will cause the bundle to fail at a load much less than that corresponding to the average fiber breaking stress.

Example 16.8. Consider a bundle of 408 glass fibers with an average diameter of 0.0003 in and average breaking strength of 450 ksi. If each of the fibers in the bundle were carrying a stress of 450 ksi and none were broken, the load in the bundle would be

$$F_{\text{bundle}} = (450E + 03)(408)\left(\frac{0.0003^2}{4}\right) = 12.9 \text{ lb}$$

But this could happen only if the breaking strengths of all fibers were identical. That just never happens.

It is reasonable to assume that the fiber breaking strength is symmetric about the mean. One-half of the fibers will break at a stress less than the mean strength, and one-half will break at a stress greater than the mean. Thus, at a stress equal to the mean, only one-half of the fibers will be still unbroken, and the load supported by the bundle will be

$$\frac{12.9}{2} = 6.45 \text{ lb}$$

The true stress, referred to the area of fibers still carrying load, is 450 ksi. But the engineering stress, referred to the original area of fibers, including the broken ones, is only 225 ksi.

Engineering stress is defined as the applied load divided by the *original* cross-sectional area. *True stress*, however, is the load divided by the area that exists at that moment. We have observed in Chapter 6 that as materials under test stretch, their cross-sectional areas decrease. The true stress becomes larger than the engineering stress.

Any discussion of the strength of a bundle of fibers, or a laminate, must be in terms of engineering stress. This is because there is no practical way to count the number of unbroken fibers, and hence their cross-sectional area, at any given load. There is no measurable narrowing or necking down. Instead, fibers break within the specimen, where the breaks cannot be seen.

An ideal bundle of fibers is shown in Figure 16.7. They are held by a fixture which somehow divides the total load so that the load on each unbroken fiber is exactly the same. As the holders are pulled steadily apart, we observe that the load

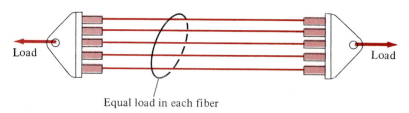

Equal load in each fiber

FIGURE 16.7
Conceptual model of a loose fiber bundle.

increases steadily to the point of fracture. The load at fracture, in other words, is the maximum load, the maximum engineering stress, that can be carried by the bundle.

The fracture of a bundle can be illustrated with a conceptual experiment. We start with zero load and increase it steadily, while keeping track of the number of surviving fibers. As fibers break, the bundle load is shared by fewer fibers. Finally a point is reached where the decrease of load-carrying capacity due to loss of fibers exceeds the increase due to greater stress. As more fibers break, the load per fiber increases. The result is a sudden fracture of the specimen.

Thus, to calculate the maximum engineering stress of a bundle is to calculate its fracture stress. From the tests on individual fibers, we know the mathematical distribution of their breaking strengths. From this, the number of survivors at any given stress is calculated. At some very small value of true stress, a few fibers will have broken. The engineering stress and true stress will have essentially the same value. As true stress increases, so does the number of broken fibers. Since most distributions of fiber breaking stress are in fact fairly symmetric about the mean, we seem to be asserting that the breaking load of a bundle would be just one-half the load that corresponds to the average breaking stress. The surprising fact is that this is not the case, as shown in Example 16.9.

Example 16.9. Assume a bundle made up of fibers whose breaking strength is uniformly distributed, with an average of $1.0E + 06$ psi and minimum breaking strength of $0.6E + 06$ psi. That is, the probability that a fiber will break at any stress between $0.6E + 06$ and $1.4E + 06$ psi is constant. A plot of the proportion of fibers broken versus stress is thus a straight line, as shown in Figure 16.8. As in Example 16.8, this distribution is symmetric about the mean.

We start with no load on the bundle, and we increase the load until the true stress in the fibers is $0.6E + 06$ psi. None have failed, so the engineering stress and true stress are identical.

The load is increased until the stress in the fibers is $0.7E + 06$ psi. We see in Figure 16.8 that one-eighth of the fibers have broken. The bundle engineering stress is

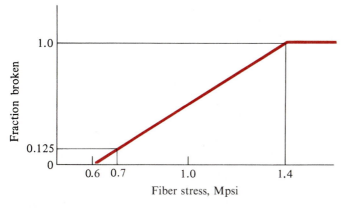

FIGURE 16.8
Cumulative-failure statistic for a hypothetical fiber.

the product of the stress in the survivors and the fraction surviving, or

$$(7E + 05)(1 - \tfrac{1}{8}) = 0.6125E + 06 \text{ psi}$$

The results of several such calculations are shown in the following table.

Fiber stress	Fraction broken	Fraction not broken	Bundle stress
0.6	0	1	0.6
0.7	$\tfrac{1}{8}$	$\tfrac{7}{8}$	0.6125
0.8	$\tfrac{1}{4}$	$\tfrac{3}{4}$	0.600
0.9	$\tfrac{3}{8}$	$\tfrac{5}{8}$	0.5625
1.0	$\tfrac{1}{2}$	$\tfrac{1}{2}$	0.500

The maximum engineering stress of the bundle is $0.6125E + 06$ psi, as any attempt to exceed this will break more fibers, producing an unstable situation with more applied load and less ability to support it. Notice that this bundle breaking strength corresponds to a true fiber stress which is only 0.7 of the average fiber breaking strength.

If the distribution of fiber fracture stress is known as a mathematical expression, the bundle fracture stress can be found by forming the product of fiber stress and area of surviving fibers and maximizing with the calculus.

A bundle of fibers which is held together by a well-bonded matrix will have a much higher fracture strength than the loose bundle of Figure 16.7. This is because the breaks in individual fibers are not usually next to one another. The matrix transfers the load, by shear stresses, from each broken fiber to its neighbors. As the load increases, so does the total number of individual breaks, and with that the number of clusters of breaks. Fracture occurs when a group of individual fiber breaks is large enough to constitute an unstable crack. As a structure becomes larger, the likelihood of such a crack forming increases. Thus we find [83] that the fracture strength of fiber-reinforced composites decreases as the size increases, as shown in Figure 16.9.

Empirical data are usually available on the strength of the lamina to be used. For example, here are typical strengths for E glass/epoxy lamina (E stands for electrical grade, as used in light bulbs):

$$S_1 = 150 \text{ ksi} \quad \text{tension in fiber direction}$$

$$S_2 = 4 \text{ ksi} \quad \text{tension perpendicular to fibers}$$

$$S_s = 4 \text{ ksi} \quad \text{shear}$$

The strength x of this lamina, when stressed at some angle θ to the 1 (fiber) direction, can be estimated with a criterion developed by Hill and Tsai [35]:

$$\frac{1}{x^2} = \frac{\cos^4 \theta}{X^2} + \left(\frac{1}{S^2} - \frac{1}{X^2} \right) \cos^2 \theta \sin^2 \theta + \frac{\sin^4 \theta}{Y^2} \tag{16.19}$$

in which X, Y, and S are the tensile, compressive, and shear strengths obtained in uniaxial tests. This is often called a *distortion energy theory*. Its predictions compare

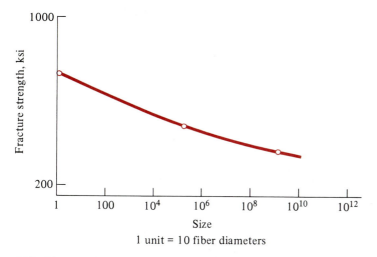

FIGURE 16.9
The effect of size on the engineering fiber stress at failure of fiberglass-reinforced structures.

well with experimental results for the E glass-epoxy lamina above, as shown in Figure 16.10.

Usually, in laminate design, the stiffness of the part will be the controlling criterion. That is, a combination of laminae must be devised which provides the stiffness required in stretching, bending, and/or compression. Then the strength of this candidate design must be checked, layer by layer, to find the lowest overall load at which

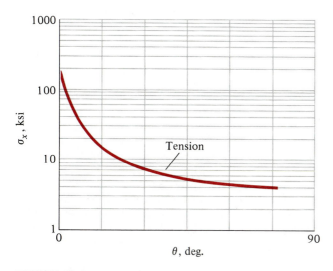

FIGURE 16.10
Prediction of the Tsai-Hill failure criterion.

failure of one of the layers is expected. If this is less than the design load as modified by the safety factor, the design must be changed.

It would seem that an optimal design would cause all components to reach failure stress at the same time, especially in a composite, where layers (laminae) can be placed to suit. However, as remarked above, lamina placement is nearly always dictated by requirements of stiffness rather than strength. If a laminate of specified stiffness does not meet strength requirements in all its layers, then both stiffness and strength will probably have to be increased.

Example 16.10. We use the design of a snow ski as an example of the optimization of a composite structure. The length of the ski, its plan form, and its beam stiffness were chosen to match those of a successful slalom design. It was known that the ski would be subjected to large flexure in use, and the ski had to be able to undergo such bending without internal damage. The criterion in design was weight minimization for quicker and easier turning. In addition, we wanted the ski to be as stiff in torsion as possible, for this was believed to correlate with the ski's ability to hold a course on an icy slope.

The physical characteristics of well-liked competitive skis were studied and expert skiers interviewed. The result was a specification for the center of the ski, where the binding is mounted. The stiffness EI was to be 2.87E + 05 lb·in², the width 2.75 in, and the thickness no more than 1 in. The required stiffness of the ski was graded from this maximum to minima at tip and tail by a synthesis of the needs for "feel" in the fore and aft parts of the ski and the desired pressure profile on the snow when the ski is flattened by the skier's weight.

This example will be limited to the optimal design of the center section of the ski. Years ago, snow skis were made of wood. The first composite skis were laminated from thin layers of wood to produce a higher-quality product than could be readily made from a solid piece. The first revolutionary change in skis came when an aircraft engineer named Howard Head built several pairs of skis of aluminum honeycomb faced with aluminum. These were far lighter and livelier than wooden skis, but the joint between the honeycomb core and the aluminum faces failed. Head then retreated to wood cores and aluminum faces, a design which dominated skiing for many years. Cores made partly or entirely of wood are still found in many skis, although the design and construction of the core have become quite sophisticated. Aluminum facings are often replaced with epoxy reinforced with fiberglass or carbon fiber. In some skis, the fiberglass laminate is molded around the core to provide a box structure. A number of skis are made entirely of fiber-reinforced plastic. Thus a wooden ski reinforced with a fiberglass skin evolved into a ski which was a fiberglass box formed over a light wood core.

For the ski of this example, with the need for minimum weight, the decision was made to use a core of expanded aluminum honeycomb, as seen in Figure 16.11. For structural integrity and torsional rigidity, the fiberglass-reinforced epoxy would be in the form of a structural box around the honeycomb, with at least two laminae literally wrapped around the ski at a 45° "helix" angle. Thus the framework for design was fixed.

The fiberglass cloth selected was a highly unidirectional weave consisting of 90 "ends" per inch of fiberglass in the 1 direction and 10 ends per inch in the transverse, or 2, direction. Each end is a bundle of fibers, analogous in every way to a strand of yarn. It looks much like a rope, including the twist. Each end consisted of 408 filaments, with an average filament diameter of 0.00026 in.

The thickness of the cloth, when impregnated with epoxy and laid up, was estimated. Although one can calculate (by assuming a perfect packing of uniform cylinders)

FIGURE 16.11
A sample of aluminum structural honeycomb, draped over a prototype of a ski built of fiberglass over a
honeycomb core.

a maximum of 91 percent fiber by volume, the practical maximum is usually closer to
60 percent. Using this guideline, we can estimate the diameter of each end as $6.78E - 05$ in.
With 90 such ends per inch, the space per end is $\frac{1}{90} = 0.011$ in, so they are not at all
crowded. The total thickness of the lamina must be greater than the diameter of an end,
because the 2 direction ends must be woven under and over the primary 1 ends. However,
there will be some squeezing together. The result of these considerations is an estimate
of the "laid-up" thickness of the lamina at 0.009 in. To calculate moduli, we model this
as a layer 0.005 in thick containing the 90 no. 1 direction ends and a layer 0.004 in thick
containing the 10 no. 2 direction ends. For the layer containing 90 ends, from Equation (16.1)

$$E_1 = 5.31E + 06 \text{ psi} \qquad E_2 = 0.643E + 06 \text{ psi}$$

These moduli are calculated in the same way for the 10-end layer [using Equation (16.1)
to calculate E_2] and the results weighted by the thickness of the layers to get overall
moduli of

$$E_1 = 5.31(\tfrac{5}{9}) + 0.422(\tfrac{4}{9}) = 3.14E + 06 \text{ psi}$$

$$E_2 = 0.643(\tfrac{5}{9}) + 1.08(\tfrac{4}{9}) = 0.837E + 06 \text{ psi}$$

Rather than take into account all the parts of the ski which contribute to its stiffness, we start this optimization by considering only the most important component—
the top and bottom skins. The core is assumed only to separate top and bottom and to

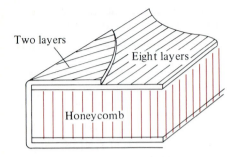

FIGURE 16.12
Cross section of a design for the ski of Figure 16.11.

transfer shear loads. With these assumptions, classic beam theory provides the first estimate of necessary top and bottom thickness:

$$t = \frac{2.87E + 05}{2(0.465)^2(2.75)(3.14E + 06)} = 0.076 \text{ in}$$

or at least 9 layers of cloth.

This is a starting point, but the original design called for two layers of cloth to be wrapped over the basic structure at 45°. We calculate the stiffness added by these layers and see if the number of longitudinal layers needs to be changed. The stiffness of the two added 45° layers in the 1 direction of the principal reinforcement fabric, longitudinally upon the ski, is calculated with Equation (16.10a):

$$E_1 = 1.88E + 06 \text{ psi}$$

Now it appears that a reasonable estimate would be eight layers of unidirectional cloth and the two layers at 45°. Figure 16.12 shows the resulting cross section. Summing three simple calculations, we estimate the total stiffness EI of the box as

$$EI = 2.95E + 05 \text{ lb·in}^2$$

which is just a bit above the desired value of 2.87E + 05. The total stiffness should now be brought into line with specifications by reducing the overall section height. Reducing the section height rather than the skin thickness has two desirable results: It reduces the stress resulting from forcing the ski into a sharp mogul (a hard-packed mound of snow), and thick skin is needed on a recreational snow ski, to hold the screws used to mount bindings and to resist the occasional thrust of a sharp-pointed ski pole. The determination of the final thickness is left as an exercise (Problem 16.10).

PROBLEMS

Section 16.1

16.1. Design a reinforced-concrete beam to withstand a moment of 36 000 lb·ft with a stress in the concrete of 3000 psi. Choose the steel reinforcement such that an appropriate number of no. 4 rods are used and the beam is not more than 14 in deep.

16.2. Repeat the design of Problem 16.1. (a) Use fiberglass as reinforcement. (b) Why would fiberglass be used instead of steel in the reinforcement of concrete? (c) How would you answer part (b) if the object being reinforced were the wall of a radar enclosure?

16.3. The manufacturer of Torlon brand polyamide-imid thermoplastic lists the following properties for two grades of its product, each reinforced with 30 percent fiber.

Grade	Tensile strength, ksi	Modulus, ksi
5030	29.7	1.61
7130	29.8	2.59

Compare these properties with those of the unreinforced plastic, shown in Table 16.1. Which is reinforced with fiberglass and which with carbon? How do these properties agree with those predicted by Equation (16.1)?

16.4. Here are properties that have been cited for a few thermoplastics reinforced with 30 percent short-fiber fiberglass. Properties for unreinforced plastic are shown in parentheses.

Base material	Tensile strength, ksi		Modulus, Mpsi		Impact, ft·lb	
ABS	14.5	(6)	1.1	(0.32)	1.4	(4.4)
Nylon 6 alloy	22	(11.8)	1	(0.4)	3.2	(1)
Polycarbonate	18.5	(9)	1.2	(0.33)	17	(60)
Polyethylene	10	(2.6)	0.9	(0.2)	1.1	(0.4)
Polystyrene	13.5	(7)	1.3	(0.45)	1	(0.45)

Comment on these in light of Equation (16.1) and the discussion of Section 16.1. Why can the presence of strong fibers lead to a reduction of the energy required for fracture?

Section 16.2

16.5. You have an application which justifies the (considerable) difficulty of fabricating a composite of long graphite fibers laid unidirectionally in a steel matrix. Estimate the principal elastic moduli of the composite. Use a reasonable fiber-volume fraction.

16.6. Pressure vessels are often made by a machine which automatically feeds a tape of unidirectional fibers (most commonly, fiberglass) onto a cylindrical mandrel. As seen earlier, the loads per unit width in the wall of a cylinder are twice as high in the "hoop" direction as they are in the "longitudinal" direction. Assuming that the fibers carry all loads (see Figure 16.4), determine the angle with the longitudinal direction at which the fibers should be placed.

16.7. The following values have been given for the properties of polyoxymethylene reinforced with 32 percent by volume short-glass fiber:

$$E_1 = 1.495E + 06 \text{ psi} \qquad G_{12} = 0.24E + 06 \text{ psi}$$
$$E_2 = 0.754E + 06 \text{ psi} \qquad \nu_{12} = 0.37$$

Write a computer program based on Equations (16.10) to calculate and print the elastic moduli of this composite as a function of the angle between the 1 direction and the direction of interest.

16.8. For the composite of Problem 16.6, calculate the stiffness invariants U_1 and U_5.

16.9. A batch of fibers was tested individually for tensile strength with these results: None of the fibers broke at less than 100 ksi, and the proportion of broken fibers increased linearly with stress, until 1100 ksi was reached with the failure of the last fiber in the lot. The average of the breaking stresses was 600 kpsi. Exactly one-half the fibers tested were able to withstand stress of 600E + 03 ksi or more.

Plot the curve of proportion broken vs. stress. Estimate the "engineering stress" (load divided by original area) at which the bundle will break, assuming that all fibers are held such that the stress in each unbroken fiber is the same. Use the method of Example 16.9, and check your result by using the calculus.

16.10. Refer to Example 16.10 on ski design. Construct and run a computer program to calculate the stiffness in bending of the ski, given inputs of total section height and width, number of "unidirectional" laminae, and number of 45°-wrap laminae.

16.11. Use the program of Problem 16.10 to finish the calculation of section height in the ski design of Example 16.10.

BOLTS, NUTS, AND JOINTS

Bolts are very important engineering members because their principal use is to clamp parts together with enough force to seal pressure within a vessel or to prevent motion between parts. It is a rare bolt which does not serve one or the other of these functions. There are dozens of examples in a car: the head-to-block joint, the main engine bearings, the connecting-rod crank bearings, the lug nuts on the wheels, and the attachment of the seat to the floor. It is said that the assembly of parts accounts for one-half of all manufacturing labor in the United States, and the majority of it is done with nuts and bolts.

17.1 TERMINOLOGY AND STANDARDS

Bolts, Screws, Studs

Our discussion concerns that class of fasteners known as *bolts*, which includes the subcategories cap screws and studs. These names are associated with the three basic clamping arrangements, shown in Figure 17.1. In the majority of cases, the bolt passes through a hole in the parts being clamped and mates with a nut, as shown in Figure 17.1a. Occasionally, the bolt mates with threads in one of the members, rather than

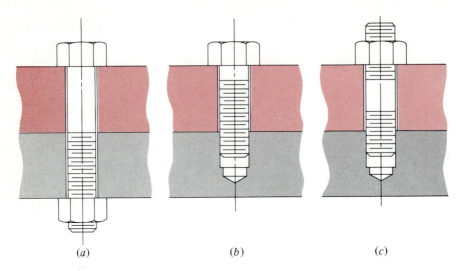

FIGURE 17.1
Basic bolting arrangements. (*a*) Bolt and nut. (*b*) Cap screw or tap bolt. (*c*) Stud.

with a nut (Figure 17.1*b*), and then the correct term is *cap screw*, *screw*, or *tap bolt*. The threaded item is in fact no different whether it screws into a nut or into one of the members. A stud is threaded on both ends and is screwed, more or less permanently, into a threaded hole in one of the parts being joined, as shown in Figure 17.1*c*. The unthreaded shank of a stud helps to locate the parts being assembled.

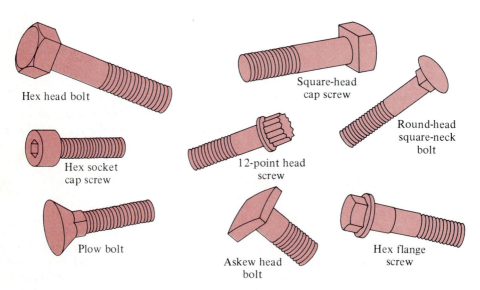

FIGURE 17.2
Bolt and capscrew types. (*Courtesy, Deere & Co. Service Training.*)

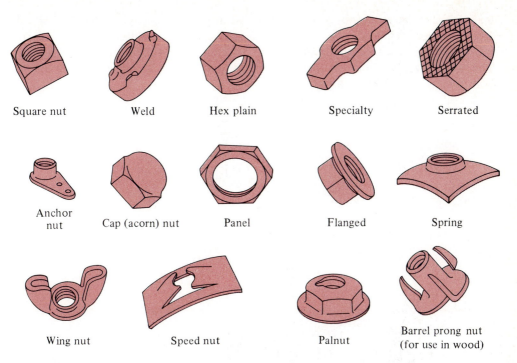

Square nut	Weld	Hex plain	Specialty	Serrated
Anchor nut	Cap (acorn) nut	Panel	Flanged	Spring
Wing nut	Speed nut	Palnut	Barrel prong nut (for use in wood)	

FIGURE 17.3
Some nut types. (Locking nuts are shown in Figure 17.4.) (*Courtesy, Deere & Co. Service Training.*)

Most bolt heads (and nuts) are hexagonal, but there are many other shapes for different purposes. A few are shown in Figure 17.2.

Nuts and Washers

Nuts are available in a large variety of styles to satisfy requirements of cost, speed of assembly, aesthetics, hand tightening, locking, stress distribution, etc. Figure 17.3 illustrates a number of types, and Figure 17.4 shows locking nuts. The latter are of two principal types. Prevailing-torque locknuts (Figure 17.4a) are designed to increase the friction between the threads. Torque is required to turn the locknuts as soon as the threads have mated. They are generally used in dynamic loading situations and represent about 90 percent of the locknut market. Free-spinning locknuts (Figure 17.4b) turn freely until seated, when the locking mechanism is activated. Either an interference fit is created, or the friction at the face is increased, or both. These locknuts have high initial holding power and are thus used where random shock loads are likely. Also shown are a jam nut and a nut secured with a cotter pin.

Some washer types are seen in Figure 17.5. Flat washers are used to improve bearing surfaces and to create a bearing surface over large clearance holes. They also spread the bolt load, which is important where the bolt is clamping relatively soft material, such as an aluminum cylinder head onto a cast-iron engine block. Conical spring washers, made of spring steel, have a nearly flat force-deflection curve and are intended to maintain bolt tension, if some deformation of the clamped parts occurs. The other washers shown are lockwashers, designed to inhibit loosening of a nut or bolt by digging into the mating surfaces.

Distorted shape

Distorted portion of nut thread produces an interference fit. Center dimple allows nut to be assembled with either side up.

Top-crimp nut is easy to start but must be properly oriented before assembly.

Thread profile is distorted, increasing interference. Complex manufacturing process increases cost.

Nut hole, forced into an out-of-round shape after initial forming, produces spring action that maintains an interference fit after assembly.

Insert

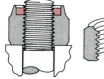 

Plastic or metal section is added to the nut to increase mating-thread friction. It distorts when the nut is installed.

Metal-insert locknut has a projecting hardened wire or pin built in to provide a ratchetlike locking action. Reuse is limited by wear of the pin tip.

(a)

FIGURE 17.4
Locking nuts. (*From Robert B. Aronson, "How Locknuts Remove the 'Unknowns' in Joint Design," Machine Design, Oct. 12, 1978, with permission.*) (*a*) Prevailing-torque locknuts. (*b*) Free-spinning locknuts. (*c*) Jam nut. (*d*) Slotted nut with securing cotter pin.

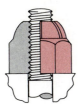

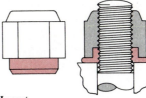

Slotted section

Spring arms formed on the domed nut top are deflected inward. When the nut is threaded on, these arms grip the bolt threads.

Insert

A plastic or metal washer built into the nut base is permanently deformed and grips the bolt threads when the nut is seated.

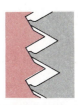

Serrated face

Serrated or grooved face of the nut digs into the bearing surface during final tightening.

Captive washer

Toothed or spring washer attached to the bearing face of the nut increases friction between the bearing surfaces.

Modified thread

Nut threads have a modified cross section which crimps the bolt-thread crests when clamp load is applied.

Spring head

When fully seated, the concave portion of the nut is forced inward and clamps against the bolt threads.

(*b*)

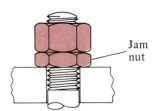

Jam nut

Standard nuts, most commonly used to create a jam lock, rely on friction between the flat bearing surfaces to increase holding power.

(*c*)

(*d*)

FIGURE 17.4   (Continued)

Certain changes in the perception of the efficacy of lockwashers and (in some cases) of the necessity of a positive lock have come about in recent years. Lockwashers, it is claimed [18], serve virtually no securing function, but merely provide a hard bearing surface at low cost. While the nuts on the connecting-rod bolts in automobile

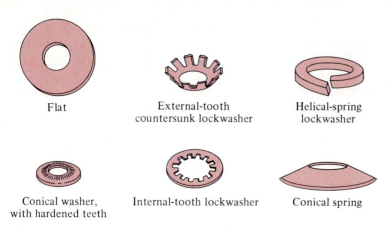

Flat   External-tooth   Helical-spring
            countersunk lockwasher   lockwasher

Conical washer,   Internal-tooth lockwasher   Conical spring
with hardened teeth

FIGURE 17.5
Washer types.

engines in the past were secured with a cotter pin, the usual practice today is to use nothing. Naturally the bolts are of high quality, and the nuts are put on very tightly.

Anaerobic (hardens in the absence of air) thread-locking products have become a popular substitute for mechanical locking methods. A drop on a $\frac{5}{16}$ bolt thread is adequate. Upon curing, the product fills the space between mating threads, thus preventing twisting or motion sideways to the thread (the latter results in twisting). Most thread lockers have good strength up to 300°F, and special formulations can increase this figure to 450°F. The product also lubricates threads, giving more uniform clamping force during bolting up.

Thread Definitions and Standards

Figure 17.6 illustrates the basic definitions associated with screw threads. The advance of the screw in one turn is called the *lead*; the thread separation, the *pitch*. Nearly

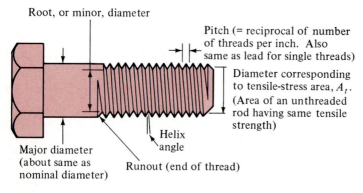

Root, or minor, diameter

Pitch (= reciprocal of number of threads per inch. Also same as lead for single threads)

Diameter corresponding to tensile-stress area, A_t. (Area of an unthreaded rod having same tensile strength)

Helix angle

Major diameter (about same as nominal diameter)

Runout (end of thread)

FIGURE 17.6
Screw-thread terminology.

all bolts and screws have a single thread, which means that one turn advances the screw an amount equal to the pitch (lead = pitch). Occasionally there may be two or three threads. The object is to cause the threaded member to advance farther per revolution and/or to bring more threads into contact per revolution. If the thread is double, the lead is twice the pitch. The most common example, before the advent of ballpoint pens, was the cap on a fountain pen. It was desired that the cap close with a half turn or so and that an adequate number of threads mate. The threads were thus double or triple.

Almost all bolts are of right-hand thread, which means that clockwise turning of the bolt will cause it to advance away from you. There are very few exceptions. For years one of the large automobile makers used left-hand (LH) threads on the wheel nuts on the left side of their cars. The idea was that if somehow the nut were restrained, forward rotation of the wheel would cause it to tighten, whereas a right-hand thread would tend to loosen. That concept was rather far-fetched, and the practice has been abandoned.

Standard Threads

Like systems of measure, international standard thread systems have evolved over the years. World War II provided an impetus to bring about interchangeability of the British and U.S. systems, the result of which was the "unified" system. There are two principal series: the Unified National Fine (UNF) and the Unified National Coarse (UNC). An extra-fine series also exists for situations where a bolt must be screwed into a thin mating part. The modifiers *coarse* and *fine* refer, naturally, to the threads; there are fewer threads per unit bolt length in the coarse series for any given size, and the threads are larger. It also follows that the helix angle, shown in Figure 17.6, is greater. This plus the fact that there are fewer threads in engagement for a given nut size means that coarse threads have a greater tendency to loosen in service. Fine threads are thus used in critical applications and where finer adjustment is required. Three classes of fit are specified: loose, standard, and close, shown with the designations in Table 17.1. The choice naturally depends on the precision which the job requires.

Tables 17.2*a* and *b* list dimensional details of UNC and UNF threads. Thread designation indicates the nominal size, the number of threads per inch, the series, the

TABLE 17.1
Thread fit and tolerance classes

| Class of fit | Unified | | Metric | |
	External	Internal	External	Internal
Loose (where joint is frequently disassembled)	1A	1B	8g	7H
Standard (general assembly)	2A	2B	6g	6H
Close (high accuracy, fine fits)	3A	3B	4g	5H

TABLE 17.2
Unified screw threads

(*a*) Diameters and areas of small unified screw threads (in inches)*

Size designation	Nominal diameter d	Coarse series—UNC		Fine series—UNF	
		Threads per inch N	Tensile-stress area $A_t{}^\dagger$	Threads per inch N	Tensile-stress area $A_t{}^\dagger$
0	0.0600			80	0.001 80
1	0.0730	64	0.002 63	72	0.002 78
2	0.0860	56	0.003 70	64	0.003 94
3	0.0990	48	0.004 87	56	0.005 23
4	0.1120	40	0.006 04	48	0.006 61
5	0.1250	40	0.007 96	44	0.008 80
6	0.1380	32	0.009 09	40	0.010 15
8	0.1640	32	0014 0	36	0.014 74
10	0.1900	24	0.017 5	32	0.020 0
12	0.2160	24	0.024 2	28	0.025 8
$\frac{1}{4}$	0.2500	20	0.031 8	28	0.036 4
$\frac{5}{16}$	0.3125	18	0.052 4	24	0.058 0
$\frac{3}{8}$	0.3750	16	0.077 5	24	0.087 8
$\frac{7}{16}$	0.4375	14	0.106 3	20	0.118 7
$\frac{1}{2}$	0.5000	13	0.141 9	20	0.159 9
$\frac{9}{16}$	0.5625	12	0.182	18	0.203
$\frac{5}{8}$	0.6250	11	0.226	18	0.256
$\frac{3}{4}$	0.7500	10	0.334	16	0.373
$\frac{7}{8}$	0.8750	9	0.462	14	0.509
1	1.0000	8	0.606	12	0.663
$1\frac{1}{4}$	1.2500	7	0.969	12	1.073
$1\frac{1}{2}$	1.5000	6	1.405	12	1.315

* American National Standards Institute B1.1-1974.

† Area of unthreaded rod of same material having same tensile strength (computed at average of pitch and minor diameters: $d_{A_t} = d - 0.974\,428/N$).

fit, and the length, e.g.,

$$\tfrac{1}{4} - 20 \text{ UNC} - 2\text{A} \times 3$$

The unified series, being based on USCS units of measure, is used primarily in the United States. (Britain uses mostly SI units now.) Metric countries and some U.S. firms utilize the ISO (International Standards Organization), i.e., metric, threads, listed in Table 17.3. They are designated by the prefix *M* for metric, the nominal size and the pitch, the fit and the length, e.g.,

$$\text{M12} \times 1.25, \; 6\text{g} \times 80$$

TABLE 17.2 (Continued)

(*b*) Hex bolt head and end thread lengths (in inches)*

| | | Thread length L_T | |
| | | Bolts 6 in | |
Nominal diameter d	Head height H_H	and shorter	Bolts over 6 in
$\frac{1}{4}$	$\frac{11}{64}$	0.750	1.000
$\frac{5}{16}$	$\frac{7}{32}$	0.875	1.125
$\frac{3}{8}$	$\frac{1}{4}$	1.000	1.250
$\frac{7}{16}$	$\frac{19}{64}$	1.125	1.375
$\frac{1}{2}$	$\frac{11}{32}$	1.250	1.500
$\frac{5}{8}$	$\frac{27}{64}$	1.500	1.750
$\frac{3}{4}$	$\frac{1}{2}$	1.750	2.000
$\frac{7}{8}$	$\frac{37}{64}$	2.000	2.250
1	$\frac{43}{64}$	2.250	2.500
$1\frac{1}{8}$	$\frac{3}{4}$	2.500	2.750
$1\frac{1}{4}$	$\frac{27}{32}$	2.750	3.000
$1\frac{3}{8}$	$\frac{29}{32}$	3.000	3.250
$1\frac{1}{2}$	1	3.250	3.500

* ANSI B18.2.1-1972.

(*c*) Bolt grades*

SAE grade	Diameter d, in	Proof load (strength)[†] S_p, ksi	Yield strength[‡] S_y, ksi	Tensile strength S_u, ksi	Identification on bolt head
1	$\frac{1}{4}$ thru $1\frac{1}{2}$	33	36	60	None
2	$\frac{1}{4}$ thru $\frac{3}{4}$	55	57	74	None
2	Over $\frac{3}{4}$ to $1\frac{1}{2}$	33	36	60	None
5	$\frac{1}{4}$ thru 1	85	92	120	
5	Over 1 to $1\frac{1}{2}$	74	81	105	
5.2	$\frac{1}{4}$ thru 1	85	92	120	
7	$\frac{1}{4}$ thru $1\frac{1}{2}$	105	115	133	
8	$\frac{1}{4}$ thru $1\frac{1}{2}$	120	130	150	

* SAE standard J429 k (1979).

[†] Permanent deformation ≤ 0.0001 in.

[‡] 0.2% offset.

TABLE 17.3
Metric (ISO) screw threads

(*a*) Diameters and areas of small metric threads (in millimeters)*

Nominal diameter d	Coarse-pitch series		Fine-pitch series	
	Pitch p	Tensile-stress area A_t[†]	Pitch p	Tensile-stress area A_t[†]
1.6	0.35	1.27		
2	0.04	2.07		
2.5	0.45	3.39		
3	0.5	5.03		
3.5	0.6	6.78		
4	0.7	8.78		
5	0.8	14.2		
6	1	20.1		
8	1.25	36.6	1	39.2
10	1.5	58.0	1.25	61.2
12	1.75	84.3	1.25	92.1
14	2	115	1.5	125
16	2	157	1.5	167
20	2.5	245	1.5	272
24	3	353	2	384
30	3.5	561	2	621
36	4	817	2	915
42	4.5	1120	2	1260
48	5	1470	2	1670

* ANSI B1.1-1974.

[†] Area of unthreaded rod of same material having same tensile strength (computed at average of pitch and minor diameters: $d_{A_t} = d - 0.938\ 194p$).

(*b*) Metric hex bolt head and thread lengths (in millimeters)*

Nominal diameter d	Head height H_H	Thread length L_T		
		≤ 125	> 125 ≤ 200	> 200
5	3.35–3.58	16	22	35
6	3.55–4.38	18	24	37
8	5.10–5.68	22	28	41
10	6.17–6.85	26	32	45
12	7.24–7.95	30	36	49
14	8.51–9.25	34	40	53
16	9.68–10.75	38	44	57
20	12.12–13.4	46	52	65
24	14.56–15.9	54	60	73
30	17.92–19.75	66	72	85
36	21.72–23.55	78	84	97
42	25.03–27.05	90	96	109
48	28.93–31.07	102	108	121

* ANSI B18.2.3.5M-1979

TABLE 17.3 **(Continued)**

(*c*) Metric bolt classes

SAE class	Diameter *d*, mm	Proof load (strength)[†] S_p, MPa	Yield strength[‡] S_y, MPA	Tensile strength S_u, MPa
4.6	5 through 36	225	240	400
4.8	1.6 through 16	310	—	420
5.8	5 through 24	380	—	520
8.8	17 through 36	600	660	830
9.8	1.6 through 16	650	—	900
10.9	6 through 36	830	940	1040
12.9	1.6 through 36	970	1100	1220

[†] SAE Standard J1199 (1979).

[†] Permanent deformation ≤ 0.0025 mm.

[‡] 0.2% offset.

When a bolt is to be threaded into soft material, a coarse thread is generally used. The threads present a greater area of resistance to shear, i.e., stripping, in the softer material. In the case of studs, the end threading into one of the clamped members is usually of coarse thread, while the end accommodating the nut has a fine thread, as shown in Figure 17.7.

Bolt and Nut Strength Grades

Bolts of the same size may be obtained in various gradations of strength. Tables 17.2*c* and 17.3*c* are a summary of the Society of Automotive Engineers (SAE) specifications of standard bolts, used in the automotive and other industries. The grade of the unified series is normally coded on the bolt head, as indicated. Nuts are graded in the same way. The American Society for Testing and Materials (ASTM), the mili-

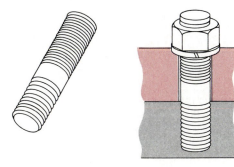

FIGURE 17.7
Studs normally have fine threads on the nut end, coarse on the other. (*Courtesy, Deere & Co. Service Training.*)

tary, and others have additional classifications. Grades in SI are termed *classes.*

The strength of bolts is based on the cross-sectional area of equivalent unthreaded rod having the same tensile strength. That area works out to be slightly greater than the area at the thread root. Tabulated in Tables 17.2*a* and 17.3*a* it is called the *tensile stress area* and is designated A_s or A_t. We will use A_t.

High-strength bolts usually do not exhibit a clear yield point; for these the yield strength is taken as the stress producing a permanent deformation of 0.02 or 0.05 percent. (This is not the same as the commonly used 0.2 percent offset yield in materials testing.)

The proof strength is the maximum tension stress which can be applied to a bolt without resulting in permanent deformation; it is about 90 to 95 percent of the yield. The proof load is the force corresponding to the proof strength applied on the tensile stress area A_t.

Thread Formation

Threads are made in one of two ways, by cutting or rolling. In production, cutting is done on a screw machine, which is an automatic lathe. When an occasional thread must be made in a shop, it is usually done by hand with a die, a cutting tool similar to the nut which will later mate with the thread. Large or special threads are turned on a lathe. Female threads are cut with a tool called a *tap.*

Rolling is a process of thread forming in which thread dies (same name, different tool) are moved (cold) under pressure across the bolt blank, which rolls, as shown in Figure 17.8. This process produces a stronger thread due to the cold working and the compressive residual stresses thus induced. In bolts which are heat-treated, rolling subsequent to the hardening produces the strongest threads.

Rolling is not used to manufacture nuts.

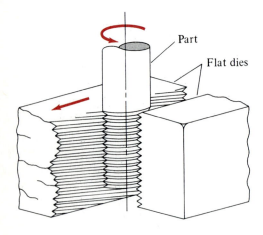

Part

Flat dies

FIGURE 17.8
Thread rolling. As one die is moved past the other, the threads are impressed in the rod. (*Reprinted with permission of General Motors.*)

17.2 LOADS AND STRESSES ON BOLTS AND NUTS

Types of Bolt Loads

We commonly think of bolts as being loaded only in tension—to clamp pieces together. Indeed, they are intended to do that, and the resulting stresses are tensile. The bolt may be subjected to other loads in addition, creating a combined and often complex stress situation.

Structural bolted joints are frequently loaded in shear, and the shear stresses produced are added to the tensile stresses due to the axial loading. The design of such joints is discussed in Section 17.9; several types are shown in Figure 17.41 (see page 508).

Tightening of a bolt is done by applying torque, resulting in shear stresses, which similarly produce a combined-stress situation with the axial tensile stresses.

If the surfaces on which the bolt head and nut bear are not exactly parallel, the bolt will bend, and the stresses created add to the others. This situation can also develop if the loading tends to pry open the joint on one side.

Stresses in Bolts and Nuts

The average tensile stress distribution along a bolt with nut is shown in Figure 17.9. It is usually assumed that the stress decreases uniformly through the thickness of the head and nut. The increased stress at the thread run-out (start of the thread) derives from the reduced cross section at the thread root. There are three locations of stress concentration, as shown in the figure, and bolts usually fail at those places.

The height of standard nuts is close to the nominal bolt diameter. In the case of cap screws, the common thread-engagement practice is shown in Table 17.4. The height of nuts, or the engagement length of tap bolts, is determined on the basis of shear strength and shear area (area at the thread root). However, the load is not distributed uniformly. The first thread carries the largest part of the load, and the first

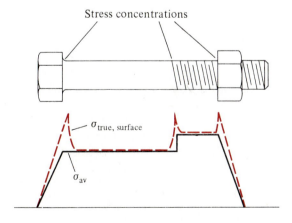

FIGURE 17.9
Stresses in a bolt under load. The points of stress concentration are the base of the head, the thread run-out, and the first thread within the nut.

TABLE 17.4
Tap-bolt thread engagements

Material	Minimum engagement
Steel	d
Brass, bronze, cast iron	$1.5d$
Aluminum and other soft materials	$2d$

d = major diameter

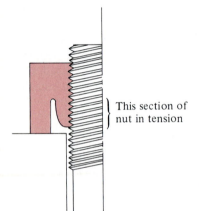

This section of
nut in tension

FIGURE 17.10
With part of the nut threads in tension, the load is more evenly
distributed among the threads.

three carry most of it. The load distribution can be improved by tapering the nut so
that the stress becomes uniform. Alternatively, the pitch of the nut may be made
slightly greater than that of the bolt, or the nut may be configured so that part of it
is in tension, as in Figure 17.10.

17.3 JOINT ANALYSIS

Preload Theory

To start the analysis, we examine the bolts holding the two parts of a joint together,
as in Figure 17.11a. It helps to consider the bolt as a tension spring, with stiffness k_b,
and the joint it clamps as a compression spring, with stiffness k_j, as shown in Figure 17.11b. The load curves (load vs. deflection, not stress vs. strain) of the bolt and
the joint members can be plotted as shown in Figure 17.11c. Extension is measured
to the right. Since the joint is in compression, the joint load line progresses leftward.
The slopes, i.e., stiffnesses, are different; usually the joint (k_j) is several times stiffer
than the bolt (k_b). When the joint is assembled, the bolt is put into an initial tension
F_i (the preload), and the joint will be under equal and opposite compression. The
two curves of Figure 17.11c can thus be put together so as to intersect at F_i (in Figure

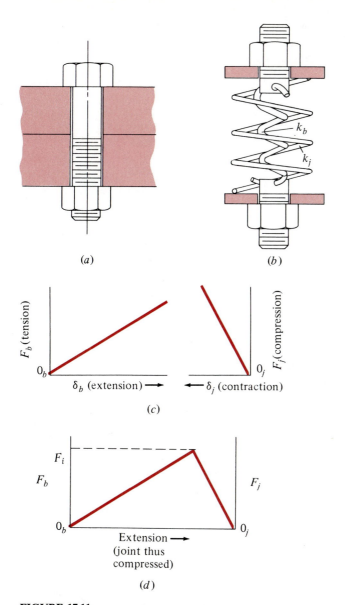

FIGURE 17.11
Bolted-joint analysis. (*a*) Bolted joint. (*b*) Spring schematic. (*c*) Bolt and joint load curves. (*d*) Joint diagram.

17.11*d*); each load line thus has its own origin. The result is called the *joint diagram*. There are other forms of this plot showing the same information, but this is the most common.

An external load *P* is now applied, causing an extension *e* of the bolt and of the joint by the same amount, as shown in the sketch of Figure 17.12*a*. The force in

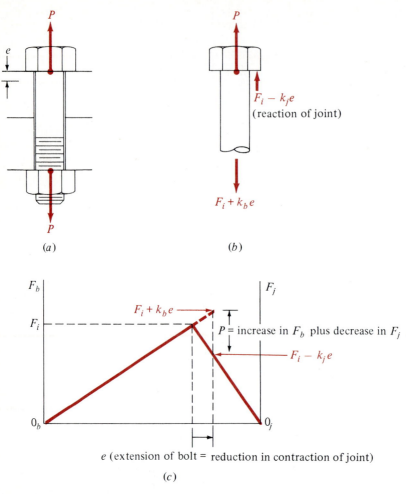

FIGURE 17.12
Application of external load to joint. (*a*) Extensions of bolt and reduction of contraction of joint are equal under load *P*. (*b*) Free-body diagram of bolt section. (*c*) Joint diagram showing result of load application.

the bolt becomes $F_i + k_b e$ and in the joint $F_i - k_j e$, as shown in Figure 17.12*b* and also on the joint diagram in Figure 17.12*c*. From equilibrium of the bolt or the similar triangles of the joint diagram, we find the load in the bolt to be

$$F_b = F_i + \frac{k_b}{k_b + k_j} P = F_i + CP \qquad (17.1a)$$

The term

$$C = \frac{k_b}{k_b + k_j} \quad \text{or} \quad C = \frac{1}{1 + k_j/k_b} \qquad (17.1b)$$

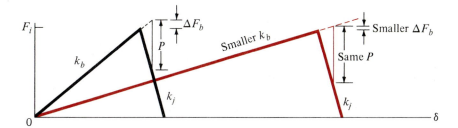

FIGURE 17.13
The smaller the stiffness of the bolt relative to the stiffness of the joint, the smaller the load in the bolt when an external load is applied.

is called the *stiffness parameter*. The load in the joint is

$$F_j = F_i - k_j e = F_i - (1 - C)P \qquad \text{compression} \tag{17.2}$$

Since the stiffness parameter C is always less than 1, expression (17.1a) indicates that only a fraction of the force P is taken by the bolt. The remainder goes to reduce the joint compression, as shown by Equation (17.2) and in Figure 17.12c. The smaller C is, the less the force on the bolt, as illustrated in Figure 17.13.

When the load P is increased to the point that the compression in the joint is reduced to zero, the joint physically separates and the bolt then assumes all the load, as seen in Figure 17.14. That is not supposed to happen in any joint. The preload to prevent separation is found from (17.2) by setting $F_j = 0$:

$$F_i = (1 - C)P \tag{17.3}$$

The joint analysis which we have laid out in the above paragraphs was based on a very unlikely loading—the external force pair was applied under the head of the bolt and under the head of the nut, shown again in Figure 17.15a. This is one extreme; the other is application of the force and reaction at the joint interface, as

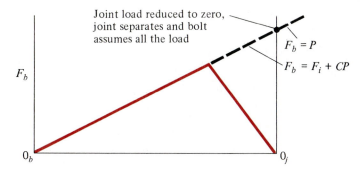

FIGURE 17.14
Separation of the joint.

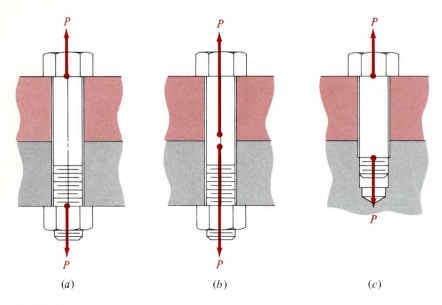

(a) (b) (c)

FIGURE 17.15
Extremes of external load application. The actual location is somewhere between the two. (a) In the analysis the forces were applied under the head of the bolt and of the nut. This is the worst case. (b) The opposite extreme is application of the forces at the joint interface. (c) Force application in a tap bolt.

shown in Figure 17.15b. To determine the true effective location of the loads in a particular application would require a finite-element or similar analysis. The assumption of loading under the bolt head and the nut is the worst case, so that design calculations on that basis are conservative.

What to do in the case of a cap screw, where there is no nut? The threaded part of the joint is, in reality, the nut, and the load can be assumed to act where the threads begin, as shown in Figure 17.15c.

Calculation of Joint Stiffness Parameter C

The stiffness parameter was defined in the preceding paragraph: $C = k_b/(k_b + k_j)$. Recall that stiffness is the slope of a load-deflection curve: $k =$ load/deflection (Figure 17.16a). This is exactly the same stiffness as that defined for springs.

BOLT STIFFNESS k_b. If a bar is loaded by an axial force P, the deflection is $\delta = PL/(AE)$. Hence the stiffness is

$$k = \frac{P}{PL/(AE)} = \frac{AE}{L}$$

If the bar is stepped, the parts must be treated as springs in series, as shown in Figure 17.16b:

$$\frac{1}{k_{\text{eff}}} = \frac{1}{k_1} + \frac{1}{k_2} + \cdots$$

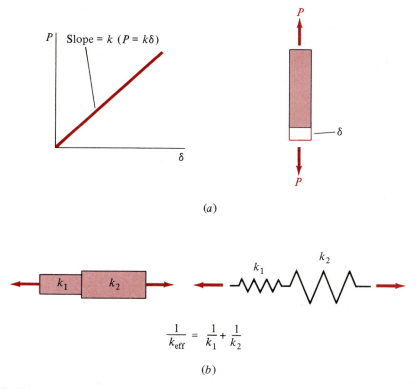

FIGURE 17.16
Calculating bolt stiffness. (*a*) Defining stiffness. (*b*) Bars in series are springs in series.

In Figure 17.17*a* the sketch of the average stress in a bolt and nut is repeated, and Figure 17.17*b* shows a model for computing the bolt stiffness. The model is stepped because of the smaller section in the threaded portion, and parts of the head and of the nut are included. (Other models use different lengths than the 0.4*d* of this scheme.) With this model, the bolt stiffness is given by

$$\frac{1}{k_b} = \frac{1}{k_1} + \frac{1}{k_2} = \frac{L_1}{A_1 E} + \frac{L_2}{A_2 E}$$

$$= \frac{1}{E}\left(\frac{L_s + 0.4d}{\pi d^2/4} + \frac{L_t + 0.4d}{A_t}\right) \tag{17.4}$$

When the application involves a tap bolt (one part of the joint threaded, no nut), the tapped part of the joint serves as the nut. If the thread engagement is about equal to the height of a nut, then one-half of the thread engagement can be included in k_b. When the engagement is greater, about one-third of that length is appropriate.

Occasionally parts of a bolt shank may be reduced to decrease the bolt stiffness, with a portion perhaps at or near the nominal size to position the parts, as sketched in Figure 17.17*c*. The several sections must be treated as springs in series.

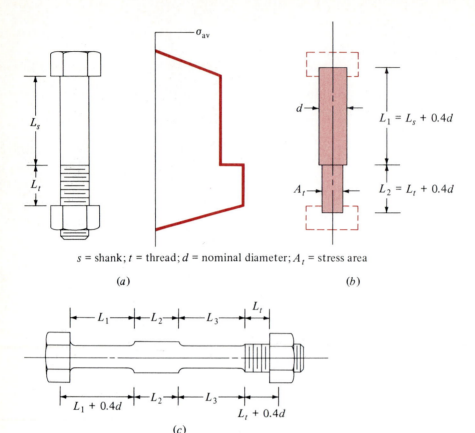

s = shank; t = thread; d = nominal diameter; A_t = stress area

(a)   (b)

(c)

FIGURE 17.17
Bolt model for calculation of K_b. (a) Average stress. (b) Stiffness model. (c) If diameters vary, each length must be taken as a separate spring in series with the others. (Those of the same diameter can be combined.)

Example 17.1. Figure 17.18a shows a section of a joint at one of the bolts. We calculate the stiffness of the bolt here as the first step in finding the joint stiffness parameter. The dimensions we need to do this are shown in Figure 17.18b. Then, filling in expression (17.4) gives

$$\frac{1}{k_b} = \frac{1}{30\text{E} + 06}\left[\frac{1.950}{\pi(0.500)^2/4} + \frac{0.450}{0.1599}\right] = 4.248\text{E} - 07$$

and $k_b = 2.354\text{E} + 06 \text{ lb/in}$

JOINT STIFFNESS k_j. Calculation of the stiffness of the joint is complicated by the fact that the area of material subjected to the bolt pressure changes with the distance into the joint, as shown in Figure 17.19. A number of models have been used to account for this variation. One of the simplest is a cylinder around the bolt with a diameter 3 times that of the bolt (Figure 17.20a). Another model consists of the

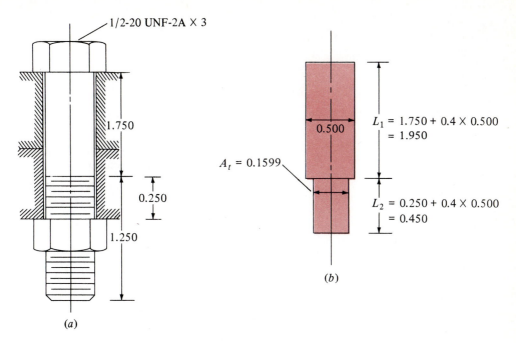

(a)

(b)

FIGURE 17.18
Joint detail and bolt stiffness model for Example 17.1.

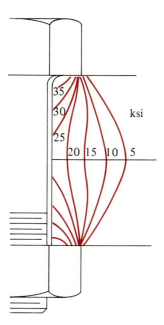

FIGURE 17.19
Lines of equal compressive stress in joint. Bolt loaded to 100 kip.
(*Reprinted from* [20], *courtesy Marcel Dekker Inc.*)

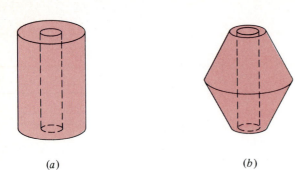

FIGURE 17.20
Joint stiffness models. (*a*) Cylindrical model. (*b*) Cone-frustrum model.

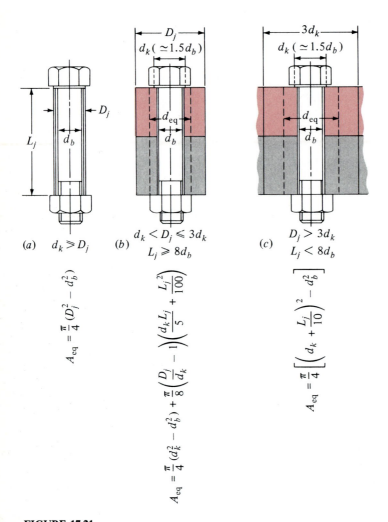

(*a*) $d_k \gg D_j$

$$A_{eq} = \frac{\pi}{4}(D_j^2 - d_b^2)$$

(*b*) $d_k < D_j \leqslant 3d_k$
$L_j \geqslant 8d_b$

$$A_{eq} = \frac{\pi}{4}(d_k^2 - d_b^2) + \frac{\pi}{8}\left(\frac{D_j}{d_k} - 1\right)\left(\frac{d_k L_j}{5} + \frac{L_j^2}{100}\right)$$

(*c*) $D_j > 3d_k$
$L_j < 8d_b$

$$A_{eq} = \frac{\pi}{4}\left[\left(d_k + \frac{L_j}{10}\right)^2 - d_b^2\right]$$

FIGURE 17.21
Equivalent joint area for stiffness calculation per VDI 2230.

frustrums of two cones (Figure 17.20*b*); the stiffness is computed through calculus [2, p. 375]. The Association of German Engineers' (VDI) directive 2230 [19] suggests an equivalent area depending on the size of the joint, as shown in Figure 17.21. The diagrams do not cover all possibilities ($D_j < 3d_k$, $L_j < 8d_b$ is lacking); when this arises, calculate the neighboring cases and use a compromise. We adopt the VDI approach.

Example 17.2. Continuing the analysis of the section of joint of Example 17.1 (Figure 17.18), we compute the stiffness of the joint and the stiffness parameter C. We see that the "diameter" of the joint D is more than 3 times the diameter of the flat under the head, and the joint length is less than 8 times the bolt diameter. So the model of Figure 17.21*c* is used:

$$A_{eq} = \frac{\pi}{4}\left\{\left[1.5(0.5) + \frac{2}{10}\right]^2 - (0.5)^2\right\} = 0.512 \text{ in}^2$$

and

$$k_j = \frac{A_{eq}E}{L} = \frac{0.512(30E + 06)}{2} = 7.687E + 06 \text{ lb/in}$$

Then since $k_b = 2.354E + 06$ (from Example 17.1), C becomes

$$C = \frac{2.354}{2.354 + 7.687} = 0.23$$

The joint stiffness calculated above is on the basis of what is termed a *hard joint*, which means that there is no gasket contributing to the stiffness. When there is a gasket, it is another spring added in series:

$$\frac{1}{k_j} = \frac{1}{k_{gasket}} + \frac{1}{k_{metal\ parts}} \tag{17.5}$$

When the gasket is an O-ring (Figure 17.22) or similar, the two parts of the joint are in contact, hence the gasket is not considered in the calculations—the joint

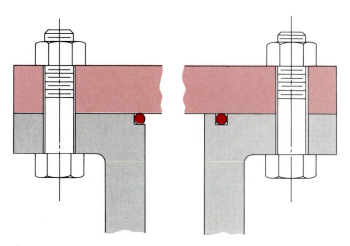

FIGURE 17.22
Hard joints sealed by an O-ring.

is a hard joint. We limit our discussion of gasketed joints to those required to conform to the ASME code (Section 17.8).

Design Analysis

First, the preload precluding joint separation was found in expression (17.3):

$$F_i = (1 - C)(\mathrm{FS}_P)P_{\max} \qquad (17.3)$$

Here P is written with the subscript "max," since the load may not be constant. We have also included a factor of safety on the load, FS_P.

Second, the force in the bolt was given in (17.1a): $F_b = F_i + CP$. A small C is desirable to keep the bolt load small, and that means minimizing the bolt stiffness relative to the joint stiffness, as we see in the second form of expression (17.1b). The joint is not amenable to a great deal of change, but the bolt can be made more compliant by reducing its shank, as we mentioned earlier. The result of doing this can be shown on the joint diagram (Figure 17.23), where the common case of an external load varying from zero to some maximum value is illustrated.

Third, fatigue failure of the bolt must be avoided. The load on the bolt under a varying load P consists of an alternating and a mean component:

$$F_{b,\mathrm{mean}} = F_i + C(\mathrm{FS}_P)P_m \qquad F_{b,\mathrm{alt}} = C(\mathrm{FS}_P)P_a$$

The stresses caused by these forces must satisfy the fatigue criterion, the Goodman relation (9.10).

The preload F_i is the principal contributor to the mean stress, determining where on a curve like that of Figure 9.14 (σ_a vs. σ_m) the bolt is operating and in consequence the permitted alternating component. A small stiffness parameter C, while reducing the mean bolt stress somewhat, has a direct effect on reducing the alternating component of the bolt stress.

The use of bolts with high fatigue strength is frequently indicated. We earlier mentioned that rolling threads after heat treatment improves strength. In fact, in some ultra high strength fasteners, it has approached the strength of the original unnotched stock.

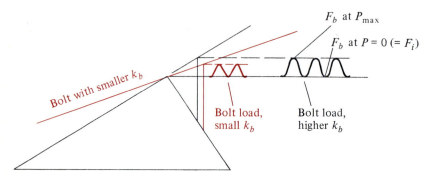

FIGURE 17.23
Effect of decreased bolt stiffness (smaller C).

Unsymmetric Joints

The joints discussed above are rarely seen in real life: cylindrical joints with bolt, load, and joint axes coincident. Figure 17.24a shows a joint in which the load is not concentric with the bolt axis, nor is the joint symmetrically situated with respect to the bolt. The joint is subjected to not only tension, but also bending. The pressure across the joint face is thus not uniform, and separation will commence on the load side, at a in the diagram. When separation starts (Figure 17.24b), there is less joint to absorb the load (to be unloaded), and the bolt load increases more rapidly. The

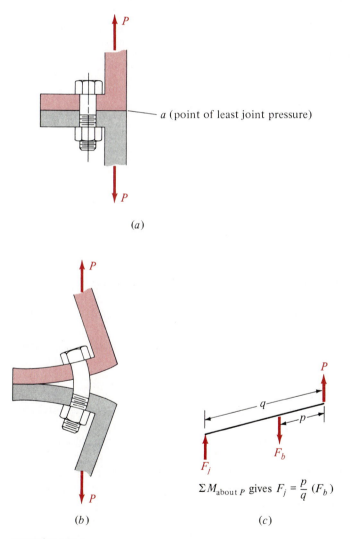

a (point of least joint pressure)

(a)

(b)

(c)

$$\Sigma M_{\text{about } P} \text{ gives } F_j = \frac{p}{q} (F_b)$$

FIGURE 17.24
Loading of eccentric joint. (a) Load, bolt, and joint axes are not aligned. (b) When separation starts, a greater fraction of the external load is transferred to the bolt, and the bolt eventually bends. (c) After the faces separate totally, a simple-lever situation results.

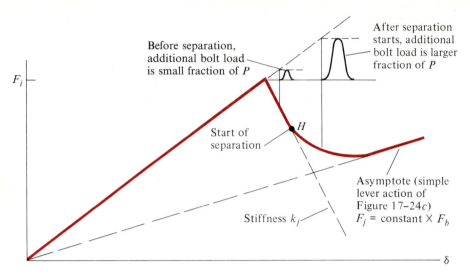

Before separation, additional bolt load is small fraction of P

After separation starts, additional bolt load is larger fraction of P

F_i

Start of separation

H

Stiffness k_j

Asymptote (simple lever action of Figure 17–24c) F_j = constant × F_b

δ

FIGURE 17.25
Joint diagram for a separating eccentric joint.

theory we developed above does not apply during this stage, because the area is changing. The limiting situation is shown in Figure 17.24c, where the loads are related through simple lever relations. Things are really out of hand at that stage. In addition to the extra tension the bolt suffers as a result of separation, it is subjected to bending, which makes matters worse.

The effect of separation can be shown on a joint diagram (Figure 17.25). As the external load P is increased, the joint is unloaded down the joint elastic line. At point H the pressure at the edge of the joint reaches zero, and separation starts. The load in the bolt rises along the bolt's elastic line, but the fraction of load P which the bolt takes increases, as shown in the diagram. The force on the joint follows a curve as shown. This continues until the joint consists of the simple lever arrangement in Figure 17.24c, when the load in the joint becomes a constant fraction of the bolt load, hence the straight-line asymptote on the joint diagram.

This analysis leads to the conclusion that separation of the joint must be prevented or minimized. That is accomplished, if possible, through a higher preload and/or improved joint geometry to more closely align the bolt, joint, and load. The effect of a higher preload is to lower the point H of incipient separation, as shown in Figure 17.26. The higher preload curves are shown in color; to better show the comparison, the diagram is extended at the bottom for the higher preload, i.e., the force scale origin is moved downward. A higher preload means a higher mean stress, and that lowers the resistance of the bolt to fatigue. However, that negative effect is more than offset by the lower cyclic stresses which the bolt sees as a result of avoiding separation. Moreover, high preload works against vibration loosening, which is probably the most frequently encountered cause of joint failure.

The stress situation in the bolt changes after the joint is bolted up and put in service. We need to examine this and to learn how preload is measured on the job.

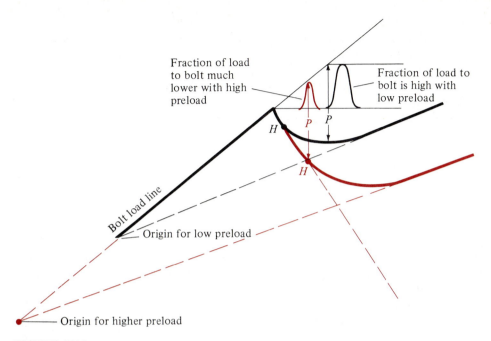

FIGURE 17.26

With higher preload the point of incipient separation H is moved downward: i.e., a greater load P is necessary to start separation. The load assumed by the bolt is smaller.

17.4   BOLT-TIGHTENING METHODS

Bolts are tightened by turning the nut or by turning the bolt. The problem is knowing when the required bolt force—the preload—has been reached. Something has to be measured. There are four basic approaches.

Torque Control

The most common method is to relate the force in the bolt to the torque applied to the nut or to the head of the bolt. A thread is like an inclined plane which the nut (or bolt) is pushed up by the torque applied to it (Figure 17.27). Resisting the motion

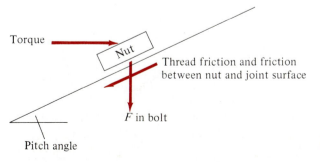

FIGURE 17.27
Force analogy of nut.

are the force in the bolt and the friction between threads and that between the nut and the surface on which it bears. These effects plus the incline of the plane (pitch of the thread) determine the torque, which can be calculated through the equations of static equilibrium. The result is very dependent on the coefficients of friction, for the frictional torque consumes 90 percent of the applied torque, and only the remaining 10 percent is used to overcome the bolt force. Unfortunately, the friction coefficients are hard to predict. An empirical approach seems to work best:

$$T = KF_i \frac{d}{12} \qquad T: \text{lb·ft, } d: \text{in, } F_i: \text{lb}$$

(17.6)

or
$$T = KF_i \frac{d}{1000} \qquad T: \text{N·m, } d: \text{mm, } F_i: \text{N}$$

Here d = nominal thread size, and K is the same in both expressions.

The friction coefficients, which depend on the nature of the sliding surfaces, are included in the constant K. For as-received steel (no lubricant), K is about 0.2. With molybdenum disulfide (MbS_2) grease it becomes about 0.137 [20, p. 429]. Needless to say, if the torque is specified as dry, then the bolt should be tightened dry; otherwise, the load will be higher than intended.

If a locknut is used which has a constant-torque drag, then that torque must be added to expressions (17.6).

Because of the unpredictability of dry friction, there is a great deal more scatter in the results with dry as compared to lubricated bolts, as shown in Figure 17.28

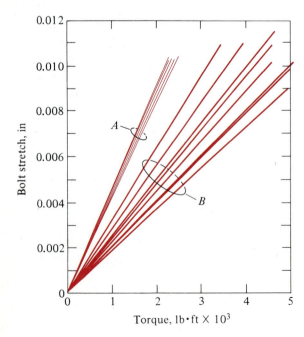

FIGURE 17.28
Results of experiments at Raymond Engineering on $2\frac{1}{4}$-in 8 × 12, B16 inner liner studs from a large steam turbine. The curves labeled A are for studs lubricated with molybdenum disulfide grease, a good thread lubricant. Curves B are for cleaned and dried studs. (*Reprinted from ref. 20, courtesy Marcel Dekker Inc.*)

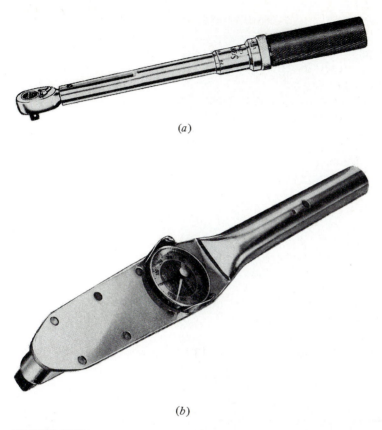

(a)

(b)

FIGURE 17.29
Torque wrenches. (*Courtesy, Snap-on Tools Corp.*) (*a*) Click type. A click is produced when the set torque is reached. Useful where a dial cannot be read. (*b*) Dial type.

[20, p. 76]. A scatter of ± 30 percent in the preload of torqued dry bolts is commonly assumed. For MbS_2-lubricated bolts, the figure is about ± 5 percent.

With a torque specified to produce a required preload, the mechanic tightens the bolt with a torque wrench, which is a tool cum instrument. Figure 17.29 shows two common manual types; nearly all auto mechanics have one or the other for tightening bearing-cap and head bolts to specification.

The use of muscle-operated torque wrenches is not practical when large torques are necessary nor economical in a production setting. Power-operated wrenches are then employed. The common hand-held types are designed to stall when a preset torque is achieved, either by controlling air pressure or through use of a slipping clutch. The impact wrenches you see in tire outlets to remove and replace wheel lug nuts are a common example. Figure 17.30 shows a hydraulic wrench being used to tighten the large bolts on a pressure vessel.

FIGURE 17.30
Hydraulic wrench is used to tighten bolts on a pressure vessel. The pump on the floor supplies fluid at 10 000 psi. (*Courtesy, Raymond Engineering, Inc.*)

Turn Control

Picture a bolt with nut against a totally rigid block, as shown in Figure 17.31a. If the nut is given a 360° twist, it advances a distance equal to the pitch of the thread p and produces that amount of stretch in the bolt. For a turn of $\theta°$, the stretch would be $p\theta/360$. In a practical setting, the block is the joint, and it is not inelastic. It deflects under the bolt load an amount Δ_j, so that the bolt stretch is

$$\Delta_b = \frac{p\theta}{360} - \Delta_j$$

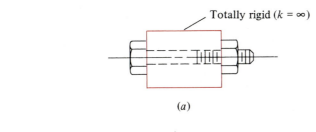

(a)

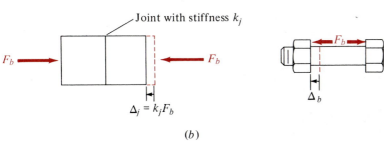

(b)

FIGURE 17.31
Calculating nut turn necessary to produce desired bolt preload. (a) A $\theta°$ turn of the nut would stretch the bolt $p\theta/360$. (b) Joint deflects under bolt load, reducing stretch in bolt by that amount to $\Delta = p\theta/360 - \Delta$.

as shown in Figure 17.31b. The desired force in the bolt and on the joint is F_i, the desired preload, and the above equation then becomes

$$\frac{F_i}{k_b} = \frac{p\theta}{360} - \frac{F_i}{k_j}$$

Solving, we find the turn angle to produce the preload:

$$\theta = \frac{(k_b + k_j)360}{k_b k_j p} F_i = \frac{360}{Ck_j p} F_i \qquad \text{degrees}$$

By measuring the nut turn, it is thus possible to produce a desired preload, if the joint and bolt stiffnesses are known. That is a calculation with some uncertainties, as we have seen, and it is the first problem with the method. The second difficulty is the question, From where do we measure the turn angle θ? When the method is used, the nut is "snugged" down, and the measurement starts there. The snugged-down position itself may be a problem, because the first part of the tightening will flatten any irregularities and produce nearly no stretch in the bolt. While there is no complete solution to this difficulty, one practice is to partly tighten the joint to remove the irregularities and then to loosen it and retighten to the final amount of turn θ.

The turn-of-nut method has been found to be more accurate than torque control in some quite sophisticated applications in the aerospace industry.

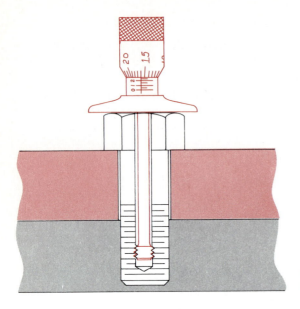

FIGURE 17.32
Measuring bolt stretch with a depth gage.

FIGURE 17.33
An ultrasonic instrument is used to check the bolt tension. (*Courtesy, Raymond Engineering, Inc.*)

Stretch Control

Turn control described above is an indirect means of determining the stretch in the bolt. That can also be done directly by measuring the bolt length before and during tightening. The bolt stiffness must be calculated, as described earlier [relation (17.4)]. Then $F_i = k_b \, \Delta L_b$. When the bolt is fully accessible, the change in length can be measured with an ordinary micrometer. If the application involves a stud or a tap bolt, where one end cannot be reached, the length change can be measured if the bolt can be drilled out and tapped to accommodate a threaded gage rod, as shown in Figure 17.32.

Ultrasonic means of measuring bolt elongation are also used. This method measures either the time it takes a burst of ultrasound to travel from one end of the bolt to the other and back or the resonant frequency of the bolt. Both depend on the bolt length and on the stress level. Figure 17.33 shows an ultrasonic instrument being used to check the tension in some large studs.

Stretch control is not well adapted to high production, but is widely used for large critical applications like pressure vessels.

Yield Control

The torque vs. turn curve for a bolt looks just like a stress-strain curve, since torque is proportional to stress and turn is proportional to strain. Such a curve is sketched in Figure 17.34. The yield is characterized by, even defined by, a rapidly changing slope. Yield-control wrenches sense that change electronically through torque and angle transducers. These wrenches can also be used to bring a bolt to a preset torque or turn of nut below the yield. The variation in preload with these devices is about ± 10 percent. Several yield-control wrenches are shown in Figure 17.35. They are also made for multiple-spindle use.

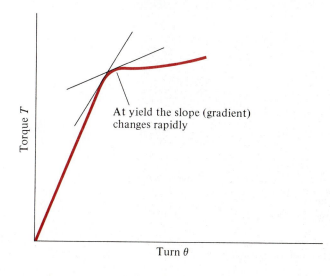

At yield the slope (gradient) changes rapidly

Torque T

Turn θ

FIGURE 17.34
Torque-turn curve.

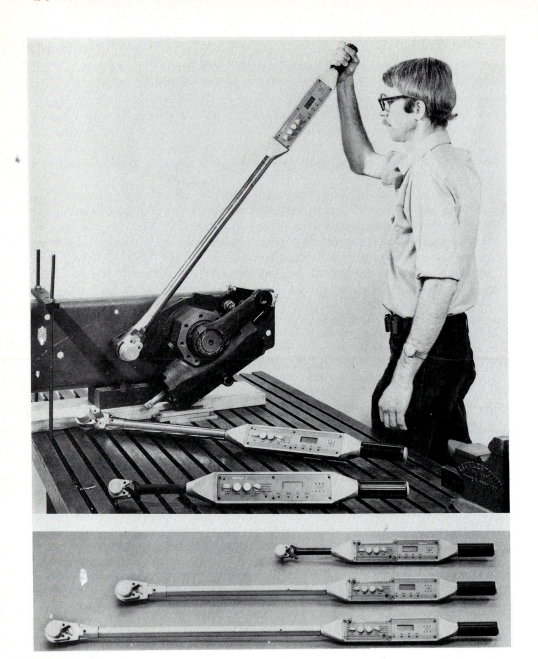

FIGURE 17.35
Yield-control wrenches. Hand-operated wrenches with torque and angle (turn) sensors that indicate when the yield point of the fastener has been reached. They can also be used to tighten to a preset torque or turn of nut.

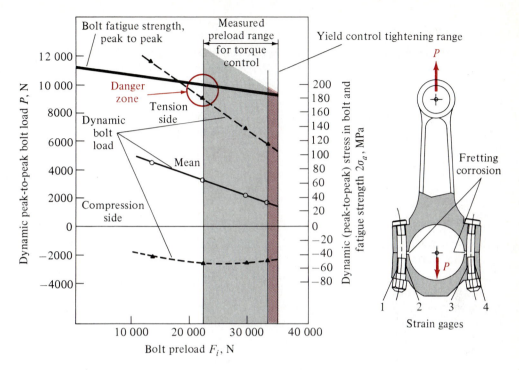

FIGURE 17.36
Connecting-rod bolt breakage analysis.

A Case History[1]

Figure 17.36 shows a production connecting rod in which the big-end bolts were failing, usually resulting in a junked engine. A stress analysis of the connecting-rod joint was undertaken, with strain gages being placed on both the inner and outer sides of the bolts, as shown. The results showed that when the actual preload was at the bottom of the torque-controlled scatter range, the maximum alternating stress in the bolt was very close to the bolt fatigue strength—the danger zone indicated on the graph. The joint was opening under these conditions, causing increased stresses in the bolt. Exacerbated by fretting corrosion (corrosion accelerated by rubbing of surfaces) on the sides of the bolt, these stresses were causing fatigue failure. This is illustrated in the figure, where the dynamic (i.e., alternating) load felt by the bolt is plotted against preload. The dark line labeled *fatigue strength, peak to peak*, is the range (twice the amplitude) of alternating stress producing failure at a particular preload. It slopes downward with increasing preload per the Goodman relation discussed in Section 9.3. (Preload is not the mean stress; rather, mean stress is preload plus the amplitude of the dynamic stress.) Three curves are shown for the bolt: the dynamic stress for the compression and tension sides of the bolt and the average dynamic

---

[1] Courtesy, SPS Technologies.

stress. The dynamic stress on the tension side, as might be expected, is the one governing fatigue failure. At the bottom of the torque-control preload scatter, the bolt peak-to-peak fatigue strength was about 200 MPa and the stress was 175 MPa, giving a factor of safety of $200/175 = 1.14$. This was inadequate. Decreasing the preload scatter through the use of yield control and then operating at the top of the previous preload scatter band, as shown on the graph, reduced the tension-side alternating stress on the bolt from over 180 to 100 MPa. The bolt fatigue strength at the high preload being about 185 MPa, the resulting factor of safety was increased to $185/100 = 1.85$.

17.5   BOLT STRESS RELAXATION

After a bolt is tightened, the force in the bolt diminishes somewhat, and fairly rapidly. The principal reason for this is plastic flow, termed *embedment*. High spots on threads or on the mating surfaces at the bolt head, the joint interface, or at the nut mean high local initial stresses and subsequent plastic deformation, which continues until enough area is created to reduce the stress below the yield. Any other places of high stress, such as a sharp corner of the hole against a fillet of the bolt head, can produce the same result. In gasketed joints, force relaxation is inevitable, since gaskets must flow to fulfill their purpose—preventing leaks.

Torque relaxation also takes place. Tightening a bolt (or its mating nut) produces a twist of the bolt and, as a result, shear stresses, which are maximum at the surface. The preload force produces tension stresses. The resulting surface stress element and its Mohr's-circle diagram are shown in Figure 17.37. In time the twist of the bolt relaxes, reducing the shear stress to near zero and leaving principal stresses and hence maximum shear stresses at a lower level. The process is, of course, accelerated if vibration is present. It is frequently said that if a bolt survives tightening, it probably will not fail in service.

The capacity of a bolt to carry the external load derives largely from the relaxation of the torsional stress. Let us see what that might mean in terms of failure theory.

In the discussion of torque control, we developed the relation between torque and bolt force: $T = KF_i d$ (torque and the bolt size are in the same length units). Substituting this in the shear stress relation

$$\tau = \frac{16T}{\pi d^3}$$

we get

$$\tau = \frac{16KF_i}{\pi d^2}$$

The tension stress is simply

$$\sigma = \frac{4F_i}{\pi d^2}$$

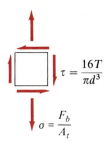

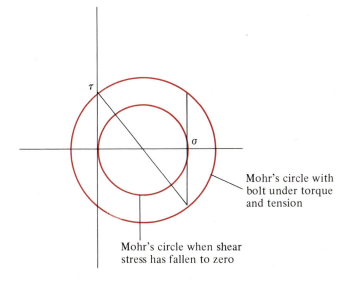

Mohr's circle with bolt under torque and tension

Mohr's circle when shear stress has fallen to zero

FIGURE 17.37
Relaxation of shear stresses.

If we use the distortion-energy failure theory and assume the bolt is tightened to the yield, the effective tension stress, relation (7.4d) (single tension plus shear), is set equal to the yield strength:

$$\sigma_{eff} = \sqrt{\sigma^2 + 3\tau^2} = \frac{4F_i}{\pi d^2}\sqrt{1 + 48K^2} = \sigma\sqrt{1 + 48K^2} = S_y$$

or

$$\sigma_{bu} = \frac{S_y}{\sqrt{1 + 48K^2}} \qquad (17.7)$$

where bu = bolting up.

When the shear stress relaxes, i.e., becomes zero, then the effective stress is $\sigma_{eff} = \sigma$. If the bolt is subsequently loaded to the yield, then the tensile stress $\sigma_{ul} = S_y$ (ul = under load). The difference between these two tensile stresses is the strength available for taking the external load:

$$\sigma_{ul} - \sigma_{bu} = S_y\left(1 - \frac{1}{\sqrt{1 + 48K^2}}\right) \qquad (17.8)$$

For dry steel this amounts to $0.42S_y$, and for lubricated threads it is $0.28S_y$. Either figure represents a respectable amount of strength for taking the external load, and the embedment relaxation will make it even greater, about another 5 percent.

Note on the Mohr's-circle diagram that when the torsional stress in the bolt has relaxed, and with it the stress governing failure, the tensile stress in the bolt and hence the force in the bolt are unaffected. If a bolt is initially brought to the yield, the tensile stress in it before and after torque relaxation is given by expression (17.7).

17.6 DESIGN APPROACH TO ECCENTRIC JOINTS

The preload necessary to preclude separation of an eccentric joint can be calculated by adding the tension and bending (prying) effects and finding the condition under which the joint pressure reduces to zero. This will, of course, be on the joint edge

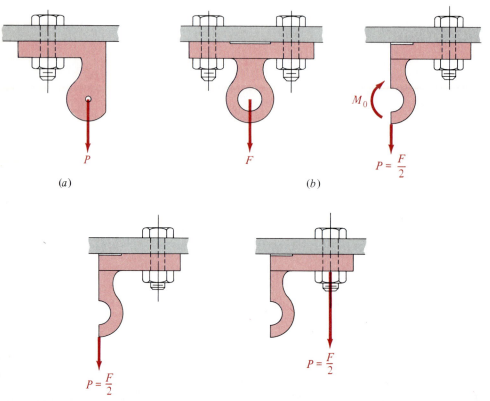

(a)   (b)

Hanger pinned in the center, $M_0 = 0$, load effectively at center

Totally stiff hanger—the load is effectively along the bolt axis

(c)

FIGURE 17.38
Location of loading force. (a) Few joints are loaded by a force only. (b) Hanger joint. (c) Extremes of hanger stiffness.

nearest the load (Figure 17.24). The problem with the procedure lies in knowing where the load P is acting.

Figure 17.38a shows a very simple joint, one of the rare cases where the location of the joint load is known for certain. In almost all joints, the load associated with each bolt consists of a force plus accompanying unknown moment, as with the hanger in Figure 17.38b. If we look at half of the hanger, as shown in the sketch, one-half of the joint load appears as a force; in addition, there is a moment, which is the reaction of the other part of the hanger. The size of the moment depends on the stiffness of the hanger, which is in essence a beam. There are two extremes. First, if the hanger were totally flexible, which would be the case if it were pinned at the center, then the moment would be zero; the joint loading would consist of the force $F/2$ only, applied at the center, as shown in Figure 17.38c. At the other extreme, if the hanger were totally stiff, the force $F/2$ would, in effect, be applied along the bolt axis, also shown in the figure. No prying of the joint would occur in the latter case. Truth lies somewhere between these extremes. The moment plus force can be replaced by a force only, positioned between the center of the hanger and the bolt. With finite-element techniques, the size of the moment could be determined and the position of the load then calculated. That implies a great deal of engineering time, which is generally not available for every joint design.

Most joints are similar to the hanger discussed above; the presence of an unknown moment means that the location of the joint load will be unknown. Most joint members are quite robust; however, even very hefty members are subject to some prying. In the absence of knowledge of the effective location of the joint load, the load is placed on the bolt axis, and the prying action is accounted for in the factor of safety.

For an analysis of eccentric joints where the load location is known, see Ref. 19.

17.7 JOINT DESIGN

Separation

The preload necessary to avoid separation from relation (17.3) is:

$$F_i \geq (1 - C)(FS_P)P_{max} \tag{17.3}$$

The preload of this expression is the in-service preload, which means that it is the load remaining after the relaxation due to embedment. (We saw in Section 17.5 that torquing shear-stress relaxation yields no decrease in the bolt load.) A figure of 5 percent is reasonable to assume for that relaxation, as noted earlier. Furthermore, this preload, being a minimum, must be calculated at the bottom of the torque scatter. We remarked earlier that under torque control this can be as high as ± 30 percent for dry bolts (± 5 percent for MbS_2-lubricated bolts). The tension stress in a bolt drawn up to the yield was given by (17.7). Including the embedment relaxation, we have

$$\sigma = \frac{0.95S_y}{\sqrt{1 + 48K^2}} \tag{17.9}$$

(If tightening is done to some other stress level, then that stress replaces S_y.)

Designating $\pm S_T$ as the torque scatter, we see the in-service minimum preload in expression (17.3) is

$$F_i = \frac{0.95 S_y}{\sqrt{1 + 48K^2}} (1 - 2S_T) A_t \geq (1 - C)(FS_P) P_{max}$$

from which

$$A_t \geq (1 - C) \frac{(FS_P) P_{max} \sqrt{1 + 48K^2}}{0.95(1 - 2S_T) S_y} \tag{17.10}$$

This gives the minimum bolt tension-stress area to avoid separation, taking into account embedment (5 percent) and stress relaxation and assuming tightening to yield (at the top of the scatter band).

Bolt Strength

STATIC LOADING. The force in the bolt was found to be

$$F_b = F_i + C(FS_P) P \tag{17.1a}$$

with an FS on the external load P. We must ensure that this force does not result in gross bolt yielding. The preload force to use in (17.1a) is that resulting from the greatest applied tightening torque, i.e., the top of the scatter. We will show P as P_{max}, since yielding must also be avoided in cases where the load P varies in time. To avoid static failure, then (17.1a), in terms of stresses, becomes

$$\frac{0.95 S_y}{\sqrt{1 + 48K^2}} + \frac{C(FS_P) P_{max}}{A_t} \leq \frac{S_y}{FS_{str}}$$

with tightening assumed to the yield.

There is a doubling up here of factors of safety with both FS_P and FS_{str} included. The yield strength of good bolts is quite well controlled, so we will not use FS_{str} (i.e., we let it be unity). Solving for the required bolt area gives

$$A_t \geq \frac{C(FS_P) P_{max}}{S_y[1 - 0.95(1 + 48K^2)^{-0.5}]} \qquad \text{yielding} \tag{17.11}$$

Expressions (17.10) and (17.11) determine the bolt size. The specified preload, that is, the mean, would be

$$F_i = \frac{A_t S_y(1 - S_T)}{\sqrt{1 + 48K^2}} \qquad \text{specified preload} \tag{17.12}$$

Example 17.3. We need to size the bolts for a hanger, such as in Figure 17.38b. The thickness of the hanger at the bolt is $\frac{3}{8}$ in, that of the frame, $\frac{1}{2}$ in. There will be a constant force of 4000 lb hung from it, hence 2000 lb per bolt. We use $FS_P = 2$. Tightening will be done dry to the yield with a manual torque wrench.

To calculate the stiffness parameter, we have to estimate the bolt size. We will assume $\frac{1}{2}$-20 UNF-2A $\times$ 1.5, grade 7 ($S_y = 115$ ksi). Using the bolt stiffness model of Figure 17.17 gives $k_b = 4.01E + 06$. For the joint stiffness, the model in Figure 17.21c gives $A_{eq} = 0.355$, so $k_j = 12.17E + 06$. Then

$$C = \frac{k_b}{k_b + k_j} = \frac{4.01}{4.01 + 12.17} = 0.25$$

We can now use relation (17.10):

$$A_t \geq (1 - 0.25) \frac{2(2000)(1.7)}{0.95[1 - 2(0.30)](115E + 03)} = 0.116$$

Since $A_t = 0.160$ for the assumed $\frac{1}{2}$-in bolts, the design passes the separation test. Let us see where we stand with the yielding criterion, (17.11):

$$A_t \geq \frac{0.25(2)(2000)}{0.44(115E + 03)} = 0.020$$

So that is no problem, and the two $\frac{1}{2}$-in UNF bolts suffice.

Now to specify the torque. With dry bolts and the scatter at ± 30 percent, the specified preload is

$$F_i = \frac{0.70S_yA_t}{1.7} = \frac{0.70(115E + 03)(0.160)}{1.7} = 7588 \text{ lb}$$

The torque, from (17.6), is

$$T = \frac{0.2(7588)(0.5)}{12} = 63 \text{ lb·ft}$$

FATIGUE LOADING. (In addition to what we develop below, the static strength requirement discussed immediately above must also be met.) There are a mean and an alternating component of stress:

$$\sigma_m = \frac{F_m}{A_t} = \frac{F_i + C(FS_P)P_m}{A_t}$$

$$\sigma_a = \frac{F_a}{A_t} = \frac{C(FS_P)P_a}{A_t}$$

(17.13)

About two-thirds of fatigue failures occur at the root of the first or second engaged thread; the others are at the fillet between head and body and at the thread run-out. These locations are points of stress concentration. Average values of K_f for the thread root and the fillet are shown in Table 17.5. The thread run-out point

TABLE 17.5
Fatigue notch factors K_f for threaded elements

SAE grade	Metric grade	Rolled threads	Cut threads	Fillet
0 to 2	3.6 to 5.8	2.2	2.8	2.1
4 to 8	6.6 to 10.9	3.0	3.8	2.3

Figures include notch sensitivity (q) and surface-finish factors C_F [2]. (Courtesy, McGraw-Hill Publishing Co.)

would have a K_f value about like that of the fillet, and the calculation would be the same. Since the nominal stress and the fatigue notch factor are higher at the thread root, that is the location to use in design. When failures occur at the other critical locations, they are due to unforeseens, like bending or a flaw.

We limit our discussion here to indefinite life. Finite-life strengths would be computed as outlined in Section 9.2. In the absence of test data from actual bolts, we apply Goodman's criterion, (9.10):

$$\frac{\sigma_a}{S_f} + \frac{\sigma_m}{S_{ut}} \le 1$$

We write this without the FS, because it is attached to the load P in the stress expressions.

By inserting the stresses (17.13), the design criterion becomes

$$A_t \ge \frac{C(FS_P)P_a}{S_f} + \frac{F_i + C(FS_P)P_m}{S_{ut}} \tag{17.14}$$

The design fatigue strength S_f may have to be estimated as discussed in Chapter 9. The worst F_i (the top of the torquing scatter) must be used. It will be the in-service value, which means the value after relaxation due to embedment. The stress at bolting up would normally be some percentage of the yield or of the proof stress. We write the expressions using the yield. Then

$$F_i = \frac{0.95S_y}{\sqrt{1 + 48K^2}} A_t$$

By inserting this in (17.14), the fatigue requirement is

$$A_t \ge \frac{C(FS_P)P_a}{S_f} + \frac{1}{S_{ut}} \left[\frac{0.95S_yA_t}{\sqrt{1 + 48K^2}} + C(FS_P)P_m \right] \tag{17.15}$$

Example 17.4. For the hanger of Example 17.3, suppose the external load of 2000 lb per bolt is applied repeatedly. Then $P_{max} = 2000$, and $P_m = P_a = 1000$.

With the $\frac{1}{2}$-in bolts chosen, separation and gross yielding were avoided for the maximum load applied, and that calculation would not change. To check the bolt fatigue strength, we first need the smooth-specimen (SS) fatigue strength S_{fSS}. Since we do not have that figure, we have to estimate it. From Table 17.2c, for the grade 7 chosen, $S_{ut} = 133$ ksi, so $S_{fSS} = S_{fRB} = 66.5$ ksi. And K_f for rolled threads from Table 17.5 is 3.0. The other correction factors are as follows:

Reliability: Since bolts are fairly well controlled, if we use a reliability factor of 90 percent, we will achieve a higher reliability than 90 percent. From Table 9.1, $C_R = 0.868$.

Finish: Included in the K_f.

Size: $C_S = 0.60$ (Section 9.1).

Impact: None, $C_I = 1$.

Then

$$S_f = \frac{0.868(0.60)(66.5E + 03)}{3.0} = 11.54E + 03$$

For dry bolts, $K = 0.2$. With the other parameters, (17.15) then becomes

$$A_t \geq \frac{0.25(2)(1000)}{11.54E + 03} + \frac{1}{133E + 03}\left[\frac{0.95(115E + 03)(0.160)}{\sqrt{1 + 48(0.04)}} + 0.25(2)(1000)\right]$$

$$= 0.124$$

The $\frac{1}{2}$-in bolts have $A_t = 0.160$, so the fatigue strength is adequate.

17.8 GASKETED JOINTS, ASME CODE

Gaskets are used to prevent leaking of a fluid or gas past a joint, often under pressure. For that reason, the design of gasketed joints frequently interfaces with public safety, and then it is regulated. The ASME established a committee in 1911 to formulate standards for the construction of steam boilers and other pressure vessels. It is now called the *Boiler and Pressure Vessel Committee*. The codes it has developed have been widely adopted as law in the United States and Canada. The gaskets prescribed at present by the code are ring gaskets which fit entirely inside the circle of bolts. Proposals are being considered for including full-flange gaskets.

The following is an introduction to code bolting requirements. The code also provides procedures for flange design, but this is beyond the scope of this text.

Gasket Seal Pressure

Since a gasketed joint must prevent leaks, a minimum (residual) gasket pressure must be maintained, and it depends on the pressure in the container:

$$p_g = mp_{max}$$

where m is called the *gasket factor*, whose value has been derived from experience with successful joints. A listing of gasket factors for a number of gasket types is shown in Table 17.6, taken from the ASME Pressure Vessel Code. As this is written, the Pressure Vessel Research Committee is preparing new gasket factors for some traditional products and for newly developed gasket materials and styles [83].

The gasket pressure p_g, as computed above, is an average value. The variation in pressure across the joint interface in effect reduces the gasket width to a value given the symbol b in the code. Computation of b is shown in Table 17.7, from the code, for various configurations.

The total bolt load (all bolts) corresponding to the minimum gasket pressure is obtained by multiplying the effective gasket area to be kept tight (computed on the basis of a width equal to twice the effective gasket width or $2b$), by the minimum required gasket pressure. Since the code treats in general gaskets of small width-to-diameter ratios, this is

$$W_{\min\ p_g} = 2\pi Gbmp_{max} \qquad (17.16)$$

TABLE 17.6
Gasket materials and contact facings*

Gasket factors m for operating conditions and minimum design seating stress y

Gasket material	Gasket factor m	Min. design seating stress y, psi	Sketches	Facing sketch and column in Table 17.7
Corrugated metal:				
Soft aluminum	2.75	3 700		(1a), (1b), (1c), (1d),
Soft copper or brass	3.00	4 500		Column II
Iron or soft steel	3.25	5 500		
Monel or 4–6% chrome	3.50	6 500		
Stainless steels	3.75	7 600		
Flat metal, jacketed asbestos filled:				
Soft aluminum	3.25	5 500		(1a), (1b), (1c),[2]
Soft copper or brass	3.50	6 500		(1d)[2], (2)[2];
Iron or soft steel	3.75	7 600		Column II
Monel	3.50	8 000		
4–6% chrome	3.75	9 000		
Stainless steels	3.75	9 000		
Solid flat metal:				
Soft aluminum	4.00	8 800		(1a), (1b), (1c), (1d),
Soft copper or brass	4.75	13 000		(2), (3), (4), (5);
Iron or soft steel	5.50	18 000		Column I
Monel or 4–6% chrome	6.00	21 800		
Ring joint:				
Iron or soft steel	5.50	18 000		(6), column I
Monel or 4–6% chrome	6.00	21 800		

Gasket material	m	y		Facing sketch and column
Stainless steels	6.50	26 000		
Elastomers without fabric or high percentage of asbestos fiber:				
Below 75A Shore Durometer	0.50	0		(1a), (1b), (1c), (1d), (4), (5); column II
75A or higher Shore Durometer	1.00	200		
Asbestos with suitable binder for operating conditions:				
$\frac{1}{16}$ in thick	2.75	3 700		(1a), (1b), (1c), (1d), (4), (5); column II
$\frac{1}{32}$ in thick	3.50	6 500		
Elastomers with cotton fabric insertion	1.25	400		(1a), (1b), (1c), (1d), (4), (5); column II
Elastomers with asbestos fabric insertion (with or without wire reinforcement):				
2-ply	2.50	2 900		(1a), (1b), (1c), (1d),
1-ply	2.75	3 700		(4), (5); column II

* This table gives a list of some commonly used gasket materials and contact facings with suggested design values of m and y that have generally proved satisfactory in actual service when you are using effective gasket seating width b given in Table 17.7. The design values and other details given in this table are suggested only and are not mandatory. Excerpted from ASME Pressure Vessel Code, Section VIII, Courtesy ASME. For other gasket types, see the code.

Note: The surface of a gasket having a lap should not be against the nubbin.

TABLE 17.7
Effective gasket width*

			Basic gasket seating width b_o	
	Facing sketch (exaggerated)		**Column I**	**Column II**
(1a)				
			$\dfrac{N}{2}$	$\dfrac{N}{2}$
(1b)†				
(1c)		$w \le N$		
			$\dfrac{w+T}{2}; \left(\dfrac{w+N_{max}}{4}\right)$	$\dfrac{w+T}{2}; \left(\dfrac{w+N_{max}}{4}\right)$
(1d)†		$w \le N$		
(2) $\frac{1}{64}$-in nubbin		$w \le N/2$	$\dfrac{w+N}{4}$	$\dfrac{w+3N}{8}$
(3) $\frac{1}{64}$-in nubbin		$w \le N/2$	$\dfrac{N}{4}$	$\dfrac{3N}{8}$
(4)†			$\dfrac{3N}{8}$	$\dfrac{7N}{16}$
(5)†			$\dfrac{N}{4}$	$\dfrac{3N}{8}$
(6)			$\dfrac{w}{8}$	$\ldots$

TABLE 17.7 (Continued)

Effective gasket seating width b

$$b = b_o, \text{ when } b_o \leq \tfrac{1}{4} \text{ in}; \quad b = \sqrt{b_o}/2, \text{ when } b_o > \tfrac{1}{4} \text{ in}$$

$$b = b_o \text{ when } b_o \leq 6.35 \text{ mm}; \quad b = 2.52\sqrt{b_o} \text{ when } b_o > 6.35 \text{ mm}$$

Location of gasket load reaction

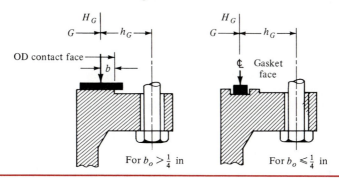

For $b_o > \tfrac{1}{4}$ in   For $b_o \leq \tfrac{1}{4}$ in

* The gasket factors listed only apply to flanged joints in which the gasket is contained entirely within the inner edges of the bolt holes.

† Where serrations do not exceed $\tfrac{1}{64}$-in depth and $\tfrac{1}{32}$-in width spacing, sketches (1b) and (1d) shall be used.

Source: From ASME Pressure Vessel Code, Section VIII, Courtesy ASME.

The end force due to the internal pressure is added to the bolt force corresponding to the minimum gasket pressure given by (17.16) to get the total required initial bolt load (all bolts):

$$W_{\text{tot}} = \frac{\pi}{4} G^2 p_{\text{max}} + 2\pi G b m p_{\text{max}}$$

$$= \pi G p_{\text{max}} \left(\frac{G}{4} + 2bm \right) \qquad \text{for minimum } p_g \qquad (17.17)$$

Our earlier joint analysis would produce a somewhat different result. The force in the joint per bolt is given by expression (17.2): $F_j = F_i - (1 - C)P$, or $W_i = (1 - C)W_p + W_j$, where the W's are the quantities for the sum of all bolts. With W_j given by expression (17.16),

$$W_i = \pi G p_{\text{max}} \left[(1 - C)\frac{G}{4} + 2bm \right] \qquad (17.18)$$

Comparing (17.17) with (17.18), we see that the code ignores the compliances of bolts and clamped members in the calculation of initial bolt forces. That is, in effect it takes $C = 0$ (joint infinitely more stiff than the bolt), which results in higher calculated bolt loads, i.e., a conservative bolt design. A factor of safety is also incorporated in the ASME allowable bolt stresses.

Gasket Seating Pressure

Since a gasket must flow to some extent to fill in the irregularities of the joint faces, a minimum "seating stress" is required when the joint is bolted up. The symbol y was adopted for it in the code. Some values are shown in Table 17.6. The area to be seated is taken on the basis of the effective gasket width b, so that the minimum total initial bolt tension for gasket seating pressure is

$$W_i = \pi bGy \qquad \text{for minimum seating pressure} \qquad (17.19)$$

There is some evidence that a greater seating pressure reduces the seal pressure necessary to prevent leakage, probably because the irregularities are better filled in. In other words, y depends on m.

To comply with the code, the initial total bolt tension must provide the minimum gasket operating pressure (17.17) and minimum seating pressure, (17.19). It is not a case of using the greater load, since the allowable bolt stresses are different, as we will see below.

Example 17.5. A 20-in cover on a pressure vessel is gasketed as shown in Figure 17.39. The pressure is static at 100 psi. A factor of safety of 1.5 is desired on the pressure. Thus $p_{max} = 150$ will be used. The gasket is copper-jacketed asbestos. The assignment is to determine the number and size of the bolts.

From Table 17.6 we find for this type of gasket that the gasket factor $m = 3.50$ and the minimum seating stress $y = 6500$ psi. It is also indicated that column II in Table 17.7 is to be used.

The gasket-face arrangement corresponds to that of sketch 1c in the table with $w = 1$, $N = 1.5$, and $T = \frac{3}{32} = 0.094$. Thus the basic gasket seating width is

$$b_o = \frac{w + T}{2} = \frac{1.094}{2} = 0.547 \text{ in}$$

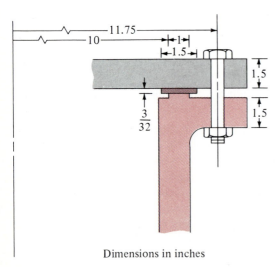

Dimensions in inches

FIGURE 17.39
Gasketed pressure-vessel cover for Example 17.5.

The limiting value of $b_o \leq (w + N)/4 = 0.62$, shown in the table, is not exceeded. Since b_o is greater than $\frac{1}{4}$ in, the effective seating width $b = \sqrt{b_o}/2 = 0.370$ in. From the left-hand sketch at the bottom of Table 17.7, we note that the gasket load reaction should be taken at a diameter $G = 22 - 2b = 21.26$ in. Entering these numbers in expressions (17.17) and (17.19), we have

$$W_i = \pi(21.26)(150)\left[\frac{21.26}{4} + 2(0.370)(3.50)\right]$$

$$= 79\ 197\ \text{lb} \qquad \text{minimum gasket pressure}$$

$$W_i = \pi(0.370)(21.26)(6500)$$

$$= 160\ 631\ \text{lb} \qquad \text{minimum seating pressure}$$

Hard Joints

The logic of the preceding paragraphs also applies for finding initial bolt loads for hard joints, i.e., where self-energizing gaskets such as O-rings are used (with the exception of certain seal configurations which produce axial loads). The minimum sealing pressure for these joints is considered to be zero ($m = 0$) as well as the minimum seating pressure ($y = 0$). Thus one expression suffices:

$$W_{\text{req}} = \frac{\pi}{4} G^2 p_{\text{max}}$$

Bolt Selection and Spacing, Static Requirements

The total cross-sectional area of bolts required for maintaining gasket pressure is obtained by dividing the required initial bolt load given by (17.17) by the allowable bolt stress at the operating temperature. To ensure proper seating, the area of bolts is the load given by (17.19) divided by the allowable stress at ordinary temperature (at bolting up). The larger resulting bolt area is used. The bolts are then chosen to produce a reasonable bolt spacing, as in the example below.

Allowable stresses for various bolt materials and grades are given in the code. The stress for temperatures from -20 to $650°F$ are about one-fourth the yield strength. Values at higher temperatures are somewhat less.

To obtain reasonably uniform gasket pressure, the bolt spacing should not be greater than a few bolt diameters. On the low side, bolt heads must not be so close together that the mechanic cannot get a wrench on the bolt head or the nut. Bolt heads are about 1.6 times the size of the bolt. If you allow $\frac{1}{8}$ in, or 3 mm, for the thickness of the wrench, then the spacing must be greater than $1.6d_b$ plus $\frac{1}{8}$ in (or plus 3 mm). On critical jobs, the exact size of the wrench can be found in a mechanic's handbook or by simply measuring it in the shop.

Example 17.5   continued. We calculated above the two bolt load requirements for this example:

$$W_i = 79\ 197 \qquad \text{for minimum gasket pressure}$$

$$W_i = 160\ 631 \qquad \text{to ensure seating}$$

Let us assume that this application will operate at 900°F. In the code a bolt speci-fied as SA-193 grade B5 (5 percent Cr, 0.5 percent Mo) has a yield strength $S_y = 80.0$ ksi and ultimate strength $S_{ut} = 100.0$ ksi. The allowable stresses are

$$S_{all} = \begin{cases} 20.0 \text{ ksi} & -20 \text{ to } 65°\text{F} \\ 10.4 \text{ ksi} & \text{at } 900°\text{F} \end{cases}$$

The required bolt areas are then

$$\frac{79\ 197}{10\ 400} = 7.62 \text{ in}^2 \qquad \text{for minimum gasket pressure}$$

$$\frac{160\ 631}{20\ 000} = 8.03 \text{ in}^2 \qquad \text{to ensure seating}$$

The second governs.

A $\frac{5}{8}$ UNF bolt has a stress area $A_t = 0.256$. The number of bolts will then be $n_b = 8.03/0.256 = 32$. Since the circumference of the bolt circle is $2(11.75\pi)$, the spacing, in bolt diameters, is $23.5\pi/(n_b d_b) = 3.7$. If $\frac{1}{2}$ UNF bolts are chosen, 51 will be needed, and the spacing will be 2.9 bolt diameters. Either choice would be satisfactory.

Bolt Selection, Fatigue Requirement

When the pressure in the vessel varies in time, the bolts will be subjected to fatigue conditions. We limit this discussion to the case where the pressure varies from zero to a maximum value; methods of accounting for other pressure variations are pro-vided in the code.

The minimum force on a bolt is the bolting-up initial force $F_i = W_i/n_b$ (to main-tain gasket pressure or to ensure seating), which we discussed above. The additional force P caused by the pressure would, according to the classic theory, be CP from (17.1a). The code takes the conservative approach of assuming that the bolt sees the entire force, that is, C is taken as unity. Thus

$$F_{\text{due to } p} = \frac{\pi G^2}{n_b(4)} p_{\max}$$

One-half of the force due to the pressure represents the alternating force on the bolt. The alternating stress is obtained by dividing that by the bolt stress area A_t and multiplying it by the fatigue notch factor K_f. The code specifies that the latter be not less than 4:

$$\sigma_a = \frac{\pi G^2}{A_t n_b(8)} p_{\max} K_f \tag{17.20}$$

The mean force on a bolt, of course, would be the initial force plus one-half of the force due to the pressure. (The stress must be increased by the effects of bending, when present.)

The ASME code provides S-N curves for the steels usually used in pressure-vessel bolts. These curves are for mean stresses between 33 percent (in the event there is bending) and 50 percent (pure tension) of the maximum allowable static stress (used for the sealing and seating calculations described earlier) for temperatures not exceeding 700°F. (The Goodman criterion is built into the curves.) We concern ourselves here only with high-cycle stressing, more than 1 million cycles. The alternating stresses allowed by the code are 12.5 ksi for steels with $S_{ut} \leq 80$ ksi and 20 ksi for steels with $S_{ut} = 115$ to 130 ksi. Interpolation between these numbers is permitted. These figures are for carbon, low-alloy, series 4XX, high-alloy steels and high-tensile steels for temperatures not exceeding 700°F.

Recognizing that the allowable stress is an upper limit, in terms of total bolt area, we see that the equality (17.20) becomes the inequality

$$n_b A_t \geq \frac{\pi G^2 p_{max} K_f}{8 S_{alt,all}} \tag{17.21}$$

Example 17.6. In Example 17.5, suppose the pressure, rather than being static, varies from 0 to 100 psi. We need to check the bolts for fatigue.

We determined in the continuation of Example 17.5 that, to ensure sealing and seating, the total bolt area necessary is $n_b A_t \geq 8.03$ in² (seating governing). With $S_{all} = 17.6$ ksi for the material of these bolts ($S_{ut} = 100.0$ ksi), K_f taken as 4, and the other numbers of the problem, expression (17.21) becomes

$$n_b A_t \geq \frac{\pi (21.26^2)(100)(4)}{8(17.6E + 03)} = 4.03 \text{ in}^2$$

Thus, the total bolt area previously determined on the basis of seating the gasket (8.03 in²) is adequate to prevent fatigue failure.

Bolt Tightening

We discussed how bolts are tightened to achieve a given bolt preload in Section 17.4. Experience indicates that a uniform pressure around the flange is very important. Not only should the bolt loads be uniform, which means using the best control available, but also the tightening sequence should be thought out. To tighten bolts by going from one to the next neighbor can warp a flange or produce a wave in the gasket, making a seal impossible. A star pattern such as suggested in Figure 17.40 can be used, and it helps, if possible, to tighten several bolts at the same time.

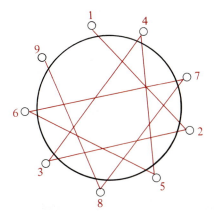

FIGURE 17.40
Typical star-pattern bolt-tightening sequence.

17.9 JOINTS LOADED IN SHEAR

When the forces on a joint tend to slide the members over each other, the joint is loaded in shear. Figure 17.41 shows joints of this type. They are most commonly found in structural-steel work: buildings, bridges, boilers. In aircraft, particularly light airplanes, there are many riveted joints in the wing ribs, spars, and elsewhere, which are also examples. The hinges of automobile doors number in the millions.

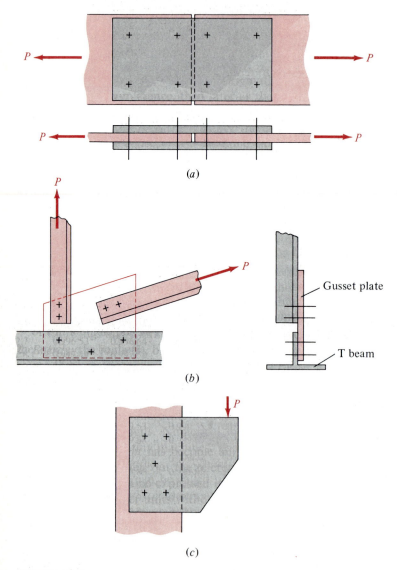

FIGURE 17.41
Joints in shear. Bolt locations are designated by a cross (+). (a) Butt splice. (b) Gusset plate. (c) Bracket.

In civil works, for reasons of public safety, the design of these connections is generally governed by codes. Most of these regulations reference the *American Institute of Steel Construction Handbook*. In the case of boilers, the Boiler Construction Code of the ASME is used.

Axial Loads

Figure 17.41*a* shows a joint secured by a symmetric group of bolts. If the load application is through the centroid of the bolt pattern, the loading is said to be *axial*, as opposed to eccentric, discussed later.

Frequently, in joints of this type, the intention is for no slippage between the plates to occur, and the joint is then termed a *slip-resistant joint*. This is the case, for example, if the joint is used for an adjustment; the bolt holes might be slotted for that purpose, as on the hood, doors, and trunk-lid hinges of a car. It would also be the case where the loads are reversing which otherwise would result in constant motion of the joint. The resistance of the joint to slip depends on the usual parameters: coefficient of friction and the normal force between the surfaces. Additionally, it depends on the number of surfaces involved. Thus

$$F_f = \mu F_{is} n_b n_s \tag{17.22}$$

where μ = friction coefficient, in this business called the *slip coefficient*
F_{is} = in-service bolt load
n_b = number of bolts
n_s = number of pairs of surfaces being clamped

Clearly, the slip coefficient depends on the nature and condition of the faying surfaces (the term for the joint surfaces), and a great deal of experimental work has been done to determine it. A large slip coefficient being desired, grease and oil should be removed as well as loose mill scale and dirt. The latter is best done by hand, since the use of a power brush tends to smooth the surface, decreasing slip resistance. For structural steel, Fisher and Struik [22, p. 76] indicate an average value of $\mu = 0.336$, with a standard deviation of 0.070. Blast cleaning results in a better value, $\mu = 0.493$, with a standard deviation of 0.074. These figures are for structural steel. In many cases the material lies around outdoors for a time before being used. Rust usually results and makes the surface appear a lot rougher, but the rust is softer and decreases the slip coefficient. To prevent rust formation, protective coatings are sometimes used, and this practice alters the coefficient.

> **Example 17.7.** The joint of Figure 17.41*a* uses four M10 × 1.5, class 5.8 bolts (per side) drawn up dry to 90 percent of the proof load (specified load). The friction coefficient is 0.4. We need the load that can be applied to the joint without slip.
>
> There are two pairs of surfaces resisting slip in this problem, so $n_s = 2$. The proof strength for class 5.8 bolts is, from Table 17.3*c*, 380 MPa, and the stress area, from Table 17.3*a*, is $A_t = 58.0$ mm². Hence the minimum bolt in-service load is
>
> $$F_{is} = \frac{0.95(1 - S_T)0.90 S_p A_t}{\sqrt{1 + 48K^2}}$$

Assuming the torquing scatter at 10 percent gives

$$F_{is} = \frac{0.95(0.90)(0.90)(380)(58.0)}{\sqrt{1 + 48(0.04)}} = 9925 \text{ N}$$

Then, using (17.22) with a slip coefficient of 0.4, we get

$$F_f = 0.4(9925)(4)(2) = 31.8 \text{ kN}$$

In the above discussion it was tacitly assumed that slip would occur before yielding or rupture of the plates. If that is in question, it must be checked. We discuss that possibility below.

While most joints are designed as friction joints, occasionally they are intended to be loaded by bearing, i.e., motion of the joint is prevented by interference between the bolts and the plates. Such connections are called *bearing joints*. A rough estimate of the strength of a bearing joint can be obtained by simply multiplying the total bolt shear area by the shear strength:

$$F_s = A n_b n_p S_s \tag{17.23}$$

where $A = A_t$ if the shear plane is in the threaded area, otherwise the shank area
n_b = number of bolts
n_p = number of shear planes
S_s = bolt shear strength

Two factors of some importance are ignored in this estimate.

1. Usually two or more bolts are used in a line to fasten the joint, such as shown in Figure 17.42. If the plates were perfectly rigid and the fits were perfect, all the bolts would be equally stressed. This is never the case, and the loads carried by the outermost bolts can be considerably greater than those seen by the inner ones. If 10 or more fasteners are used in a line, the difference can be a factor of 2 [22,

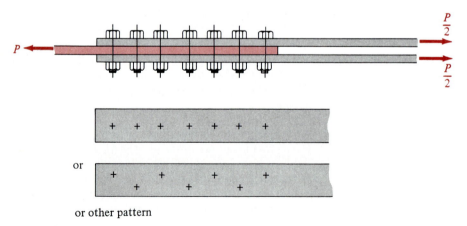

or other pattern

FIGURE 17.42
Joints usually have several bolts in line or in an array.

p. 96]. Additionally, stress concentrations exist. Empirical relations have been developed to account for these effects.

2. The loading in the bolts is not pure shear. Tension also exists in the form of the bolt load F_{is}. Fisher and Struik [22, p. 54] reported that for high-strength bolts at rupture

$$2.60\tau^2 + \sigma^2 = S_{ut}^2 \tag{17.24}$$

Both stresses are on the basis of A_t, the stress area. (Whether the threaded portion or the shank fell at the shear plane made no difference in their tests.)

The empirical relation (17.24) bears considerable resemblance to the distortion-energy theory result for yield failure [relation (7.4d) for single tension plus shear]:

$$3\tau^2 + \sigma^2 = S_y^2 \tag{17.24'}$$

which can be used for yield failures.

Example 17.8. The connection of Figure 17.42 is a bearing joint bolted up dry with $\frac{5}{16}$ − 24 UNF grade 5.2 bolts to a specified load corresponding to the proof stress. A yield-control wrench ($S_T = 0.10$) is used. For these bolts S_{ut} is 120 ksi, $S_p = 85$ ksi, and $A_t = 0.058$ in^2. At what load do we rupture a bolt?

We assume that the outer bolts see 50 percent more than the average load. There are 7 bolts and 2 pairs of shear surfaces, so that on one of the outer bolts

$$\tau = 1.5 \frac{P}{7(2)(0.058)} = 1.85P$$

The minimum in-service tensile stress is

$$\sigma_{is} = \frac{0.95(0.90)(85E + 03)}{1.71} = 42.8E + 03$$

Then, from (17.24),

$$2.60(1.85^2)P^2 + (42.8E + 03)^2 = (120E + 03)^2$$

whence $$P = 37.6 \text{ klb}$$

On the basis of shear only and the tensile stress area A_t, the failure load from (17.23) is

$$F_s = \frac{0.058}{1.5}(7)(2)(120E + 03)(0.577) = 37.5 \text{ klb}$$

This is remarkably close. (The 0.577 on the right converts S_{ut} to S_{us} per the distortion energy theory—an approximation only.)

Rupture of the bolts is not the only manner in which a connection can fail. This varies somewhat from joint to joint; the basic possibilities are shown in Figure 17.43.

At (*a*) the failure is of the plate ("pl") in tension

$$F = S_{ut,pl}A_{pl}$$

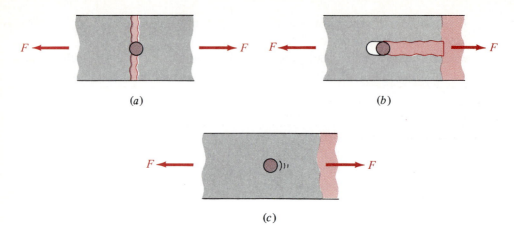

FIGURE 17.43
Types of plate failures. (*a*) Tension failure. (*b*) Shear failure. (*c*) Crushing failure.

where A_{pl} is the cross-sectional area which parts. Naturally, the smallest area in the connection is used.

Failure by shear of the plate is shown at (*b*); again the worst spot must be chosen:

$$F = S_{s,pl}A_{pl} \qquad S_{s,pl} \simeq 0.58S_{ut,pl}$$

Failure by crushing of the plate at a bolt location is shown at (*c*):

$$F = S_{y,pl}d_{bolt}t_{pl}$$

where $d_{bolt}t_{pl}$ is the projected bearing area of the bolt against a single plate.

Eccentric Loading

A joint is eccentrically loaded when the line of action of the load does not pass through the bolt-pattern centroid. The attachment of the gusset plate to the T beam in Figure 17.41*b* is an example, as well as the bracket at (*c*).

In a slip-resistant connection, the maximum load, i.e., the slip load, causes the joint to translate and to rotate. These two motions are the equivalent of a rotation about an instantaneous center, and the friction force of each bolt is thus perpendicular to its radius of rotation. If the further assumption is made that the plate is totally stiff, then all the bolts slip simultaneously, and the friction forces are all equal.

We limit our discussion to symmetric connections, such as the bracket of Figure 17.41*c*, whose force diagram at the moment of slip is shown in Figure 17.44. The direction of each bolt friction force depends on the location of the instantaneous center, which is situated on a line passing through the centroid of the bolt group and perpendicular to the direction of the load, as in the figure. The position of the IC on that line is determined by the equations of static equilibrium. An xy system is set up at the centroid of the bolt pattern. The slip resistance of each bolt is $f_s = \mu F_i$. The summation of forces in the x direction is satisfied by the symmetry, independently

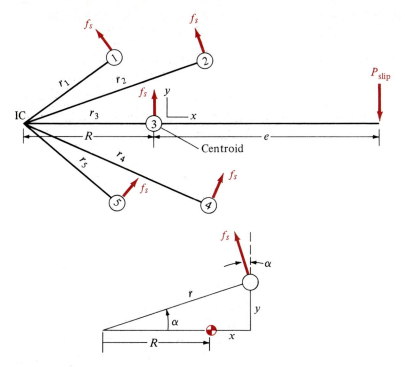

FIGURE 17.44
Forces on bracket of Figure 17.41c at instant of slip.

of the location of the IC. Applying the other static equilibrium conditions gives

$$\sum F_y: \qquad \sum_{i=1}^{n} f_s \cos \alpha - P = f_s \sum_{i=1}^{n} \frac{R + x_i}{\sqrt{(R + x_i)^2 + y_1^2}} - P = 0 \qquad (17.25a)$$

$$\sum M_{\text{(about IC)}}: \quad f_s \sum_{i=1}^{n} r_i - P(R + e) = f_s \sum_{i=1}^{n} \sqrt{(R + x_i)^2 + y_i^2} - P(R + e)$$

$$= 0 \qquad (17.25b)$$

To find R, solve for P from (17.25a) and substitute it in (17.25b). This gives

$$\sum_{i=1}^{n} \sqrt{(R + x_i)^2 + y_i^2} - (R + e) \sum_{i=1}^{n} \frac{R + x_i}{\sqrt{(R + x_i)^2 + y_i^2}} = 0 \qquad (17.26)$$

Thus the instantaneous center does not depend on the size of the forces f_s or on the size of the applied force P. Naturally, it does depend on the location of P, the distance e. The summations in expression (17.26) make an explicit solution for R impossible. A recursion solution could be set up by solving for the R which stands in the parentheses with e; however that solution, in some instances at least, is divergent. So expression (17.26) must be solved by trial and error, which is readily done with a desk computer, by interval halving or other means. Once R is known, then P or f_s, depending on which is the unknown, can be found from (17.25a) or (17.25b).

Example 17.9. Dimensions for the bolt pattern and the eccentricity of the bracket of Figure 17.41c are shown in Figure 17.45. The bracket must support a load of 35 kN, with the factor of safety included. We need to size the bolts. The slip coefficient is 0.4.

A recursion solution of (17.26) for this problem is divergent. The expression was solved by interval halving to give $R = 10.0$ mm. Expression (17.25a) or (17.25b) can now be used to find f_s and the required preload; (17.25b) looks easier:

$$f_s = \frac{P(R + e)}{\sum\limits_{i=1}^{5} \sqrt{(R + x_i)^2 + y_i^2}}$$

The summation in the denominator works out to be 212.5, so we have

$$f_s = \mu F_{is} = 0.4 F_{is} = \frac{(35E + 03)(10.0 + 130)}{212.5}$$

whence
$$F_{is} = 57.6 \text{ kN}$$

If a class 5.8 bolt is used, $S_p = 380$ MPa, and if the bolts are tightened (nominally, dry) to this stress with a torquing scatter of 10 percent, then

$$\frac{0.95(380)(0.90)}{\sqrt{1 + 48(0.04)}} A_t = 57.6E + 03$$

whence
$$A_t = 303 \text{ mm}^2$$

An M24 × 3 bolt ($A_t = 353$ mm^2) meets the requirement.

When an eccentric joint is a bearing joint, i.e., motion is prevented by interference between bolts and plates, the load is shown transferred to the centroid of the bolt group as a force plus a moment, as illustrated in Figure 17.46. No frictional forces are assumed. The effect of each is then calculated separately, as follows (P is assumed to incorporate a factor of safety).

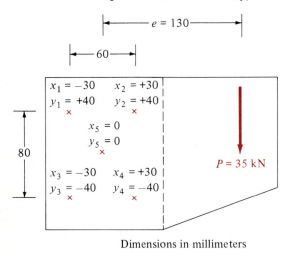

$e = 130$

60

$x_1 = -30 \quad x_2 = +30$
$y_1 = +40 \quad y_2 = +40$

$x_5 = 0$
$y_5 = 0$

80

$x_3 = -30 \quad x_4 = +30$
$y_3 = -40 \quad y_4 = -40$

$P = 35$ kN

Dimensions in millimeters

FIGURE 17.45
Dimensions and load for eccentric bracket in Example 17.9.

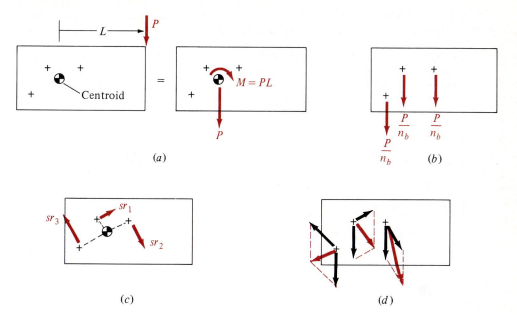

FIGURE 17.46
Bolt loading in eccentric bearing joint. (a) Load is transferred to the centroid of bolt group. (b) Direct shear is divided equally among bolts. (c) Effect of moment is perpendicular to, and in proportion to, the radius from the centroid. (d) Component forces are added to give the total force on each bolt.

1. The direct shear is divided equally among the bolts:

$$V_P = \frac{P}{n_b} \qquad (17.27)$$

This is shown in Figure 17.46b.

2. The force on each bolt due to the moment is assumed to be proportional to its distance from the centroid and acting perpendicular to the radius from the centroid, as seen in Figure 17.46c. Thus

$$V_{M,i} = sr_i$$

for any bolt. The constant s is determined by writing the total moment:

$$M = s(r_1^2 + r_2^2 + \ldots) = PL$$

Substituting s into the preceding relation for $V_{M,i}$ gives

$$V_{M,i} = \frac{PLr_i}{r_1^2 + r_2^2 + \cdots} \qquad (17.28)$$

The total force acting on each bolt is the resultant of the direct shear force and that due to the moment, shown in Figure 17.46d. The bolt with the largest force is critical, and the stresses on it can be determined as detailed earlier for axial bearing joints.

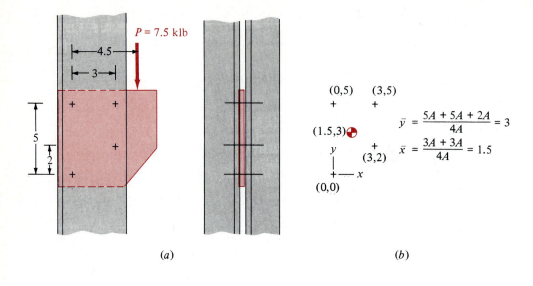

(a)

$$\bar{y} = \frac{5A + 5A + 2A}{4A} = 3$$

$$\bar{x} = \frac{3A + 3A}{4A} = 1.5$$

(b)

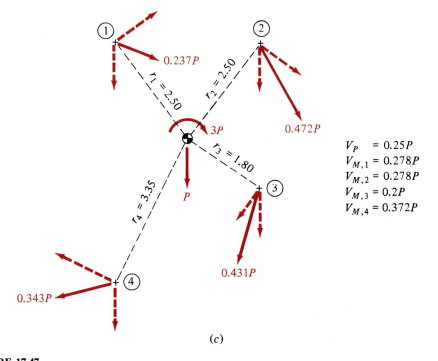

$$V_P = 0.25P$$
$$V_{M,1} = 0.278P$$
$$V_{M,2} = 0.278P$$
$$V_{M,3} = 0.2P$$
$$V_{M,4} = 0.372P$$

(c)

FIGURE 17.47
Eccentric bearing plate for Example 17.10. (a) Plate sketch and dimensions (in inches). (b) Location of bolt-pattern centroid. (c) Equivalent loading on bolts for one pair of shear planes.

Example 17.10. We need to size the bolts for the bearing joint of Figure 17.47, with failure based on yielding, not rupture. Note that there are two pairs of shear planes. The easiest way to handle that is to analyze one pair only, using one-half the indicated load. The FS on the load of 5 klb is to be 3, so the load to use in the calculation is $P = 7500$ lb.

The first step is to locate the centroid of the bolt group; this is done in Figure 17.47b. The equivalent force at the centroid and the moment are shown in (c).

The direct average shear load per bolt is the load divided by 4, the number of bolts:

$$V_P = 0.25P$$

The shear due to the moment, per (17.28), is

$$V_{M,i} = \frac{3Pr_i}{(6.25 + 6.25 + 11.25 + 3.25)} = \frac{Pr_i}{9}$$

These forces are listed for the individual bolts in Figure 17.47c. They are shown as vectors on the loading schematic, and the resultant for each bolt is indicated.

Our concern is with the bolt with the greatest force, which is the one at the upper right, number 2, with a resultant force $V_{bolt\ 2} = 0.472P$.

We might expect to use a grade-5 bolt for the application, which has

$$S_p = 85 \text{ kpsi} \qquad S_{yt} = 92 \text{ ksi}$$

If we assume the bolts are tightened (nominally, dry) to the proof load with a torquing scatter of 10 percent, then the in-service bolt tensile stress is

$$\sigma_{is} = \frac{0.95(85E + 03)(0.9)}{\sqrt{1 + 48(0.04)}} = 42.5E + 03$$

Then the distortion-energy yield criterion, expression (17.24'), becomes

$$3\left[\frac{0.472(7.5E + 03)}{A_t}\right]^2 + (42.5^2)E + 06 = (92^2)E + 06$$

from which

$$A_t = 0.0751 \text{ in}^2$$

A $\frac{3}{8}$-16 UNF bolt ($A_t = 0.0775$) can be used.

17.10   CLOSURE

This chapter has presented the basics of joint design, which would be adequate in most applications for producing a prototype. For design of pressure connections to code, you may be required to follow the ASME procedures. The AISC *Handbook* provides a great deal of information on shear-joint design.

PROBLEMS

Section 17.1

17.1. What do the following designations mean?
   (a) $\frac{3}{8}$-16 UNC − 2A × $2\frac{1}{2}$
   (b) M8 × 1.25, 8g × 40

(c) 1-12 UNF − 3A × 5

(d) M14 × 2, 4g × 60

17.2. If a bolt is threaded into aluminum, say, is a coarse or fine thread usually used?

17.3. What is the advantage of rolled threads?

17.4. In the case of a stud, is the nut end usually a fine or a coarse thread? What thread is usually used on the other end?

17.5. Define tensile stress area A_t. Define proof strength and proof load.

Section 17.2

17.6. How is the load distributed among the threads of a nut? What steps are sometimes taken to improve the situation?

17.7. Sketch the Mohr's-circle diagram for the critical-stress element of a bolt which is loaded in shear.

17.8. Sketch the Mohr's-circle diagram for the critical-stress element of a bolt while it is being tightened.

Section 17.3

17.9. A $\frac{1}{2}$-13 UNC × 3-in bolt with nut is used to grip a 2-in joint. Compute the stiffness of the bolt k_b.

> Answer: $k_b = 2.29E + 06$ lb/in

17.10. An M20 × 1.5 × 60 bolt is used as a tap bolt to secure a 35-mm-thick member to a larger casting. Compute the stiffness of the bolt k_b.

17.11. The clamped members of the joint of Problem 17.9 are steel. Compute the joint stiffness k_j and the stiffness parameter C. In Figure 17.21, assume $D_j = 3$.

> Answer: $K_j = 7.68E + 06$, $C = 0.230$

17.12. The members of the joint of Problem 17.10 are steel. Find the joint stiffness k_j and the stiffness parameter C. Assume $D_j = 100$.

17.13. The end plate of a pneumatic cylinder, with O-ring seal, is bolted to a flange, such as in Figure 17.22. The plate and flange each have a thickness of 20 mm. M10 × 1.25, 6g × 60 bolts are used. The bolt spacing is about 3 bolt diameters. Find the stiffness parameter.

> Answer: $C = 0.25$

17.14. We have a flanged connection as shown in Figure P17.14. Four $\frac{3}{8}$-24 UNF-2A grade 7 bolts are used. The diameter of the shaft is $\frac{3}{4}$ in and of the flange is 3 in. Each flange

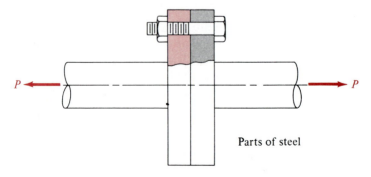

Parts of steel

FIGURE P17.14

thickness is $\frac{3}{8}$ in. If the preload in the bolts corresponds to the proof strength, what force P will cause separation of the joint?

Section 17.4

17.15. Calculate the value of the constant K in expressions (17.6) from Figure 17.28. Assume that the length which stretches is the 12-in shank of the $2\frac{1}{4}$-in studs.
  Answer: A: $K \simeq 0.12$, B: $K \simeq 0.23$

17.16. A 1–8 UNC grade 8 bolt is to be tightened to 80 percent of proof strength (lubricated with molybdenum disulfide grease) with a torque wrench. Find the nominal torque which must be applied.

17.17. An M16 × 1.5 bolt is to be tightened to achieve a force of 57 kN. Find the nominal torque if the bolt is tightened dry.
  Answer: 183.4 N·m

17.18. The bolt of Problem 17.9 is to be tightened to a preload of 10 klb. Find the turn of nut necessary.

17.19. A joint uses $\frac{5}{8}$-18 UNF bolts; $k_b = 4.387\text{E} + 06$, and $k_j = 13.11\text{E} + 06$ lb/in. Find the turn of nut to produce a preload of 15 330 lb.
  Answer: 30.2°

Section 17.7

17.20. The diameter of the flange in Figure P17.14 is 4 in; the bolts are on a 3-in diameter, and the shaft size is $1\frac{1}{2}$ in. The flanges are $\frac{1}{2}$ in thick. There are two $\frac{1}{2}$-13 UNC × $1\frac{1}{2}$ grade 7 bolts, preloaded dry with a manual torque wrench to the yield. What minimum force P can be expected to cause separation of the joint?

17.21. An O-ring is used in the design of the cap for a cylinder, as shown at the left in Figure 17.22. The pressure in the vessel will be constant at 8 MPa. The cap and flange each have a thickness of 20 mm. The diameter at the O-ring is 200 mm, and the bolt pattern diameter is 280 mm. Determine a suitable bolt for this application, such that the bolt spacing will be 3 to 4 bolt diameters. A manual torque wrench will be used, and the bolts will be brought up to yield.
  Find the specified preload F_i and from it the specified torque. What will be the resulting range of preload? Assume no lubrication. Use a factor of safety of 1.6.

17.22. The connection of Problem 17.20 is subjected to a load P which varies from 0 to 6000 lb. If the bolts (grade 7) are tightened (MbS$_2$-lubricated) to the proof load with a manual wrench, find the FS on the load P.
  Answer: FS = 1.13

17.23. Repeat Problem 17.21, except the pressure varies continuously from 3 to 9 MPa.

17.24. Repeat Problem 17.21, except a yield-control torque wrench is used. While the instrument will detect the yield, what torque will it apply, and what will be the resulting preload?
  Answer: M6 × 1 × 50 class 10.9, 47 bolts

Section 17.8

17.25. The joint shown in Figure P17.25 is bolted up so as to achieve the minimum seating pressure. What internal pressure would the ASME code permit?
  Answer: 2737 psi

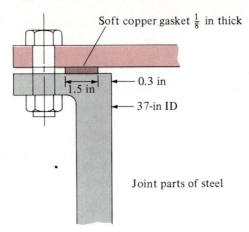

Soft copper gasket $\frac{1}{8}$ in thick

0.3 in

1.5 in

37-in ID

Joint parts of steel

FIGURE P17.25

17.26. What pressure, according to the ASME code, is permitted for the vessel sketched in Figure P17.26? The bolts are drawn up to produce the required gasket seating pressure.

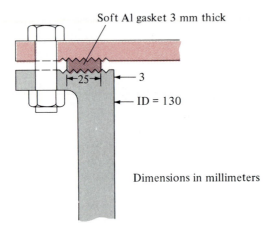

Soft Al gasket 3 mm thick

25

3

ID = 130

Dimensions in millimeters

FIGURE P17.26

17.27. The pressure in the vessel shown in Figure P17.25 is static at 80 psi. Calculate the total area of bolts required by the ASME code if SA 325 carbon-steel bolts with a minimum S_{ut} = 120 ksi and an allowable stress of 23.0 ksi for ambient temperatures are used.
Answer: 30.52 in^2

17.28. The pressure in the vessel shown in Figure P17.26 is static at 600 kPa, and the operating temperature is 850°F.
(a) Find the total area of bolts required by the ASME code if SA 193 grade B7 bolts are used, which have a minimum S_{ut} = 125.0 ksi and an allowable stress of 25.0 ksi from −20 to 650°F and 17.0 ksi at 850°F.

(b) Choose a bolt size to give reasonable bolt spacing. (Dimension the bolt circle to suit.)

17.29. A gasketed cylindrical steel (AISI 1040: $S_{yt} = 380$ MPa) pressure vessel in Figure P17.29 is to hold a static pressure of 5 MPa.

(a) The maximum stress in the head can be estimated from the relation

$$\sigma = \frac{3(3 + v)}{32} \frac{d^2}{t^2} p$$

where v is Poisson's ratio (0.3), d is the diameter of the head, and t is the thickness. Dimension the head, using a factor of safety of 4. Use the diameter to the outside of the nubbin for d.

  Answer: $t = 20.2$ mm

(b) Assume a bolt diameter and choose a bolt circle such that the bolts are far enough out to provide clearance between the nuts and the cylinder wall. (You must be able to get a wrench on the nuts.)

  Answer: $d \simeq 177$ mm

(c) Calculate the ASME code-required bolt size and the number of bolts, assuming the use of bolts with a minimum $S_{ut} = 120$ ksi and an allowable stress of 23.0 ksi. Operation is at ambient temperature.

  Answer: $A_{\text{bolts}} = 421$ mm^2

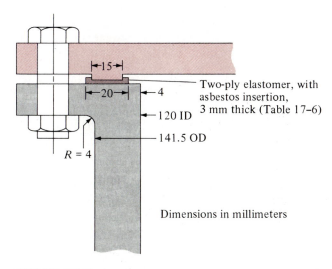

FIGURE P17.29

17.30. Suppose the pressure in Problem 17.27 varies from 0 to 80 psi. Check for fatigue of the bolts.

17.31. In Problem 17.28 the pressure varies from 0 to 600 kPa. Check for fatigue of the bolts.

  Answer: Design meets code.

17.32. The pressure in the vessel of Problem 17.29 varies from 0 to 250 kPa. Are the bolts adequate?

Section 17.9

17.33. The friction joint of Figure P17.33 is assembled with $\frac{1}{2}$-20 UNF, grade 5 bolts, which are drawn up dry with a torque wrench to a nominal 70 percent of the proof load. The joint must resist a force of 20 klb. What is the factor of safety?
Answer: FS = 1.33

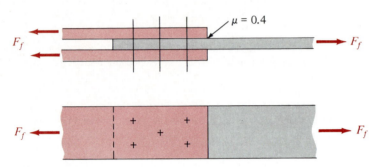

FIGURE P17.33

17.34. Rework Example 17.9, leaving out the bolt at the center of the pattern. Use class 5.8 bolts.

17.35. The bracket shown in Figure P17.35 must support a load of 1000 lb with a factor of safety $FS_P = 1.7$. Dimensions shown are inches. Size the bolts (grade 5) for a friction joint on the basis of the bolts being brought up (nominally) to 90 percent of the yield. A yield-control wrench will be used ($S_T = 10$ percent). The slip coefficient is 0.35. Specify the torque for lubricated bolts.
Answer: $\frac{7}{16}$-14 UNC, $T = 44.0$ lb·ft

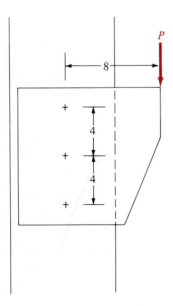

FIGURE P17.35

17.36. The bracket of Figure P17.35 is to support a load of 5 kN, with $FS_P = 2$. Dimensions shown are centimeters. Size the bolts (class 5.8) for a bearing joint on the basis of bolt rupture and bolting up being nominally to the proof load with a yield-control wrench ($S_T = 10$ percent). Specify the torque for lubricated bolts.

17.37. The bearing joints in Figure P17.37 are made of 1040 steel, cold-drawn 20 percent (Appendix D). They are bolted up dry to (nominally) 85 percent of the proof strength with a manual torque wrench. Find the failure modes (based on rupture where applicable) and failure loads.

$\qquad$ *Answer:* (*a*) $F = 18.1$ kN, $\qquad$ (*b*) $F = 30.4$ klb

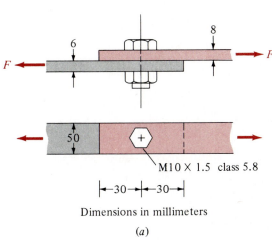

M10 × 1.5 class 5.8

Dimensions in millimeters

(*a*)

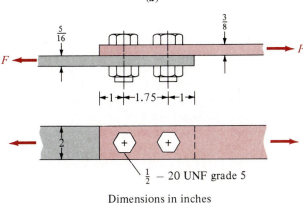

$\frac{1}{2} - 20$ UNF grade 5

Dimensions in inches

(*b*)

FIGURE P17.37

CHAPTER
18

SPRING DESIGN*

The function of mechanical springs is to produce a pulling or pushing force or a torque when displaced and/or to store energy. Functional and physical restraints result in an infinitude of configurations, some shown in the display of Figure 18.1. The basic types are sketched in Figure 18.2. The push function is provided by helical compression springs, spring washers, volute springs, and beam springs [(a) through (d)]. In applications involving large deflections, helical springs are usually chosen. The washers are generally used for small deflections associated with motion along a bolt or other guide. Volute springs have high inherent damping and good resistance to buckling; they are, however, expensive. Beam springs can push or pull and may have any of the various end conditions and shapes usually discussed in mechanics-of-materials courses.

Helical extension springs [(e) and (f)] and constant-force springs [(g)] provide a pull. The latter are formed by prestressed flat material wound on an arbor. Since the unwinding is in effect the incremental bending of a uniform beam, they have the characteristic of a nearly flat force-deflection curve. Typewriter carriage-return springs are a common example.

The twisting function is provided by helical torsion springs [(h)] and spiral springs (coils in the same plane) [(i)]. The former are frequently used for counter-

* We are grateful to Associated Spring, Barnes Group Inc. for permission to use information and figures from their *Design Handbook, Engineering Guide to Spring Design*, 1987 edition, © Barnes Group Inc., [82] in this chapter.

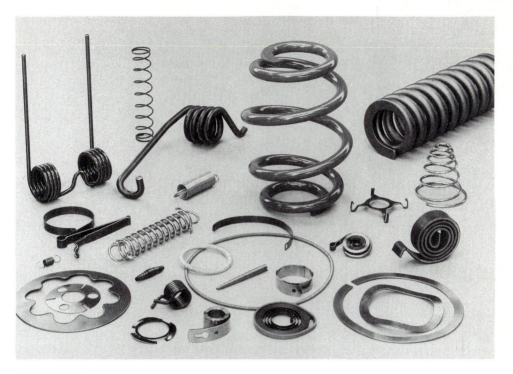

FIGURE 18.1
A few of the many spring configurations [82].

balance of such things as doors which rotate about a horizontal edge. Clothespins and mousetraps are other examples. Hairsprings are used in instruments and mechanical clocks and watches; one of their characteristics is low hysteresis (small energy loss). Brush springs hold motor and generator brushes against their commutators. Power springs [(j)] are used to supply rotational energy for clocks and mechanical toys. A variation is the prestressed power spring [(k)], which is coiled against its preformed spiral configuration to produce large energy-storage capacity. Seat-belt retractors are an application. The constant-torque spring motor [(l)] provides level torque.

Most springs are formed cold with a subsequent heat treatment to reduce residual stresses and to stabilize dimensions. Some shapes make the use of prehardened material impossible. The spring is then formed of annealed material and heat-treated to attain the desired properties. When the spring is to be formed of large bar (wire) [$\simeq$ 16-mm ($\frac{5}{8}$-in) diameter], it may be necessary to wind the spring hot. The required properties are then achieved through heat treatment. Vehicle suspension springs are an example.

Manufacturing companies normally purchase springs from outside, as opposed to making them in-house. The engineer with a spring problem should be acquainted with the basics of spring design and be able to specify springs for common applications. The following centers on the helical spring. A discussion of the characteristics

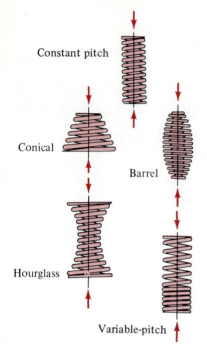

Constant pitch

Push—wide load and deflection range—constant rate.

Conical

Barrel

Push—wide load and deflection range. Conical spring can be made with minimum solid height and with constant or increasing rate. Barrel, hourglass, and variable-pitch springs used to minimize resonant surging and vibration.

Hourglass

Variable-pitch

(a) **Helical compression**

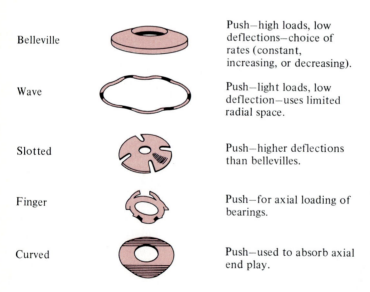

Belleville

Push—high loads, low deflections—choice of rates (constant, increasing, or decreasing).

Wave

Push—light loads, low deflection—uses limited radial space.

Slotted

Push—higher deflections than bellevilles.

Finger

Push—for axial loading of bearings.

Curved

Push—used to absorb axial end play.

(b) **Spring washer**

FIGURE 18.2
Basic types of springs [82].

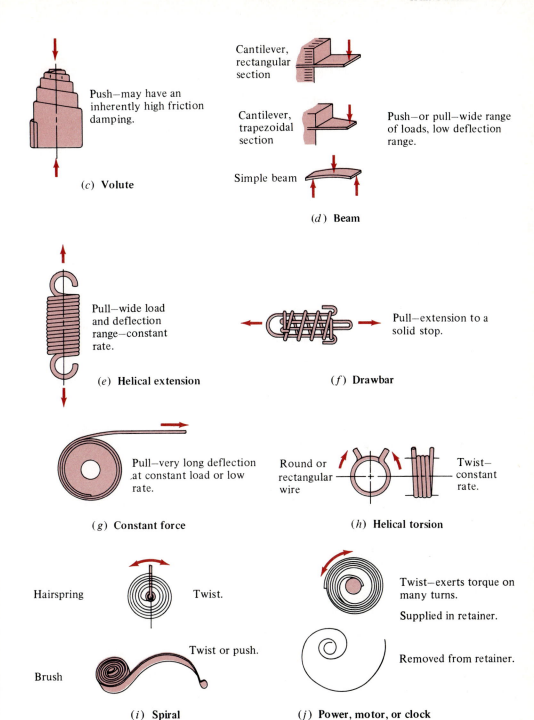

Push—may have an inherently high friction damping.

(*c*) **Volute**

Cantilever, rectangular section

Cantilever, trapezoidal section

Push—or pull—wide range of loads, low deflection range.

Simple beam

(*d*) **Beam**

Pull—wide load and deflection range—constant rate.

(*e*) **Helical extension**

Pull—extension to a solid stop.

(*f*) **Drawbar**

Pull—very long deflection .at constant load or low rate.

(*g*) **Constant force**

Round or rectangular wire

Twist—constant rate.

(*h*) **Helical torsion**

Hairspring

Twist.

Brush

Twist or push.

(*i*) **Spiral**

Twist—exerts torque on many turns.

Supplied in retainer.

Removed from retainer.

(*j*) **Power, motor, or clock**

FIGURE 18.2 (Continued)

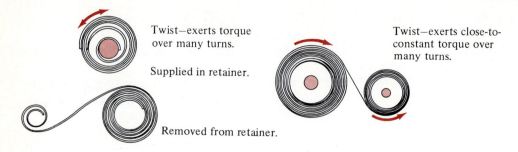

(k) Prestressed power

(l) Constant-torque spring motor

FIGURE 18.2 (Continued)

of the interesting Belleville washer is given at the end of the chapter. Chapter 14 deals with beams as springs.

18.1 STRESS ANALYSIS OF HELICAL SPRINGS

Despite its appearance, a helical spring is a type of torsion bar—a curved (rather than straight) component subjected to twisting plus a certain amount of direct shear. In the free-body diagram of Figure 18.3, the load P results in a combination of a direct force P and a moment T at the cross section. The shear stress due to the torsion is given by expression (13.1):

$$\tau_{\text{tors}} = \frac{16T}{\pi d^3} = \frac{8PD}{\pi d^3} \tag{13.1}$$

The average direct stress is

$$\tau_{\text{direct}} = \frac{P}{A} = \frac{4P}{\pi d^2}$$

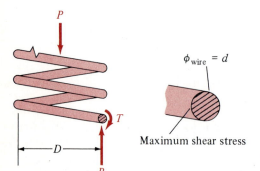

$\phi_{\text{wire}} = d$

Maximum shear stress

FIGURE 18.3
Free-body diagram of helical spring.

Adding the two gives

$$\tau = \frac{8PD}{\pi d^3}\left(1 + \frac{0.5}{D/d}\right) = \frac{8PD}{\pi d^3}K_s = \frac{8PC}{\pi d^2}K_s \qquad (18.1)$$

where K_s can be termed a *combined loading factor*. It is equal to the term in parentheses at the left. $C = D/d$ is a commonly used parameter in the spring business known as the *spring index*.

Now, the fact that the torsion bar analyzed above is curved causes the stress to be higher. A correction factor K_w, known as the *Wahl factor* [40], accounts for this increase and incorporates the combined loading effect (direct shear and torsion, which we computed above, K_s):

$$\tau = \frac{8PD}{\pi d^3}K_w$$

where

$$K_w = \frac{4C - 1}{4C - 4} + \frac{0.615}{C} \qquad (18.2a)$$

or

$$K_w = \frac{C^2 + 0.365C - 0.615}{C(C - 1)} \qquad (18.2b)$$

Analytically, K_w in the form given by Equation (18.2) is sometimes awkward to use. Many such expressions can be closely approximated by exponential relations of the form $y = Ax^b$. K_w can be so represented as

$$K_w = 1.60C^{-0.140} \qquad (18.2c)$$

which Johnson [3] states to be accurate within 2 percent for the range $5 \leq C \leq 18$.

The increase in stress due to the curvature is a highly localized phenomenon at the inside of the coil; in static loading it is usually relieved by yielding of the material at that point, and so the stress is computed on the basis of K_s. In fatigue loading, the notch sensitivity of these hard, strong materials is nearly unity, hence the stress concentration owing to the curvature of the wire is treated as the fatigue notch factor $K_f = K_{curv}$, increasing the stress or decreasing the strength, as explained in Section 9.1. We will see that K_{curv} always appears in the strength expression multiplied by K_s (after rearranging), as the Wahl factor ($K_w = K_{curv}K_s$).

18.2 SPRING MATERIALS AND FAILURE CRITERIA

Round wire is the form most often used for helical springs. Occasionally wire of rectangular or keystone cross section is employed; such springs can store more energy in the available space than their round counterparts.

Medium- and high-carbon steels are the most common spring materials. Stainless steel is used in corrosive environments, while phosphor bronze and beryllium copper provide electrical conductivity. When high strength at elevated temperatures is necessary, nickel alloys are often indicated. Surface finish is especially important in fatigue applications. Table 18.1 shows preferred sizes and tolerances of spring wire.

TABLE 18.1
Spring steel wire sizes and tolerances

(*a*) Preferred spring steel wire sizes

Metric sizes, mm			English sizes, in	
First preference	Second preference	Third preference	First preference	Second preference
0.10			0.004	
	0.11		0.005	
0.12			0.006	
	0.14		0.008	
0.16				0.009
	0.18		0.010	
0.20				0.011
	0.22		0.012	
0.25				0.013
	0.28		0.014	
0.30				0.015
	0.35		0.016	
0.40				0.017
	0.45		0.018	
0.50				0.019
	0.55		0.020	
0.60				0.021
	0.65		0.022	
	0.70		0.024	
0.80			0.026	
	0.90		0.028	
1.0			0.030	
	1.1			0.031
1.2				0.033
		1.3	0.035	
	1.4		0.038	
1.6			0.040	
	1.8			0.042
2.0			0.045	
		2.1		0.047
	2.2		0.048	
		2.4	0.051	
2.5			0.055	
		2.6	0.059	
	2.8		0.063	
3.0			0.067	
		3.2	0.072	
	3.5		0.076	
		3.8	0.081	
4.0			0.085	
		4.2	0.092	
	4.5		0.098	
		4.8		0.102
5.0			0.105	
	5.5		0.112	

TABLE 18.1 (Continued)

Metric sizes, mm			English sizes, in	
First preference	Second preference	Third preference	First preference	Second preference
6.0				0.120
	6.5		0.125	
	7.0			0.130
		7.5	0.135	
8.0				0.140
		8.5	0.148	
	9.0			0.156
		9.5	0.162	
10.0				0.170
	11.0		0.177	
12.0			0.192	
	13.0			0.200
14.0			0.207	
	15.0			0.218
			0.225	
16.0			0.250	
				0.262
			0.281	
				0.306
			0.312	
			0.343	
			0.362	
			0.375	
			0.406	
			0.437	
			0.469	
			0.500	

Source: Ref. 82.

(*b*) Tolerances for spring wire

Diameter, mm (in)		Tolerance, mm (in)	Maximum out-of-roundness, mm (in)
0.51–0.71	(0.020–0.028)	±0.010 (0.0004)	0.010 (0.0004)
0.71–2.00	(0.028–0.078)	±0.015 (0.0006)	0.015 (0.0006)
2.00–3.00	(0.078–0.118)	±0.020 (0.0008)	0.020 (0.0008)
3.00–6.00	(0.118–0.240)	±0.030 (0.00118)	0.030 (0.0012)
6.00–9.00	(0.240–0.354)	±0.050 (0.00197)	0.050 (0.002)
9.50–16.00	(0.375–0.625)	±0.070 (0.00276)	0.070 (0.0028)

Most spring wires can be purchased to tighter tolerances. Music wire and most nonferrous materials are regularly made to closer tolerances.

Source: Ref. 82.

Tensile Strength

Spring wire is made by drawing an originally heavy wire through progressively smaller dies. High tensile strength results from the cold work. Figure 18.4 shows the variation in strength with size of common wires. Several of the curves on this semilog plot are nearly straight lines, so that strength can be written as $S_{ut} = Ad^m$. For example,

For ASTM A228 (music wire):

$$S_{ut} = \begin{cases} 172d^{-0.186} & \text{ksi} & d \text{ in inches, size } 0.012\text{--}0.250 \\ 2170d^{-0.186} & \text{MPa} & d \text{ in millimeters, size } 0.3\text{--}6.5 \end{cases} \tag{18.3a}$$

For ASTM A227 (hard-drawn wire):

$$S_{ut} = \begin{cases} 137d^{-0.192} & \text{ksi} & d \text{ in inches, size } 0.020\text{--}0.590 \\ 1760d^{-0.192} & \text{MPa} & d \text{ in millimeters, size } 0.5\text{--}15 \end{cases} \tag{18.3b}$$

These expressions are sometimes more convenient than a chart like Figure 18.4.

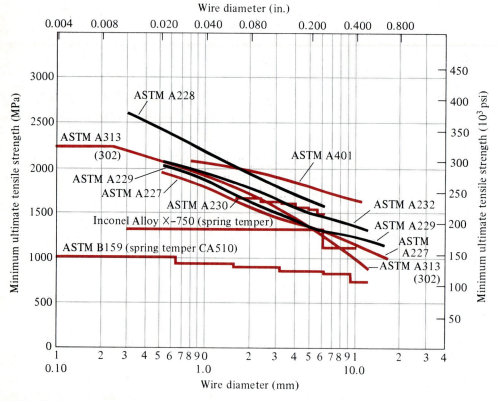

FIGURE 18.4
Minimum tensile strengths of spring wire [82].

Static Applications

In static applications, residual stresses which enhance the load-carrying capacity of compression springs are often created by making the spring longer than the designed free length and then compressing it to solid. Plastic deformation occurs, mainly at the inside of the coil, resulting in stresses of the opposite direction upon release. The length of the spring naturally decreases in this process—the loss of length should be a minimum of 10 percent for the procedure to be effective. The process is called *removing set*. It is not always cost-effective, especially with small springs.

The allowable shear stress with and without set removal for static applications is shown in Table 18.2*a*. We label this stress the *yield strength*, although in the spring business the concern is more with the elastic limit than the yield point. Thus, for static safety, from (18.1)

$$\frac{8PD}{\pi d^3} K = \frac{8PC}{\pi d^2} K \leq \frac{S_{ys}}{FS} \tag{18.4}$$

Before set removal (yielding), K in (18.4) is the Wahl factor K_w; after set removal, K_s is used.

TABLE 18.2
Allowable stresses for helical compression springs

(*a*) Maximum allowable torsional stresses in static applications (bending or buckling stresses not included)

	Maximum percentage of tensile strength	
Materials	**Before set removed**	**After set removed**
Patented and cold-drawn carbon steel	45	60–70
Hardened and tempered carbon and low-alloy steel	50	65–75
Austenitic stainless steels	35	55–65
Nonferrous alloys	35	55–65

(*b*) Maximum allowable alternating torsional stresses

	Percentage of tensile strength			
	ASTM A228, Austenitic Stainless steel and nonferrous		**ASTM A230 and A232**	
Fatigue life, cycles	**Not shot-peened**	**Shot-peened**	**Not shot-peened**	**Shot-peened**
10^5	18	21	21	25
10^6	17	20	20	24
10^7	15	18	19	23

This information is based on the following conditions: no surging, room temperature, and noncorrosive environment (stress ratio $R = 0$).

Source: Adapted from [82].

Fatigue Limits

Table 18.2b shows shear fatigue strength as a function of tensile strength for a number of spring materials. Shot-peening improves the fatigue performance.

Using expression (18.1) and remembering to decrease the fatigue strength by the curvature effect K_c, we see that the fatigue strength criteria [(9.14a) and (9.14b)] become, with $K_c K_s = K_w$,

$$\frac{8P_{\text{cyc}}D}{\pi d^3} K_w \le \frac{S_{f,\text{sh}}}{\text{FS}}, \text{ if } \frac{8P_m D}{\pi d^3} \le \frac{1}{\text{FS}} \left(\frac{S_{sy}}{K_s} - \frac{S_{f,\text{sh}}}{K_w} \right) \qquad (18.5)$$

(fatigue failure)

Expression (18.4) above, if $\dfrac{8P_m D}{\pi d^3} > \dfrac{1}{\text{FS}} \left(\dfrac{S_{sy}}{K_s} - \dfrac{S_{f,\text{sh}}}{K_w} \right)$

(static failure)

One of the above criteria must be assumed and then checked afterward. You cannot always be sure beforehand which is valid, for K_w and K_s depend on the final dimensions.[1]

18.3 SPRING RATE k

The load producing unit deflection P/δ is termed the *spring rate, constant,* or *stiffness*. The calculation is normally based on the torsional deflection only of the spring wire; the deflection due to bending and direct shear is neglected.

For one coil of the spring, the distance moved by the load during its application is, for small motion, $D\alpha/2$, as sketched in Figure 18.5. The angle α is given by expression (13.3) with $\theta = \alpha$, $L = \pi D$, and $T = PD/2$, so that the deflection is

$$\delta_{\text{per coil}} = \frac{8PD^3}{d^4 G}$$

The spring rate $k = F/\delta$ for a spring of N_c active coils is then

$$k = \frac{Gd^4}{8D^3 N_c} \qquad (18.6)$$

Generally N_c, the number of active coils, differs somewhat from the total number of coils, as we see in the following section. Note in (18.6) that the stiffness rises with

---

[1] These fatigue-failure relations are based on the torsional alternating fatigue strength being unaffected by a mean torsional stress (Figure 9.16). There is, of course, some effect. To account for it, a pseudo-Goodman relation may be used:

$$\frac{S_{ut}(\tau_{\text{max}} - \tau_{\text{min}})}{S_{ut} - \tau_{\text{min}}} = S_G$$

where S_G is the "fatigue strength" from tests in which the stress varies from zero to a maximum value (stress ratio = 0). The torsional fatigue strength values of Tables 18.2b and 18.4b are one-half S_G.

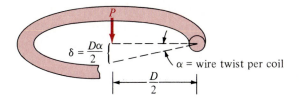

$\delta = \dfrac{D\alpha}{2}$

α = wire twist per coil

$\dfrac{D}{2}$

FIGURE 18.5
Spring-deflection analysis.

fewer coils. If you cut a spring in half, each part deflects half as much under a given load as the original and is thus twice as stiff.

18.4   COMPRESSION SPRINGS

Compression springs are used in a multitude of applications, e.g., the valve springs of internal-combustion engines; the suspension springs of road, off-road, and rail vehicles; inner-spring mattresses; and retractable pens.

Types of Ends

Figure 18.6 shows the basic types of spring ends. Squared and/or ground ends naturally allow better seating of the spring; valve springs are squared and ground. With these types of ends, a certain amount of spring does not deflect when the load is applied, i.e., there are inactive coils. The number of inactive coils for the several end

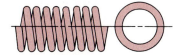

Plain ends
(coiled right-hand)
Inactive coils = 0

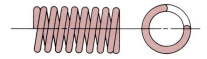

Squared and ground ends
(coiled left-hand)
Inactive coils = 2

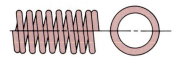

Squared or closed ends
(not ground, coiled right-hand)
Inactive coils = 2

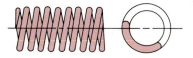

Plain ends ground
(coiled left-hand)
Inactive coils = 1

FIGURE 18.6
Compression spring end types. Ground ends are not recommended with wire diameters less than about 0.5 mm (0.020 in), a spring index greater than 12, or low spring rates [82].

types is indicated in the figure. A common example of plain ends is the helical suspension springs of automobiles. The cups in which they seat are formed to fit. Plain ends allow for more active coils in the given space and are cheaper. Also shown in the figure are left- and right-hand windings, defined as for bolts and screws. This is important mainly in applications where springs are nested one within the other to achieve greater stiffness. The hands should then be opposite, to prevent interference.

Spring Index Range

If the spring index $C = D/d$ is lower than about 4, the spring will be difficult to make (small spring of large wire). When the index is larger than about 12, springs tend to tangle (large spring of small wire), especially when a number are packaged together. Thus the preferred range for C is about 4 to 12 [82]. Many special-purpose springs will fall outside this range, however, especially on the high side.

Buckling

A compression spring is in essence a column—it resists a compressive load along its axis, and it may become unstable, i.e., buckle. Buckling conditions for compressive springs are shown in Figure 18.7. The parameters are the ratios of free length to mean spring diameter (analogous to L/ρ for columns) and of deflection to free length (analogous to P_{cr} for columns), as indicated in the figure. The two curves correspond to the two common end conditions shown. Occasionally the buckling problem may be solved by placing the spring over a rod or in a tube.

Springs of Variable Diameter and/or Pitch

In Figure 18.2a several compression springs with special shapes are shown. The conical, hourglass, and barrel shapes allow a smaller collapsed height than corresponding springs of cylindrical shape. If the coils nest within one another, the solid height for a conical spring can be as little as one wire diameter. The stiffness of these springs generally increases with deflection, since coils at the bottom are progressively taken out of action. The variable stiffness helps minimize vibration problems, discussed later in Section 18.5. A constant rate can be achieved by making the spring pitch (the coil separation) greater toward the base, so that the coils do not bottom out when the spring is deflected.

   The spring rate may be calculated by using expression (18.6) to find the rate for each coil, or a portion of a coil, and then adding the effects as springs in series:

$$\frac{1}{k_{spr}} = \sum_{n=1}^{N} \frac{1}{k_n} = \frac{8N_c}{Gd^4N}(D_1^3 + D_2^3 + \cdots + D_N^3)$$

where N is the number of units used in the calculation; $N = N_c$ if whole coils are used.

   If the taper is uniform (spring shape is a true cone), then the stiffness can be found through calculus:

$$k = \frac{d^4G}{2N_c(D_L + D_S)(D_L^2 + D_S^2)} \tag{18.6'}$$

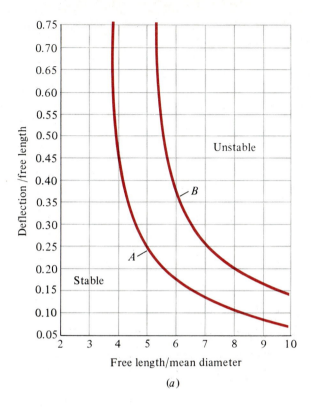

Free length/mean diameter

(a)

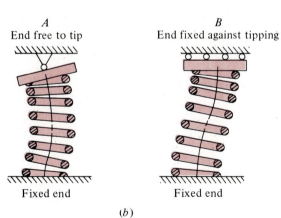

A	B
End free to tip	End fixed against tipping
Fixed end	Fixed end

(b)

FIGURE 18.7
Buckling of compression springs [82]. *(a)* Buckling conditions. *(b)* End conditions for *(a)*.

where the subscripts refer to the large and small ends. Proof of this expression is left as an exercise (Problem 18.2). The rate for barrel and hourglass springs can be found by considering them as two conical springs in series.

Stress calculations for springs of variable diameter are based on the largest active coil, since the stress is greatest there.

18.5 DESIGN FOR MINIMUM MASS[1]

Fatigue Loading

In this problem we seek the minimum mass for a spring to have a specified deflection δ under the cyclic part of a load. The mass of a spring is

$$M = \rho \frac{\pi d^2}{4} \pi D(N_c + Q) = \frac{\rho \pi^2}{4} C d^3(N_c + Q) \qquad (18.7)$$

in which N_c, as before, is the number of active coils and Q is the number of inactive coils. Any wire used to form special ends, such as a hook (the development is applicable to extension as well as to compression springs), is not included.

The fatigue-stress safety relation (18.5) in terms of the spring index is

$$\frac{8P_{\text{cyc}}C}{\pi d^2} K_w \le \frac{S_{f,\text{sh}}}{\text{FS}}$$

which gives

$$d = \left[\frac{8P_{\text{cyc}}CK_w(\text{FS})}{\pi S_{f,\text{sh}}} \right]^{1/2} \qquad (18.8)$$

As the minimum mass will clearly be achieved with the wire diameter d at its smallest value, the inequality has been replaced with an equal sign.

The spring deflection due to the cyclic part of the load is P_{cyc}/k. With k given by (18.6), this becomes

$$\delta = \frac{8C^3 N_c P_{\text{cyc}}}{Gd}$$

whence

$$N_c = \frac{\delta Gd}{8C^3 P_{\text{cyc}}} \qquad (18.9)$$

Substituting (18.8) and (18.9) in (18.7) and clearing the right side of all terms except C and K_w (a function of C), we get

$$\frac{4}{\sqrt{\pi}\rho Q} \left[\frac{S_{f,\text{sh}}}{(\text{FS})8P_{\text{cyc}}} \right]^{3/2} M = BK_w^2 + K_w^{3/2}C^{5/2} \qquad (18.10)$$

where

$$B = \frac{\delta G}{Q(8\pi P_{\text{cyc}}S_{f,\text{sh}}/\text{FS})^{1/2}} \qquad (18.11)$$

Since we are seeking a minimum for the mass M, we take the derivative of (18.10) with respect to C and set it to zero:

$$\frac{dM}{dC} = 0 = 2BK_w \frac{dK_w}{dC} + \frac{3}{2}C^{5/2}K_w^{1/2}\frac{dK_w}{dC} + \frac{5}{2}C^{3/2}K_w^{3/2}$$

---

[1] This method was published by R. T. Hinkle and I. E. Morse, Jr., in Ref. 80.

whence
$$B = -\frac{C^{3/2}}{4K_w^{1/2}}\frac{3C\,dK_w/dC + 5K_w}{dK_w/dC} \qquad (18.12)$$

From expression (18.2b)

$$\frac{dK_w}{dC} = \frac{-1.365C^2 + 1.230C - 0.615}{(C^2 - C)^2} \qquad (18.13)$$

By substituting expressions (18.2b) and (18.13) in (18.12), and after a good deal of algebra, there results

$$B = \frac{C^3(C^3 - 0.635C^2 - 0.980C + 0.615)^{1/2}(5C^3 - 7.27C^2 - 1.21C + 1.23)}{4(1.365C^4 - 0.732C^3 - 0.673C^2 + 0.981C - 0.378)}$$

The index C which satisfies this relation results in a spring of minimum mass. Of course, the expression can be solved only by numerical means, or graphically. The latter solution is shown in Figure 18.8. In calculating B, a wire diameter must be assumed, to get the fatigue strength. Note that B is dimensionless, hence the chart of Figure 18.8 can be used with either USCS or SI units.

A spring of minimum mass will be the most efficient in terms of material and money. This can be important when large production is involved. A minimum-mass spring will also have the highest possible natural frequency, discussed later in this section.

Example 18.1. We are to design a minimum-mass spring of ASTM A228 wire. The spring will not be shot-peened. It is to have a life of 10^6 cycles, and $P_{cyc} = 70$ lb, $\delta = 0.5$ in, FS = 1.4, and $Q = 2$.

From Table 18.2b, $S_{f,sh,10^6} = 0.17S_{ut}$. With S_{ut} given by expression (18.3a),

$$S_{f,sh,10^6} = 27.2d^{-0.186}$$

With a guess of $d = 0.2$, this gives

$$S_{f,sh,10^6} = 39.4E + 03$$

The constant B from (18.11) is then

$$B = \frac{0.5(11.5E + 06)}{2[8(70\pi)(39.4E + 03)/1.4]^{1/2}} = 409$$

For this value of B, we read in Figure 18.8 a value for the spring index $C = 6.2$. The Wahl factor (18.2a) is then

$$K_w = \frac{4(6.2) - 1}{4(6.2) - 4} + \frac{0.615}{6.2} = 1.24$$

The wire diameter is, from (18.8),

$$d = \left[\frac{8 \times 70 \times 5.5 \times 1.24 \times 1.4}{\pi(77E + 03)}\right]^{1/2} = 0.1487 \text{ in}$$

This is quite far removed from our guess of 0.2, so we try the standard size $d = 0.148$ (Table 18.1). The strength $S_{f,sh,10^6}$ is then 41.7 ksi, and $B = 397$. The spring index is about the same, $C = 6.2$, and we settle on this wire size.

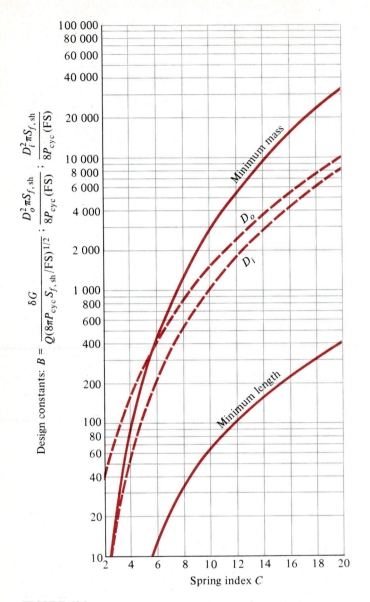

FIGURE 18.8
Curves of spring index for minimum mass or length or with specified outside or inside diameter [80].

The mean spring diameter is

$$D = Cd = 6.2(0.148) = 0.918 \text{ in}$$

and the number of active coils from (18.9) is

$$N_c = \frac{0.5(11.5\text{E} + 06)(0.148)}{8(6.2^3)(70)} = 6.4$$

The minimum mass (18.7) is then

$$M = \frac{0.28\pi^2}{4}(6.2)(0.148^3)(6.4 + 2) = 0.12 \text{ lb}$$

We indicated in Section 18.4 that the preferred range for the spring index C is 4 to 12. When the problem parameters result in an index which falls outside this range, the limiting value for C should be used in expression (18.8) to determine the required wire size. Then the spring diameter $D = Cd$, and the number of coils is calculated from (18.9).

Static Loading

Most of the above development may be used to determine the minimum mass of a spring having a given deflection under a static load. Relation (18.12) is valid if K_s replaces K_w. In the definition of B, (18.11), P_{cyc} is replaced by P and $S_{f,\text{sh}}$ by S_{ys}:

$$B = \frac{\delta G}{Q(8P\pi S_{ys}/\text{FS})^{1/2}}$$

Then, with

$$\frac{dK_s}{dC} = \frac{d}{dC}\left(1 + \frac{0.5}{C}\right) = \frac{-0.5}{C^2}$$

there results

$$B = \frac{C^3(5C + 1)}{2\sqrt{C + 0.5}}$$

To calculate the parameter B, the shear yield strength, which depends on the wire diameter, must be estimated. The value of C in the last relation is then found most easily by trial and error. It may be necessary to recalculate B with a new wire diameter and run through the cycle again.

If the spring index C falls outside the preferred range, then the limiting value of C is used in (18.4) to find the required wire size. The spring diameter $D = Cd$, and N_c is determined from (18.9).

Example 18.2. The load for the spring data in Example 18.1 is 100 lb steady. A spring of minimum mass is desired. Set will be removed.

From Table 18.2a, the yield is 60 to 70 percent of S_{ut}. Taking the mean of these figures and using the tensile strength given by (18.3a), we get

$$S_{ys} = 112d^{-0.186} \quad \text{ksi}$$

After one or two tries, we find that a wire diameter of 0.092 in satisfies the above relations. For this size, $S_{ys} = 175$ ksi, and

$$B = \frac{0.5(11.5\text{E} + 06)}{2[800\pi(175\text{E} + 03)/1.4]^{1/2}} = 162$$

This results in an impractical spring, since $C = 3.5$. By taking the spring index at the low end of the preferred range $C = 4$, the wire diameter is found from (18.4) with $K_s = 1.13$

and the expression for S_{ys} substituted:

$$d = \left[\frac{8 \times 100 \times 4 \times 1.13 \times 1.4}{\pi(112E+03)}\right]^{1/1.816} = 0.097$$

We round this to the preferred size of 0.098. Then the mean diameter $D = 4(0.098) = 0.392$ in.

The number of coils is

$$N_c = \frac{Gd\delta}{8C^3P} = \frac{(11.5E+06)(0.098)(0.5)}{8(4^3)(100)} = 11.0$$

The minimum mass is then

$$M = \frac{0.28\pi^2}{4}(4)(0.098^3)(11.0+2) = 0.034 \text{ lb}$$

Natural Frequencies

Springs have a natural frequency of vibration in the longitudinal direction, i.e., along the length of the spring, called *spring surge*. Operation of a spring near its natural frequency (resonance) can result in stress levels above those predicted by the stress relations. The spring may also jump away from its end plates, resulting in loads less than calculated. To avoid these problems, a natural frequency a minimum of 13 times the operating frequency is recommended.

If the mass of the spring could be considered as concentrated at the moving end, the natural frequency would be simply the square root of the stiffness divided by the mass. But the mass, being the spring itself, is distributed along the length, and this raises the natural frequency. For a helical spring fixed at both ends, the natural frequency is

$$f = \frac{1}{2}\sqrt{\frac{kg}{M}} \qquad \text{Hz}$$

M is the mass of the active part of the spring; i.e., it is calculated on the basis of N_c.

Since the spring stiffness k is normally specified, the highest fundamental frequency is achieved by minimizing the mass. If the problem specifications are expressed as limit constraints, as in the example following, then the approach of finding the minimum mass through differentiation is cumbersome.

Rather than calculate the minimum mass, it is more direct and just as simple to maximize the frequency. We express the frequency of the spring in terms of each constraint and see how it varies with the wire size. The several curves so constructed will enclose a space of possible solutions with an indication of the optimum. The method is best shown with the specifics of an example.

The force on the spring will vary between 3 and 13 lb. The spring is to have a rate $k = 30$ lb/in and a mean coil diameter $D < 0.5$ in. The number of active coils $N_c \geq 8$. Material is to be music wire (ASTM A228). The spring will be shot-peened. We use FS = 1.5. A life of 10^7 cycles is desired.

First, we insert the stiffness and the expression for the mass of the spring ($M = \rho\pi^2 d^2 D N_c/4$), with the physical properties of steel, into the natural-frequency expression. The result is

$$f = \frac{64.73}{d\sqrt{DN_c}} \quad \text{Hz} \tag{18.14}$$

which is the criterion function.

Now to get an expression for frequency as a function of diameter, while maintaining the specified spring rate, we eliminate N_c from (18.14), using (18.6) solved for N_c: $N_c = (4.79\text{E} + 04)d^4/D^3$. Thus

$$f = \frac{0.296D}{d^3} \quad \text{Hz} \tag{18.15}$$

Inserting the limit $D < 0.5$ gives

$$f < \frac{0.148}{d^3} \quad D \text{ limited} \tag{A}$$

An upper limit on frequency, this is the curve labeled A in Figure 18.9, a plot of f vs. d. In like manner, eliminating D from (18.14) by using (18.6), we get

$$f = \frac{10.74}{N_c^{1/3}d^{5/3}}$$

which, with $N_c \geq 8$, gives

$$f \leq 5.37d^{-5/3} \quad N_c \text{ limited} \tag{B}$$

This is the curve labeled B in the figure, also an upper limit on frequency.

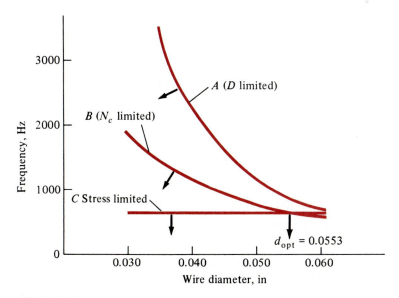

FIGURE 18.9
Spring frequency optimum.

With $P_a/P_m = \frac{5}{8}$, the situation is clearly fatigue, not static. Thus expression (18.5) applies:

$$\frac{8P_{cyc}D}{\pi d^3} K_w \leq \frac{S_{f,sh}}{FS}$$

The shear fatigue strength [(18.3a) and Table 18.2b] is

$$S_{f,sh} = (31E + 03)d^{-0.186}$$

By substituting this and the Wahl approximation (18.2c), the strength relation becomes

$$\frac{8(5D)}{\pi d^3} (1.60)\left(\frac{D}{d}\right)^{-0.140} \leq \frac{(31E + 03)d^{-0.186}}{1.5}$$

or

$$D \leq (31.3E + 03)d^{3.109} \tag{18.16}$$

Inserting this limit in (18.15) gives an upper limit on frequency due to the stress:

$$f \leq 920d^{0.109} \qquad \text{stress limited} \tag{C}$$

Relation (C) is also shown in Figure 18.9, in which it is evident that the optimum occurs at the intersection of relations (B) and (C), i.e., the upper limit on stress and the lower limit on the number of coils. Setting the two relations equal, we see the optimum wire diameter is $d = 0.0553$, which we round to the standard size 0.055. The frequency from (B) or (C) is $f = 671$ Hz.

The mean coil diameter can be computed from the strength relation (18.16): $D = 0.380$ in, resulting in a spring index $C = 0.380/0.055 = 6.9$. The number of coils is the minimum required: $N_c = 8$.

Usually, the number of turns is specified so that after the required ends are made, the desired effective number of turns N_c will be obtained. In our example, we have $N_c = 8.0$ turns. Suppose we decided to get the best possible seat for the end of this high-speed spring by closing the end coils and grinding them square. The total number of coils should be increased by 2, to a total of 10.0 turns. The allowance for the inactive coils is an approximate one; we might end up with anything between 7.8 and 8.2 active coils. In addition, the commercial tolerance on wire diameter in the 0.055 size is 0.0006 in (Table 18.1). That is, we might expect to get something between 0.0544 and 0.0556 in. Finally, the coil diameter is subject to tolerance. We have specified a coil diameter of 0.380 in, or spring index $C = 6.9$. The tolerance on the coil diameter for our values of C and d is ±0.004 in. These are variables over which the designer has little control. In Problem 18.6, the effect of these tolerances on the performance of our spring is to be checked.

18.6   DESIGN FOR MINIMUM SOLID LENGTH

Solid means "fully compressed." The solid length is given by

$$L = (N_c + Q)d$$

The procedure is the same as for the minimum-mass spring of the previous section. The constant B is defined as before in (18.11); the index C for fatigue loading is found in the chart of Figure 18.8.

When the spring index falls outside the preferred range, the spring specifications are computed as for the case of the minimum-mass spring, discussed in Section 18.5.

Example 18.3. For the specifications of Example 18.1, the minimum solid-length spring is desired.

The value of the parameter B for a guess of $d = 0.2$ is $B = 409$. From Figure 18.8, $C = 20+$, which is outside the preferred range. Taking C at the top of the range, $C = 12$, gives

$$K_w = \frac{4(12) - 1}{4(12) - 4} + \frac{0.615}{12} = 1.12$$

and

$$d = \left[\frac{8 \times 70 \times 12 \times 1.12 \times 1.4}{\pi(39.4E + 03)} \right]^{1/2} = 0.292 \text{ in}$$

The nearest standard size is $d = 0.306$. Then $S_{f,\text{sh}}$ is 36.4 ksi, and d becomes 0.304. Thus we use 0.306. The mean spring diameter is $D = Cd = 12(0.306) = 3.672$ in, and the number of coils is

$$N_c = \frac{0.5(11.5E + 06)(0.306)}{(8 \times 12^3)(70)} = 1.8$$

Depending on the application, a spring with this few coils may not be practical. The minimum solid length is $(1.8 + 2)0.306 = 1.16$ in.

18.7 OUTSIDE OR INSIDE DIAMETER SPECIFIED

Frequently the spring is to fit over a rod or inside a hole, which implies a fixed inside or outside diameter.

Clearance

The minimum diametral clearance between spring and hole or spring and rod should be [82]

$$0.05D \qquad \text{when } D_{\text{hole or rod}} > 13 \text{ mm } (0.512 \text{ in})$$

$$0.10D \qquad \text{otherwise}$$

Outside Diameter Specified

The outside diameter is given by

$$D_o = D + d = d(C + 1) \tag{18.17}$$

which, with the substitution of relation (18.8), gives

$$D_o = \left[\frac{8P_{\text{cyc}}K_w C(\text{FS})}{\pi S_{f,\text{sh}}} \right]^{1/2} (C + 1)$$

Substituting the Wahl relation (18.2*b*) and squaring give

$$\frac{D_o^2 \pi S_{f,sh}}{8P_{cyc}FS} = \frac{C^4 + 2.365C^3 + 1.115C^2 - 0.867C - 0.615}{C - 1} \qquad (18.18)$$

This expression, for fatigue loading, is shown in Figure 18.8.

The result for static loading is

$$\frac{D_o^2 \pi S_{ys}}{8P(FS)} = (C + 0.5)(C + 1)^2 \qquad (18.19)$$

The value of C satisfying (18.19) is found most easily by trial and error, as in Example 18.4.

For both the fatigue and static cases, the wire size can be found from (18.17) or (18.8), or both, as a check. A value must be assumed to start, since the strength depends on it.

If the clearance is small, the increase in diameter of the spring when it is compressed must be accommodated. A spring's diameter when compressed solid is [82]

$$OD_{solid} = \sqrt{D^2 + \frac{p^2 - d^2}{\pi^2}} + d \qquad (18.20)$$

In this expression p is the pitch, the distance between the coils.

Example 18.4. A compression spring is required which must exert a force $P_1 = 255$ N at a height $L_1 = 57$ mm and $P_2 = 600$ N at a height $L_2 = 40$ mm. The spring must fit in a hole of 30-mm diameter. The application is essentially static. ASTM A229 (oil-tempered carbon-steel) wire will be used. The factor of safety is 1.2.

With the load and height values specified, the spring stiffness must be

$$k = \frac{600 - 255}{57 - 40} = 20.29 \text{ N/mm}$$

The deflection under the larger load is then

$$\delta = \frac{P}{k} = \frac{600}{20.29} = 29.57 \text{ mm}$$

With a (guessed) wire diameter of 4.5 mm, the tensile strength is 1400 MPa (Figure 18.4). After set removal, the permitted torsional stress S_{ys} is 65 to 75 percent of S_{ut} (Table 18.2), say, 70 percent. Then $S_{ys} = 0.70 \times 1400 = 980$ MPa. If we allow 2-mm clearance, then $D_o = 28$ mm. With these numbers and using the larger load, expression (18.19) becomes

$$\frac{0.028^2(\pi)(980E + 06)}{8(600)(1.2)} = 419 = (C + 0.5)(C + 1)^2$$

For a few trial values of C, the right side is

$$C = 6.6: \qquad 7.1 \times 7.6^2 = 410$$

$$C = 6.7: \qquad 7.2 \times 7.7^2 = 426$$

$$C = 6.8: \qquad 7.3 \times 7.8^2 = 444$$

For this first iteration, then, $C = 6.7$. With an outer diameter of 28 mm and 4.5-mm wire, the mean diameter of the spring $D = 28.0 - 4.5 = 23.5$ mm. Then, using the value of C just found, we get a new estimate of the wire diameter: $d = D/C = 23.5/6.7 = 3.51$ mm. The nearest preferred size is 3.5 mm, which has $S_{ut} = 1480$ and $S_{ys} = 0.70 \times 1480 = 1036$ MPa. Then (18.19) becomes

$$\frac{0.028^2(\pi)(1036E + 06)}{8(600)(1.2)} = 443$$

for which $C = 6.8$. To check the wire size as before, $D = 28.0 - 3.5 = 24.5$ mm, and $d = 24.5/6.8 = 3.6$ mm. Thus 3.5-mm wire is chosen. The number of coils is computed from the spring-rate expression (18.6):

$$N_c = \frac{(79.3E + 06)(0.0035^4)}{8(0.0245^3)(20.29E + 03)} = 5.0$$

The free length of this spring is the 40.0-mm length under the larger load (the smaller load could be used as well) of 600 N plus $600/k = 600/20.29 = 29.6$, giving $L_{\text{free}} = 69.6$ mm. The number of coils is 2.0 (inactive end coils) plus $N_c = 5.0$, for a total of 7.0.

We can also check the minimum clearance of the spring in the hole by computing relation (18.20). The pitch $p = L_{\text{free}}/N_{\text{total}} = 69.6/7.0 = 9.94$. So

$$\text{OD}_{\text{solid}} = \sqrt{24.5^2 + \frac{9.94^2 - 3.5^2}{\pi^2}} + 3.5 = 28.18 \text{ mm}$$

The clearance then is $30.0 - 28.18 = 1.82$ mm (6 percent of the hole diameter), which is satisfactory. Of course, the spring may never be compressed solid.

Inside Diameter Specified

When the inside diameter is specified, we have, paralleling the above,

$$D_i = D - d = d(C - 1) = \left[\frac{8P_{\text{cyc}}K_w C(\text{FS})}{\pi S_{f,\text{sh}}}\right]^{1/2}(C - 1) \qquad (18.21)$$

giving

$$\frac{D_i^2 \pi S_{f,\text{sh}}}{8P_{\text{cyc}}\text{FS}} = C^3 - 0.635C^2 - 0.98C + 0.615$$

for fatigue loading. This is also plotted in Figure 18.8. When the loading is static,

$$\frac{D_i^2 \pi S_{ys}}{8P(\text{FS})} = (C + 0.5)(C - 1)^2$$

The wire size can be found from (18.21) or (18.8), or both as a check.

Example 18.5. For the same spring data as in Example 18.1, suppose the inside diameter is 1.25 in. Guessing $d = 0.18$, we get $S_{f,\text{sh}} = 78E + 03$. Then

$$\frac{D_i^2 \pi S_{f,\text{sh}}}{8P_{\text{cyc}}(1.4)} = \frac{125^2(\pi)(77E + 03)}{8(70)(1.4)} = 482$$

From Figure 18.8 a value $C = 8.0$ is found. Then K_w is

$$K_w = \frac{4(8.0) - 1}{4(8.0) - 4} + \frac{0.615}{8.0} = 1.18$$

The wire size is given by (18.8):

$$d = \left[\frac{8(70)(8.0)(1.18)(1.4)}{\pi(78E + 03)} \right]^{1/2} = 0.1738$$

which can be checked with (18.21):

$$d = \frac{D_i}{C - 1} = \frac{1.25}{7} = 0.1786$$

The difference between the two values is, of course, attributable to error in reading the chart. We would round the wire size to the nearest standard size, $d = 0.177$ in. (A re-computation with this size results in no change.)

For both cases, inside or outside diameter specified, if the spring index C falls above the preferred range, then the limiting value ($C = 12$) can be placed in $D_o = d(C + 1)$ (outside diameter specified) or $D_i = d(C - 1)$ (inside diameter specified) to determine the wire size. The result will be an economy of stress; i.e., the material will not be stressed to $S_{f,\text{sh}}/\text{FS}$ or S_{ys}/FS, as the case may be. (The curves of Figure 18.8 presuppose the stress to be at those limits.) When the value of C is below the preferred range, it will not be possible to design the spring to specification (with the chosen wire material), because the resulting stress will be excessive.

18.8   EXTENSION SPRINGS

Helical extension springs differ from compression springs in three respects: the direction of the force the spring generates, the provision of end hooks or loops to attach the spring to the load, and their manufacture, usually close wound (collapsed) with initial tension.

Initial Tension

When the spring is wound, if the wire is twisted in the same direction as it will be when the spring is later stretched, the stresses created will remain in the wire because the spring cannot contract beyond its collapsed position to remove them. The result is that the force on the spring must exceed an initial value before there is any

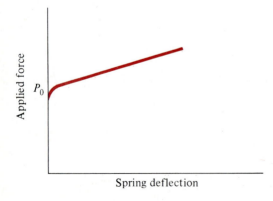

FIGURE 18.10
Initial tension in a helical extension spring.

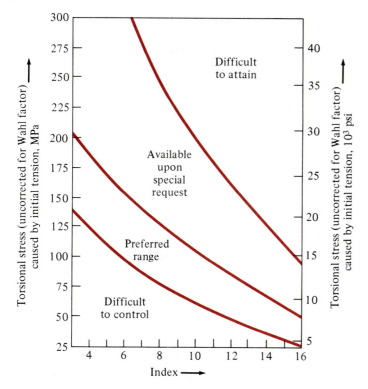

FIGURE 18.11

Torsional stress due to initial tension as a function of index in helical extension springs [82].

deflection, as shown in the sketch of Figure 18.10. The amount of feasible initial torsional stress depends on the spring index, as shown in Figure 18.11.

Types of Ends and Stresses

Some common end configurations for extension springs are displayed in Figure 18.12. The stresses in these ends involve bending, tension, and torsion with stress concentration, and they are usually greater than the stresses in the spring body.

A "twist loop" is shown in Figure 18.13. At point A the stress is due to the tension force P and to the bending moment PR_1:

$$\sigma = \frac{4P}{\pi d^2} + K_1 \frac{32PR_1}{\pi d^3} \tag{18.22}$$

The stress concentration factor K_1 is given by

$$K_1 = \frac{4C_1^2 - C_1 - 1}{4C_1(C_1 - 1)} \qquad [\text{Ref. 81}] \tag{18.23}$$

where

$$C_1 = \frac{2R_1}{d}$$

Type	Configurations	Recommended length* minimum–maximum
Twist loop or hook		0.5–1.7 ID
Cross-center loop or hook	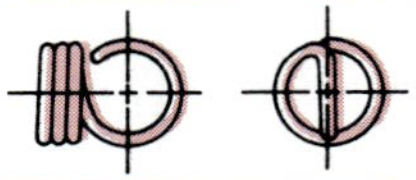	ID
Side loop or hook		0.9–1.0 ID
Extended hook	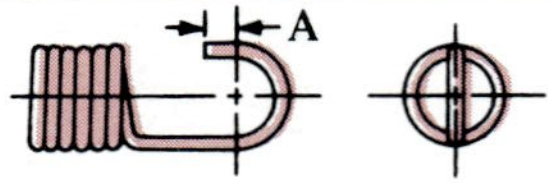	1.1 ID and up, as required by design
Special ends	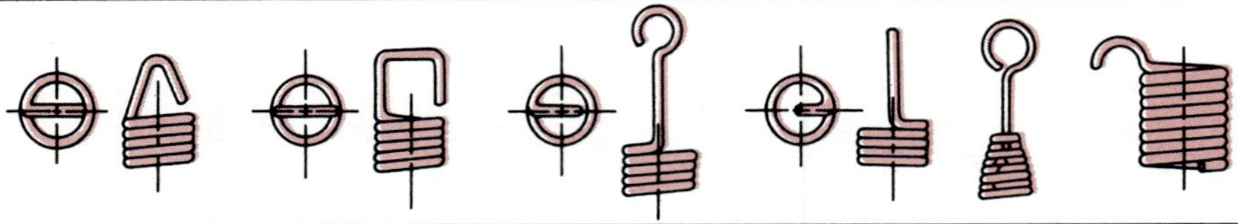	As required by design

*Length is distance from last body coil to inside of end. ID is inside diameter of adjacent coil in spring body.

FIGURE 18.12

Common end configurations for helical extension springs [82].

550

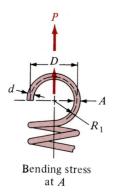

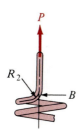

Bending stress
at A

Torsion stress
at B

FIGURE 18.13
Location of maximum bending and torsion
stresses in twist loops [82].

At point B the effect is mainly torsion, the twisting moment being $PD/2$:

$$\tau = K_2 \frac{8DP}{\pi d^3} \qquad [\text{Ref. 81}] \tag{18.24}$$

where

$$K_2 = \frac{4C_2 - 1}{4C_2 - 4} \quad \text{with } C_2 = \frac{2R_2}{d} \tag{18.25}$$

In a practical sense K_2 is meaningless for $C_2 \le 4$. (It becomes infinite at $C_2 = 1$.) C_2 must be a reasonable size, 4.2 or greater, implying $R_2 > 2.1d$, which would normally be the case.

Another expression sometimes used for K_1 and K_2 is

$$K_{1,2} = \frac{R_m}{R_i} \tag{18.26}$$

where R_m is the mean radius of the bend and R_i is the inner radius. The value of K_1 per (18.26) is within 4 percent (on the high side) of K_1 per (18.23) for $R_1 > 4d$. The value of K_2 from (18.26) is within 14 percent (on the high side) of K_2 per (18.25) for $R_2 > 2d$.

Reduction of the stress at end loops may be achieved by increasing the turn radius at point A in Figure 18.13 and/or by reducing the coil diameter gradually at the end. In the latter case, while the coil curvature is thus reduced, the moment arm for the force is reduced also. End turns utilizing these principles are shown in Figure 18.14.

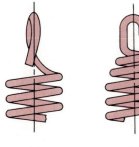

(a)

(b)

FIGURE 18.14
End hooks for reducing stresses. (a) Increased radii. (b) Reduced coil diameter.

TABLE 18.3
Permitted stresses for helical extension springs

(a) Maximum allowable stresses (K_s corrected) in static applications*

Materials	Percentage of tensile strength		
	In torsion (shear)		In bending plus tension (tensile)
	Body	End	End
Patented, cold-drawn or hardened and tempered carbon and low-alloy steels	45–50	40	75
Austenitic stainless steel and nonferrous alloys	35	30	55

* Set not removed and low-temperature heat treatment applied. For springs that require high initial tension, use the same percentage of tensile strength as for end.

(b) Maximum allowable alternating stresses for ASTM A228 and type 302 stainless steel

Number of cycles	Percentage of tensile strength		
	In torsion (shear)		In bending plus tension (tensile)
	Body	End	End
10^5	18	17	26
10^6	17	15	34
10^7	15	14	29

Stress-relieved springs with low levels of initial tension, not shot-peened. Ambient environment; stress ratio = 0; no surging.

Source: Adapted from [82].

Permitted Stresses

In static applications the allowed stress levels for the body of the spring are about the same as those for compression springs, as seen in Table 18.3a. Fabrication of the ends stretches and otherwise distorts the wire, so the permitted end stresses are somewhat lower. Table 18.3b gives allowable stresses for two types of wire under cyclic conditions. The tensile stresses in (18.22) will consist of mean and alternating components, requiring application of Goodman's relation (9.10), as in Example 18.7.

Dimensions and Active Coils

The free length of an extension spring is the distance between the points of attachment, thus between the inner surfaces of the ends. The body length is given by $d(N_{body} + 1)$, to which is added the length of the ends (Figure 18.12).

The number of active coils is approximately the number of coils in the body, plus the effect of the ends, especially hooks and loops. A figure of $0.1N_a$ is sometimes used for the latter.

Example 18.6. An extension spring is to be designed with twist loops corresponding to Figure 18.13 ($R_1 = 0.5D$, $R_2 = 0.1D$). The spring must supply a load $P_1 = 20$ N at $L_1 = 33.4$ mm and $P_2 = 33$ N at $L_2 = 38.4$ mm. ASTM A228 wire is used. The mean diameter of the spring should be about 10 mm. Operation is cyclic, and a life of 10^7 cycles is required. FS = 1.2.

We can expect the strength in the loop to be critical. At point A (Figure 18.13), the stresses are tensile, mean, and alternating. Since $P_a = 6.5$ and $P_m = 26.5$,

$$\sigma_m = \frac{26.5}{6.5}\,\sigma_a = 4.08\sigma_a$$

From Table 18.3b, $S_f = 0.29S_{ut}$, so applying Goodman's criterion yields

$$\frac{\sigma_a}{0.29S_{ut}} + \frac{4.08\sigma_a}{S_{ut}} \le \frac{1}{FS}$$

which, with $S_{ut} = 2170d^{-0.186}$, gives $\sigma_a \le 240d^{-0.186}$. Using a guess of the standard wire size $d = 1.2$ gives $\sigma_a \le 232$ MPa (allowed).

The stress at point A is given by expression (18.22), for which $R_1 = D/2$, and hence $C_1 = C$. Our interest lies in the alternating component ($P = P_a = 6.5$). With $d = 1.2$, $C = 10/1.2 \simeq 8.3$. Thus

$$\sigma_a = \frac{(6.5)(4)}{(1.2^2)(\pi)}\left[1 + \frac{4(8.3^2) - 8.3 - 1}{7.3}\right] = 215 \text{ MPa}$$

which is below the allowed level. (The next smaller size is overstressed.)

Note in the above procedure that the stress-concentration factor is applied to both the mean and the alternating components of stress. In Chapter 9 we used it with the latter only, with the argument that yielding in a cycle or two of loading would relieve the stress. In this instance, that could distort the loop. Application of the factor to both stresses is also more conservative, i.e., it boosts the factor of safety slightly.

At point B (Figure 18.13), $R_2 = 0.1D$, thus C_2 for expression (18.24) is $0.2C$, and we have

$$\tau = \frac{(0.8C - 1)C(8P_a)}{(0.8C - 4)\pi d^2}$$

which, with $C = 8.3$, is $\tau = 204$ MPa. From Table 18.3b, the permitted level is 14 percent of S_{ut}/FS, or 294 MPa, so that point is safe.

To find the stress in the body of the spring, we need the Wahl factor for $C = 8.3$, which is $K_w = 1.18$. Then

$$\tau = \frac{8P_a CK_w}{\pi d^2} = \frac{(8)(6.5)(8.3)(1.18)}{(\pi)(1.2^2)} = 113 \text{ MPa}$$

The allowed stress is 15 percent of S_{ut}/FS, or 315 MPa.

The stiffness of this spring is $k = (33 - 20)/(5E - 03) = 2600$ N/m. Then, using (18.6), we get

$$N_c = \frac{Gd}{8C^3 k} = \frac{(79.5E + 09)(1.2E - 03)}{(8)(8.3^3)(2600)} = 8.0$$

The free length is

$$L_{\text{free}} = 2D_i + (N_c + 1)d = 2(1.2)(8.3 - 1) + 9(1.2) = 28.32 \text{ mm}$$

The deflection under, say, P_1 is

$$\delta_1 = L_1 - L_{\text{free}} = 33.4 - 28.32 = 5.08 \text{ mm}$$

The initial tension in the spring is then

$$P_{\text{init}} = P_1 - k\delta_1 = 20 - 2600(5.08\text{E} - 03) = 6.79 \text{ N}$$

The stress (uncorrected for the Wahl factor) under the initial load is

$$\tau = \frac{8PD}{\pi d^3} = \frac{8PC}{\pi d^2} = \frac{8(6.79)(8.3)}{\pi(1.2^2)} = 99.7 \text{ MPa}$$

We see in Figure 18.11 that this is acceptable.
    Finally, the mean diameter $D = Cd = 8.3(1.2) = 9.96$ mm.

18.9   BELLEVILLE SPRING WASHERS

These are spring-steel washers formed to a conical shape, as shown in Figure 18.15. As may be intuitively evident, increasing the ratio h/t of the spring height to the thickness increases the stiffness. The stiffness is, in general, not constant throughout the deflection range; the shape of the load-deflection curve depends on the ratio h/t, as shown in Figure 18.16. At $h/t = 1.41$, the curve exhibits a portion which is nearly horizontal. A washer with this ratio can thus be used to maintain a constant force in a mechanical part which wears or otherwise relaxes with time.
    Belleville washers can be stacked to achieve greater or less stiffness. In the series stack (Figure 18.17), the deflection is the sum of the deflections of the individual washers, while the load is that of one washer. The spring rate is thus reduced by the factor of the number in the stack. In the parallel stack, the load is increased, while the deflection is that of one washer, i.e., the stiffness is increased. These observations are simply statements of the laws of (ordinary) springs in series or in parallel.

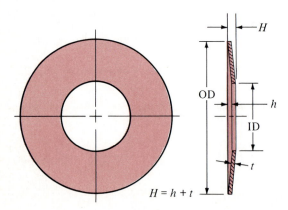

$H = h + t$

FIGURE 18.15
Belleville spring washer [82].

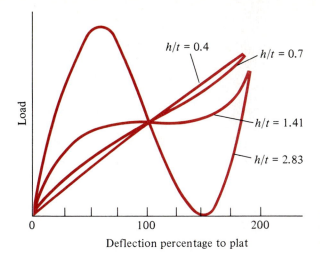

FIGURE 18.16
Load-deflection curves for Belleville washers with various h/t ratios [82].

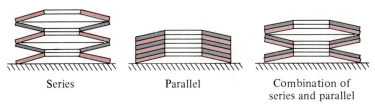

Series   Parallel   Combination of series and parallel

FIGURE 18.17
Stacks of Belleville washers.

In the series stack, if any washers have a negative spring rate at some point in their deflection curve, the possibility exists of their snapping through to their mirror image, resulting in a discontinuous deflection curve. To prevent this, the ratio h/t should be less than 1.3 [81].

Deflections and loads intermediate between a pure-series and a pure-parallel arrangement can be achieved through a combination stack, such as shown in the figure. For that illustrated, the stiffness is the same as for a single washer, but the stack can be deflected twice as far.

PROBLEMS

Section 18.2

18.1. Develop relations in the USCS and SI units approximating the strengths shown in Figure 18.4 for the following steels.
(*a*) ASTM A401
   Answer: $S_{ut} = 2060d^{-0.0938}$ (mm, MPa)
(*b*) ASTM A232
(*c*) ASTM A229
   Answer: $S_{ut} = 148d^{-0.180}$ (in, ksi)

Section 18.4

18.2. Derive expression (18.6′).

18.3. The eight active coils of a conical compression spring (Figure 18.2*a*) have a diameter D of 1.2 in at the base and 0.5 in at the top, varying linearly in between. The wire has diameter $d = 0.105$ in. Calculate the spring rate by dividing the spring into four parts and per relation (18.6′).
    Answer: 30.7, 30.4 (1 percent difference)

18.4. The active part of a conical compression spring (Figure 18.2*a*) tapers uniformly from a diameter D of 15 mm at the top to 30 mm at the bottom. There are 10 active coils. The wire size is 2.0 mm. Determine the spring rate.

Section 18.5

18.5. This problem is similar to that of Section 18.5. We wish to maximize the frequency of a spring, with the following restrictions: $k = 30$ lb/in, $3 \le P \le 13$ lb, $N_c \ge 7$, $C = 8$, and FS = 1.5. ASTM A228 wire is used, and the spring will be shot-peened.
    Answer: $d = 0.076$, $f = 407$

18.6. Check the effect of wire-size, coil-diameter, and coil-number tolerances on the frequency of the spring designed in Section 18.5.

18.7. A minimum-mass spring is required to meet the following specifications: Life 10^7 cycles, $P_{\text{cyc}} = 100$ N, $\delta = 5$ mm, FS = 1.6, and $Q = 2$. The material is ASTM A230, and the spring will be shot-peened.
    Answer: $d = 2.8$, $D = 15.12$, $M = 40.0$ g

18.8. We must design a spring for indefinite life. The material will be ASTM A228 wire, and the spring will be shot-peened. The specifications are $k = 100$ lb/in, FS = 1.4, $N_c = 8$, $f \ge 1500$ Hz, $D \le 0.75$ in, and $11 \ge C \ge 5$.
    We want to maximize the cyclic force we can put on the spring P_{cyc}. To get started, write (18.5), solve for P_{cyc} (as an equality), and go from there. Specify D, d, f, and P_{cyc}.

18.9. We wish to design a spring with the following constraints: $k = 18$ kN/m, $36 \le P \le 116$ N, $D = 18$ mm, $5 \le C \le 9$, $N_c = 8$, and $f > 1700$ Hz. Material is to be ASTM A229; the spring will not be shot-peened.
    The object is to maximize the reliability, i.e., tthe factor of safety, for a service life of 10^5 cycles. *Hint:* Be careful of units; the strength expression (18.3*a*) is in millimeters!
    Answer: FS = 1.96, $d = 2.35$

Section 18.6

18.10. We need a compression spring with a minimum solid length. The spring will normally operate at $P_{\text{cyc}} = 400$ N with a deflection under that load of 32 mm. Life is to be 10^7 cycles, $Q = 2$, and FS = 1.9. Material is ASTM A232. Determine the specifications for the spring (not shot-peened) and the solid length.
    Answer: $d = 9$, $L = 55.3$ mm, $N_c = 4.1$

Section 18.7

18.11. A spring made of ASTM A230 wire must fit into a 1.45-in hole. It must exert a force $P_1 = 52$ lb at a height $L_1 = 2.2$ in and $P_2 = 84$ lb at $L_2 = 1.7$ in. These forces are applied repeatedly. The spring will be peened; a life of 10^6 cycles is required. Use FS = 2. Determine the specifications for the spring.

18.12. Repeat Problem 18.11, but the application is for a few cycles only. The spring will have set removed. (This is a hardened carbon steel.)
    Answer: $d = 0.105$, $D = 1.275$, $N_c = 1.33$

18.13. Repeat Problem 18.11, but the hole size is 0.8 in.

18.14. Repeat Problem 18.11, except that the spring must fit over a $\frac{1}{2}$-in rod, rather than in a hole. Use both (18.8) and (18.21) to find the wire size.

  Answer: $d = 0.105$, $D = 0.655$, $N_c = 9.9$

18.15. A spring is to be guided by a rod of diameter 40 mm. It must exert a repeated force $P_1 = 5050$ N at a height $L_1 = 100$ mm and $P_2 = 7000$ N at $L_2 = 80$ mm. Life is to be indefinite; FS = 1.7. Use ASTM A232 wire. The spring will be shot-peened. Determine the specifications for the spring.

18.16. We need a helical compression spring and have settled on music wire (ASTM A228); the spring will be unpeened. The ends will be squared and ground. The force will vary repeatedly from 10 N (at a spring length of 23 mm) to 60 N (at a spring length of 18 mm). The safety factor is included. Indefinite life is desired. Minimize the OD of this spring while ensuring that it will not buckle. Specify the free length, wire size, OD, and number of coils. *Hint:* Figure 18.8 indicates the spring should have as small an index as possible. Use (18.2c) for K_w and (18.3a) for S_{ut}.

(*a*) The spring ends are restrained from tipping.

(*b*) One end is free to tip.

  Answer: (*a*) OD = 6.66 m,   (*b*) $D = 7.45$ mm

Section 18.8

18.17. Rework Example 18.7, but change the loads to $P_1 = 15$ N at $L_1 = 41.2$ mm and $P_2 = 19$ N at $L_2 = 46.2$ mm.

18.18. We have a spring with full-loop ends (Figure 18.13), $R_2 = 0.25$ in. The wire is 0.170-in-diameter ASTM A313 (austenitic) steel. The spring has 60 coils of $1\frac{1}{2}$-in mean diameter. The preload shear stress in the spring is 12 ksi. Application is essentially static.

(*a*) Find the free length of the spring.

(*b*) Find the spring rate.

(*c*) Find the preload force.

(*d*) How great a force may be applied to the spring, if FS = 1.3?

(*e*) What is the length of the spring under the condition off (*d*)?

  Answer: (*b*) 5.4 lb/in,   (*d*) 47.7 lb

CHAPTER
19

DESIGN OF
WELDED
COMPONENTS

A *weld* can be defined, broadly, as a localized union accomplished by applying heat and/or pressure, with or without extra material being added. The strength of a weld can be as great as that of the materials being joined, although prudence usually requires that a reduced strength be assumed.

When the thing to be made is large, or when only a few copies are needed, it is usually more practical to join simple pieces by welding, bolting, or riveting than to create a single entity by casting and/or machining. Such welded assemblies are called *weldments*. Today's automobile body is a weldment, a single unit combining the functions of body and frame. Steel ships are also weldments. In both cases, the superior properties and economy of sheet metal are realized by using welding to produce a complex article. Welding has also become more economical for high-volume production with the use of robots, which can be programmed to produce a complex series of perfect welds.

Welding is done in a number of ways. At one extreme, two surfaces of identical contour and intimate proximity can form a weld through interatomic bonding forces. The fit between the surfaces can be the result of careful preparation or can be achieved by the combination of heat and pressure. Thus, people have long welded bronze and wrought iron by heating the pieces to a plastic state and then hammering them into union. Forge welding, as this is called, is still a viable technology, often seen in the fabrication of gold jewelry.

At the other extreme, two parts can be bonded together by a third material which is applied in melted form to their mating surfaces. The general definition of welding thus includes the processes commonly known as *welding, brazing,* and *sol-*

dering. The process is called welding if no filler metal is used, or if the composition and melting point of the filler are near those of the metal being welded. Brazing and soldering both use a filler having a lower melting point than those of the parts being joined. The problems of warpage and change in properties due to overheating are thus much less severe in brazing than in welding. Steel parts, for example, can be finish-machined before they are assembled with a bronze filler, because the brazing temperature is normally not high enough to melt or distort the steel. The operation must be carried out with some care, because the brazing temperature may be high enough to reduce hardness or to warp parts that contain high residual stresses. Customarily, the process is called brazing if the temperature required is above 425°C (800°F) and soldering if it is below that figure.

Nonmetals are welded by processes entirely analogous to the above. Although the materials used and the names of the processes are different, the remarks about types of joints and relative strengths apply. Indeed, many wood glues are stronger than the materials they fasten.

19.1   FUSION WELDING

In fusion welding, the union is formed by bringing the mating edges to a molten state, usually by localized application of heat. Filler metal may be added to the weld.

Fusion welds can be as strong and tough as the materials being welded. They can, however, be afflicted by cracking, warping, and poor strength, regardless of the method used to produce the molten zone. These problems arise from the fact that the molten zone is achieved in a very short time and is thus localized. The molten zone cools very rapidly because it is surrounded by a relatively huge volume of conductive material at low temperature. Rapid cooling, or quenching, hardens most steels. With hardness comes a loss of ductility. If the part being welded or the filler is such a steel, then the finished joint will be brittle. The large temperature difference also produces stress, much of which may remain when the joint has cooled.

Brittleness and residual stress can be ameliorated by using a ductile metal for the filler, by pre- and postheating the weld zone, and by careful design of the sequence in which the parts of the weld are made.

Thin sheets of metal pose a special problem in fusion welding, because the torch or electric arc used to provide the heat can easily overheat the metal, resulting in holes instead of welds. Resistance welding avoids much of this problem. In the technique known as spot welding, the sheets are welded by passing a heavy electric current through water-cooled copper electrodes on either side. Enough pressure is used to confine the current, and the heating, to a small area. Spot welding is usually not used when the strength of the joint must approach that of the original metal, because the weld is not continuous. Continuous resistance welds can be made by replacing the electrode of spot welding with a roller electrode.

Localized union without filler metal can also be accomplished by forging, as described above, heavy pressure, or induction heating. Such joints can be of very high quality, and some metals cannot be welded in any other way. The fine Saracen, Toledo, and Samurai swords of antiquity were made by hammer-forging many sheets of wrought iron into a single piece.

Torch Welding

Torch welding is a familiar fusion technique. The flame, usually oxygen-acetylene, is adjusted to a narrow point, at which the temperature is over 6000°F. The oxygen content of the flame can be adjusted to be slightly deficient, neutral, or excessive (reducing, neutral, or oxidizing), depending on the metals being welded. A high level of skill is needed to achieve fusion throughout the thickness of the joint, but no farther, and to move filler into the melt zone at the proper rate. The very great advantage of torch welding is that the equipment is inexpensive, portable, and independent. That is, no connection to electricity or other utility is needed.

Torch welding can be done with a bare metal wire for filler. Many an automobile exhaust pipe has been thus repaired, by using a wire coathanger. But cleaner, denser, and stronger welds result when the filler wire contains deoxidants, such as manganese and silicon, and has a coating which vaporizes to form a protective shielding atmosphere.

Electric Arc Welding

SHIELDED METAL ARC WELDING. "Stick" welding is the prevailing technology for structural steel and for the joints in steel automobile bodies. As with torch welding, the consumable electrode or "stick" and its protective coating should be formulated for the application. In addition to providing a shield of vapor, the protective coating contains a flux which floats slag out of the weld. The slag forms a solid blanket over the weld to prevent oxidation while the weld cools.

The flux coating on welding electrodes is a brittle material, and for this reason, electrodes are commonly supplied in short lengths. Electrode is also available with the flux contained within the rod. It can be handled in large coils and fed automatically to the weld.

SUBMERGED ARC WELDING. With the right equipment, a continuous blanket of flux can be fed onto the arc. Nearly all the heat, smoke, and fumes which ordinarily result are thus trapped. With more complete shielding than provided by other techniques, high-quality welds result. Submerged arc welding can be used only when the welding gun can be above the weld.

GAS METAL ARC WELDING. This technique, also known as *metal inert-gas* (MIG) or CO_2 *welding*, uses a continuous, bare-metal wire electrode. The welding zone is shielded by an inert gas, such as CO_2. Gas metal arc welding is well suited to automatic applications and requires a relatively low level of operator skill.

GAS TUNGSTEN ARC WELDING. This is also known as *tungsten inert-gas* (TIG), *heliarc*, and *heli welding*. A nonconsumable tungsten electrode is shielded by an inert gas such as helium or argon, and a separate filler rod is used. Gas tungsten arc welding (GTAW) is normally a manual process requiring high skill. It is suitable for both steel and other metals and is better for welding thin sections than other techniques.

PLASMA ARC WELDING. This is similar to GTAW, but the arc is constricted either between the electrode and the work (transferred arc) or between the electrode and the constricting nozzle (nontransferred arc). In the latter case, the shielding gas leaves the nozzle as a superheated plasma—hence the name.

19.2   STRENGTH OF WELDED JOINTS

In this section, we discuss the factors which affect the strength of a weld, including standard weld joints, the attainable properties of commercial filler rods, and the recommendations of the American Welding Society (AWS) and the American Institute of Steel Construction (AISC).

Some Widely Used Weld Joints

Nearly all fusion welds combine fusion of the base with addition of material from the welding rod. The latter is called a *fillet* and the end product a *fillet weld*. The usual categories of fillet weld joints are shown in Figure 19.1.

For stress analysis, as for the single lap joint shown in Fig. 19.2, the cross section of the fillet is considered to be a right isosceles triangle. The width t of the fillet is taken to be the length of the side of the triangle. Fillet welds nearly always fail in shear at the minimum cross section, or "throat," of the fillet. In the joint of Figure 19.2, the load is taken by two fillets, so the total stress area is $2(0.707t)$ per unit length of weld.

Butt

Tee

Corner

Lap

Edge

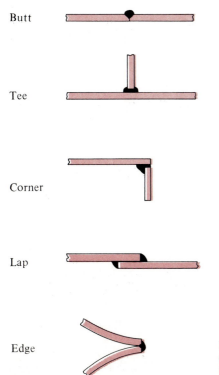

FIGURE 19.1
Commonly used weld joints. (*Courtesy James F. Lincoln Arc Welding Foundation.*)

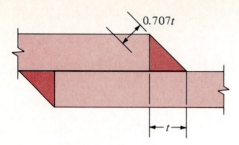

FIGURE 19.2
A single lap joint showing effective thickness in shear.

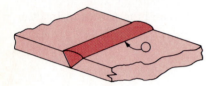

Reinforced butt weld: $k = 1.2$

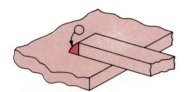

Transverse fillet weld: $k = 1.5$

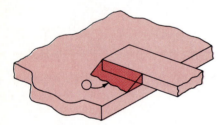

Parallel fillet weld: $k = 2.7$

FIGURE 19.3
Stress-concentration factors in welds.

TABLE 19.1

Recommended size of weld fillets (AWS)

Thickness of thicker plate, in	Minimum fillet width,* in
$<\frac{1}{4}$	$\frac{1}{8}$
$\frac{1}{4}$ to $\frac{1}{2}$	$\frac{3}{16}$
$\frac{1}{2}$ to $\frac{3}{4}$	$\frac{1}{4}$
$\frac{3}{4}$ to $1\frac{1}{2}$	$\frac{5}{16}$
$1\frac{1}{2}$ to $2\frac{1}{4}$	$\frac{3}{8}$
$2\frac{1}{4}$ to 6	$\frac{1}{2}$
Over 6	$\frac{5}{8}$

* Weld size need not exceed the thickness of the thinner part joined.
Generally, the fillet width should be $\frac{1}{16}$ in less than the thickness
of the material.

In the design of a fillet weld which will be subjected to fatigue loadings, the stress concentrations which occur at the edge of the weld must be taken into account. The stress pattern in these fillet welds can be very complex. A few examples, with stress-concentration factors commonly used, are shown in Figure 19.3. The locations of the maximum local stresses are indicated. The effect of these stress concentrations decreases as the size of the smaller member being joined approaches that of the larger member.

Fillet sizes recommended by the American Welding Society for welding steel sheet are given in Table 19.1.

Welding is often chosen for economy. To realize the full economic potential of welding, it is important to specify only the amount of weld metal required and to design the weld pattern to suit the type of welding to be done. Some examples are shown in Figure 19.4. When a choice is possible, the weld should be made in the

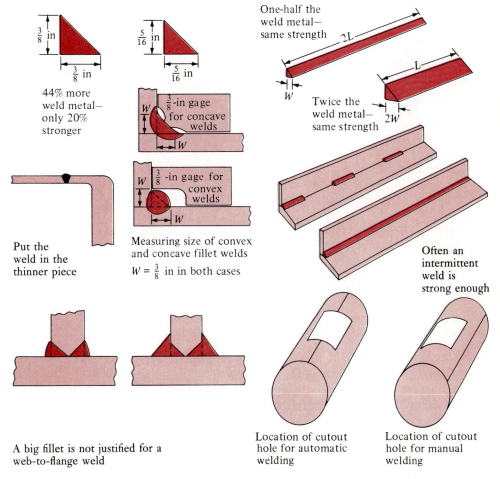

FIGURE 19.4
Examples of oversized weld fillets. (*Courtesy James F. Lincoln Arc Welding Foundation.*)

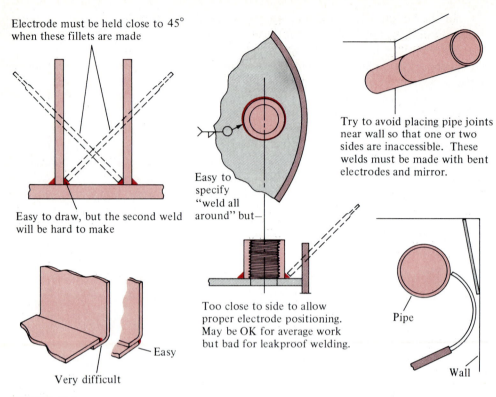

Electrode must be held close to 45° when these fillets are made

Easy to draw, but the second weld will be hard to make

Very difficult

Easy

Easy to specify "weld all around" but—

Too close to side to allow proper electrode positioning. May be OK for average work but bad for leakproof welding.

Try to avoid placing pipe joints near wall so that one or two sides are inaccessible. These welds must be made with bent electrodes and mirror.

Pipe

Wall

FIGURE 19.5
Weld joints should be designed for ease of access. (*Courtesy James F. Lincoln Arc Welding Foundation.*)

thinner material. With manual welding, it may be more economical to use intermittent fillet welds than continuous ones; the opposite applies for automatic welding.

Economical weldment design will avoid specifying welds that are difficult to make properly. Several examples are shown in Figure 19.5.

Welding Rods for Steel

Filler rod for welding should be chosen to match the alloy content and strength of the pieces to be welded. The identification number used by the American Welding Society for welding rod indicates the type of welding it is intended for, its minimum yield strength, and the composition of the deposit. For example, E6010 is an electrode for arc welding. The first two or three numbers indicate the minimum ultimate strength, here 60 kpsi. The third or fourth number (here: 1) indicates that it may be used for all welding positions (2 = horizontal, fillet, and flat positions only; 4 = flat, horizontal, overhead, and vertical down). The last number identifies the type of coating and welding current, here a cellulose sodium coating, and that it should be used with direct current with the electrode positive. It is produced as Hobart 10 by Hobart

Brothers Co. who provide these specifications as typical:

Carbon	0.12%
Manganese	0.54%
Silicon	0.16%
Phosphorous, sulfur	0.018% each
Ultimate strength	79 500 psi
Yield strength	66 500 psi
Elongation	34%

This is a ductile, low-carbon steel. Manganese and silicon act as both deoxidants and strength improvers.

AMERICAN WELDING SOCIETY. Table 19.2 is a sample of AWS electrode specifications. The carbon content of these rods varies from 0.04 to 0.12 percent. They are all quite ductile.

Allowable stresses recommended by the AWS for fillet welds made with various combinations of base metals and electrode/flux combinations are listed in Table 19.3.

TABLE 19.2
American Welding Society specifications for welding electrodes*

AWS no.	Yield, ksi	Principal alloy	Remarks
6010	66.5	Mn, Si	Piping, ships, structures
7018	73.7	Mn, Si	For harder steels
8010	80–90	Mn, Ni	Pipelines
10018	95.9	Mn, Ni, Mo	To weld manganese steel
308-15	65	Cr, Ni, Mn	For stainless steel
316-L15	68	Cr, Ni, Mn, Mo	For stainless steel

* Courtesy Hobart Brothers Company.

TABLE 19.3
Fillet weld strength (AWS)

Base metal (ASTM)	Yield stress, ksi	Electrode	Allowable shear stress, ksi
A500 A		AWS E60XX	18.0
A36, 53B	36	AWS E70XX	21.0
A572	50	AWS E80XX	24.0
A514 ($>$2.5 in)	100	AWS E90XX	27.0
A514 ($>$2.5 in)		AWS E100XX	30.0
A514 ($<$2.5 in)		AWS E110XX	33.0

TABLE 19.4
Allowable stress on welds (AISC)

Type of weld	Allowable stress
Complete-penetration groove welds	
Tension	Same as base metal
Compression	Same as base metal
Shear	$0.30S_{ut}$, except shear stress on base metal $< 0.4S_y$ (yield strength of base metal)
Partial-penetration groove welds	
Tension normal to effective area	$0.30S_{ut}$, except tensile stress on base metal shall not exceed $0.60S_y$
Fillet welds	
Shear on effective area	$0.30S_{ut}$, except shear stress on base metal shall not exceed $0.40S_y$

All stresses relate to effective area.

AMERICAN INSTITUTE FOR STEEL CONSTRUCTION. As shown in Table 19.4, the AISC allows designers to utilize the ultimate strength S_{ut} of the base metal in tension or compression, providing that filler metal of matching composition and strength is used.

Aluminum

Almost all aluminum alloys can be welded. Those which are non-heat-treatable retain about 90 percent of their ultimate strength and 60 percent of their yield strength after welding. Heat-treatable alloys lose much of their strength in welding, and for this reason they are usually joined mechanically. Welds in heat-treated aluminum alloy can be rehardened, but the elongation and toughness of the weld are much reduced. For example, Reynolds Metals Co. states that the yield strength of 6061-T6 is reduced from 40 ksi before welding to 17 ksi after. The elongation at break is reduced from 17 to 10 percent. Re–heat treatment brings the yield strength back to 38 ksi, but the elongation drops to 3 percent.

WELDING RODS FOR ALUMINUM. Aluminum welding rods are classified by the same four-digit system used to designate cast and wrought aluminum alloys. Of these, only the 1XXX (Al) and 4XXX (Al-Si) series are recommended for torch welding. The 5XXX (Al-Mg) rods can be used for arc welding of 5XXX, 6XXX, and 7005 alloys. Welds made with this alloy possess highest as-welded strength and ductility.

Welds may crack because the aluminum alloy in the weld is not ductile enough at welding temperatures. The condition is known as *hot shortness*. It is best avoided by choosing a filler metal with a lower melting temperature than that of the alloy being welded. Usually this means a higher alloy content than the base metal. For example, 6061 alloy, which contains 0.6 percent Si, is crack-sensitive when it is welded with 6061 filler. The use of 4043 filler, which contains 5 percent Si, solves the problem. The 4043 filler has a lower melting point and remains more plastic than the base metal during cooling. However, this filler should *not* be used to weld high-magnesium alloys such as 5083, 5086, or 5456, because weld cracking would result. You should choose a filler metal for a particular welding application only after consulting a good reference on the subject, such as the AWS *Welding Handbook.*

STRENGTH OF ALUMINUM WELDS. In any fusion weld, the properties of the joint depend on the response of the metal to the temperature history of the joint. Most aluminum alloys are strengthened by precipitation hardening after quenching or by cold work. During welding, the metal in the weld zone will be annealed and then quenched. It will no longer be hard. For this reason, the minimum annealed tensile strength of the base metal is generally considered as the design strength for butt joints in these alloys. A major exception is the 7XXX series (Al-Zn) alloys which age naturally to high strength after welding.

Generally, with higher welding speeds, the time at high temperature is less and the final properties are better. Since thinner sheets can be welded at higher speeds, weld properties are generally better in thin sheets. Some typical as-welded tensile properties of gas-shielded arc welds are given in Table 19.5. The tensile properties of a given joint may vary, as discussed above.

As with welds in steel, the critical stress in aluminum fillet welds is shear on the throat—a plane at 45° to the legs of the fillet. Table 19.6 gives typical shear strengths in fillet welds made with several electrodes.

TABLE 19.5
Typical tensile properties of fillet welds in aluminum alloys

Base	Filler	Yield, ksi	Elongation
	Non-heat-treatable alloys		
1060	1260	5	29% in 2 in
3003	1100	7	24
5050	5356	12	18
5086	5356	19	17
5456	5456	23	14
	Heat-treatable alloys		
2014-T6	4043	28	4
6061-T4	4043	18	8
6061-T6	5356	19	11
7005-T53	5356	30	10

Courtesy of the Aluminum Association.

TABLE 19.6
Shear strength of fillet welds on aluminum

| Electrode | Shear strength, ksi | |
	Longitudinal	Transverse
ER1100	11	13
ER4043	16	21
ER5356	21	34
ER5183	24	39
ER5556	26	41

Courtesy of the Aluminum Association.

Spot Welds

In the main, spot welding is used for production fastening where the loads on the joint are not large compared with the strength of the sheet being fastened. Spot welds should normally be used to carry only shear loads, as shown in Figure 19.6, not tensile or peel loads. The load-carrying ability of a spot weld can be estimated by assuming that failure consists of the shearing of a circular "plug," shown in Figure 19.7, whose diameter is 1.5 times that of the weld and whose thickness is that of the

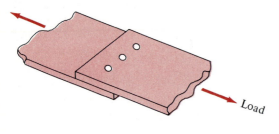

Load

FIGURE 19.6
Spot welds should be loaded only in shear.

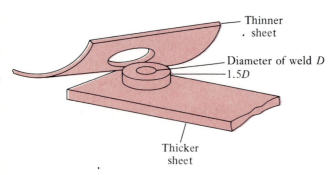

Thinner sheet

Diameter of weld D
1.5D

Thicker sheet

FIGURE 19.7
A model for the fracture of a spot weld.

thinnest sheet. The allowable stress used can be that of the thinnest sheet, if that sheet is in the soft condition. However, if the sheet has been hardened, the allowable stress should be that which results from the heating associated with the welding process. A precipitation-hardened sheet may revert to the as-quenched condition at a spot weld, and thus the stress used in estimating the strength of the weld should correspond to that condition.

Example 19.1. Aluminum sheet can vary in tensile yield strength from less than 20 to over 50 ksi. Estimate the failure strength of $\frac{3}{16}$-in spot welds in aluminum sheet of 0.050-in thickness for several yield strengths, and compare with military specifications.

The effective weld diameter is $(3/16)1.5 = 0.28$ in. The failure strengths are as follows:

Original sheet tensile yield strength, ksi	20	42	65
Shear yield strength "as quenched," ksi	10	12	14
Calculated failure load, lb	440	528	616
Military specification (minimum average, three tests), lb	370	540	585

Fatigue-Resistant Design

The road to fatigue failure is often paved with good intentions—a cliché, but true. Too often a change intended to add stiffness or strength only succeeds in creating an origin of fatigue failure.

In the discussion of stress concentrations, we used the example of an abrupt change in cross section. In fact, it is the abrupt change in stiffness which causes the intensification of stress. "Reinforcing," by introducing discontinuities of stiffness and concentrations of stress, can lead to early failure. An example is shown in Figure 19.8. Transverse members attached to a plate reduced its fatigue strength. Three weld beads which meet in a corner, as shown in the first example of Figure 19.9, can result in a condition of triaxial stress, with brittle failure possible even in a static loading situation. The other examples in Figure 19.9 illustrate two rules: (1) Place weld beads in low-stress locations. (2) Avoid abrupt changes in stiffness.

Effect of transverse attachments on fatigue strength

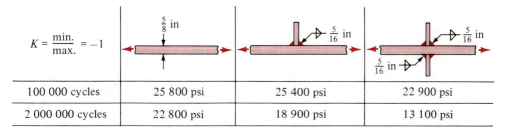

$K = \dfrac{\text{min.}}{\text{max.}} = -1$	$\frac{5}{8}$ in	$\frac{5}{16}$ in	$\frac{5}{16}$ in $\frac{5}{16}$ in
100 000 cycles	25 800 psi	25 400 psi	22 900 psi
2 000 000 cycles	22 800 psi	18 900 psi	13 100 psi

FIGURE 19.8
Fatigue strength is reduced by discontinuities.

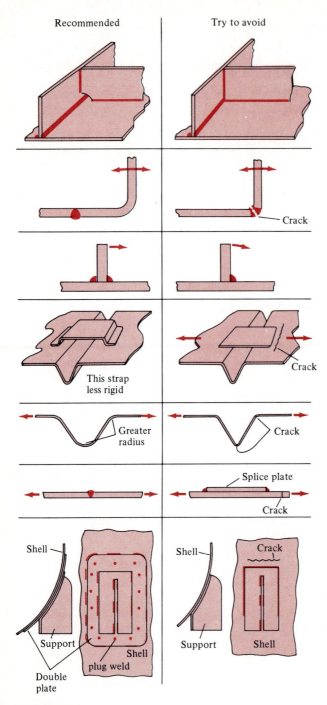

FIGURE 19.9
Design practice to avoid fatigue cracking.

Joints under Eccentric Loads

Figure 19.10*a* shows a bracket which has been welded to a larger structure to support a load at its end. The effect of the load on the weld can be analyzed by resolving it into an identical load through the centroid of the weld pattern and a twisting moment. This is shown in Figure 19.10*b*. The total shear stress in the weld metal is the vector sum of the shear stress P/A caused by the load applied at the centroid and the torsional stress

$$\tau = \frac{Mr}{J} \tag{13.1}$$

caused by the moment M. The maximum distance to any point of the weld is designated r, and J is the polar moment of the weld pattern area about its centroid.

The first step in this stress analysis is to find the centroid of the weld metal pattern. As the throat width is usually very small compared with other dimensions, the centroid and area moment of the pattern can be computed as though the throat area were in fact concentrated on a line. With the centroid located, the moments of

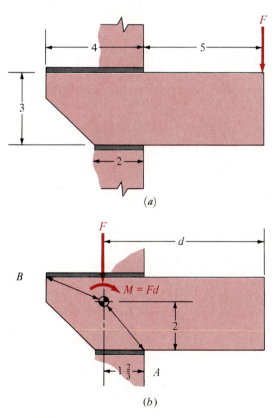

(a)

(b)

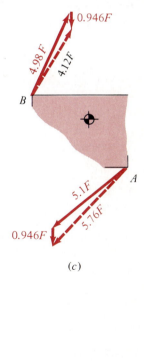

(c)

FIGURE 19.10
A bracket welded to a thicker member.

inertia I_x and I_y of the weld area about the centroid are calculated and summed to give the polar moment J. The total stress is then computed as the vector sum of the torsional and direct components.

Example 19.2. Consider the bracket of Figure 19.10 to be $\frac{1}{2}$-in steel plate welded to a heavier structure. Using Table 19.1, we decide on a $\frac{1}{4}$-in fillet, whose throat area is $0.25(0.707) = 0.176$ in^2/in of weld. With the origin of coordinates at the right end of the lower weld, the centroid is found to be at $x = -1.67$, $y = 2$. The moment of inertia of the weld throat area about an x axis through the centroid is

$$I_x = [2(2^2) + 4(1^2)](0.176) = 2.113 \text{ in}^4$$

In the same fashion,

$$I_y = [(\tfrac{1}{12})4^3 + 4(\tfrac{1}{3})^2 + (\tfrac{1}{12})2^3 + 2(\tfrac{2}{3})^2]0.176 = 1.29 \text{ in}^4$$

These are summed to give $J = 3.403$ in^4.

The torsional component of stress reaches a maximum, as stated earlier, at the weld edge farthest from the centroid of the pattern. Figure 19.10b shows two extreme points, of which A is farther (2.603 in) from the centroid than B (2.538 in). The torsional component of stress at these points is $\tau_A = F(5 + 1\tfrac{2}{3})(2.603)/3.403 = 5.1F$, and $\tau_B = 4.98F$. The total throat area is $0.176(4 + 2) = 1.056$ in^2, and thus the direct component of shear stress is $\tau = 0.946F$.

The vector diagrams of Figure 19.10c show that the vector sum of the direct and torsional components of shear stress is much larger at A than at B. The magnitude of this resultant stress is

$$\tau = F \sqrt{\left[0.946 + 5.1\left(\frac{1.67}{2.6}\right)\right]^2 + \left[5.1\left(\frac{2}{2.6}\right)\right]^2} = 5.76F$$

On this structural-steel application, the 6010 electrode is indicated. Its minimum tensile yield stress is 50 ksi. With a factor of safety of 2.5, the design stress for shear becomes

$$\frac{S_y}{2FS} = \frac{50E + 03}{5} = 10 \text{ ksi}$$

Equating, we find the allowable load to be 1739 lb.

The allowable stress used in this example can be checked against the other standards given here, if we keep in mind that they incorporate an FS which is probably not the same as that used in Example 19.1. In Table 19.3, we find tabulated, for the E60XX electrode, an allowable shear stress of 18 ksi. In Table 19.4 for fillet welds, an allowable shear stress of $0.30S_{ut} = 18$ ksi, or not more than $0.4S_y = 20$ ksi, is recommended. From this, we would infer that both tables include a factor of safety of $25/18 = 1.4$.

19.3   STRUCTURAL STIFFNESS

Bending

Frequently a weldment is made far stronger than it needs to be, in order to obtain the needed stiffness. If that stiffness is not obtained in the most effective way, the weldment will be needlessly heavy and costly. Here we discuss the stiffness of a welded

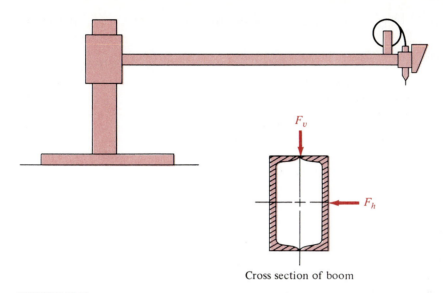

FIGURE 19.11

An automatic welding head mounted on a long boom. (*Courtesy James F. Lincoln Arc Welding Foundation.*)

structure in bending. It is also important to make sure that the most important stiffness is emphasized, as pointed out in this cautionary tale, courtesy of the Lincoln Arc Welding Foundation:

> A fabricating plant has set up an automatic welding head on a boom [shown in Figure 19.11]. The allowable vertical deflection of the welding head is set at $\frac{1}{8}$ in. Knowing the weight of the welding head, etc., and recognizing a simple cantilever beam, the engineer selects a welded box beam, deeper than it is wide. When the welding head is mounted to the end of the beam, it deflects downward $\frac{1}{8}$ in, and there it remains. During the operation of the machine, the fixture may vibrate slightly, and it will vibrate a greater distance horizontally than vertically because it is less stiff in that direction.
>
> ... The real design problem here is to maintain proper stiffness against possible movement of the boom, which is greatest in the horizontal direction. The result would probably look like the original cross section, but rotated 90 degrees.

For bending loads, the best ratio of stiffness to weight is usually obtained by using the maximum depth consistent with resistance to buckling and other considerations. Blodgett [84, 1.3-3] points out that while an increase in beam depth increases stiffness, it can decrease the strength. The U-shaped section in Figure 19.12 is nearly one-half again as stiff in bending as the flat bar, with only one-quarter the mass. But for any given bending load, the maximum stress at the ends of the legs in the U is twice as great as in the bar. The U section is stiffer, but weaker. To avoid costly mistakes, do not place thin sections far away from the neutral axis, and always check maximum stress when the shape of the cross section is changed.

Consider a beam fabricated as a rectangular hollow box, shown in Figure 19.13. The largest moment of inertia of the cross section will be about an axis I_{xx} perpendicular to the long side h. If the thickness of the box is small compared to its other

Section property	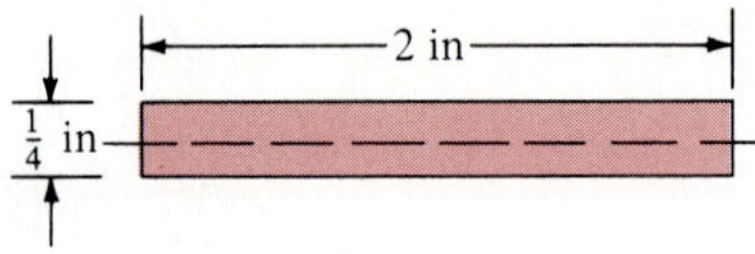 2 in, 1/4 in	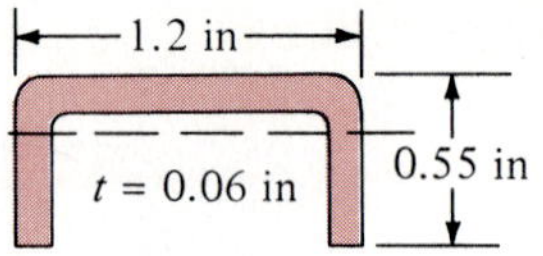 1.2 in, t = 0.06 in, 0.55 in
I	1	1.4
S	1	0.44
W_i	1	0.26

FIGURE 19.12
Thin cross sections should not be placed far from the neutral axis. (*Courtesy James F. Lincoln Arc Welding Foundation.*)

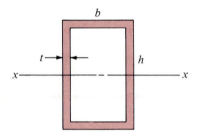

FIGURE 19.13
A rectangular beam cross section with constant thickness.

dimensions, that moment of inertia can be expressed as

$$I_{xx} = \frac{th^2}{6}(h + 3b) \tag{19.1}$$

which is a measure of its bending stiffness. The moment of inertia divided by the volume per unit length is

$$\frac{I}{V} = \frac{h^2}{12}\frac{h + 3b}{h + b} \tag{19.2}$$

This is simplified by denoting the ratio b/h as n:

$$\frac{I}{V} = \frac{h^2}{12}\frac{1 + 3n}{1 + n} \tag{19.3}$$

The stiffness of the box in bending relative to its bulk, I/V, is related to the square of its depth, but not at all to its thickness. Of course, a thicker box will be stiffer—but its bulk will increase exactly with the stiffness.

As the width/height ratio n approaches zero, the shape approaches a plate on edge, and its relative stiffness approaches that of a solid beam: $h^2/12$. With $n = 1$, we have a square tube, whose I/V of $h^2/6$ is slightly better than that of a circular tube. The relative stiffness increases with n to a limit of $h^2/4$. In sum, the deepest box is

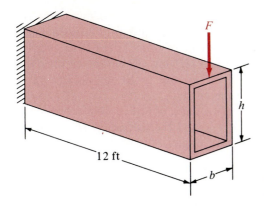

FIGURE 19.14
A rectangular tube acting as a cantilever beam.

most efficient for bending stiffness. For any given depth, it is more efficient to use a wider box than one with thicker walls.

The degree to which a structural box can be optimized by increasing its depth or width is limited in practice because the large, thin sides may buckle or may not be robust enough to provide attachment points. Blodgett [84] suggests that the largest allowable ratio of panel height to thickness may vary from $2400/\sqrt{S_y}$ to $10\,000/\sqrt{S_y}$, depending on restraint. For a steel with 40-kpsi yield strength, this guideline would allow panel height/thickness ratios of 12 to 50.

Example 19.3. A steel tube 12 ft long with rectangular cross section is loaded as a cantilever beam, as shown in Figure 19.14. It must carry a load of 2000 lb with no more than 0.06-in deflection. The mass of the beam should be minimum.

For sidewall stability, we set the minimum thickness ratio between the extremes given above at $h/t = 30$, and since we want maximum values for both h and b, they are set equal. Equation (19.1) becomes

$$I = \tfrac{1}{45}h^4$$

The deflection of a cantilever beam to a load at its end is, from Appendix G.9, $y = PL^3/(3EI)$. Solving gives

$$h = b = 14.94 \text{ in} \qquad t = 0.498 \text{ in}$$

Panels may be stiffened against buckling by increasing their effective thickness. This is done by corrugation or by welding stiffeners to the sheet. Many early aircraft used corrugated skin; modern aircraft use stiffeners bonded or riveted to the inside surface of the skin. The choice between a fairly complex but weight-efficient weldment and a simpler box made of heavier material is usually made on the basis of cost. The complex structure is lighter and requires less material, but the simpler, heavier structure uses less labor and costs less to maintain.

Torsional Stiffness of Built-up Sections

We discussed in Chapter 13 the torsion of tubes, i.e., long members of uniform cross section. However, the width and depth of many weldments are the same magnitude as their length, and their cross sections are anything but uniform. A simple example is shown in Figure 19.15. Instead of representing torsion by a pair of moments, it is more realistic to show the frame acted on by two pairs of equal forces P, or couples. The figure shows the resulting deflection of members AD and CD relative to the other sides of the frame. Member AD has been twisted relative to BC by an angle θ_{AB}, while CD has been twisted relative to BA by the angle θ_{BC}.

We can regard the forces P as the sum of two sets of forces. One set of forces P' twists members AB and CD through angle θ_{AB}, and the second set $(P - P')$ acts only on BC and AD to cause the twist θ_{BC}. These angles of twist are related by the deflection Δ. That is,

$$\Delta = \theta_{AB}W = \theta_{BC}L \tag{19.4}$$

Equation (13.3) relates the twist in a rod to the torque T:

$$\theta = \frac{TL}{GJ_T} \tag{13.3}$$

So,

$$\theta_{AB} = \frac{P'WL}{2GJ_{T,AB}} \qquad \theta_{BC} = \frac{(P - P')WL}{2GJ_{T,BC}}$$

These equations are combined with expression (19.4) to get an expression for the deflection of the frame:

$$\Delta = \frac{PW^2L^2}{2G(WJ_{T,BC} + LJ_{T,AB})} \tag{19.5}$$

Equation (19.5) expresses the torsional resistance of a built-up frame as the sum of the torsional resistances of both its longitudinal members and its transverse members. The members of a real frame bend as well as twist, and this will increase

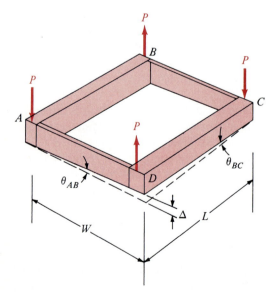

FIGURE 19.15
A welded frame in torsion. (*Courtesy James F. Lincoln Arc Welding Foundation.*)

deflection. Blodgett [84] cites an experiment showing that bending increased the deflection of a simple frame 10 percent.

BRACING OF LARGE WELDMENTS. In the previous paragraph, we calculated the torsional rigidity of a frame as the sum of the torsional rigidities of its members. In Chapter 13, it was shown that the torsional rigidity of any open cross-sectional tube could be increased hundreds of times by closing the cross section. To that we add that closing the end of the tube also increases the torsional rigidity.

As the size of a frame increases, cross bracing becomes necessary for many reasons, and of course we seek a way of placing that bracing to most effectively increase the stiffness of the frame. A simple approach is to place the braces parallel to one end of the frame. However, as we saw in developing Equation (19.5), each parallel brace contributes only its own torsional rigidity to the whole. The torsional rigidity of a thin plate was shown in Chapter 13 to be small. Parallel braces are thus not an effective use of material.

A clue to the most effective bracing can be found by examining the stress in the main members of a frame under twisting loads, which is shear. A member like *AB* in Figure 19.15 is shown in Figure 19.16 at *a*, with the shear stress indicated on

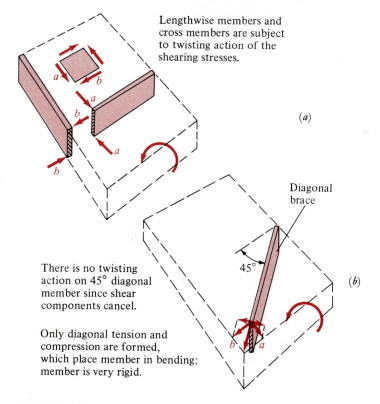

Lengthwise members and cross members are subject to twisting action of the shearing stresses.

(*a*)

Diagonal brace

45°

(*b*)

There is no twisting action on 45° diagonal member since shear components cancel.

Only diagonal tension and compression are formed, which place member in bending; member is very rigid.

FIGURE 19.16
Braces set at 45° see only bending loads and are therefore much stiffer than braces at other angles, which see torsional loads. (*Courtesy James F. Lincoln Arc Welding Foundation.*)

the cut section. Since every pair of shear stresses *b-b* must be accompanied by an equal set *a-a*, the cross member is subjected to torsional loads, to which its stiffness is very low.

As shown at *b* in the figure, the principal stresses corresponding to the torsion in the side members are tension and compression at 45° to the members. A brace placed at this angle to the frame will thus be subjected only to bending loads, to which it is very stiff.

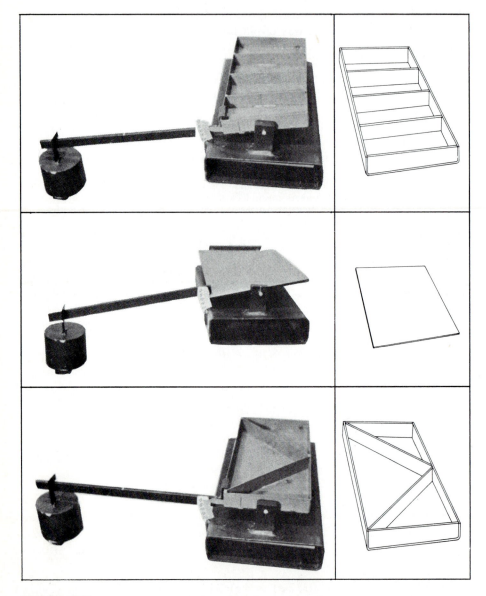

FIGURE 19.17
A model test verifying the stiffness of 45° bracing. (*Courtesy James F. Lincoln Arc Welding Foundation.*)

While a rigorous derivation of the stiffness of diagonally braced frames is beyond the scope of this text, Figure 19.17 shows the result of model tests on three versions of a frame. The thickness of material was the same in all three models. The model with parallel bracing proved to be only 10 percent stiffer than the plain sheet, while the diagonally braced frame, using 6 percent less material, was 36 times stiffer.

The use of models to evaluate candidate frame designs is surprisingly accurate— and the most practical procedure for one-of-a-kind construction, where the cost of a computerized finite-element analysis cannot be justified.

19.4 BRAZING AND SOLDERING

Brazing

The temperatures required for brazing and soldering are well below the melting points of the metals being joined. Three benefits result: Lower temperatures are easier to work with, problems of loss of heat treatment and warpage are minimized, and dissimilar metals can be joined. The disadvantage is that the joint may be less resistant to heat than the base metal.

Brazed and soldered joints derive their strength from the fact that when two materials are brought within a few angstrom units ($1 \text{ Å} = 1E - 10$ m) of each other, a bond results. The process is usually called brazing if the temperature is above 425°C (800°F). In a brazed joint, an alloy can form between the braze and the base metal at the interface. This is possible because even though the base metal does not melt, a good deal of interchange between base metal and filler can take place by diffusion. The yield strength of the resulting alloy can be superior to that of the base metal, even at high temperature. Alloys which are already tempered generally lose little temper to brazing. Fully hardened heat-treatable alloys are usually tempered, i.e., softened, by the brazing cycle. In many cases, these heat-treatable alloys can be hardened and tempered after brazing.

The filler metal, braze or solder, can be placed in several ways. It may be added from outside the joint, such as when copper plumbing is soldered. The pieces to be joined can be assembled in a rack or jig and then dipped in the molten solder, as electronic components are often fastened to circuit boards. Or the joints can be designed with recesses in which the filler is placed before the joint is heated. With the braze material inside the joints, the assembly is simply placed in a furnace until fusion has occurred. Control of the atmosphere in the furnace can produce remarkably clean results. Large assemblies can thus be bonded in a single operation.

When the braze metal is used to form a fillet or bead, the process is called *braze welding*. The technology of braze welding is much like that of torch welding.

A flux is used in both brazing and soldering to loosen and float away anything on the metal surface that would interfere with the desired bond. In designing joints for brazing, it is necessary to provide an avenue by which both the liquid and vaporized flux can escape.

For best results, joints to be brazed or soldered should be designed so that the clearance between them is appropriate for the braze or solder to be used. With the right clearance, capillary action will carry the molten filler into the entire joint. In this way joints up to 24 in high can be filled from the bottom. Some recommended

TABLE 19.7
Recommended clearances for brazing aluminum

Joint width, in	Clearance, in
Dip and vacuum brazing	
<0.25	0.002–0.004
>0.25	0.002–0.025
Torch, furnace, and induction brazing	
<0.25	0.004–0.008
>0.25	0.004–0.025

Courtesy the Aluminum Association.

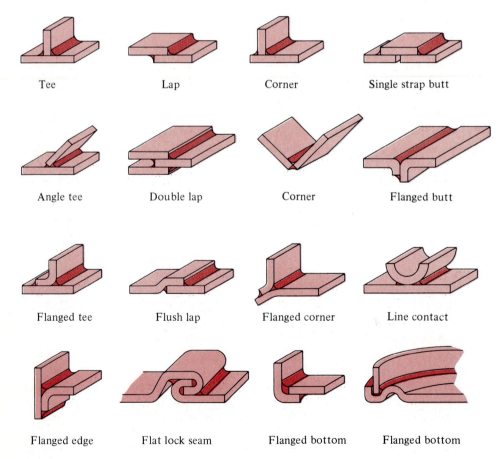

| Tee | Lap | Corner | Single strap butt |

| Angle tee | Double lap | Corner | Flanged butt |

| Flanged tee | Flush lap | Flanged corner | Line contact |

| Flanged edge | Flat lock seam | Flanged bottom | Flanged bottom |

FIGURE 19.18
Typical brazed joint designs. (*Courtesy of the Aluminum Association.*)

TABLE 19.8
Representative brazing alloys

Alloy no.	Use with:	Brazing temperature, °F	Yield strength, kpsi
Royal 110	Aluminium	1100	35
Crown 255	Cast iron		55
Sil 45	All except zinc	1145	60
Sil-Cop	Copper alloys	1460	
Royal 310	Ferrous to nonferrous	arc	35
Crown 120FC	Ferrous to nonferrous	1300–1750	76
Crown 95	Solder	425	10

Courtesy of Crown Alloys Co.

clearances are shown in Table 19.7. A number of representative brazing joints are shown in Figure 19.18.

STRENGTH OF BRAZED JOINTS. Brazed joints should be stronger than the parts they join. This requires design and preparation so that the braze metal is in a thin layer between the parts being joined, and primarily subjected to shear stresses.

Braze welding is often used for repair. A fillet of braze metal is used to join two parts which fit each other only approximately, such as those shown in Figure 19.18. These joints can be quite strong, as indicated by the physical properties of the sample of brazing alloys shown in Table 19.8.

PROBLEMS

Section 19.1

19.1. Figure P19.1 shows two details of an I beam fastened to a vertical column. State the advantages of the welded connection and of the bolted connection.

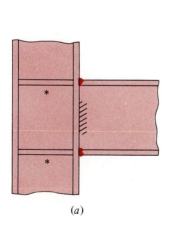
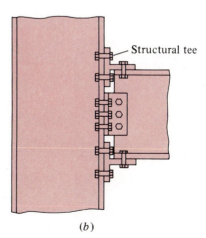

Structural tee

(a)   (b)

FIGURE P19.1
An I beam fastened to a vertical column by welding and by bolting.

19.2. In Figure P19.1, what is the function of the extra piece, indicated by an asterisk, that has been welded in between the webs of the column?

Section 19.2

19.3. As shown in Figure P19.3, a steel strap $\frac{3}{8}$ in thick by 8 in wide is welded to a heavier piece of steel. Both pieces are AISI 1040 HR steel. The joint will be loaded in tension, as indicated, and the surface of the larger piece is vertical.
(a) Specify the appropriate electrode and fillet size.
(b) Calculate the largest static load that can be applied, by AISI standards. Compare with the load calculated by using the AWS specifications.
(c) Assume the load is fully reversed. Compute the largest completely reversed fatigue load that can be supported.

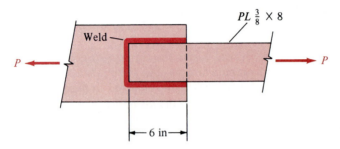

FIGURE P19.3
A steel strap welded to a heavier piece of steel.

19.4. Repeat Problem 19.3, except that AISI 1040 CR (cold-rolled) steel is used.
    Answer: 117.5E + 03 lb; 24.3E + 03 lb

19.5. The bracket shown in Figure P19.5 is AISI 1040 HR steel $\frac{3}{8}$ in thick, welded to a heavier piece and loaded as shown. Find the factor of safety for both (a) static and (b) fully reversed dynamic loads.

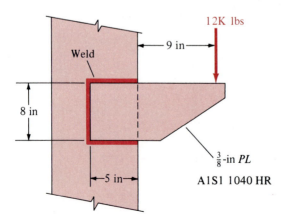

FIGURE P19.5
A steel bracket welded to a massive steel column.

19.6. The dogleg beam shown in Figure P19.6 is welded from $\frac{3}{8}$-in AISI 1040 HR steel plate. For static loading and a factor of safety FS = 3, find the rectangular cross section for the entire beam which uses the least amount of steel. Don't worry about how the beam is fastened to the wall at the left.

Answer: width, 10.9; depth, 13.3 in

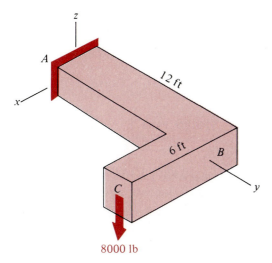

FIGURE P19.6
A hollow beam fabricated by welding steel plate.

19.7. The L-shaped member shown in Figure P19.6 is a weldment of $\frac{3}{8}$-in AISI 1040 HR steel with square cross section, each side 20 in. Design a flange of the same steel that can be welded onto the structure at end *A*. The flange will be held to the wall at left with bolts. Use FS = 3.

19.8. The bracket shown in Figure P19.8 is to be welded from $\frac{1}{2}$-in AISI 1040 HR plate, and to the column, using E70 electrode.

(a) Determine the factor of safety of the design for a static load, using recommended fillet sizes and stresses.

(b) Evaluate the design, assuming the load to be cyclic, fully reversed. If it is not satisfactory, engineer the most economical acceptable design.

Answer: FS = 4.22; FS = 0.96

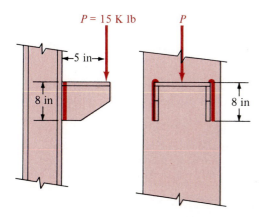

FIGURE P19.8
A bracket welded from steel plate.

19.9. Figure P19.9 shows a circular steel shaft welded to a heavy steel base. Determine the safety factor for static loading, using E70 electrode.

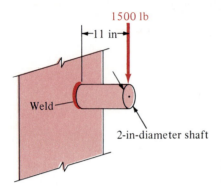

1500 lb

11 in

Weld

2-in-diameter shaft

FIGURE P19.9
A solid steel shaft welded to a massive column.

19.10. The joint in Figure P19.10 consists of a double lap joint 6-in wide of AISI 1040 HR steel. Find the maximum allowable force, if E70 electrode is used, for (a) static, and (b) fully reversed loading.
   Answer: 127.2E + 03 lb; 23.24E + 03 lb

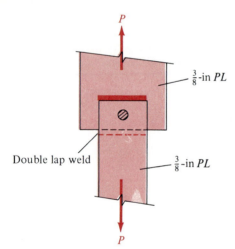

P

$\frac{3}{8}$-in PL

Double lap weld

$\frac{3}{8}$-in PL

P

FIGURE P19.10
A lap joint held together by a plug weld.

19.11. Figure P19.11 shows two members joined by fillet welds. Material: AISI 1040 HR
   (a) Determine the maximum static load that can be applied if E70 electrode has been used.
   (b) Repeat for a cyclic, fully reversed load.

19.12. A shaft is welded to a hub assembly, as shown in Figure P19.12. Assume AISI 1040 HR steel and E60 electrode. The shaft rotates, so this is a fatigue situation. Evaluate the design for a factor of safety of 1.5. If it is not satisfactory, propose an economical alternative.
   Answer: FS < 1; not satisfactory

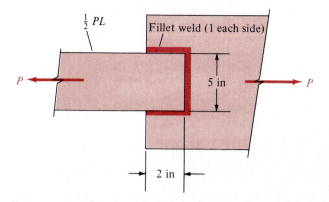

FIGURE P19.11
A lap joint held by three fillet welds.

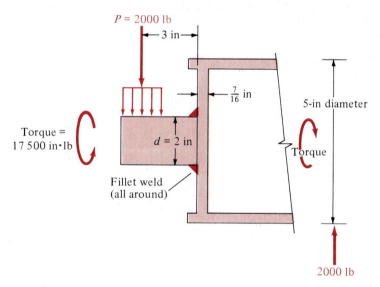

FIGURE P19.12
A solid circular shaft welded to a circular hub.

19.13. Figure P19.13 shows a steel channel section of AISI 1020 HR which will be welded to a thick steel wall. As indicated, the weld fillet will be on the outside of the channel only. Select the least expensive fillet that will provide an FS of 3 for the static load shown.

19.14. A beam is formed of two channels, as shown in Figure P19.14, and welded to the wall as a cantilever. The fillet size is specified as $\frac{3}{16}$ in. How great can the force F be, if the allowable shear stress in the weld is 15 kpsi?
   Answer: $F = 5E + 03$ lb

19.15. A steel platform for lifting a person in a wheelchair is supported by two 4-in × 7.5-lb/ft steel (AISI 1040 HR) channels which are welded to the platform using E70 electrodes. What is the factor of safety for static loading?

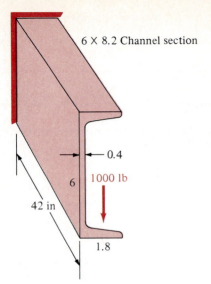

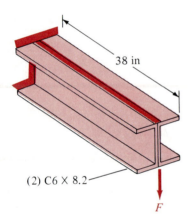

6 × 8.2 Channel section

0.4

6

1000 lb

42 in

1.8

FIGURE P19.13
Cross section of a 6 × 8.2 (6-in-deep, 8.2-lb/ft) channel section.

38 in

(2) C6 × 8.2

F

FIGURE P19.14
Two steel channels welded together to form a beam.

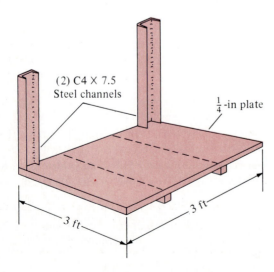

(2) C4 × 7.5
Steel channels

$\frac{1}{4}$-in plate

3 ft

3 ft

FIGURE P19.15
A platform consisting of a thick steel plate held by two channels welded to one edge.

Section 19.3

19.16. Assuming a constant cross section as in Problem 19.7, calculate the deflection at end C due to the 8000-lb load.

19.17. Redesign the structure of Problem 19.6 for minimum weight. Do not use steel less than $\frac{1}{4}$-in thick. Assume AISI 1040 HR and FS = 1. You can use any cross section you like for either leg, so long as the perimeter is not more than 72 in and the cross section is the same throughout the leg.

CHAPTER
20

BEARING SELECTION

The development of bearings closely parallels that of transport. Prior to the wheel, heavy loads were dragged in one way or another over the ground, usually against rather large frictional forces. The wheel dramatically altered the scene. It is supposed to have originated in the delta of the Tigris and Euphrates rivers, where specimens have been found estimated to be 5000 years old. Figure 20.1 shows a cart, whose construction has not changed for centuries. The wheels and axletree are a solid unit rotating between pegs. Because this arrangement was poor for making turns, it was often replaced by a fixed axle on which the wheels turned independently. The bearings of both these designs resulted in severe wear problems, which can be seen in Figure 20.1. At some stage it was discovered that the wheel turned more easily and the axle and wheel lasted longer if a lubricant, such as animal fat, was added. In time the wood bearing surfaces were lined with metal for longer wear.

Replacement of sliding bearing surfaces by shapes which roll is not a modern idea. Indeed, the Egyptians are known to have moved heavy stone slabs by rolling them on logs. Heavy machinery is still moved today in this manner, but usually on pipes. Figure 20.2 shows a Celtic cart which was built over 2000 years ago. The wheel bearings consist of cylindrical surfaces separated by wood rollers.

Although the invention of the wheel and the development of bearings were very important for transport, many other wheel-axle combinations required bearing mountings: waterwheels, mills, potter's wheels, spinning wheels, etc. A turntable believed to have been used for a large statue in the time of Emperor Caligula (A.D. 12–41) was mounted on spherical wheels with pins inserted through them that served to confine

(a)

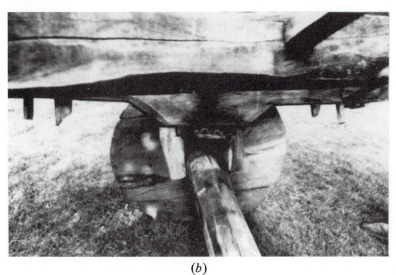

(b)

FIGURE 20.1
Two-wheel cart, found at the site of Troy, Turkey. Probable age is 100 to 150 years. (*Courtesy, Prof. Manegold, Göttigen, Germany.*)

the motion. Leonardo da Vinci (1452–1519) utilized rolling bearings in water screws and in the rolls of a small sheet-lead rolling mill. Figure 20.3 shows bearings from a French patent of 1802 for the mounting of the king post of a horse-drawn merry-go-round.

The evolution of the bicycle into a mass-produced vehicle sparked a parallel development in rolling bearings. In Great Britain alone about 900 patents for bicycle

(a)

(b)

FIGURE 20.2
Celtic cart, 300 to 100 B.C., with detail of wood-roller wheel bearings. (*Courtesy, Danish National Museum, Copenhagen.*)

ball bearings were filed up to 1905. The first ball-bearing catalog was published in Germany in 1900. Roller bearings date from a decade or two after that.

We first describe what has become of sliding-element bearings and then turn to rolling-component bearings.

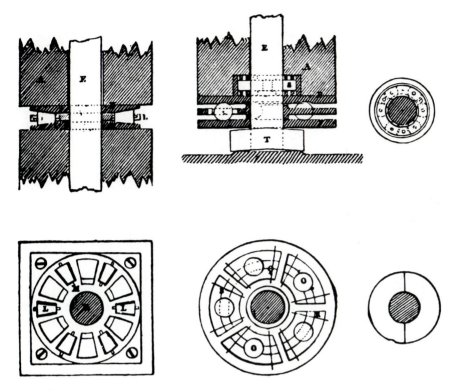

FIGURE 20.3
Drawing from a French patent of 1802 showing a tapered roller thrust bearing (left) and a thrust ball bearing (right), both with cages to space the rolling elements. Bearings intended for the king post of a horse-drawn merry-go-round.

PART A
SLIDING-CONTACT BEARINGS

A shaft which rotates within another member with sliding contact between the two is termed a *journal bearing*. The journal is the shaft, as depicted in Figure 20.4. Such an arrangement must be lubricated. The function of the lubricant is to maintain a separation of the surfaces and to produce a low resistance to motion, i.e., a low coefficient of friction. The nature of the lubricant varies from oil under pressure in turbine-generator units and internal-combustion engines to water in wooden ship sea bearings (where the propeller shaft enters the water) to air in high-speed instruments. Magnetic fields have also been used for support. Materials with inherently low coefficients of friction, such as nylon and teflon, are often used in light-load applications as bearing materials, with no other lubrication. Nylon impregnated with molybdenum disulfide and porous metallurgy products impregnated with oil are similarly employed.

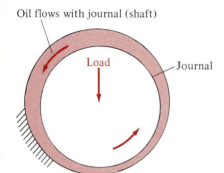

Oil flows with journal (shaft)

Load

Journal

FIGURE 20.4
Oil film generated by rotating journal bearing. Because of the oil pressure, the shaft is displaced slightly to the right.

The load-bearing capability of the lubricant derives from a wedge of it, under pressure, created by the rotation, as shown in Figure 20.4. The lubrication is hydrodynamic, i.e., due to a moving fluid.

In small bearings, such as lawn-mower engine connecting-rod bearings, the parts themselves are used as the bearing surfaces. More commonly a liner, called *the insert*, is fitted to the housing. The liner is usually a backing of steel with a thin overlay of soft material, often lead-tin. The latter absorbs any small dirt particles, which otherwise could score the journal. Worse, the bearing could seize on the journal and then rotate within its housing and destroy itself.

If the end of a journal is accessible, the bearing can usually be installed complete. This is the case with fractional-horsepower motors, for example, where a bushing is installed in each end of the motor housing prior to assembly. In larger installations, the journal is generally not accessible in this manner, and then the bearing must be installed in halves. The most common examples are the main and connecting-rod bearings of automobile engines. The latter are illustrated in Figure 20.5.

No bearing operates totally free of friction, so a certain amount of heat is generated. In small installations, such as fractional-horsepower electric motors, the whole unit is cooled by a fan mounted on the motor shaft which blows air through the motor, and this is sufficient to prevent overheating of the bearings. In large machines, such as steam-turbine units, the heat is removed by the lubricating oil, which is subsequently cooled. The heat is usually more than that generated in the bearings, since the entire machine is often at high temperature. This is the case with automotive engines—the oil is close to the temperature of the engine parts it contacts, which are in turn cooled by the radiator-pump system.

Removing heat from lubricating oil is also a means of supplementing the engine cooling system. This is not generally done in street vehicles for economic reasons, though some cars and motorcycles have fins on the oil sump to enhance cooling. Hot-rodders sometimes add an oil cooler, and trucks and airplanes frequently have such units. One of the engines in the globe-circling plane *Voyager* was cooled in the upper cylinder area by traditional coolant and in the lower end by a continuous spray of oil.

Oil loses viscosity with increasing temperature, and this can lead to an unstable condition in journal bearings. Figure 20.6 shows the relation between the coefficient

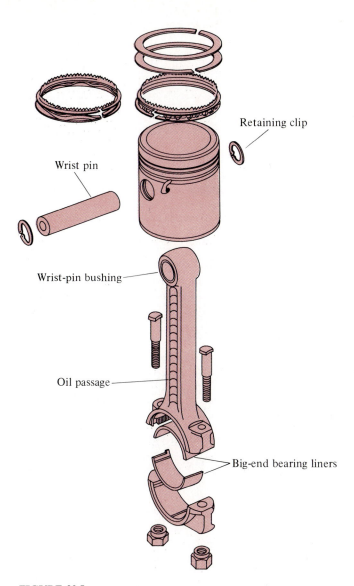

Retaining clip

Wrist pin

Wrist-pin bushing

Oil passage

Big-end bearing liners

FIGURE 20.5
Components of piston–connecting-rod assembly. Oil under pressure from the crankshaft lubricates the big-end bearing as well as the wrist-pin bushing through a passage in the shank of the connecting rod. (*Courtesy, Jaguar Cars, Ltd.*)

of friction in such bearings and a parameter known as the *bearing characteristic* $\mu n/P$ [62]. Here μ is the oil viscosity, n the journal speed, and P the load. To the right of the low point A, the lubrication is stable; to the left, unstable. To understand this, suppose the bearing operation is at point B, in the stable region, and the load becomes smaller. This causes an increased bearing characteristic and a movement up the curve to the right. A higher friction results, and consequently greater heat production and

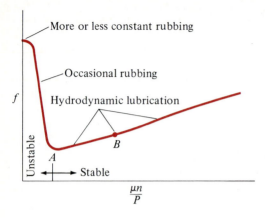

FIGURE 20.6
Coefficient of friction f as function of bearing characteristic (f is ratio of tangential friction force to bearing load).

temperature ensue. The viscosity decreases and with it the bearing characteristic, and the operating point moves back down the curve. Thus the situation is stable. To the left of point A, the opposite happens—the lubrication is unstable, and there may be incipient rubbing. Although bearings are not intended to operate in that region, it is unavoidable at start-up when the speed n is low. To avoid damage, we rely on the toughness of the residual oil film.

There are numerous sliding-contact bearings not involving rotating journals, e.g., ball joints in the steering articulations of vehicles, most cam-follower combinations, and bushings at the ends of steering racks in automobiles. Even skis, ice skates, and sleds could be called sliding-contact bearings. In fact, hydrodynamic lubrication does take place.

The above few paragraphs are intended to acquaint you with the components and functioning of journal bearings. Their design relies on the application of fluid flow and thermodynamic principles. It also leans heavily on experience and experimentation.

PART B
ROLLING-COMPONENT BEARINGS

20.1 BEARING TYPES

These bearings are commonly referred to by the generic term *rolling bearings* or by the names of the two main classes, *ball bearings* and *roller bearings*. They are sometimes called *antifriction bearings* by the industry. The small turning resistance of these bearings is due to the ease with which a sphere or cylinder rolls on a hard surface. Rolling bearings have several advantages compared to sliding-component bearings:

1. The starting torque is considerably lower. It is, in fact, not much greater than the running torque.

2. The bearings are internationally standardized, making for more economical designs and easier replacement.
3. Rolling bearings can be designed to take both radial and axial loads.
4. They can be arranged so as to position very precisely the parts they connect, both radially and axially.

Ball Bearings

The components of ball bearings are shown in Figure 20.7*a*. There are four main parts. The inner and outer races or rings provide the pathway on which the balls move. The cross-sectional radius of the pathway grooves is slightly greater than that of the balls, to avoid rubbing contact. The retainer or cage maintains separation of the balls, and there is often a shield (as shown) to keep dirt from entering or a seal to retain lubrication.

There are two methods of assembling ball bearings, as seen in Figure 20.7. In the method shown at (*b*), called the *Conrad type*, the balls are placed between the two rings, and then the separator is fitted. The balls fill a little more than half the space. In the second method, called the *filling-notch type*, a notch is ground into each

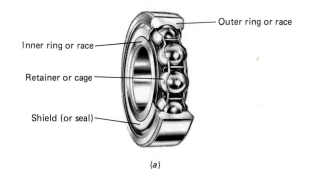

(*a*)

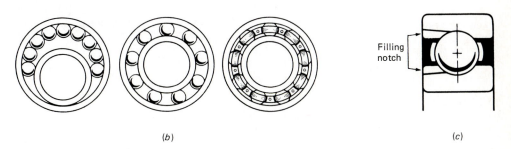

(*b*)   (*c*)

FIGURE 20.7
Components of ball bearings and methods of assembly. (*Courtesy, New Departure Hyatt, General Motors.*) (*a*) Components of ball bearings. (*b*) Assembly of Conrad deep-groove bearing. (*c*) Filling-notch method of assembly.

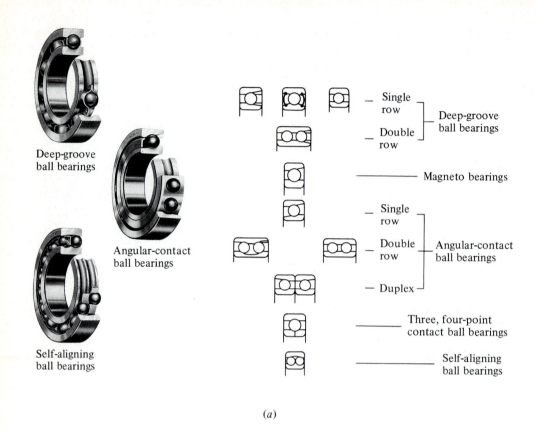

Deep-groove
ball bearings

Angular-contact
ball bearings

Self-aligning
ball bearings

Single row
Double row
} Deep-groove ball bearings

Magneto bearings

Single row
Double row
Duplex
} Angular-contact ball bearings

Three, four-point contact ball bearings

Self-aligning ball bearings

(*a*)

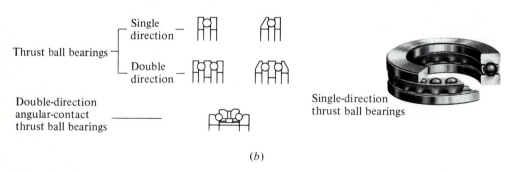

Thrust ball bearings
{ Single direction
Double direction }

Double-direction angular-contact thrust ball bearings

Single-direction thrust ball bearings

(*b*)

FIGURE 20.8
Types of ball bearings. (*Courtesy, NSK Corp.*) (*a*) Radial bearings. (*b*) Thrust bearings.

ring, which, when the two are aligned, allows passage of the balls (Figure 20.7c). More of them can be inserted in this manner than in the other method, allowing greater bearing loadings; hence they are called *maximum-capacity bearings*. The notches, however, limit the thrust (axial-load) capability.

Conrad-type bearings are sometimes known as *deep-groove bearings*. However, the term *deep groove* is often applied to both Conrad and filling-notch types, as in the figure discussed below.

Some of the numerous types of ball bearings are shown in Figure 20.8. The deep-groove bearings are radial bearings; i.e., they are primarily intended to take radial loads. Magneto bearings are small bearings (4- to 30-mm ID) used in magnetos, gyroscopes, and instruments. They are usually mounted in pairs. Since the outer ring has no shoulder on one side, it may be removed, which is sometimes helpful for installation.

All ball bearings can withstand a certain amount of thrust, but if the thrust is sufficiently high to preclude the use of a deep-groove bearing, an angular-contact bearing may be used. If the bearing is single-row, the thrust, of course, can be resisted in one direction only. Double-row angular-contact bearings allow thrust in both directions.

If shaft misalignment is seen as a problem, a self-aligning bearing may be used. The outer raceway surface in these bearings is spherical.

Double-row bearings are indicated where loads are heavy and rigidity of the shaft is important. They may be obtained in the deep-groove style or as an angular-contact bearing. In the latter case, the angles are generally in opposite directions, which ensures axial rigidity of the shaft. Occasionally, to get greater thrust capacity, a pair of deep-groove bearings is used rather than a single double-row bearing, which may be available only in the filling-slot construction.

If the load is pure thrust, bearings such as shown in Figure 20.8b are indicated.

Almost all types of ball bearings are available with a shield or seal on one or both sides. When a double seal is provided, the bearing is given lifetime lubrication at the factory. Occasionally provision is made for adding lubricant.

When the shaft is parallel to the surface on which the bearing is to be mounted, a housed bearing, called a *pillow block*, is frequently used. An example is shown in Figure 20.9a. The bolt holes are often slotted, as shown, to facilitate adjustment. If the mounting surface is perpendicular to the shaft, a flanged cartridge, seen at (b), may be used. In both of these mounting styles, the bearing insert is commonly spherical, to mate with a spherical bore in the housing, allowing self-alignment.

Figure 20.10 shows various special-purpose adaptations of ball bearings.

Roller Bearings

Roller bearings have the same essential components as ball bearings, with rollers replacing the balls. Generally the parts are separable, so they are not usually made with integral seals. Because the contact between a cylinder and the surface on which it rolls is a line rather than a point, roller bearings can withstand higher loads, other things being equal.

(a)

(b)

FIGURE 20.9
Common bearing mountings. (*Courtesy, Fafnir Bearings Divisions, The Torrington Co.*) (*a*) Pillow block. (*b*) Flanged cartridge.

The common types of roller bearings are shown in Figure 20.11. The cutaway photograph at the top shows a single-row bearing. It can take no thrust, but permits axial positioning of the shaft. The addition of shoulder(s) on the outer race, as shown in the sketches at the right, allows for a small amount of thrust.

Small cylindrical roller bearings are termed *needle bearings*. They may or may not have a cage. Because of the full complement of needles, the capacity of these bearings in relation to the space they occupy is very high.

Tapered roller bearings are intended to take considerable thrust loads and are most often used in pairs. A common example is the front-wheel bearings of rear-drive

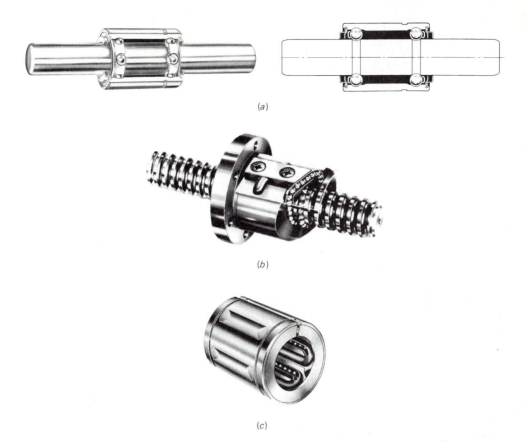

(a)

(b)

(c)

FIGURE 20.10
Special adaptations of ball bearings. (a) Automotive fan and water-pump shaft bearing. Here the inner race is the shaft itself. (*Courtesy, New Departure Hyatt, General Motors.*) (b) Cutaway view of a ball screw. The endless row of balls separates the nut from screw and allows rotary motion to be converted to linear movement with a minimum of friction. This scheme is used in many vehicle steering mechanisms. (*Courtesy, Thompson Saginaw Ball Screw Co.*) (c) Linear ball-bearing guide. These units also utilize endless rows of balls. Coefficients of friction as low as 0.002 can be achieved. (*Courtesy, NSK Corp.*)

vehicles, as shown in Figure 20.12. (Front-wheel-drive cars generally have enclosed double-row ball bearings.) The cup (outer race) of tapered roller bearings is separable; the cone (inner race) may or may not separate from the roller-cage assembly. The raceways and the roller surfaces meet at a common apex, which ensures pure rolling.

Spherical roller bearings, as with ball bearings of the same type, have spherically shaped outer raceways, allowing for some misalignment. The rollers are shaped to fit the spherical outer raceway; they are no longer cylindrical or conical.

Several types of roller thrust bearings are illustrated in Figure 20.11b.

No bearing has infinite life. In many applications the down time associated with changing bearings can be very costly, and the design must take into account this

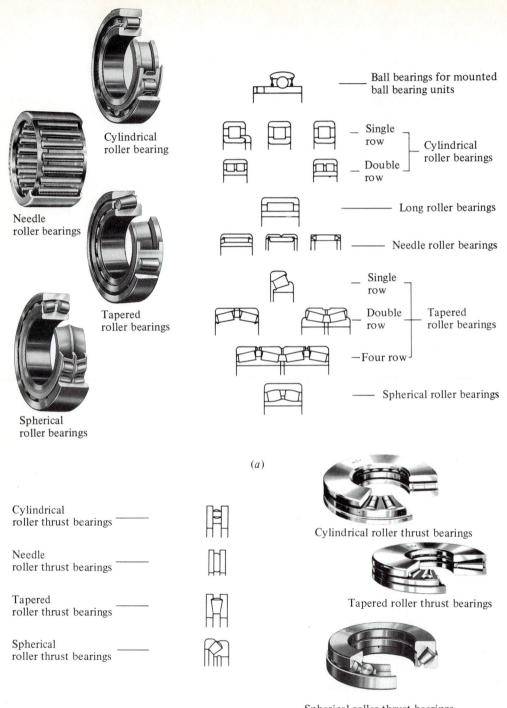

Cylindrical
roller bearing

Needle
roller bearings

Tapered
roller bearings

Spherical
roller bearings

Ball bearings for mounted
ball bearing units

Single
row

Double
row

Cylindrical
roller bearings

Long roller bearings

Needle roller bearings

Single
row

Double
row

Four row

Tapered
roller bearings

Spherical roller bearings

(a)

Cylindrical
roller thrust bearings

Needle
roller thrust bearings

Tapered
roller thrust bearings

Spherical
roller thrust bearings

Cylindrical roller thrust bearings

Tapered roller thrust bearings

Spherical roller thrust bearings

(b)

FIGURE 20.11
Roller-bearing types. (*Courtesy, NSK Corp.*) (*a*) Radial bearings. (*b*) Thrust bearings.

600

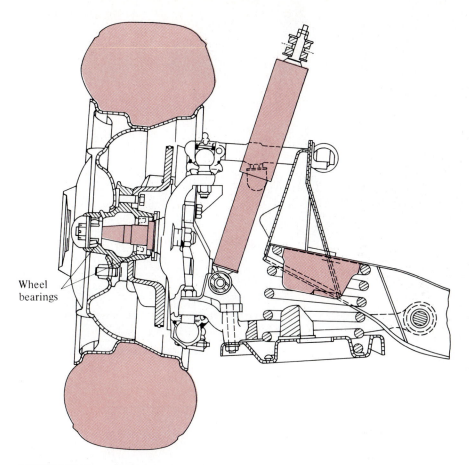

Wheel
bearings

FIGURE 20.12
Section through front-wheel hub and suspension showing wheel bearings. Since the road reaction (for straight-ahead driving) is vertical through the center of the tire, the in-board bearing takes nearly all of it and is therefore larger. (*Courtesy, Jaguar Cars, Ltd.*)

important economic fact. Imagine, for example, the cost of having the excavator shown in Figure 20.13*a* idle and the labor involved in disassembling major parts for bearing change. The bearings on the bucket shaft and its drive are split double-row spherical roller bearings, such as shown at (*b*). The bore of these bearings may be as great as 14 in, and all the components, including a seal, are split into two parts to allow the easiest possible replacement.

Figure 20.14 shows special adaptations of roller bearings.

Summary of Characteristics

Chart 20.1 is a summary of the principal bearing types and their characteristics.

(a)

(b)

FIGURE 20.13
Large split bearings find use in heavy machinery. (*Courtesy, FAG Bearings Corp.*) (*a*) Bucketwheel excavator for a daily stripping rate of 240 000 m³. (*b*) All components of this bearing are in two parts. Note the bolt for joining the roller-cage parts. Bearings like this are used in the excavator shown and other heavy applications.

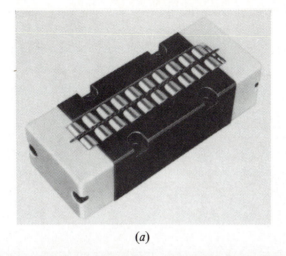

(*a*)

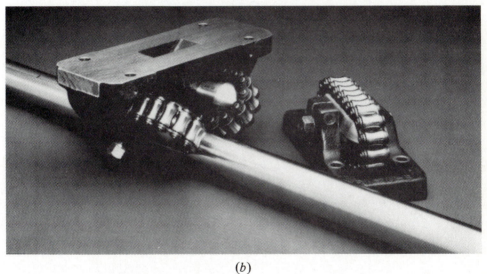

(*b*)

FIGURE 20.14
Special adaptations of roller bearings. (*a*) Linear roller bearing. (*Courtesy, NSK Corp.*) (*b*) Roller bearing for linear motion on rod. (*Courtesy, Thomson Industries.*)

20.2 BEARING SIZES

Most rolling-component bearings have traditionally been made in metric sizes, probably because of the European origin of the first standardized bearings. Tapered roller bearings and needle bearings have developed more in USCS. There is a movement with all these toward SI units, because large suppliers are to be found in all parts of the globe. Bearing sizes, methods of rating, etc., have been standardized

Bearing Type / Features	Deep Groove Ball Bearings	Magneto Bearings	Angular Contact Ball Bearings	Double Row Angular Contact Ball Bearings	Duplex Angular Contact Ball Bearings	Self-Aligning Ball Bearings	Cylindrical Roller Bearings	Double Row Cylindrical Roller Bearings	Cylindrical Roller Bearings	Cylindrical Roller Bearings with Angle Ring
Load Carrying Capacity — Radial Load	○	∘	⊙	⊙	⊙	○	⊙	◎	⊙	⊙
Axial Load	↔	← ∘	← ⊙	↔ ⊙	↔ ⊙	↔ ∘	×	×	←	↔
Combined Load	○	∘	⊙	⊙	⊙	∘	×	×	○	○
High Speeds	◎	⊙	◎	○	⊙	⊙	◎	⊙	⊙	⊙
High Accuracy	◎		◎		◎		◎	◎		
Low Noise and Torque	◎						⊙			
Rigidity					⊙		⊙	◎	⊙	⊙
Angular Misalignment	⊙	∘	∘	∘	∘	◎	○	⊙	○	○
Self-Aligning Capability						Yes				
Ring Separability		Yes					Yes	Yes	Yes	Yes
For Use on Fixed-end	Yes			Yes	Yes	Yes				Yes
For Use on Free-end	Yes*			Yes*	Yes*	Yes*	Yes	Yes		
Tapered Bore of Inner Ring						Yes		Yes		
Remarks		Two bearings are usually mounted in opposition.	Contact angle 15° 30° 40°. Two bearings are usually mounted in opposition. Clearance adjustment is necessary.			Any arrangement of the pair is possible.				

Legend · ◎ Excellent ⊙ Good ○ Fair ∘ Poor × Impossible
— One direction only ↔ Two directions

* Can be used as free-end bearings if tap fit allows axial motion.

CHART 20.1
Types and characteristics of rolling bearings. (*Courtesy, NSK Corp.*)

by the Anti-Friction Bearing Manufacturers Association (AFBMA) and the International Organization for Standardization (ISO).

In the most common application, a bearing is required for a shaft, which has been dimensioned on the basis of strength. Thus the starting point for bearing selection is often the bore diameter. Figure 20.15 shows, from left to right, the proportions of the extra light, light, and medium series of ball bearings, which are the most

Needle Roller Bearings	Tapered Roller Bearings	Double, Row Tapered Roller Bearings	Spherical Roller Bearings	Thrust Ball Bearings	Thrust Ball Bearings with Aligning Seats	Angular Contact Thrust Ball Bearings	Cylindrical Roller Thrust Bearings	Tapered Roller Thrust Bearings	Spherical Roller Thrust Bearings
⊙	⊙	◎	◎	×	×	×	×	×	○
×	⊙ ↔	⊙ ↔	○ ↔	⊙ ←	⊙ ←	⊙ ↔	◎ ←	◎ ←	◎ ←
×	⊙	◎	⊙	×	×	×	×	×	○
⊙	○	○	○	×	×	○	○	○	○
	⊙			⊙		◎			
⊙	⊙	◎				⊙	◎	◎	
○	○	○	◎	×	◎	×	×	×	◎
			Yes		Yes				Yes
Yes	Yes	Yes		Yes	Yes	Yes	Yes	Yes	Yes
		Yes	Yes						
Yes		Yes*	Yes*						
			Yes						
	Two bearings are usually mounted in opposition. Clearance adjustment is necessary.						Needle roller thrust bearings are available.		This type should have oil lubrication.

commonly used (L is this manufacturer's designation for the 100 series). The size is denoted by three-digit numbers. The first indicates the width series. The second and third digits multiplied by 5 give the bore size in millimeters, from size 04 up. For example, no. 214 is a light series with a bore of $5 \times 14 = 70$ mm. A further digit before the size designation indicates the type of bearing. For example, the 6 in a 6205 bearing indicates a single-row deep-groove bearing; 7306 is a single-row, angular-contact

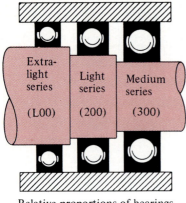
Relative proportions of bearings with same outside diameter

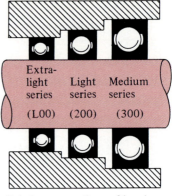
Relative proportions of bearings with same bore dimension

FIGURE 20.15
Ball-bearing size proportions. (*Courtesy, New Departure Hyatt, General Motors.*)

bearing, etc. Manufacturers frequently add other digits and letters before and after the series code to indicate various features and details, such as the type of seal or the presence of a snap ring. Bearings lighter and heavier than the three series of Figure 20.15 are also available in many sizes.

A standardized metric dimension plan has also been developed by ISO for single-row tapered roller bearings. Since it is fairly new, it is not in universal use. The bearings are grouped according to the amount of taper. Width and diameter series are designated within each group.

20.3 BEARING TESTING

When a bearing is turned under load, the balls or rollers and the raceways are subjected to repeating stresses, which, because of the small contact area, are generally quite high. Under these fatigue conditions, failure occurs with the separation of scale-like particles from the surface. This is known as *flaking* or *spalling*. A given area of spalled surface, usually 0.6 mm^2, is the criterion of bearing failure in manufacturers' tests.

If a group of seemingly identical bearings is tested under specified speed, load, and environment in the laboratory, the life distribution will take a form similar to that shown in Figure 20.16. This is not a normal distribution, but rather the Weibull discussed in Section 2.10. The rating life used by the industry is that corresponding to the point where 10 percent of the group has failed, expressed in revolutions or hours. It is given the symbol L_{10} and is called the "L-10 life." It represents, for the conditions of the tests, a 90 percent reliability. That is, the chance that a bearing under the load in question will last as long as L_{10} is 90 percent.

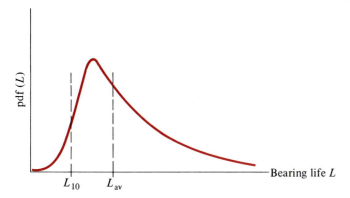

pdf (L)

Bearing life L

L_{10} L_{av}

FIGURE 20.16
Life pdf for bearings.

20.4   DYNAMIC BEARING LOAD RATINGS

In general, the life of ball bearings varies inversely with the third power of the load. In the case of roller bearings, the exponent is 10/3. Thus

$$\frac{L_1}{L_2} = \left(\frac{P_2}{P_1}\right)^s \qquad s = \begin{cases} 3 & \text{for ball bearings}[1] \\ 10/3 & \text{for roller bearings} \end{cases} \qquad (20.1)$$

To bring uniformity to bearing ratings, a *basic load rating* was established by the AFBMA. It is the constant radial load (axial load for pure-thrust bearings) which a group of identical bearings with stationary outer ring can endure for a rating life of 1 million revolutions of the inner ring. There is a certain amount of fiction in this rating, for the load to cause failure in only 1 million cycles would be very high in almost any bearing and would cause yielding of the surfaces before the bearing ever turned. It is a cataloging convenience and a calculating convenience, as we will see below, if lives are expressed in millions of revolutions.

The manufacturer converts laboratory results to the above standard by writing Equation (20.1) as follows:

$$\frac{L_{10, \text{ lab conditions}}}{1\,000\,000} = \left(\frac{C_{\text{basic load}}}{P_{10, \text{ lab conditions}}}\right)^s$$

The symbol C is used by most manufacturers for ratings, here the basic load rating, sometimes also termed the *basic dynamic capacity* or the *dynamic load rating*.

The designer's task is to select the bearing which will give the life required under the loading conditions of the problem. To find the bearing in a catalog, its basic

---

[1] Some manufacturers use 10/3 for both types.

load rating C must be known. In Equation (20.1), then,

$$L_2 = 1\,000\,000$$

$$P_2 = C \text{ (the basic load rating, to be calculated)}$$

$$L_1 = L \text{ (life required)}$$

$$P_1 = P \text{ (actual load)}$$

Substituting gives $$C = P\left(\frac{L}{1\,000\,000}\right)^{1/s} \tag{20.2}$$

If L is expressed in millions of revolutions, this simplifies to

$$C = PL^{1/s} \tag{20.3}$$

Frequently we have the life indirectly in terms of a number of hours H at a specified speed. Then $L = H \times 60 \times \text{rpm}/(1E + 06)$ in Equation (20.3), giving

$$C = P\left(\frac{H \times 60 \times \text{rpm}}{1E + 06}\right)^{1/s} \tag{20.4}$$

Some manufacturers rate their bearings at a given speed and number of hours, which work out to a number of revolutions different from the standard 1 million. Frequently 500 rpm for 3000 h is used, which comes to 90 million revolutions. In those cases, the denominator of (20.2) and of (20.4) is that figure. Thus expressions (20.3) and (20.4) become

$$C = P\left(\frac{L}{90}\right)^{1/s} \tag{20.3'}$$

and
$$\left.\begin{array}{l}\text{rated at 500}\\\text{rpm for 3000 h}\end{array}\right\}$$

$$C = P\left(\frac{H \times 60 \times \text{rpm}}{9E + 07}\right)^{1/s} \tag{20.4'}$$

Example 20.1. A ball bearing is required to resist a radial load of 6 kN with an L_{10} life (reliability of 90 percent) of 20 000 h at 1800 rpm. We need to find the basic load rating, so as to choose a bearing from a catalog.

These numbers are entered in expression (20.4) with $s = 3$ for a ball bearing, to give

$$C = 6\left[\frac{(2E + 04)(60)(1800)}{1E + 06}\right]^{1/3} = 77.6 \text{ kN}$$

Life and Load Rating Factors

A reliability of more than 90 percent is required in most applications, and other conditions may be different from those for which the bearing ratings are listed. Manufacturers account for these by modifying the life from expression (20.3) with "adjustment factors":

$$L = a_1 a_2 a_3 \left(\frac{C}{P}\right)^s \qquad L \text{ in } 10^6 \text{ r} \tag{20.5}$$

The factor a_1 accounts for reliability and could be calculated from the Weibull distribution, but the approximation

$$a_1 = 4.48 \left\{ \ln \frac{100}{R} \right\}^{2/3}$$

is usually used, where R is the desired reliability in percent. For commonly required reliabilities, the values of a_1 per the above formula are shown in Table 20.1.

The factor a_2 is an adjustment for bearing material. The steel regularly used in bearing manufacture is now superior to the standard on which a_2 was based, and the factor is absorbed by many makers in their published ratings, since it makes their bearings look better competitively. Table 20.2 displays typical ratings, which incorporate the a_2 factor, for single-row deep-groove ball bearings of the 100, 200, and 300 series.

Environmental conditions different from those of the manufacturer's tests (which are optimum) are accounted for in a_3. The conditions resulting in low values of a_3 are inadequate lubrication (including contamination, especially water), slow rotation, high temperature, and misalignment. Of these, lubrication is usually the most important. Since the value of a_2 is known to depend somewhat on operating conditions as well, the two factors are often combined and labeled a_{23}. Appropriate values of a_{23} derive mainly from experience. Manufacturers' catalogs and/or their applications engineers should be consulted in the absence of in-house experience.

Other factors may be included. For example, one manufacturer notes that bearings can be used for a certain period beyond the point where the bearing is taken to have failed (appearance of 0.6-mm^2 spalled area) and accounts for it with a further adjustment factor.

An application factor f_{app} may also be applied to the load to account for such unquantified effects as variation in speed, shock and vibration, environment, etc. Values of the factor derive from experience. In the absence of experience, some guidance may be found in Table 20.3

Thus the life expression becomes

$$L = a_1 a_2 a_3 \left(\frac{C}{P f_{app}} \right)^s \qquad L \text{ in } 10^6 \text{ r} \qquad (20.6)$$

TABLE 20.1
Adjustment factor for reliability a_1

Reliability R, %	L_n	Adjustment factor a_1
90	L_{10}	1
95	L_5	0.62
96	L_4	0.53
97	L_3	0.44
98	L_2	0.33
99	L_1	0.21

TABLE 20.2
Typical ratings for 100, 200, and 300 series deep-groove (Conrad) ball bearings*

Bearing number	Bore in	Bore mm	Outside diameter in	Outside diameter mm	Width in	Width mm	Dynamic load rating lb	Dynamic load rating N	Static load rating lb	Static load rating N
9100[†]	0.3937	10	1.0236	26	0.3150	8	1 140	5 100	440	1 960
200	0.3937	10	1.1811	30	0.3543	9	1 500	6 550	600	2 650
300	0.3937	10	1.3780	35	0.433	11	2 000	9 000	850	3 750
9101	0.4724	12	1.1024	28	0.3150	8	1 270	5 600	500	2 240
201	0.4724	12	1.2598	32	0.3937	10	1 700	7 500	680	3 000
301	0.4724	12	1.4567	37	0.472	12	2 080	9 150	850	3 750
9102	0.5906	15	1.2598	32	0.3543	9	1 400	6 200	560	2 500
202	0.5906	15	1.3780	35	0.4331	11	1 930	8 650	780	3 450
302	0.5906	15	1.6535	42	0.512	13	2 900	13 200	1270	5 600
9103	0.6693	17	1.3780	35	0.3937	10	1 500	6 700	630	2 800
203	0.6693	17	1.5748	40	0.4724	12	2 360	10 600	1 000	4 400
303	0.6693	17	1.8504	47	0.551	14	3 350	15 000	1 460	6 550
9104	0.7874	20	1.6535	42	0.4724	12	2 320	10 400	1 000	4 400
204	0.7874	20	1.8504	47	0.5512	14	3 200	14 300	1 400	6 200
304	0.7874	20	2.0472	52	0.591	15	4 000	17 600	1 760	7 800
9105	0.9843	25	1.8504	47	0.4724	12	2 500	11 000	1 120	5 000
205	0.9843	25	2.0472	52	0.5906	15	3 450	15 600	1 560	6 950
305	0.9843	25	2.4409	62	0.669	17	5 850	26 000	2 750	12 200
9106	1.1811	30	2.1654	55	0.5118	13	3 350	14 600	1 560	6 950
206	1.1811	30	2.4409	62	0.6299	16	4 800	21 600	2 280	10 000
306	1.1811	30	2.8346	72	0.748	19	7 500	33 500	3 550	15 600
9107	1.3780	35	2.4409	62	0.5512	14	4 000	17 600	1 900	8 500
207	1.3780	35	2.8346	72	0.6693	17	6 400	28 500	3 050	13 700
307	1.3780	35	3.1496	80	0.827	21	9 150	40 500	4 500	20 000
9108	1.5748	40	2.6772	68	0.5906	15	4 400	19 600	2 280	10 000
208	1.5748	40	3.1496	80	0.7087	18	8 150	36 000	4 000	17 600
308	1.5748	40	3.5433	90	0.906	23	11 000	49 000	5 600	24 500
9109	1.7717	45	2.9528	75	0.6299	16	5 200	23 200	2 750	12 200
209	1.7717	45	3.3465	85	0.7480	19	8 150	36 000	4 000	17 600
309	1.7717	45	3.9370	100	0.984	25	13 200	58 500	6 700	30 000
9110	1.9685	50	3.1496	80	0.6299	16	5 400	24 000	2 900	13 200
210	1.9685	50	3.5433	90	0.7874	20	8 800	39 000	4 500	19 600
310	1.9685	50	4.3307	110	1.063	27	15 300	68 000	8 000	35 500
9111	2.1654	55	3.5433	90	0.7087	18	6 950	31 000	3 800	17 000
211	2.1654	55	3.9370	100	0.8268	21	10 800	48 000	5 600	25 000
311	2.1654	55	4.7244	120	1.142	29	18 000	80 000	9 500	41 500
9112	2.3622	60	3.7402	95	0.7087	18	7 350	32 500	4 150	18 300
212	2.3622	60	4.3307	110	0.8661	22	12 900	58 500	6 950	31 000
312	2.3622	60	5.1181	130	1.220	31	20 400	90 000	10 800	48 000
9113	2.5591	65	3.9370	100	0.7087	18	7 650	33 500	4 400	19 600
213	2.5591	65	4.7244	120	0.9055	23	14 300	63 000	7 800	34 000
313	2.5591	65	5.5118	140	1.299	33	23 200	102 000	12 500	56 000
9114	2.7559	70	4.3307	110	0.7874	20	9 500	42 500	5 500	24 500
214	2.7559	70	4.9213	125	0.9449	24	15 600	69 500	8 500	37 500
314	2.7559	70	5.9055	150	1.378	35	26 000	116 000	14 300	63 000

* The a_2 life adjustment factor is incorporated in these figures.

[†] This manufacturer designates the 100 series as the 9100 series.

Courtesy, Fafnir Bearings Division, The Torrington Co.

TABLE 20.3
Guide for application factor f_{app}

Operating conditions	Applications	f_{app}
Smooth operation free from shock	Electric motors, machine tools, air conditioners	1–1.2
Normal operation	Air blowers, compressors, elevators, cranes, paper-making machines	1.2–1.5
Operation accompanied by shock and vibration	Construction machines, crushers, vibration screens, rolling mills	1.5–3.0

Courtesy NSK Corp.

and the basic (equivalent) load rating is

$$C = f_{app}P\left(\frac{L}{a_1 a_2 a_3}\right)^{1/s} \qquad L \text{ in } 10^6 \text{ r} \tag{20.7}$$

If the manufacturer rates bearings at 500 rpm for 3000 h, add 90 to the string of a factors, per Equation (20.3′).

Axial and Radial Loading

Many applications involve combinations of axial and radial loading. It is then necessary to compute an equivalent dynamic load. This is that radial load which would result in the same life as the actual loading. It is computed as the sum of part of the radial load and part of the axial load:

$$P = XF_r + YF_a \tag{20.8}$$

The X and Y factors vary with the type of bearing and can be found in manufacturers' catalogs. In the case of single- and double-row deep-groove ball bearings, the factors depend on the ratio of the axial load F_a to the static load rating C_o (discussed in the section immediately following). The amount of clearance between the balls and the raceways also has an influence. Table 20.4 displays values of X and Y for bearings

TABLE 20.4
Equivalent dynamic bearing load factors for deep-groove bearings with normal fit and clearance

F_a/C_o	e	$F_a/F_r \le e$		$F_a/F_r > e$	
		X	Y	X	Y
0.025	0.22	1	0	0.56	2
0.04	0.24	1	0	0.56	1.8
0.07	0.27	1	0	0.56	1.6
0.13	0.31	1	0	0.56	1.4
0.25	0.37	1	0	0.56	1.2
0.5	0.44	1	0	0.56	1

mounted with normal fits. The axial load has no effect on life below a certain ratio of F_a/F_r (called e in the table), as seen in the X and Y values in the second and third columns.

20.5 STATIC LOAD RATINGS

In some applications, the shaft motion is intermittent or oscillatory, and there are cases where a bearing may have a large load on it, even though the machine is shut down. A gear shaft in an automobile transmission when the car is left in gear on a hill is an example of the latter. When, in such instances, there is no motion of the bearings, the stresses between balls and raceway are static and may become quite high.[1] Permanent deformation, called *brinelling* because of its likeness to the indentation produced in a Brinell hardness test, is a possible result, which may adversely affect bearing performance and life. The basic static load rating was established by the AFBMA as that radial load which would cause a total permanent deformation at the rolling-element/race contact of 0.0001 of the diameter of the ball or roller.

Values of static load capacity are tabulated for ball bearings along with the dynamic load capacity. With roller bearings, static effects are not as acute, and the static load capacity is not generally listed in the catalogs. A method of calculating it is usually provided, however.

For bearings loaded radially, the static load is simply the force on the bearing. When an axial load is also present, an equivalent static load must be computed from expression (20.8). For deep-groove bearings $X = 0.6$ and $Y = 0.5$; the figures for other bearings can be found in catalogs. If the equivalent static load thus computed is less than the radial load, the radial load should be used.

20.6 LIFE UNDER VARYING LOADS AND SPEEDS

In many applications the loads and/or speeds are not constant. While bearing selection can be based on the maximum load, it is more meaningful to use a weighted life, if the fraction of time at the various load conditions can be estimated. The calculation then follows Miner's rule, discussed in Section 9.4. Equation (9.15) as it applies here is

$$\frac{\text{Revs}_1}{L_1} + \frac{\text{Revs}_2}{L_2} + \cdots = 1$$

Here "Revs$_1$" means the revolutions under the speed and load conditions of segment 1 of the work cycle, etc, and L_1 is the life under those conditions. If the revolutions are expressed as $FR(L)$, where FR is the fraction of the total life (in revolutions)

---

[1] Some years ago this problem became quite severe with the axle bearings of automobiles transported on railway cars. The weight of the car on the bearings was exacerbated by the fatigue loading produced by the vibration during transport.

turned in each segment, then we have

$$\frac{FR_1(L)}{L_1} + \frac{FR_2(L)}{L_2} + \cdots = 1$$

Substituting expression (20.6) for L_1, L_2, etc., we get

$$\frac{1}{a_1 a_2 a_3} \left(\frac{f_{app}}{C}\right)^s (FR_1 P_1^s + FR_2 P_2^s + \cdots) = \frac{1}{L}$$

which can be solved to estimate the life of a chosen bearing

$$L = a_1 a_2 a_3 \left(\frac{C}{f_{app}}\right)^2 \frac{1}{FR_1 P_1^s + FR_2 P_2^s + \cdots} \qquad (20.9)$$

or to find the basic load rating for a required bearing life

$$C = f_{app} \left[(FR_1 P_1^s + FR_2 P_2^s + \cdots) \frac{L}{a_1 a_2 a_3} \right]^{1/s} \qquad (20.10)$$

Here L is in millions of revolutions. As before, if the bearing is rated at 500 rpm for 3000 h, add 90 to the string of a factors.

When the speeds of the several segments of a bearing history are different, the segments are usually expressed in fractions of clock time rather than as fractions of revolutions turned. It is then necessary to compute the values of FR, as in the following example.

Example 20.2. We wish to know the L_{10} life of a 205 bearing used in an application with the following radial load cycle;

$$P_1 = 1500 \text{ lb at } 1800 \text{ rpm for 20 percent of time}$$

$$P_2 = 1000 \text{ lb at } 1200 \text{ rpm for 35 percent of time}$$

$$P_3 = 800 \text{ lb at } 800 \text{ rpm for 45 percent of time}$$

The computation of each FR is done most easily on the basis of one minute's running:

Interval no.	Time, min	rpm	r	FR
1	0.20	1800	360	360/1140
2	0.35	1200	420	420/1140
3	0.45	800	360	360/1140
			$\overline{1140}$ r in 1 min = average rpm	

This bearing has a dynamic load rating $C = 3450$ lb, so then

$$L_{10} = 3450^3 \frac{1140}{360(1500^3) + 420(1000^3) + 360(800^3)} = 25.73E + 06 \text{ r}$$

Example 20.3. In Example 20.2, suppose the required life is 1000 h with a reliability of 95 percent. The assignment is to choose a bearing.

Since the average speed of the bearing is 1140 rpm the required life is

$$L = 1140 \times 60 \times 1000 = 68.4 \text{ million r}$$

In Table 20.1 we find the reliability correction factor $a_1 = 0.618$. Filling in the numbers in (20.10). we have

$$C = \left\{ [360(1500^3) + 420(1000^3) + 360(800^3)] \frac{68.4}{0.618(1140)} \right\}^{1/3} = 5611 \text{ lb}$$

Using the same bore diameter as previously, 25 mm, we find that this application would require a 305 bearing, whose dynamic load rating is 5850 lb. The bearing is larger than in the earlier calculation, since the required life is some 4 times greater.

20.7 THRUST CONSIDERATION, TAPERED ROLLER BEARINGS

Tapered roller bearings are intended to take considerable thrust. The following method of taking account of the effect of thrust includes the essentials of the Timken calculation. It follows the AFBMA procedure, as do the methods of other manufacturers, detailed in their catalogs. Calculations similar to these are also specified by some manufacturers for pairs of angular-contact ball bearings.

Tapered roller bearings most often are mounted in pairs. Figure 20.17 illustrates an indirect or back-to-back mounting above the shaft centerline and a direct or face-to-face mounting below it. The line of action of the force between the roller and cup

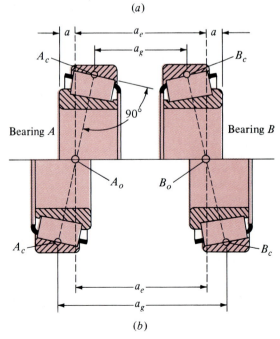

FIGURE 20.17
Tapered roller-bearing geometry. (*Courtesy, The Timken Co.*) (*a*) Indirect or back-to-back mounting. (*b*) Direct or face-to-face mounting.

of bearing A is at the center of the roller-cup contact and perpendicular to the raceway surface, as shown. It intersects the shaft centerline at A_o. The intersection of bearing B is at B_o. The bearing reactions on the shaft should therefore be taken as acting at A_o and B_o. The capability of a bearing pair to resist a moment on the shaft depends directly on the distance between those two points, termed the *effective spread* a_e. (The distance a in the figure is listed among the catalog data.) Because indirect mounting requires the least amount of space for the same spread, it is more commonly used.

Because of the taper in the design of the bearing, a radial load will induce a thrust reaction, and this must be opposed so that the bearing will not separate. The ratio of the radial load to the thrust load and the degree of taper of the bearing determine the number of rollers in contact. It is customary in calculating loads and ratings to consider that one-half of the rollers are in contact, i.e., the load is spread over 180° of the bearing. The induced bearing thrust then works out to be

$$F_a = \frac{0.47F_r}{K}$$

The subscript a means "axial" (thrust) and r, "radial." K is the ratio of the radial rating of the bearing to the thrust rating; it is listed with the bearing data in the catalog.

The basic dynamic radial load rating C of a tapered roller bearing is the radial load-carrying capacity with one-half the rollers in contact. When the thrust load on a bearing exceeds the induced thrust, a dynamic equivalent radial load must be used to calculate bearing life. This is that radial load which, if applied to a bearing, will result in the same life as the bearing will attain under the actual loading conditions. In Figure 20.18 indirect and direct mounting arrangements are shown where an axial (thrust) force is present as well as radial forces. In both arrangements bearing A provides the reaction to the axial force. The total thrust seen by bearing A is the axial force F_a plus the induced axial thrust of bearing B:

$$\text{Thrust}_A = F_{aA} = \frac{0.47F_{rB}}{K_B} + F_a \tag{20.11}$$

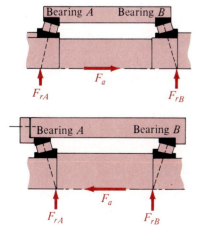

FIGURE 20.18
Tapered roller-bearing loading, including an axial force. (*Courtesy, The Timken Co.*)

The thrust on bearing B is the induced axial thrust of bearing A less the axial force:

$$\text{Thrust}_B = F_{aB} = \frac{0.47F_{rA}}{K_A} - F_a \tag{20.12}$$

The dynamic equivalent radial loads for the bearings are given by expressions paralleling (20.8):

$$P_A = 0.4F_{rA} + K_A F_{aA} \qquad P_B = 0.4F_{rB} + K_B F_{aB} \tag{20.13}$$

If either of these computed equivalent radial loads is less than the actual radial load on the bearing, then the actual radial load is used. In making the calculations, values for K_A and K_B must be assumed and then checked after the bearing is chosen.

Example 20.4. Figure 20.19 shows a helical gear shaft with direct-mounted tapered bearings. The speed of the shaft is 1100 rpm. The design life $L_{10} = 5000$ h (90 percent reliability), and the desired shaft diameter $d = 30$ to 32 mm. The assignment is to select bearings for this application. The various adjustment factors a_1, a_2, etc., and the application factor f_{app} we will assume to be 1.0.

The bearing reactions intersect the shaft centerline at points A_o and B_o, as in Figure 20.17. To find the radial forces on the bearings at those points, we use the equations of static equilibrium. A coordinate system is shown at the left side of the sketch in Figure 20.19. Summing moments about A_o and taking counterclockwise as plus, we have

$$M_z = 0 = F_{aG}\frac{D_{pG}}{2} - F_{rG}(105) - F_{yB}(155)$$

$$= 3720(100) - 3550(105) - 155F_{yB}$$

whence

$$F_{yB} = -5 \text{ N}$$

$$M_y = 0 = F_{tG}(105) + F_{zB}(155)$$

$$= 8360(105) + 155F_{zB}$$

whence

$$F_{zB} = -5663 \text{ N}$$

Moments of forces about the x (axial) axis do not concern us, because they result in no forces on the bearings—they have to do with the torque transmitted by the shaft.

The two forces F_{yB} and F_{zB} are in the plane of bearing B; their resultant is equal to the radial force on that bearing:

$$F_{rB} = \sqrt{5^2 + 5663^2} = 5663 \text{ N}$$

Its direction does not concern us. The radial load on bearing A is obtained most easily by summing forces (it can also be found by taking moments about point B):

$$\sum F_y = 0 = F_{yA} + F_{yB} + F_{rG}$$

whence

$$F_{yA} = -3550 + 5 = -3545 \text{ N}$$

$$\sum F_z = 0 = F_{tG} + F_{zB} + F_{zA}$$

whence

$$F_{zA} = 5663 - 8360 = -2697 \text{ N}$$

The radial force on bearing A is then

$$F_{rA} = \sqrt{3545^2 + 2697^2} = 4454 \text{ N}$$

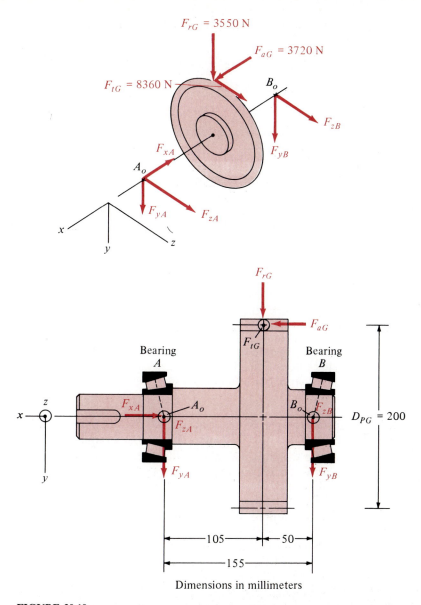

FIGURE 20.19
Helical gear shaft for Example 20.4. (*Courtesy, The Timken Co.*)

In addition to the above, the axial (thrust) force on the bearing pair is the axial force on the gear: $F_a = F_{aG} = 3550$ N, directed to the left. Hence bearing A takes the thrust. Values of $K_A = K_B = 1.5$ are reasonable as a start. The thrust on the two bearings is then given by (20.11) and (20.12):

$$F_{aA} = \frac{0.47(5663)}{1.5} + 3550 = 5324 \text{ N}$$

and
$$F_{aB} = \frac{0.47(4454)}{1.5} - 3550 = -2154 \text{ N}$$

The dynamic equivalent radial loads are found with expressions (20.13):

$$P_A = 0.4(4462) + 1.5(5324) = 9771 \text{ N}$$

$$P_B = 0.4(5663) + (1.5)(-2154) = -966 \text{ N}$$

The dynamic equivalent load for bearing B being less than the actual radial load $F_{rB} = 5663$, we use

$$P_B = 5663 \text{ N}$$

We must now calculate the dynamic radial rating C for the bearings. The L_{10} life is calculated by this manufacturer on the basis of 90 million r (3000 h at 500 rpm), so we use expression (20.4′):

$$C_A = 9771\left[\frac{5000(60)(1100)}{9\text{E} + 07}\right]^{3/10} = 9771(1.477) = 14\,428 \text{ N}$$

$$C_B = 5663(1.477) = 8362 \text{ N}$$

Figure 20.20 shows the page from the Timken catalog listing bearings of the required bore (30 to 32 mm). For bearing A several choices would be satisfactory. The seventh bearing on the page, as well as the fourth from the bottom, has $C = 14\,500$. For both of these, the radial/thrust ratio $K = 1.07$, which is less than the assumed value of 1.5. A recalculation of the equivalent rating would not be necessary then. Of course, price and availability would be determining factors in the selection.

For bearing B, the penultimate bearing in Figure 20.20, having a load rating $C = 9460$ lb, satisfies the requirement.

The catalog page used in Example 20.4 is from a U.S. bearing manufacturer, hence most of the basic dimensions are in USCS units, though the SI equivalents are listed. Thus the seventh bearing on the page, a candidate for A, has a bore of $1.1875 = 1\frac{3}{16}$ in. Some entries are standard SI sizes, e.g., the second, third, and fourth, which are indicated by shading.

Example 20.5. In Figure 20.21 a bearing pair is shown which is subjected to thrust load only. Clearly, bearing A supports the thrust and is thus a tapered bearing. Bearing B sees no thrust, but is nonetheless a tapered bearing, for positioning purposes. The speed is 1000 rpm. Design life is to be $L_{10} = 15\,000$ h, and the desired shaft diameter $d \simeq 30$ mm. Application and other factors will be taken as 1.

The radial loads on both bearings are nil, hence we do not have combined loading to deal with. The load rating for bearing A is obtained directly with Equation (20.4′), with $P_A = 5700$ N:

$$C_A = 5700\left[\frac{(15\,000)(60)(1000)}{9\text{E} + 07}\right]^{3/10} = 11\,373 \text{ N} \qquad \text{thrust rating}$$

In Figure 20.20, the second bearing in the metric group (third on the page) has a thrust rating of 11 300 N, and it would be a good choice. If an extra margin of safety is desired, there are several others which more than meet the requirement.

Bearing B presumably has no load, radial or axial, and is thus selected on the basis of economy. It must, however, be a reasonable size. The first bearing in the SI group

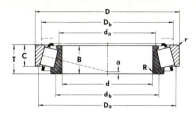

			Cone		Cup

Bore d	Outside diameter D	Width T	One-row radial N lbf	Thrust N lbf	Factor K	Eff. load center a^2	Cone	Cup	Max. shaft fillet radius R^1	Width B	d_b	d_a	Max. housing fillet radius r^1	Width C	D_b	D_a
			Rating at 500 rpm for 3000 hours L_{10}				**Part numbers**		**Max. shaft fillet**		**Backing shoulder diameters**		**Max. housing fillet**		**Backing shoulder diameters**	
30.000	69.012	19.845	13100	8570	1.53	−4.3	14117A	14276	3.5	19.583	42.5	39.5	1.3	15.875	60.0	63.0
1.1811	2.7170	0.7813	2950	1930		−0.17			0.14	0.7710	1.67	1.56	0.05	0.6250	2.36	2.48
30.000	72.000	20.750	15400	8320	1.85	−5.8	◆30306	◆30306	1.5	19.000	38.0	35.5	1.5	16.000	64.0	66.0
1.1811	2.8346	0.8169	3470	1870		−0.23			0.06	0.7480	1.50	1.40	0.06	0.6299	2.52	2.60
30.000	72.000	28.750	21000	11300	1.85	−10.7	◆32306	◆32306	1.5	27.000	40.5	37.0	1.5	23.000	62.0	66.0
1.1811	2.8346	1.1319	4730	2550		−0.42			0.06	1.0630	1.59	1.46	0.06	0.9055	2.44	2.60
*30.000	*72.000	29.370	20900	19500	1.07	−5.6	JHM88540	JHM88513	1.3	27.783	44.5	42.5	3.3	23.020	58.0	69.0
*1.1811	*2.8346	1.1563	4700	4390		−0.22			0.05	1.0938	1.75	1.67	0.13	0.9063	2.28	2.72
30.112	62.000	19.050	12100	7280	1.67	−5.8	15116	15245	0.8	20.638	36.0	35.5	1.3	14.288	55.0	58.0
1.1855	2.4409	0.7500	2730	1640		−0.23			0.03	0.8125	1.42	1.40	0.05	0.5625	2.17	2.28
30.162	62.000	16.002	10400	6800	1.53	−3.6	17119	17244	1.5	16.566	37.0	34.5	1.5	14.288	54.0	57.0
1.1875	2.4409	0.6300	2330	1530		−0.14			0.06	0.6522	1.46	1.36	0.06	0.5625	2.13	2.24
30.162	64.292	21.433	14500	13500	1.07	−3.3	M86649	M86610	1.5	21.433	41.0	38.0	1.5	16.670	54.0	61.0
1.1875	2.5312	0.8438	3250	3040		−0.13			0.06	0.8438	1.61	1.50	0.06	0.6563	2.13	2.40
30.162	69.850	23.812	20100	9410	2.14	−8.6	2558	2523	2.3	25.357	40.0	36.5	1.3	19.050	61.0	64.0
1.1875	2.7500	0.9375	4520	2120		−0.34			0.09	0.9983	1.57	1.44	0.05	0.7500	2.40	2.52
30.162	72.626	30.162	22700	13000	1.76	−10.2	3187	3120	0.8	29.997	39.0	38.5	3.3	23.812	61.0	67.0
1.1875	2.8593	1.1875	5110	2910		−0.40			0.03	1.1810	1.54	1.52	0.13	0.9375	2.40	2.64
30.213	62.000	19.050	12100	7280	1.67	−5.8	15118	15245	3.5	20.638	41.5	35.5	1.3	14.288	55.0	58.0
1.1895	2.4409	0.7500	2730	1640		−0.23			0.14	0.8125	1.63	1.40	0.05	0.5625	2.17	2.28
30.213	62.000	19.050	12100	7280	1.67	−5.8	15119	15245	1.5	20.638	37.5	35.5	1.3	14.288	55.0	58.0
1.1895	2.4409	0.7500	2730	1640		−0.23			0.06	0.8125	1.48	1.40	0.05	0.5625	2.17	2.28
30.213	62.000	19.050	12100	7280	1.67	−5.8	15120	15245	0.8	20.638	36.0	35.5	1.3	14.288	55.0	58.0
1.1895	2.4409	0.7500	2730	1640		−0.23			0.03	0.8125	1.42	1.40	0.05	0.5625	2.17	2.28
30.226	69.012	19.845	13100	8570	1.53	−4.3	14116	14274	0.8	19.583	37.0	36.5	3.3	15.875	59.0	63.0
1.1900	2.7170	0.7813	2950	1930		−0.17			0.03	0.7710	1.46	1.44	0.13	0.6250	2.32	2.48
30.226	69.012	19.845	13100	8570	1.53	−4.3	14116	14276	0.8	19.583	37.0	36.5	1.3	15.875	60.0	63.0
1.1900	2.7170	0.7813	2950	1930		−0.17			0.03	0.7710	1.46	1.44	0.05	0.6250	2.36	2.48
30.226	72.085	22.385	13100	8570	1.53	−4.3	14116	14283	0.8	19.583	37.0	36.5	2.3	18.415	60.0	65.0
1.1900	2.8380	0.8813	2950	1930		−0.17			0.03	0.7710	1.46	1.44	0.09	0.7250	2.36	2.56
30.955	64.292	21.433	14500	13500	1.07	−3.3	M86648A	M86610	1.5	21.433	42.0	38.0	1.5	16.670	54.0	61.0
1.2187	2.5312	0.8438	3250	3040		−0.13			0.06	0.8438	1.65	1.50	0.06	0.6563	2.13	2.40
31.750	58.738	14.684	7610	6170	1.23	−1.3	08125	08231	1.0	15.080	37.5	36.0	1.0	10.716	52.0	55.0
1.2500	2.3125	0.5781	1710	1390		−0.05			0.04	0.5937	1.48	1.42	0.04	0.4219	2.05	2.17
31.750	59.131	15.875	9460	6680	1.42	−3.0	LM67048	LM67010	SPCL	16.764	42.5	36.0	1.3	11.811	52.0	56.0
1.2500	2.3280	0.6250	2130	1500		−0.12				0.6600	1.67	1.42	0.05	0.4650	2.05	2.20
31.750	62.000	18.161	12100	7280	1.67	−4.8	15123	15245	SPCL	19.050	42.5	36.5	1.3	14.288	55.0	58.0
1.2500	2.4409	0.7150	2730	1640		−0.19				0.7500	1.67	1.44	0.05	0.5625	2.17	2.28

[1] These maximum fillet radii will be cleared by the bearing corners.
[2] Minus value indicates center is inside cone backface.
† For standard class **only**, the maximum metric size is a whole millimetre value.
* For "J" part tolerances—see metric tolerances, page 73, and fitting practice, page 65.
◆ ISO cone and cup combinations are designated with a common part number and should be purchased as an assembly. For ISO bearing tolerances—see metric tolerances, page 73, and fitting practice, page 65.

FIGURE 20.20
Timken catalog page for Example 20.4. (*Courtesy, The Timken Co.*)

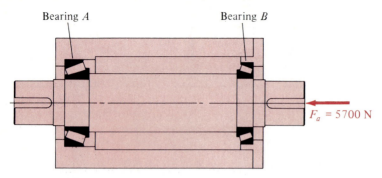

FIGURE 20.21
Bearing pair for Example 20.5. (*Courtesy, The Timken Co.*)

with the same bore as bearing A, an OD $= 72$ mm, and a thrust rating of 8320 N would suit.

20.8   PRELOAD

The thrust calculation of paired tapered roller bearings above assumed that the bearings were just drawn up, i.e., no thrust load was put on the bearings at bolting up. There are cases where it is quite important that there be no axial or radial play of the rotating part, as in precision machine tools. Bearings in such installations are preloaded. Preloading places an additional load on bearings and hence lowers the life. The bearing manufacturer should be consulted before designing in a preload.

Ball bearings may be purchased in matched pairs wherein a predetermined preload is established by a small gap between the inner or the outer rings of the pair. Figure 20.22 shows the arrangement with a pair of angular-contact ball bearings.

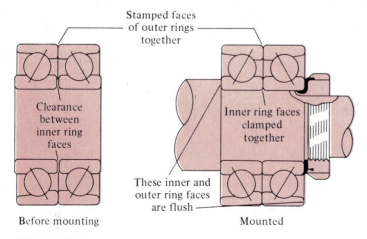

FIGURE 20.22
Matched pairs of ball bearings may be obtained to achieve a predetermined preload. (*Courtesy, Fafnir Bearings Division, The Torrington Co.*)

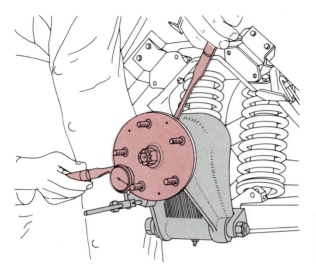

FIGURE 20.23
Checking rear-wheel hub end float of independent rear-suspension car. (*Courtesy, Jaguar Cars, Ltd.*)

When the mounting nut is drawn up tight, the bearings are preloaded. The gap is very small, so the sides to be placed together are identified by a stamp.

Many, if not most, bearing installations are subjected to a temperature rise. If the expansion of the parts is not exactly the same, an additional thrust load, a "preload," can be placed on the bearings. The engineer must be aware of this problem and account for it in design and bearing selection. Figure 20.23 shows the rear-wheel hub of an automobile with independent rear suspension. The housing for the bearings is aluminum; the wheel axle itself, steel. Aluminum has a greater coefficient of thermal expansion than steel, and with the back-to-back mounting, increased temperature results in less clearance between bearings. Thus, when the axle is assembled, 0.004 in of axial play or end float is specified to allow for differential expansion when the temperature rises as the car is driven. The figure shows a mechanic measuring the end float.

20.9 BEARING MOUNTING

There are many variations of bearing-mounting arrangements, and nearly every application presents some new challenge. The following are a few general principles.

It is customary to press-fit the rotating ring of a bearing. The tolerances are critical, because the slight change in size of the ring will reduce the small clearance between the balls or rollers and the raceway. That clearance is such that a press fit on both rings cannot be used. Hence the stationary ring is tap-fitted, a snug, sliding fit such that the member can be installed by light tapping with a hammer. The tap fit is often used to allow for axial motion of the ring to accommodate thermal expansion. Permitted interferences for press and tap fits are listed in manufacturers' catalogs. The most common bearing application is that of a rotating shaft, in which case the inner ring is press-fitted onto the shaft. Often the press fit is considered sufficient to hold the ring in position; otherwise, it may be secured with a nut.

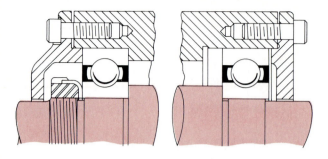

FIGURE 20.24
Fixed (left) and floating (right) bearing mountings. (*Courtesy, New Departure Hyatt, General Motors.*)

Two other aspects of the selection and arrangement of bearings are very important. First, the design should minimize manufacturing costs. Thus economy can be realized at times in choosing bearings of the same size when it facilitates a single boring operation of the housing. Second, the arrangement must permit disassembly for repair. For example, the outer ring of a roller bearing should not be placed in a blind hole without provision for its removal.

Figure 20.24 shows bearings in fixed and floating mountings. On the left, the bearing is secured to the shaft by a nut with locking ring and in the housing by the end cap. The small clearance between the cap and the housing face ensures a positive clamp. This fixed bearing mounting positions the shaft with respect to the housing. The bearing at the right is free to float axially, and it can thus accommodate thermal expansion and accumulated manufacturing tolerances of the shaft or housing.

The transmission shown in Figure 20.25 utilizes two single-row bearings on the sleeve riding the central shaft at the right. The bearings have the same outside diameter, so that the housing can be through-bored. The bearing bores, however, are different,

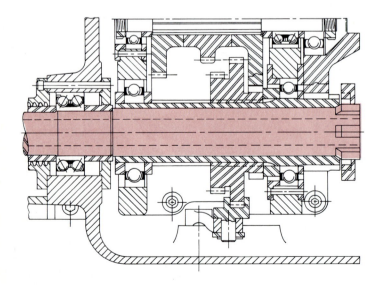

FIGURE 20.25
Transmission-bearing mountings. (*Courtesy, New Departure Hyatt, General Motors.*)

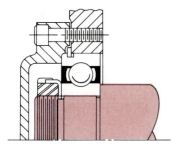

FIGURE 20.26
Snap ring is used to position bearing. (*Courtesy, New Departure Hyatt, General Motors.*)

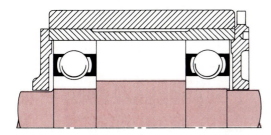

FIGURE 20.27
If end play is not critical, both bearings may be float-mounted. (*Courtesy, New Departure Hyatt, General Motors.*)

to permit easy mounting on the stepped sleeve. The bearing at the right is mounted in a fixed position, to locate the sleeve and to resist thrust from either direction. The bearing at the left is float-mounted in the housing. Note the angular-contact bearings, which provide rigidity of relatively short rotating parts.

A machined shoulder for fixing a bearing in a through-bored housing is not necessary if a snap-ring bearing is used, as shown in Figure 20.26. Here again, enough clearance must be provided between the cap and the housing face to ensure that the snap ring is clamped in place. This installation is suitable if thrust loads, especially to the right, are light.

If shaft end play is not critical, both bearings may be float-mounted, as shown in Figure 20.27. Axial movement is limited by the caps.

The angular rigidity of a shaft can be enhanced by increasing the effective spread between the bearings. In Figure 20.28, which shows a precision spindle, the two angular-contact bearings, mounted back to back, are separated by spacers, thus increasing the resistance of the shaft to applied moments. The spacers must have exactly equal lengths so as to maintain the predetermined preload on the matched pair of bearings.

20.10 LUBRICATION

The majority of small ball bearings are integrally sealed and lubricated for life. Common examples are the bearings in street-vehicle alternators and in small electric motors. Frequently bearings which turn relatively slowly are grease-lubricated, as in automobile wheels. Bearings which are internal to machinery are lubricated by a system which must provide a constant supply of lubricant, usually oil. Such systems include splash, bath, wick-feed, drip-feed, air-oil mist, and oil jets. The choice depends

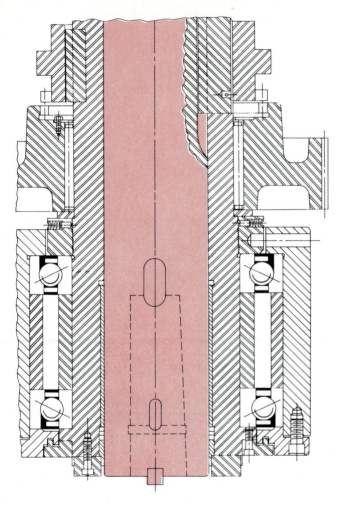

FIGURE 20.28
Angular rigidity of a shaft can be improved by increasing the effective spread of the bearings. (*Courtesy, New Departure Hyatt, General Motors.*)

on several factors, including, of course, economics and space. High speed applications generally require air-oil mist. Oil-jet lubrication is sometimes needed to remove heat coming to the bearing from elsewhere in the machinery. Recommendations for lubrication of bearings are found in manufacturers' catalogs.

20.11 CLOSURE

This chapter has provided information on the rating, selection, and application of bearings. The choice of bearings must be made from manufacturers' catalogs. Most suppliers have applications engineers to assist designers in working out particular problems, and you should not hesitate to use them.

PROBLEMS

In the problem sets below, assume factors other than those specifically mentioned to be unity.

Section 20.4

20.1. How great an increase in the radial load on a ball bearing will cause the life to be halved? Is your answer valid for all loads and lifetimes?

   Answer: 26 percent

20.2. The sketch shows a chain drive. Choose suitable ball bearings from Table 20.2 if a life of 5000 h at 400 rpm is required with a reliability of 95 percent. Desired shaft size is $1\frac{1}{2}$ to 2 in.

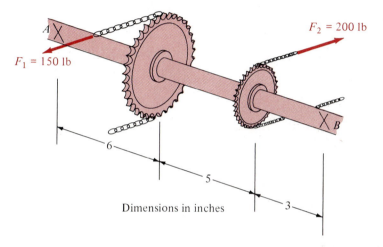

FIGURE P20.2

20.3. Figure P20.3 is a schematic of a vertically mounted motor armature. The bearings are no. 203. The sum of the belt tensions is a steady 150 lb at a speed of 945 rpm. For 90 percent reliability, estimate the time in hours to the first bearing failure.

   Answer: 46 362

20.4. Calculate the life expectancy in hours of a 208 deep-groove ball bearing which turns at 1050 rpm under a radial load of 3000 N. Operation is "normal," and reliability is to be 97 percent.

20.5. Repeat Problem 20.4, but the bearing is subjected, in addition, to an axial force of 1600 N (95 percent reliability).

   Answer: 1946

20.6. Select a bearing from Table 20.2 which will give 30 000 h of service at 900 rpm with 98 percent reliability under a radial load of 600 lb. The application factor is 1.4. Preferred ID is 1.75 in or greater.

20.7. Redo Problem 20.6, but a 320-lb axial load is also present and 10 000 h of service is required.

   Answer: No. 311

20.8. Select bearings (not necessarily the same at both positions) for the motor armature of Figure P20.3. The sum of the belt tensions is 300 lb. Required life is 30 000 h at 1050 rpm with 90 percent reliability.

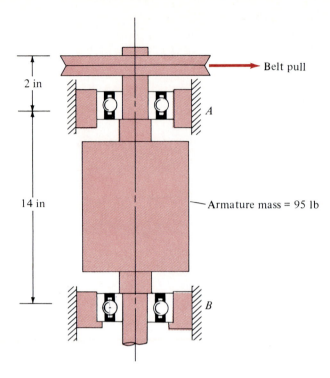

2 in

14 in

Belt pull

A

Armature mass = 95 lb

B

FIGURE P20.3

Section 20.6

20.9. Figure P20.9 is a schematic of a shaft which is driven at the left through a universal joint (pure torque) and which drives a pair of spur gears on the right. The shaft speed is a fairly constant 1500 rpm. The gear separating force F varies in time as follows:

$$500 \text{ N about 50 percent of time}$$

$$700 \text{ N about 25 percent of time}$$

$$900 \text{ N about 25 percent of time}$$

Calculate the required bearing ratings and choose appropriate bearings for $L_{05} = 30\,000$ h. The ID should be about 1 in.

Answer: $C_A = 14\,113$ (no. 106); $C_B = 25\,404$ (no. 305)

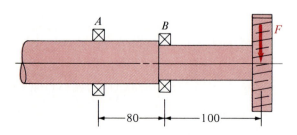

A

B

F

80 — 100

FIGURE P20.9

20.10. Redo Problem 20.9, but the speed decreases under greater loads:

> 500 N at 1500 rpm about 40 percent of time
>
> 750 N at 1300 rpm about 45 percent of time
>
> 925 N at 1150 rpm about 15 percent of time

Calculate the required bearing ratings.

20.11. In Problem 20.10, bearing A is a no. 205 deep-groove ball bearing (Table 20.2), and bearing B is a no. 304. Estimate the time in hours to the first bearing failure for 95 percent reliability.

> *Answer:* 9296

20.12. Repeat Problem 20.3, but the sum of the belt tensions is

> 150 lb at   945 rpm for about 50 percent of time
>
> 122 lb at 1230 rpm for about 20 percent of time
>
> 93 lb at 1790 rpm for about 30 percent of time

There are light shock conditions. Estimate the time in hours to the first bearing failure for 90 percent reliability.

20.13. For the armature shaft shown in Figure P20.3 and the loading conditions of Problem 20.12, choose suitable bearings for a life of 10 000 h with a reliability of 98 percent.

> *Answer:* A: no. 302, B: no. 303

Section 20.7

20.14. The bearings shown in Figure P20.14 are Timken no. 33213 (Table P20.14). Calculate the life of the unit, based on a reliability of 95 percent and "normal operation."

20.15. Select bearings for the application shown in Problem 20.14. A life of 40 000 h at 560 rpm at a reliability of 98 percent is required. There is a moderate amount of shock and vibration.

> *Answer:* A: no. 32213, B: no. 30213

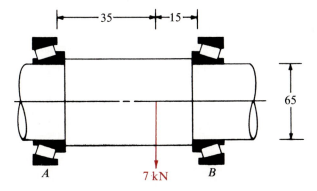

Dimensions in millimeters

TABLE P20.14

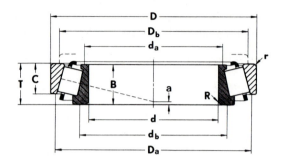

Cone   Cup

Bore d	Outside diameter D	Width T	One row radial N lbf	Thrust N lbf	Factor K	Eff. load center a²	Part numbers Cone	Part numbers Cup	Max. shaft fillet radius R¹	Width B	Backing shoulder diameters d_b	Backing shoulder diameters d_a	Max. housing fillet radius r¹	Width C	Backing shoulder diameters D_b	Backing shoulder diameters D_a
65.000	100.000	23.000	23600	18500	1.27	−0.3	◆32013X	◆32013X	1.5	23.000	73.0	71.0	1.5	17.500	91.0	97.0
2.5591	3.9370	0.9055	5300	4160		−0.01			0.06	0.9055	2.87	2.80	0.06	0.6890	3.58	3.82
65.000	100.000	27.000	25700	15300	1.68	−6.1	◆33013	◆33013	1.5	27.000	74.0	70.0	1.5	21.000	91.0	96.0
2.5591	3.9370	1.0630	5780	3450		−0.24			0.06	1.0630	2.91	2.76	0.06	0.8268	3.58	3.78
*65.000	*105.000	24.000	26000	20200	1.29	−0.3	JLM710949C	JLM710910	3.0	23.000	77.0	71.0	1.0	18.500	96.0	101.0
*2.5591	*4.1339	0.9449	5840	4540		−0.01			0.12	0.9055	3.03	2.80	0.04	0.7283	3.78	3.96
*65.000	*110.000	28.000	33900	23300	1.45	−3.3	JM511946	JM511910	3.0	28.000	78.0	72.0	2.5	22.500	99.0	105.0
*2.5591	*4.3307	1.1024	7610	5240		−0.13			0.12	1.1024	3.07	2.83	0.10	0.8858	3.90	4.13
65.000	110.000	34.000	40100	26600	1.51	−7.9	◆33113	◆33113	1.5	34.000	77.0	73.0	1.5	26.500	99.0	106.0
2.5591	4.3307	1.3386	9010	5980		−0.31			0.06	1.3386	3.03	2.87	0.06	1.0433	3.90	4.17
65.000	120.000	24.750	31400	21700	1.44	−1.0	◆30213	◆30213	2.0	23.000	75.0	72.0	1.5	20.000	109.0	113.0
2.5591	4.7244	0.9744	7050	4890		−0.04			0.08	0.9055	2.95	2.83	0.06	0.7874	4.29	4.45
65.000	120.000	32.750	41700	28900	1.44	−5.6	◆32213	◆32213	2.0	31.000	77.0	73.0	1.5	27.000	108.0	114.0
2.5591	4.7244	1.2894	9380	6500		−0.22			0.08	1.2205	3.03	2.87	0.06	1.0630	4.25	4.49
*65.000	*120.000	39.000	53500	30900	1.73	−10.7	JH211749	JH211710	3.0	38.500	80.0	74.0	2.5	32.000	107.0	114.0
*2.5591	*4.7244	1.5354	12000	6950		−0.42			0.12	1.5157	3.15	2.91	0.10	1.2598	4.21	4.49
*65.000	*120.000	39.000	53500	30900	1.73	−10.7	JH211749A	JH211710	7.0	38.500	88.0	74.0	2.5	32.000	107.0	114.0
*2.5591	*4.7244	1.5354	12000	6950		−0.42			0.28	1.5157	3.46	2.91	0.10	1.2598	4.21	4.49
65.000	120.000	41.000	53100	35500	1.50	−11.2	◆33213	◆33213	2.0	41.000	79.0	74.0	1.5	32.000	107.0	115.0
2.5591	4.7244	1.6142	11900	7970		−0.44			0.08	1.6142	3.11	2.91	0.06	1.2598	4.21	4.53
65.000	140.000	36.000	53300	31500	1.69	−7.6	◆30313	◆30313	3.0	33.000	80.0	75.0	2.5	28.000	125.0	131.0
2.5591	5.5118	1.4173	12000	7080		−0.30			0.12	1.2992	3.15	2.95	0.10	1.1024	4.92	5.16
*65.000	*140.000	53.975	77200	42900	1.80	−19.3	J6392	J6327	3.0	56.007	83.0	77.0	3.3	44.450	119.0	136.0
*2.5591	*5.5118	2.1250	17300	9640		−0.76			0.12	2.2050	3.27	3.04	0.13	1.7500	4.69	5.35
65.000	140.000	51.000	72600	42900	1.69	−16.8	◆32313	◆32313	3.0	48.000	83.0	77.0	2.5	39.000	123.0	131.0
2.5591	5.5118	2.0079	16300	9640		−0.66			0.12	1.8898	3.27	3.03	0.10	1.5354	4.84	5.16

The "Rating at 500 rpm for 3000 hours L_{10}" heading spans the radial and thrust columns.

[1] These maximum fillet radii will be cleared by the bearing corners.
[2] Minus value indicates center is inside cone backface.
† For standard class **ONLY**, the maximum metric size is a whole millimetre value.
* For "J" part tolerances—see metric tolerances, page 73, and fitting practice, page 65.
◆ ISO cone and cup combinations are designated with a common part number and should be purchased as an assembly. For ISO bearing tolerances—see metric tolerances, page 73, and fitting practice, page 65.

20.16. Figure P20.16 shows an intermediate shaft in a gear reducer. The two gears are helical, hence have axial force components, as shown. The shaft speed is 1300 rpm. The required life is 5000 h at 95 percent reliability, and the desired shaft diameter $d = 65$ mm. Select tapered roller bearings for this application. That portion of the catalog page listing Timken standard metric bearings of 65-mm bore is shown in Table P20.14.

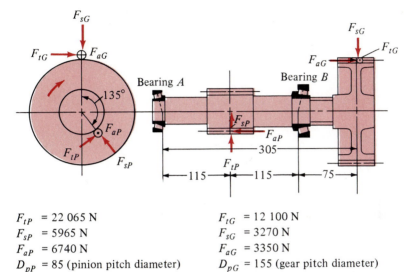

F_{tP} = 22 065 N

F_{sP} = 5965 N

F_{aP} = 6740 N

D_{pP} = 85 (pinion pitch diameter)

F_{tG} = 12 100 N

F_{sG} = 3270 N

F_{aG} = 3350 N

D_{pG} = 155 (gear pitch diameter)

FIGURE P20.16

CHAPTER
21

BELT
DRIVES

Belts of various kinds are probably people's oldest means of transmitting power. The primitive who threw a vine over a tree limb used some of the same principles found in modern belt drives. Indeed, modern automobile supply catalogs list an emergency belt which is simply wound around the pulleys of an automobile engine, just as our ancestors used a vine.

Innovation continues in drive belts; new materials and assembly technology seem to appear daily. Drive belts have many advantages. They can be fabricated from widely available materials, by using primitive techniques and tools. The shafts they connect can be far apart and at large angles to one another. The same belt can drive several shafts, wrapping one way around one pulley and the opposite way around the next pulley. That is, one belt can drive shafts in opposite directions. Belt drives are quiet and need no lubrication, and large speed ratios can easily be handled with high efficiency.

Many specialized types of belts have been developed. We discuss three categories: flat belts, V belts, and toothed or "synchronous" belts. Flat belts are chosen where belt speed is high, where a single belt must drive several shafts, or where pulleys (often called *sheaves*) must be small. V belts run in grooved pulleys for a better grip. They are the workhorses of belting, carrying heavy loads and tolerant of abuse. Toothed belts are used primarily where precise register of driving and driven shafts is required—hence the name *synchronous*.

21.1   FLAT BELTS

A flat belt is shown in Figure 21.1, wrapped around an ungrooved pulley. A net torque $R(T_1 - T_2)$ is transmitted to the pulley by the difference in tension between the tight and loose sides of the belt. A free-body diagram of a small part of the belt

630

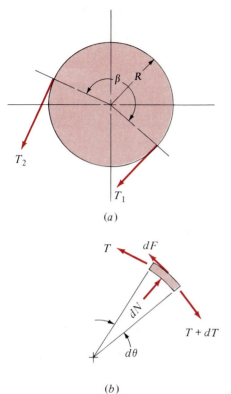

(a)

(b)

FIGURE 21.1
(a) A flat belt on a pulley. (b) An element of a flat belt on a pulley.

is shown in Figure 21.1b. It is in equilibrium under tension T and the reaction of the pulley. The reaction is shown as a normal component dN and a frictional component dF. The torque that can be transmitted to the pulley will be maximum when the belt is on the point of slipping. The friction component is then

$$dF = \mu \, dN \qquad (21.1)$$

Summing the normal components of the forces gives

$$dN = (2T + dT) \sin \frac{d\theta}{2}$$

The sine of an infinitesimal angle can be taken equal to the angle, and the product of the two infinitesimals dT and $d\theta$ can be neglected. Doing that and incorporating (21.1) produce:

$$dT = \mu T \, d\theta$$

or

$$\frac{dT}{T} = \mu \, d\theta$$

This is integrated over the wrap angle $\beta = \theta_1 - \theta_2$:

$$\frac{T_1}{T_2} = e^{\mu\beta} \tag{21.2}$$

The power transmitted by a flat belt is the product of the net tension and the belt speed V. Expressing power in horsepower and using units of pounds, feet and minutes, we have

$$\text{Power} = \frac{(T_1 - T_2)V}{33\,000} \quad \text{hp} \tag{21.3}$$

The difference between tight-side tension T_1 and slack-side tension T_2 is usually called the *effective tension*. Equations (21.2) and (21.3) can be combined to give

$$\text{Power} = \frac{T_1 V(1 - e^{-\mu\beta})}{33\,000} \quad \text{hp} \tag{21.4}$$

Thus the power capability of a flat belt is a function of the angle of wrap β and the permissible speed and belt tension. Normally the maximum permissible tight-side tension T_1, or *working tension*, is specified by the manufacturer.

Example 21.1. A flat rubber-covered belt, such as the one shown in Figure 21.1, is driving a pulley of 12-in diameter, with an angle of wrap β of π rad, or 180°. The minimum, or slack-side, tension is 10 lb, and the speed is 600 rpm. Assume a coefficient of friction of $\mu = 0.8$. How much horsepower could be transmitted, with the belt at the point of slippage?
From (21.2)

$$T_1 = 10e^{0.8\pi} = 123.5$$

Thus, expression (21.3) gives as the power transmitted

$$\frac{113.5(600)(\pi)(1)}{33\,000} = 6.48 \text{ hp}$$

Belts are not usually loaded to the point of slippage; the heat generated by the slippage would soon destroy the belt. Instead, allowable belt power transmission ratings are based on the lifetime and size of the smallest pulley. As a belt bends around a pulley, it is subjected to flexural and shear strains just as any other beam. These strains are less severe with larger pulleys, and therefore manufacturers' power ratings increase with the size of the smallest pulley.

Smooth belts will tolerate occasional slippage. This fact is exploited to make an inexpensive clutch for small engines such as those in lawn mowers. An extra pulley called an *idler* is moved in, to increase tension and cause the belt to drive, as shown in Figure 21.2. When the idler is retracted, the belt tension drops and the engine is effectively declutched from the load.

A modern endless flat belt is usually a single ply of cloth impregnated with rubber. Commonly, the cloth is polyester, and the rubber is a synthetic called *neoprene*, which has excellent resistance to attack by most petroleum solvents. Neoprene

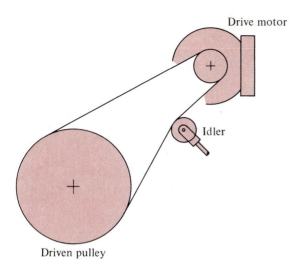

Drive motor

Idler

Driven pulley

FIGURE 21.2
A belt drive used as a clutch.

is a trademark which has become generic in usage. *Elastomer* is the general term for materials commonly known as "rubber."

The smooth side of the belt is run against the pulley in clean conditions and the textured side in oily conditions. The rating chart for such a belt is shown in Table 21.1. Notice that the rated power increases with pulley diameter, as remarked earlier. As we would expect, rated power also increases with speed. However, the increase of power is less than proportional to speed. That is, the rating for twice the speed is less than twice the original power. The reason? As the speed increases, the normal force dN and thus the effective tension, decrease due to centrifugal forces. Also, if belt ratings are based on an expected lifetime measured in hours, then for a given lifetime, greater speed means a greater number of flexings of the belt.

> **Example 21.2.** In the belt installation of Example 21.1, the belt speed is 1885 ft/min. That speed is not listed, so Table 21.1 is entered with the next greatest speed of 2000 ft/min. The table lists ratings only for pulleys up to a diameter of 3 in, but the instructions for use of the table allow us to increase that rating by 7 percent for each inch of pulley diameter. Thus
>
> $$\text{Rated power} = 2.7 + 2.7(12 - 3)(0.07) = 4.40 \text{ hp}$$
>
> Equation (21.3) can be entered with a slack-side tension of 10 lb, to find the working tension of 87 lb. PBI 300 belting is rated at 140 lb/in of width, and thus the required belt width is $87/140 = 0.621$ in. We would probably specify a width of 1 in. This belt is available off the shelf in widths up to 6 in.

The pulley of a flat belt is not really flat across the outer side of its rim, but *crowned*, to keep the belt on the pulley. When the belt moves off the crown for whatever reason, as shown in Figure 21.3, the belt tension tends to pull the "off" (in this illustration the *left*) side of the belt tight against the pulley face. This bends the belt toward the center of the pulley, and it moves toward center position.

TABLE 21.1
Rating chart for flat belts

Moderate to ultra high speed drives such as computer disc drives or grinding and polishing applications are prime applications for this belt.

SPECIFICATIONS

Tensile: 1400 lbs tensile strength per inch of width in endless form.
Working Tension: Up to 140 lbs per inch of width.
Maximum Belt Speed: 15,000 FPM.
Minimum Pulley Diameter: 5/8 inch.
Nominal Thickness: .035" — .040".
Belt Construction: 1 ply polyester cord woven endless.
Saturant: Black Neoprene.
Saturant Properties: Anti-static; excellent oil and abrasion resistance. Good heat resistance and non-marking properties.

Belt Surfaces: Standard construction offers two drive surfaces — smooth inside, textured outside. Normal installation would run the smooth surface to the pulley — in cases of oil or contamination run rough side to the pulley.
Normal Elongation: 0.5% maximum.
Widths Available: To 6 inches —over 6 inches. contact factory.
Lengths Available: 4 to 175 inches.
Standard Width Tolerances: Under 5 inches ±1/32"— Over 5 inches ±1/16".
Standard Length Tolerances: ±0.5%. Special tolerances can be accommodated upon request.

PBI 300 HORSEPOWER RATING CHART*
(Based On 180° Arc Of Contact On Small Pulley)
Small Pulley Diameter

Speed ft/min	1/2"	1"	2"	3"
1000	.5	.9	1.2	1.4
2000	1.3	2.0	2.4	2.7
3000	1.7	2.7	3.4	4.0
4000	2.1	3.5	4.5	5.1
5000	2.5	4.0	5.3	6.1
6000	2.7	4.1	6.0	6.8
7000	2.9	4.9	6.5	7.5
8000	3.0	5.3	7.0	8.1
9000	3.1	5.5	7.3	8.4
10,000	3.2	5.6	7.5	8.7

Correction Factor For Arc of Contact on Small Pulley

45° =	.30	140° =	.83
90° =	.60	150° =	.87
110° =	.70	160° =	.91
120° =	.74	170° =	.96
130° =	.78	180° =	1.00

*Add 7% to horsepower values for each inch of small pulley diameter above 3".

PBI 400

The PBI 400 woven endless belt is a heavy duty drive belt. Its two ply construction was designed for high tension loads run with flanged pulleys. Because of the cushioning effect of this construction, edge wear is kept to a minimum. Due to the high tensile ratings this belt maintains, it is recommended for many high tension low speed applications.

SPECIFICATIONS

Tensile: 1600 lbs tensile strength per inch of width in endless form.
Working Tension: Up to 160 lbs per inch of width.
Maximum Belt Speed: 8,000 FPM.
Minimum Pulley Diameter: 7/8 inch.
Nominal Thickness: .065" — .070".
Belt Construction: 2 ply polyester and cotton woven endless.
Saturant: Black Neoprene.
Saturant Properties: Anti-static; excellent oil and abrasion resistance. Good heat resistance and non-marking properties.

Belt Surfaces: Standard construction offers two drive surfaces — smooth inside, textured outside. Normal installation would run the smooth surface to the pulley — in cases of oil or contamination run rough side to the pulley.
Normal Elongation: .5% maximum.
Widths Available: To 6 inches — over 6" contact factory.
Lengths Available: 4 to 175 inches.
Standard Width Tolerances: Under 5 inches ±1/32"–Over 5 inches ±1/16".
Standard Length Tolerances: ±0.5%.
Special Tolerances can be accommodated upon request.

PBI 400 HORSEPOWER RATING CHART*
(Based On 180° Arc Of Contact On Small Pulley)
Small Pulley Diameter

Speed ft/min	1/2"	1"	2"	3"
1000	.8	1.4	1.8	2.1
2000	2.0	3.0	3.6	4.1
3000	2.6	4.1	5.1	6.0
4000	3.2	5.3	6.8	7.7
5000	3.8	6.0	8.0	9.2
6000	4.1	6.2	9.0	10.2
7000	4.4	7.4	9.8	11.3
8000	4.5	8.0	10.5	12.2
9000	4.7	8.3	11.0	12.6
10,000	4.8	8.4	11.3	13.1

Correction Factor For Arc of Contact on Small Pulley

45° =	.30	140° =	.83
90° =	.60	150° =	.87
110° =	.70	160° =	.91
120° =	.74	170° =	.96
130° =	.78	180° =	1.00

*Add 7% to horsepower values for each inch of small pulley diameter above 3".

Courtesy Pacific Belting Industries, Inc.

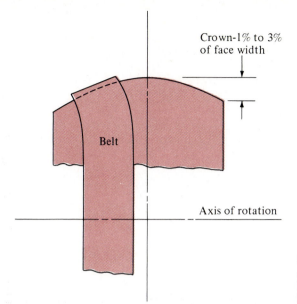

Crown-1% to 3% of face width

Belt

Axis of rotation

FIGURE 21.3
A flat belt running off center will be bent by the crown of the pulley and forced back into line.

21.2   V BELTS

It is likely that pulleys were originally grooved to keep drive ropes in place. In addition the wedging action of a grooved pulley or sheave increases the friction and thus the power that can be transmitted. The positive locating action of the groove makes V belts a popular choice as clutches, as in Figure 21.2, and variable-speed drives, as shown in Figure 21.4. Because the belt does not fill its groove completely, these drives are quite tolerant of dirty environments.

The modern V belt is the result of encasing the tension members in rubber and shaping the rubber to interact with the sides of the pulley groove for best performance. The power transmission capability of a V belt can be analyzed, as for the flat belt. The radial force dN can be expressed in terms of the forces F_2 on each side of the V, as shown in Figure 21.5. In converting the free-body diagram of a flat belt to that of a V belt, we have replaced the single normal force dN by a pair of inclined forces F_2. The normal force becomes

$$dN = 2F_2 \sin \frac{\phi}{2}$$

or

$$\frac{2F_2}{dN} = \frac{1}{\sin (\phi/2)} \tag{21.5}$$

The ratio v of the frictional force on the V belt to that of a flat belt is the same as the ratio of normal forces:

$$v = \frac{F_v}{F} = \frac{2F_2}{dN} = \frac{1}{\sin (\phi/2)} \tag{21.6}$$

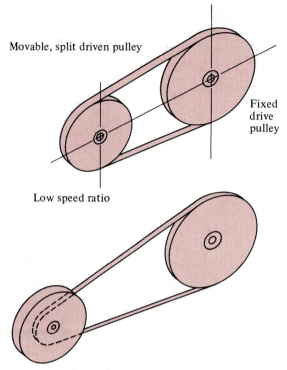

Movable, split driven pulley

Fixed
drive
pulley

Low speed ratio

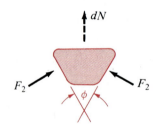

High speed ratio

FIGURE 21.4
A variable-speed V-belt drive.

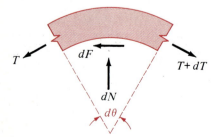

FIGURE 21.5
Forces acting on an element of a V belt.

And the ratio of tight-side to slack-side tension becomes

$$\frac{T_1}{T_2} = e^{v\mu\beta} \qquad (21.7)$$

For an angle ϕ of 38°, typical of industrial V belts, the relation becomes

$$\frac{T_1}{T_2} = e^{3.0\mu\beta} \qquad (21.8)$$

As we shall see, this leads to very large ratios of tight- to slack-side tension. Belt manufacturers, to provide a safety margin against slip, recommend smaller tension ratios: 5 to 8 when the belt is adjusted manually, and 8 when the adjustment is automatic.

Example 21.3. For the conditions of Example 21.1, what power could be transmitted by a V belt without slipping? Assuming that the wedge angle is 38°, we get

$$\frac{T_1}{T_2} = e^{3\pi 0.5} = 111.3$$

The maximum tight-side tension without slip is thus

$$111.3(10) = 1113 \text{ lb}$$

and the corresponding power is

$$\frac{(1113 - 10)(1)(\pi)(600)}{33\,000} = 63 \text{ hp}$$

Using the recommended value of $T_1/T_2 = 8$, we get

$$\text{Power} = \frac{(8 - 1)(10)(600)(\pi)(1)}{33\,000} = 4 \text{ hp}$$

Note that this is nowhere near the load that would cause this belt to slip. As remarked earlier, belts are not rated at the point of slippage.

Manufacturers of belt drives have devised rating schemes that take into account the service to be seen by the belt and the life expected. It is not unusual to see, along with the usual tables of rated power, an offer to run a computer simulation of the life to be expected of a proposed belt installation.

A typical procedure (courtesy of Goodyear Power Transmission Products, Lincoln, Nebraska) starts with the determination of a service factor, which may vary from 1.0 for electric motors driving "smooth" loads (such as blowers in intermittent service) to 1.8 for high-torque motors driving crushers, to 2.0 when the load is "chokable" (i.e., the resistance may exceed the torque of the drive motor). The horsepower required is multiplied by the service factor to give the design horsepower.

Next, the size of belt is chosen from an application chart, such as that shown in Figure 21.6. As you might expect, smaller belts such as the A size (0.5 in wide) are recommended for light loads and high speeds, and larger belts (D size, 1.25 in wide) for heavy loads and low speeds.

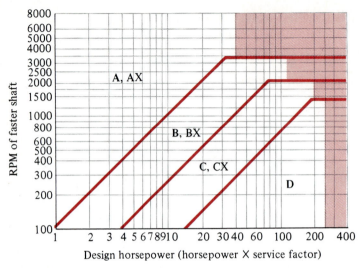

RPM of faster shaft

Design horsepower (horsepower X service factor)

The area indicators are belt sizes   Shaded area, refer to factory

FIGURE 21.6
Typical rating chart for V-belt drives. (*Courtesy Goodyear Industrial Products Division.*)

The minimum recommended size of the smaller pulley or sheave is taken from a chart provided by the National Electric Manufacturers Association (NEMA). It is shown here as Table 21.2. Again, higher power and lower speed correlate with larger sheaves. The power ratings given in the table are related to the diameter in inches of the smaller sheave d and its speed $n = \text{rpm}/1000$ by a formula fitted to the results of life testing:

$$\text{Hp} = dn \left[K_1 - \frac{K_2}{d} - K_3(dn)^2 - K_4 \log(dn) \right] + K_2 n \left(1 - \frac{1}{K_{SR}} \right) \qquad (21.9)$$

The coefficients K_1, K_2, K_3, and K_4 are specific to the type and size of belt (Table 21.3), and K_{SR} depends on the speed ratio. The parameter K_{SR} can be calculated from:

$$K_{SR} = \frac{1}{1 + 0.35 \log \dfrac{1 + 10^{-(1/0.35)[1-(1/SR)]}}{2}} \qquad (21.10)$$

TABLE 21.2
Minimum recommended sheave O.D. when driver is an electric motor

Motor RPM										Motor HP											
	½	¾	1	1½	2	3	5	7½	10	15	20	25	30	40	50	60	75	100	125	150	200
575	2.5	3.0	3.0	3.0	3.65	4.5	4.5	5.3	6.0	6.9	8.0	9.0	10.0	10.0	11.0	12.0	14.0	18.0	20.0	22.0	22.0
695	2.5	2.5	2.5	3.0	3.0	3.65	4.5	4.5	5.3	6.0	6.9	8.0	9.0	10.0	10.0	11.0	13.0	15.0	18.0	20.0	22.0
870	2.2	2.4	2.4	2.4	3.0	3.0	3.8	4.4	4.4	5.2	6.0	6.8	6.8	8.2	8.4	10.0	10.0	12.0	...	...	...
1160	...	2.2	2.4	2.4	2.4	3.0	3.0	3.8	4.4	4.4	5.2	6.0	6.8	6.8	8.2	8.2	10.0	10.0	12.0	...	...
1750	...	...	2.2	2.4	2.4	2.4	3.0	3.0	3.8	4.4	4.4	4.4	5.2	6.0	6.8	7.4	8.6	8.6	10.5	10.5	13.2
3450	...	...	...	2.2	2.4	2.4	2.4	3.0	3.0	3.8	4.4	4.4	...	...	...	...	...	...	...	...	...

Note: for internal combustion engine drivers, use the next larger size sheave.

TABLE 21.3
Constants for the belt horsepower formula

Cross section	K_1	K_2	K_3	K_4
A	1.3948	2.6198	0.00029043	0.27041
B	2.2149	5.8478	0.00047867	0.41948
C	3.6653	13.7060	0.00081326	0.66836
D	6.7891	39.352	0.00156760	1.1898

In similar fashion, experience in the setting of belt tension was distilled into this formula:

$$T_i = \frac{(63\,030)(\text{hp})(2A_R - 1)}{Ndn} + \frac{W(dn)^2}{1.69E + 06} \tag{21.11}$$

N is the number of belts in the drive, n the rpm, W the weight of the belt per foot, and A_R a tabulated factor related to the arc of contact on the small sheave. A_R varies from 1.25 for an arc of 180° to 1.91 when the arc is 83°.

Example 21.4. Suppose we need a V-belt drive from a 5-hp 1750-rpm electric motor to a hoist which will run intermittently at 440 rpm. The service factor is 1.5, and thus the design horsepower is $5(1.5) = 6.5$ hp. From the chart of Figure 21.4, we determine that the most appropriate belt to use is size A, and in Table 21.2 we find that the smallest recommended sheave is 3 in in diameter. Inserting in Equation (21.9) the parameters listed for size-A belts, we find that the rated horsepower per belt is

$$\text{Rated power} = dn\left[1.3948 - \frac{2.6198}{d} - 0.00029(dn)^2\right.$$

$$\left. - 0.2704 \log dn\right] + 2.6198n\left(1 - \frac{1}{1.116}\right) = 2.15 \text{ hp}$$

The total rating of 3 belts is thus 6.45 hp—close enough.

The installed tension in the belt can be calculated from Equation (21.11):

$$T_i = \frac{(63\,030)(5)(2.5 - 1)}{3(3)(1750)} + \frac{0.07[(3)(1750)]^2}{1.69E + 06}$$

$$= 29 \text{ lb}$$

When the motor starts, the belt will tighten on one side of the belt and slacken on the other. With the ratio of tight- to loose-side tension at 8/1, and the sum of the tensions the same as when the unit is not running, we find that the tensions are $T_1 = 81.8$ lb and $T_2 = 10.2$ lb.

The power can be recalculated from the difference in tensions and the size of the sheave:

$$\text{Hp} = \frac{45(\frac{3}{12})(\pi)(1750)}{33\,000} = 1.87 \text{ hp per belt}$$

FIGURE 21.7
Positive drive belts can be used in very small situations. (*Courtesy Belting Industries.*)

21.3   POSITIVE DRIVE BELTS

Also called *synchronous* or *timing belts*, these belts have formed teeth which mesh with teeth in the pulleys or sprockets. They can thus maintain exactly the position of the driven shaft relative to the driver. In this sense, positive drive belts perform the same function as the older chain drives. Chain drives have been replaced in many applications by these one-piece molded composite belts, which are much lighter and smoother in operation. The very high strength-to-weight ratio of composite belts makes high operating speed feasible, and the thinness of the load-carrying portion of the belt makes it possible to operate over very small sprockets. Figure 21.7 shows such a belt.

Positive drive belts owe much to the technology of rubber tires, in that they comprise a rubber body molded over a layer of strong cords. The cords are made of steel, fiberglass, kevlar, or polyester. The rubber is urethane or neoprene. The molded gear teeth which mesh with the teeth on the sprocket are usually covered with nylon fabric.

Belt Pitch

By design, a positive drive belt moves over its sprocket as though it were a very thin smooth belt running on a smooth pulley. The diameter of this imaginary pulley is called the *pitch diameter* of the real sprocket. As shown in Figure 21.8, the pitch circle thus defined lies beyond the sprocket, on the reinforcing cords of the belt.

The pitch diameter is chosen to minimize any relative motion between belt and sprocket due to flexing of the belt. Positive drive belts are classified by pitch—the length per tooth. Standard pitches are given in Table 21.4, with typical working tension and weight per foot for a belt 1 in wide. The working tension of a positive drive belt is taken to be 5 to 10 percent of its breaking strength.

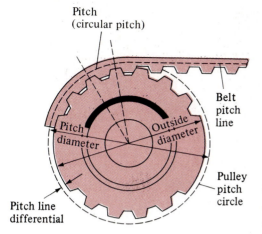

Pitch
(circular pitch)

Pitch
diameter

Outside
diameter

Belt
pitch
line

Pulley
pitch
circle

Pitch line
differential

FIGURE 21.8
The concept of pitch diameter of a positive
drive sprocket. (*Illustration courtesy Goodyear
Industrial Products Division etc.*)

TABLE 21.4
Positive drive belts

Pitch, in	Working tension, lb/in width	Weight, lb/ft (1 in wide)
0.08	32	0.016
0.20	41	0.046
$\frac{3}{8}$	55	0.064
$\frac{1}{2}$	140	0.090
$\frac{7}{8}$	191	0.210
$1\frac{1}{4}$	234	0.27

Courtesy Goodyear Tire and Rubber Company

Positive Drive Belt Selection

Slippage is not an issue in the design of positive drive belt installations, and therefore these belts are selected on the basis of their rated tension. The selection of a positive drive belt cannot be done by formula—tables and charts are necessary. This is because both the belt and the sprockets must have an integral number of teeth, and only a finite number of combinations are available from stock. It follows that the final design of a positive drive system must be made with up-to-date information from vendors.

The first step, as with V belts, is to choose the size of belt suitable for the power to be transmitted and the operating speed. A selecting chart is reproduced here as Figure 21.9. As you can see, the power capability of the belt is determined by the working tension and the size of the smallest sprocket. Heavier loads and lower speeds correlate, as usual, with heavier belts of greater pitch.

To use Figure 21.9, we need a size for the smallest sprocket or pulley. We can get a rule of thumb for minimum sprocket size by fitting a parabola to the numbers given in the literature.

$$\text{Minimum sprocket diameter} = 12(\text{pitch})^2 \tag{21.12}$$

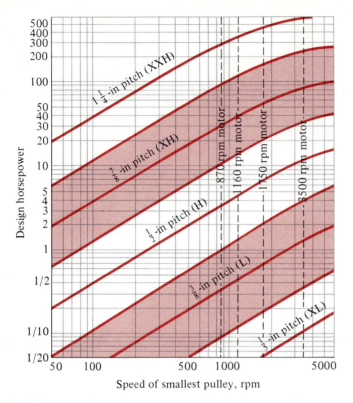

FIGURE 21.9
Typical guide for selection of positive drive belts. (*Courtesy Goodyear Industrial Products.*)

When designing with these rules of thumb, we must use a somewhat larger belt tension than that corresponding to the power requirement, because part of the working belt tension will be taken up by centrifugal force. The selection chart of Figure 21.9 included this correction. The *effective tension* T_e, that which corresponds to the power transmitted, is

$$T_e = T_1 - \frac{WV^2}{g} \tag{21.13}$$

in which T_1 is the working tension, W the weight per unit length, V the belt velocity, and g the acceleration of gravity. The horsepower rating per inch of belt width is then calculated with Equation (21.3).

> **Example 21.5.** A 10-hp 3500 rpm electric motor is to drive a centrifugal pump at 2450 rpm. The service is intermittent, and the desired center distance 25 to 27 in.
> The application chart (Figure 21.9) shows that the $\frac{1}{2}$-in-pitch belt should be used. The service factor for this application is 1.5, and thus the design horsepower is 15. A reasonable first estimate of sprocket size is 5-in diameter, well above the minimum of 2.5 for this belt. This size sprocket, at 3500 rpm, produces a belt speed

of
$$V = \frac{5\pi(3500)}{12} = 4582 \text{ ft/min} = 76.3 \text{ ft/s}$$

The operating tension and weight for this size belt are, from Table 21.4, 140 lb and 0.09 lb/ft. The effective tension is, by Equation (21.13),
$$T_e = T_1 - \frac{(0.09)(76.3)^2}{32.2} = 123.7 \text{ lb}$$

And the horsepower per inch of belt, by Equation (21.3), is
$$\text{hp} = \frac{123.7(4582)}{33\,000} = 17.2$$

So a belt of $\frac{1}{2}$-in pitch and 1-in width will be adequate. A solution to this problem using the tables provided by Goodyear Power Transmission products yielded the following:

$\frac{1}{2}$-in pitch, 1-in width

Sprocket diameters: 4.456, 6.366 in

Center distance: 26.49 in

PROBLEMS

Section 21.1

21.1. Design a belt drive to transmit 2 hp from a driving shaft running at 3600 rpm to a drill which must run at 10 800 rpm. Use the data of Table 21.1.

21.2. A flat belt is needed to drive two shafts in opposite directions. One of the shafts needs 0.65 hp at 3600 rpm, and the other needs 2.3 hp at 7200 rpm. Choose pulley sizes and belt width, using Table 21.1, and make a schematic of the drive. Evaluate the angle of wrap and limiting horsepower on each pulley.
    Answer: large pulley, 2 in diameter; belt, $1\frac{1}{4}$ in wide

Section 21.2

21.3. A centrifugal "squirrel cage" fan will be driven by a 50-hp 1750-rpm electric motor. The fan should run at 715 rpm. Use a service factor of 1.2. Choose suitable pulleys and select an appropriate size and number of V belts.

21.4. An internal-combustion engine is used to drive a fan in continuous operation. The engine speed is 2000 rpm, and it is rated at 75 hp for continuous service. The desired fan speed is 1100 rpm, and the desired center distance is 43 in. Use a service factor of 1.3, and select a suitable V-belt drive.
    Answer: 5 belts

21.5. An 80-hp, 2200-rpm diesel engine is to drive a rotary compressor continuously at 1000 rpm. The load peaks at 100 hp twice each day. Use a service factor of 1.2. Select a suitable V-belt drive.

21.6. The design for a very quiet, efficient airplane calls for the propeller to be driven by V belts at a speed of 625 rpm while the engine is turning at 2500 rpm. The maximum horsepower requirement, at 3000 engine rpm, is 150 hp. The engine output shaft is 1.5 in in diameter, and the distance between centers should be kept to a minimum. Design the drive, using Figure 21.6 or contemporary catalog information.
    Answer: Driver pulley, 6 in diameter; 19 belts (probably not practical)

Section 21.3

21.7. Design a drive for the requirements of Problem 21.6, but use a synchronous belt drive.

21.8. Design a drive for the requirements of Problem 21.3, using a synchronous belt.

   Answer: 9 in diameter; belt 3 in wide

21.9. The twin camshafts in a new racing engine will be driven by a synchronous belt. The drive sprocket for the belt will be fastened to the crankshaft. The maximum rpm expected is 12 000; and, of course, each camshaft will run at one-half that speed. The torque required to turn each camshaft has been measured at 8 ft·lb. Select the belt and sprockets for the camshaft drive. Assume that the camshafts will be 8 in apart and 16 in from the crankshaft centerline.

THEORETICAL STRESS-CONCENTRATION FACTORS, K_t*

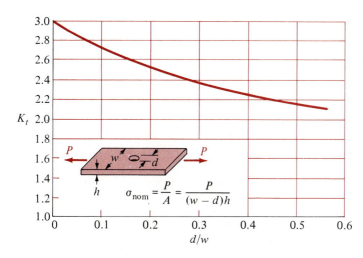

FIGURE A.1
Bar with central hole with axial load.

---

* Figures A.1 to A.13 from R. E. Peterson, *Design Factors for Stress Concentration*, published Feb.–Jul. 1951 in *Machine Design.* Reproduced with permission. These factors and others are also to be found in ref. [8]. Fig. A.14 from ref. [2].

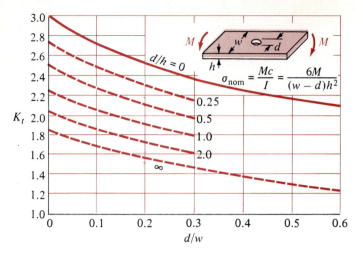

FIGURE A.2
Bar with central hole in bending.

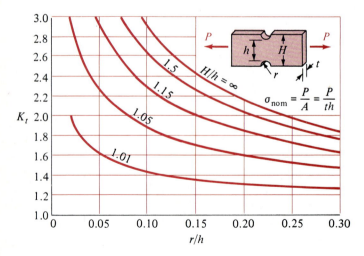

FIGURE A.3
Notched bar with axial load.

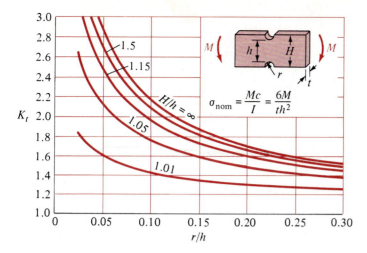

FIGURE A.4
Notched bar in bending.

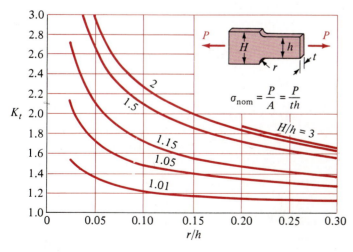

FIGURE A.5
Filleted bar with axial load.

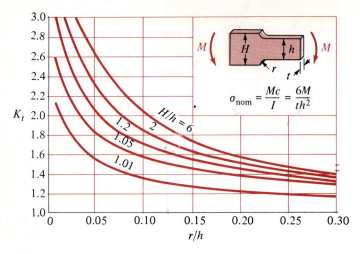

FIGURE A.6
Filleted bar in bending.

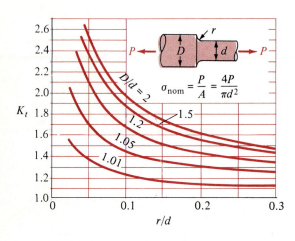

FIGURE A.7
Filleted shaft with axial load.

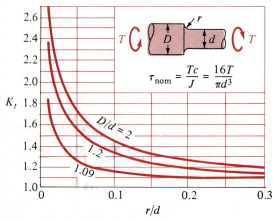

FIGURE A.8
Filleted shaft in torsion.

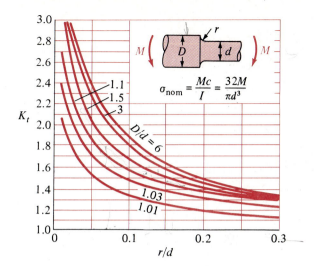

FIGURE A.9
Filleted shaft in bending.

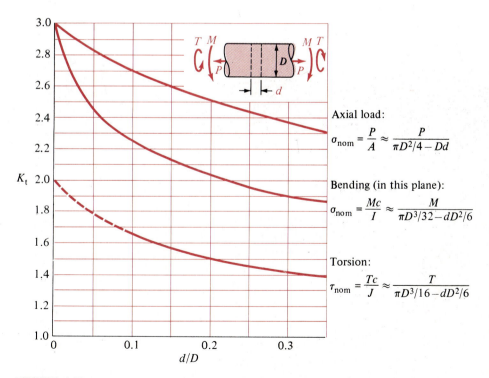

Axial load:

$$\sigma_{nom} = \frac{P}{A} \approx \frac{P}{\pi D^2/4 - Dd}$$

Bending (in this plane):

$$\sigma_{nom} = \frac{Mc}{I} \approx \frac{M}{\pi D^3/32 - dD^2/6}$$

Torsion:

$$\tau_{nom} = \frac{Tc}{J} \approx \frac{T}{\pi D^3/16 - dD^2/6}$$

FIGURE A.10
Shaft with radial hole.

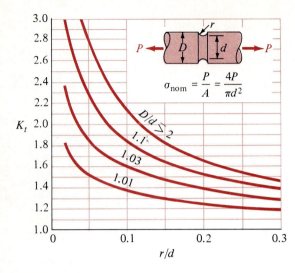

FIGURE A.11
Grooved shaft with axial load.

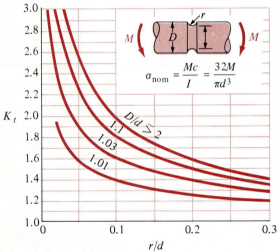

FIGURE A.12
Grooved shaft in bending.

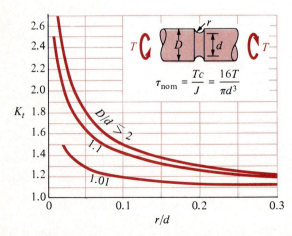

FIGURE A.13
Grooved shaft in torsion.

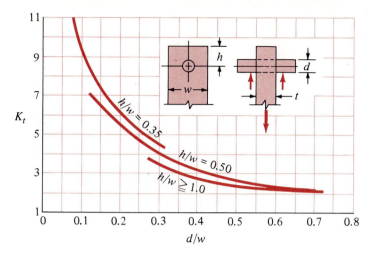

FIGURE A.14

Plate loaded in tension by a pin through a hole; $\sigma_{\text{nom}} = F/A$, where $A = (w - d)t$. When clearance exists, increase K_t by 35 to 50 percent [71].

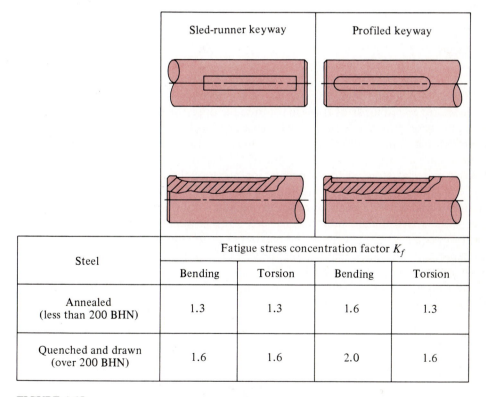

Steel	Fatigue stress concentration factor K_f			
	Bending	Torsion	Bending	Torsion
Annealed (less than 200 BHN)	1.3	1.3	1.6	1.3
Quenched and drawn (over 200 BHN)	1.6	1.6	2.0	1.6

FIGURE A.15

Fatigue stress-concentration factor K_f for keyways [1, p. 76] (with permission). Use total shaft section for nominal stress.

APPENDIX
B

CALCULATION OF EFFECTIVE DIAMETER FOR FATIGUE SIZE FACTOR C_S

As pointed out in the text, the size factor C_S depends on the stress gradient, and values are suggested for size ranges in rotating bending, torsion, and axial applications for round bars. If a part has a round cross section but is subjected to alternating bending rather than rotating bending, or if the cross section is other than round, some guidelines are needed to establish a value of C_S. The observation has been made [53] that the material stressed to 95 percent or more of the maximum determines the fatigue resistance. Thus the proposition can be advanced that, say, a rectangular beam in bending and a round beam in rotating bending would have the same fatigue resistance if the areas of each subjected to such stresses were equal. The area subjected to stresses equal to 95 percent of the maximum or greater is designated $A_{0.95}$. Once the diameter of the equivalent rotating beam has been established, the size factor can be chosen from the listing in Section 9.1.

We need $A_{0.95}$ for a round bar in rotating bending and then $A_{0.95}$ for the section of interest. The stress in a beam is given by the classical formula

$$\sigma = \frac{My}{I}$$

which, for a round bar, becomes

$$\sigma = \frac{32MD}{\pi D_o^4} \tag{B.1}$$

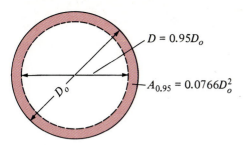

$D = 0.95D_o$

$A_{0.95} = 0.0766D_o^2$

FIGURE B.1
Area of rotating-bending specimen subjected to stress equal to 95 percent of $\sigma_{\max}$ or more.

where D is the diameter at the point of interest and D_o is the outside diameter. The maximum stress occurs at $D = D_o$, and so

$$0.95\sigma_{\max} = \frac{0.95 \times 32MD}{\pi D_o^4} \tag{B.2}$$

The diameter where $\sigma = 0.95\sigma_{\max}$ is obtained by equating (B.1) and (B.2), which results in

$$D = 0.95D_o \qquad \sigma = 0.95\sigma_{\max}$$

The cross-sectional area over which the range of stress $\sigma \geq 0.95\sigma_{\max}$ exists is then

$$A = \frac{\pi}{4} D_o(1 - 0.95^2) = 0.0766D_o^2 \tag{B.3}$$

This area is that shown shaded in Figure B.1. It is an annulus because the beam rotates, i.e., the material all the way around is stressed to the $\geq 0.95\sigma_{\max}$ level.

Suppose the loading in question involves bending of a beam of rectangular cross section, as shown in Figure B.2. The stress in such a beam is

$$\sigma = \frac{My}{I} = \frac{My(12)}{bh^3} \tag{B.4}$$

Then

$$0.95\sigma_{\max} = 0.95\frac{M(6)}{bh^2} \tag{B.5}$$

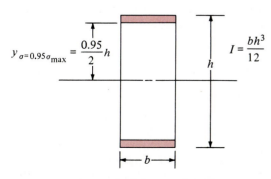

$y_{\sigma=0.95\sigma_{\max}} = \frac{0.95}{2}h$

$I = \frac{bh^3}{12}$

$A_{0.95} = $ shaded area $= 0.05bh$

FIGURE B.2
Area $A_{0.95}$ for beam of rectangular cross section in reverse bending.

Equating (B.4) and (B.5) yields

$$\frac{12yM}{bh^3} = 0.95\frac{6M}{bh^2}$$

whence

$$y = \frac{0.95h}{2} \qquad \sigma = 0.95\sigma_{max}$$

The area subjected to $\sigma \geq \sigma_{max}$ is shown shaded in Figure B.2 and is

$$A_{0.95} = 2b\left(\frac{h}{2} - \frac{0.95h}{2}\right) = 0.05bh$$

This result shows that $A_{0.95}$ is the same with bending about either principal axis.

Equating the areas for the round rotating beam of diameter D_o to the above result, we get

$$0.0766D_o^2 = 0.05bh$$

which gives

$$D_o = 0.81\sqrt{bh} = 0.81\sqrt{A}$$

This is the result shown in the text. It is the diameter of a round bar subjected to rotating bending which has the same fatigue resistance as a beam of cross-sectional dimensions b by h, subjected to reverse bending.

If a round bar is subjected to reverse bending, without rotation, the distance from the neutral axis at which the stress is 95 percent of the maximum is calculated as it was for the rotating case, with the same result: $D = 0.95D_o$. The $A_{0.95}$ area is not an annulus in this case, however, but is as shown in Figure B.3. It is given by

$$A_{0.95} = \frac{D_o^2}{2}\left(\arcsin\sqrt{1 - 0.95^2} - 0.95\sqrt{1 - 0.95^2}\right) = (1.046E - 02)D_o^2$$

This must now be equated to $A_{0.95}$ for the rotating bending case (B.4):

$$0.0766D_{eq}^2 = (1.046E - 02)D_o^2$$

giving

$$D_{eq} = 0.37D_o$$

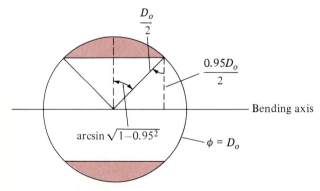

$A_{0.95}$ = shaded area = $1.046E - 02\ D_o^2$

FIGURE B.3
Area $A_{0.95}$ for bar in bending (no rotation).

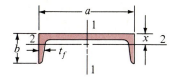

$$A_{0.95,1-1} = 0.05ab \quad (t_f > 0.025a)$$
$$A_{0.95,2-2} = 0.05xa + 0.1t_f (b-x)$$

(a)

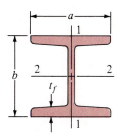

$$A_{0.95,1-1} = 0.10at_f$$
$$A_{0.95,2-2} = 0.05ba$$
$$(t_f > 0.025a)$$

(b)

FIGURE B.4
Area $A_{0.95}$ for channels and I beams, after Shigley and Mitchell [2]. (a) Channel. (b) Wide-flange or I beam.

the diameter of the equivalent rod in rotating bending. Other shapes are handled in the same way. Shigley and Mitchell [2, p. 294] have calculated $A_{0.95}$ for channels and I beams, as shown in Figure B.4.

APPENDIX
C

MONOTONIC STRESS-STRAIN PROPERTIES

Material	BH	Process description	S_u, ksi/MPa	S_y, ksi/MPa	σ_f, ksi/MPa	RA, %	ε_f	n	K, ksi/MPa
Steel									
RQC-100[a]	290	HR[b] plate	135/931	128/883	193/1331	67	1.02	0.06	170/1172
1005–1009	125	CD[c] sheet	60/414	58/400	122/841	64	1.02	0.05	76/524
1005–1009	90	HR sheet	50/345	38/262	123/848	80	1.6	0.16	77/531
1015	80	Normalized	60/414	33/228	105/724	68	1.14	0.26	
1020[d]	108	HR plate	64/441	38/262	103/710	62	0.96	0.19	107/738
1045[e]	225	Q&T[f]	105/724	92/634	178/1227	65	1.04	0.13	166/1145
1045[e]	410	Q&T	210/1448	198/1365	270/1862	51	0.72	0.08	302/2082
5160	430	Q&T	242/1669	222/1531	280/1931	42	0.87	0.06	308/2124
9262	260	Annealed	134/924	66/455	151/1041	14	0.16	0.22	253/1744
9262	410	Q&T	227/1565	200/1379	269/1855	32	0.38	0.06	283/1951
950X	150	HR plate	77/531	48/311	145/1000	72	1.24	0.19	131/903
Aluminum									
2024-T351		Solution treated, strain-hardened	68/469	55/379	81/558	25	0.28	0.03	66/455
2024-T4		Solution treated and RT age	69/476	44/303	92/634	35	0.43	0.20	117/807
7075-T6		Solution treated and artificially aged	84/579	68/469	108/745	33	0.41	0.11	120/827

[a] Tradename, Bethlehem Steel Corp. Roller quenched and tempered carbon steel. Used in structural, heavy-machinery applications.

[b] Hot-rolled.

[c] Cold-drawn.

[d] Low-carbon, common machining steels.

[e] Bar stock, medium carbon; higher-strength machining steel.

[f] Quenched and tempered.

Source: SAE J1099 [67]. Reprinted with permission.

656

APPENDIX
D

TYPICAL MECHANICAL PROPERTIES

Steel

Material[a]	Form	Condition[b]	S_{yt}, ksi/MPa	S_{ut}, ksi/MPa	RA, %	ε_f	S_{fss}, ksi/MPa[c]	S_{yc}^d, ksi/MPa[d]	S_{uc}^d, ksi/MPa[d]	S_{ys}^a, ksi/MPa[d]	S_{us}^d, ksi/MPa[d]	E, Mpsi/GPa	G, Mpsi/GPa; K_{Ic} ksi√in/MPa√m
1016		CD 0%	40/275	66/455	70	1.20	35/240[e]						
		CD 30%	85/585	90/620	62	0.97	46/315[e]						
		CD 60%	88/605	102/710	54	0.78	51/350[e]						
		CD 80%	96/660	115/790	26	0.30	53/365[e]						
1020	1-in bar or plate	HR	48/331	65/448	59	0.89	35/241[e]						
1030	1-in bar or plate	WQ + 1200F	64/441	85/586	70	1.20	43/296			35/241		29.6/204	12.0/83
1040	1-in bar	Annealed	51/352	75/517	57	0.84	39/269						
		HR	60/414	90/621	50	0.69	43/296[a]						
		CD 20%	97/670	117/805	44	0.58	54/370[e]						
		CD 50%	124/855	140/965	25	0.33	60/410[e]						
1050	1-in bar	Annealed	53/365	92/634	40	0.51	53/365						
1060		CD 20% + s.r. 2 h 900F	101/696	127/876	31	0.37	62/427[e]						11.4/79
4130	1-in bar	WQ + 1200F	102/703	118/814	64	1.02	71/490	105/724					
4340	1-in bar	OQ + 1000F	170/1172	183/1262	52	0.73	97/669	190/1310		109/752	127/876	30.0/207	11.7/81 $K_{Ic}=100/110$
		OQ + 800F	200/1379	222/1531	47	0.63	68/469	220/1517		124/855	146/1007		$K_{Ic}=68/75$
18% Ni maraging													
200	L plate	Aged 482C	215/1480	225/1540	55	0.80	100/690						
250	L plate	Aged 482C	237/1630	256/1760	62	0.97	100/690						
300	L plate	Aged 482C	279/1920	288/1980	50	0.69	110/760						
Cast iron, gray, medium section													
#20				20-25/138-172			9.5-10/66-69		80-105/552-724		26/179	9.6-14/66-97	3.9-5.6/27-39
#40				40-48/276-331			17.5-19.5/121-134		140-143/965-986		54-57/372-393	16.0-20.0/110-138	6.4-7.8/44-54
#60				60-66/414-455			24.5-29.5/169-203		170-187/1172-1289		76-88/524-607	20.4-23.5/141-162	7.8-8.5/54-59

Material	Condition								
Ferritic malleable grade 32510		32.5/224	50–58/345–400	28/193	208/1434	19–24/131–165	54–58/372–400	25/172	12.5/86
Aluminum									
2024	T3	50/345	70/483	20/138[f]		41/283		10.6/73	
7075	T6, T651	73/503	83/572	23/159[f]		48/331		10.3/71	
Titanium									
6-4	Annealed[g]	130/896	140/965	60/414(A)[h]	140/965	80/552			K_{Ic} = 19/21
	Aged	160/1103	170/1172	50/345(A)[h]	155/1069	110/758			
Copper alloys[i]									
60/40 (Cu/Zn) (Muntz metal)	Annealed	14/95	46/315	12/85					
70/30 (Cu/Zn) (Cartridge brass)	Half-hard rod	70/483	52/359	22/152[j]					
Phosphor bronze (90% Cu/10% Sn; flat products 0.040 in thick)	Half-hard	83/572	68/469	30/207[k]					

[a] A description of the materials and typical uses follows the table.

[b] CD = cold-drawn (the percentages are reduction in area); HR = hot-rolled; OQ = oil-quenched; WQ = water-quenched (temperature following is the tempering temperature); s.r. = stress-relieved.

[c] Smooth-specimen rotating-beam results, unless noted A (=axial).

[d] S_{yc} = compressive yield strength, S_{uc} = compressive ultimate strength, S_{ys} = shear (torsional) yield strength, S_{us} = shear ultimate strength, K_{Ic} = fracture toughness.

[e] 10^6 cycles.

[f] 5×10^8 cycles.

[g] 2 h at 1300 to 1600°F.

[h] $R = 0.1$; 10^7 cycles.

[i] *Metals Handbook*, vol. 1, ASM.

[j] 50×10^6 cycles.

[k] 1×10^8 cycles.

Source: Except as noted, this table and following notes are drawn from the *Structural Alloys Handbook*, published by the Metals and Ceramics Information Center, Battelle Memorial Institute, Columbus, Ohio, 1985, with whose permission it is reproduced.

Description of Materials and Typical Uses

1020 (1016)

This ordinary carbon steel is widely used for general structural applications. It is commonly used in the carburized condition where wear resistance rather than core strength is important. Due to its low carbon content, it is readily welded. It can be fully hardened only in very thin sections. As with other plain-carbon steels, cold work produces sizable increases in hardness and strength. It is available in plate, sheet, strip, bars, billets, tubes, and strip wire. Its typical applications include fan blades, chain, clamps, cams, sprockets, gears, bolts, pins, machinery components, case-hardened parts, and forming tools.

1030

This water-hardening medium-carbon steel is commonly used for small machinery parts of moderate strength. It may be used in annealed sheet form for deep drawing; however, permissible wall thickness reductions are not as great as for 1020 or 1010. It provides medium strength and toughness at low cost. It has low hardenability and can be hardened fully by water quenching only in very thin sections. Cold work can produce sizable increases in hardness and strength. It can be welded with either the gas or the arc process. It is available in bars, billets, tubes, sheets, plates, and rods. Typical uses include shear blades, shafts, axles, pins, gears, and bolts.

1040

This medium-carbon general-purpose steel has higher strength than 1030 and is used for machinery parts such as gears, sprockets, and crankshafts. Its low manganese content results in a lower hardenability than is obtainable with low-alloy steels such as 1340. This steel may be welded; however, preheating and postheating may be necessary because of its carbon content. It may be induction- or flame-hardened; however, water quenching produces fully hardened material only in thin sections. It is available in bars, billets, tubes, plates, sheets, castings, and forgings. Typical uses are hand tools, screwdrivers, wrenches, pliers, cold-drawn bolts, shafts, axles, gears, wire, bulldozer edges, and medium-strength heat-treated forgings.

1050

This heat-treatable carbon steel is used for parts requiring medium strength. It can be flame- or induction-hardened for parts requiring a wear-resistant surface. Its low hardenability allows for a flame-hardened zone of shallow depth, which in turn reduces the tendency for surface cracks. As with other high-carbon steels, it is very difficult to weld due to an increased susceptibility to weld cracking. It is available in bars, billets, tubes, plates, sheets, and rods. Typical uses include shafts, connecting rods, couplings, cams, heavy machinery parts, gears, bolts, nuts, hand tools, axles, and helical springs.

1060

This alloy, as other high-carbon steels, is more restricted in application than the lower-carbon steels. Items fabricated from high-carbon steels are more costly due to their decreased machinability, poor formability, and poor weldability. They are also more brittle when heat-treated. This steel distorts more than low-carbon steel when heat-treated, and surface decarburization can occur. It is available in strips, rods, bars, and forgings. Typically it is used for heavy-machinery parts, shafts, springs and torsion bars, blades for agricultural and earth-moving equipment, and tools requiring good strength and wear resistance.

4130

This is a heat-treatable water-hardening steel of low to intermediate hardenability. It possesses uniform results when tempered and retains much of its strength at high tempering temperatures. It retains good tensile, fatigue, and impact properties at cryogenic temperatures. It is frequently used in the normalized condition for applications requiring higher tensile strengths than are obtainable from low-carbon steels. It can be readily fusion-welded; however, resistance welding is not recommended. Due to its hardenability properties, the section thickness must be considered when it is heat-treated to high strength. It is not subject to temper embrittlement and responds to nitriding. It is available in sheets, plates, strips, bars, billets, pipes, and tubes. Typically it is used for automotive connecting rods, aircraft shapes, and tubing.

4340

This alloy is considered the "standard" to which other ultra-high-strength steels are compared. It is a triple-alloy steel that has excellent hardenability, ductility, and resistance to impact. It is used where severe service conditions exist and where high strength in heavy sections is required. As with other low-alloy steels, this alloy will decarburize as a result of normal heat treating, which produces a detrimental effect on the fatigue strength. Hydrogen embrittlement is a problem if it is heat-treated to ultimate-strength values greater than about 200 ksi. This material also exhibits extremely poor stress corrosion properties when tempered to tensile strengths of 220 to 280 ksi. It is available in plates, sheets, strips, bars, and billets. Typical uses include axle shafts, heavy truck gears, industrial machinery, earth-moving equipment, aircraft gears and other components, piston rods, and heavy-duty crankshafts.

18% Nickel maraging steels

These steels have excellent fracture toughness, good weldability, and a simple heat treatment. The relatively low heat-treatment temperature (age at 900°F) results in an exceptionally low amount of distortion or dimensional change. Section size has no effect on the hardenability of these steels, and tempering is not required, which is in contrast to the technology of quenched and tempered steels. These steels are not intended to be used for extended times at elevated temperatures. They are readily machined (similar to 4340) and have a low rate of work hardening, allowing for good cold-working properties. They are free from decarburization, without the use of a protective atmosphere. They have excellent weldability and also possess a low coefficient of thermal expansion. They are available in bars, billets, rods, sheets, plates, and wire. Typically they are used for rocket-motor cases, extrusion dies, mandrels, flexible shafts, precision-machine components, shear blades, punches, pins, collets, and chuck parts.

Gray cast iron

The main advantages of gray iron are its low cost, low shrinkage, best castability of any ferrous material, excellent machinability, highest damping capacity of any common metal, good wear resistance, low notch sensitivity, and high compressive strength.

Ferritic (or standard) malleable iron

This product is produced by the full annealing of suitable white cast irons. The anneal must be long enough to allow all or at least most of the cementite to decompose into iron and free "temper" carbon. Ferritic malleable iron is characterized by lower strength and higher impact resistance than the other malleable irons and also by the dark color (black heart) it possesses when fractured. Unlike gray iron, ferritic malleable iron exhibits stress-strain properties with a definite yield point similar to that of low-carbon steel. The modulus of elasticity and fatigue strength of ferritic malleable iron are about double those of gray iron. Grade 32510 can be cast in sections as little as $\frac{1}{32}$ in thick and is used primarily for automotive, agricultural, and electrical products. Ferritic malleable iron is considered the most machinable and free-cutting of any of the irons.

2024

This is the most universally used high-strength alloy of aluminum. It is available in sheet, plate, tube, extruded shapes, rod, bar, wire, and rivets. Typical uses are aircraft structures, screw-machine products, and truck wheels.

7075

This aluminum alloy has very high strength and hardness. Its fatigue properties are comparable to those of 2014 and 2024, which have somewhat lower static strengths. At high temperatures, 7075 loses its strength advantage over 2024 even under static conditions. In general, 7075 should be formed in the annealed condition and then heat-treated to develop its high-strength properties. Alloy 7075 in the T651 condition has about the greatest plane-stress and plane-strain fracture toughness of any similarly tempered aluminum alloy. It is available as sheet, plate, tube, rod, bar, wire, extrusion, forgings, and rivets. Typical uses are high-strength aircraft structures, such as jumbo-jet wing stringers, and door keys.

Ti-6Al-4V

This alloy is the most used of any of the titanium materials, the "work horse" of the titanium alloys. At least a dozen varieties of 6-4 are produced that vary slightly from one another in composition, but these variations can be significant in terms of properties. The kinds of processing and heat treatment are quite numerous and also lead to differences in properties. As indicated in the tabulation, the aged condition is a higher-strength, lower-ductility, lower-toughness condition than the annealed. When all the possible variations in composition, processing, and heat treatment are considered, the 6-4 material can be tailored to adjust its properties to meet a wide variety of service conditions. This commonly is the practice, and much of the 6-4 sold is customized. Given also that the properties can vary with the mill product form, the difficulty of generally characterizing the 6-4 alloy in a selection chart is severe. It is therefore recommended that producers be consulted when a potential use of 6-4 is identified.

The 6-4 alloy can be formed into billet, bar, plate, sheet, strip, foil, extrusions, tubing, rod, and wire. It finds many aerospace uses in gas-turbine engines, forgings for airframes, fasteners, wheels, etc. In marine equipment, it is used for deck fittings, sailing masts, shafts, and pumps, and in industrial equipment its uses vary from blades for steam turbines to cryogenic handling equipment.

60/40

Brasses are known for their corrosion resistance and machinability. The 60/40 is called *Muntz metal* and is particularly resistant to the corrosion of seawater and freshwater. It is used for marine hardware and in condenser tubing in power plants.

70/30

Because of its ductility, this brass is particularly suited to severe cold-forming operations. One of these is deep drawing, such as in the manufacture of gun cartridges, hence the name *cartridge brass*. It has good strength in the worked condition and is also corrosion-resistant. Cartridge brass is used mainly in cold-formed products, such as light sockets, flashlights, and rifle shells.

Phosphor bronzes

These copper-tin alloys (small amounts of phosphorous are used as a deoxidizer) are known for their high tensile strength, toughness, formability, fatigue strength, and corrosion resistance. They are formed by commonly used methods such as blanking, forming, and bending. They are also readily machined. Phosphor bronzes are available as strip, rod, wire, and tube. They are used in the electrical and chemical industries as well as in bridge bearing plates. They also find use as bearings, bushings, gears, and valve plates.

APPENDIX
E

CYCLIC STRESS-STRAIN PROPERTIES (AXIAL TESTS)

Material	BH	Process description	S'_y, ksi/MPa	n'	K', ksi/MPa	σ'_f, ksi/MPa	b	ε'_f	c
			Steel						
Gainex[a]		HR[b] sheet	54/372	0.11	114/786	117/807	−0.07	0.86	−0.65
RQC-100[c]	290	HR plate	87/600	0.14	208/1434	180/1241	−0.07	0.66	−0.69
RQC-100[c]		Tempered air-cooled plate	74/510	0.09	127/876	159/1096	−0.08	2.26	−0.77
1005–1009	125	CD[d] sheet	36/248	0.11	71/490	78/538	−0.07	0.11	−0.41
1005–1009	90	HR sheet	33/228	0.12	67/462	93/641	−0.11	0.10	−0.39
1012		HR sheet	32/223	0.19	107/738	370/2552	−0.22	0.76	−0.59
1015[e]	80	Normalized	35/241	0.22	137/945	120/827	−0.11	0.95	−0.64
1020[e]	108	HR plate	35/241	0.18	112/772	130/896	−0.12	0.41	−0.51
1045[f]	225	Q&T[g]	60/414	0.18	195/1344	178/1227	−0.10	1.00	−0.66
1045[f]	410	Q&T	120/827	0.15	335/2310	270/1862	−0.07	0.60	−0.70
1541F	290	Q&T forging grade	95/655	0.17	255/1758	185/1276	−0.08	0.68	−0.65
5160	430	Q&T	145/1000	0.15	335/2310	280/1931	−0.07	0.40	−0.57

[a] Tradename, Armco Steel Corp. HSLA: high-strength, low-alloy.

[b] Hot-rolled.

[c] Tradename, Bethlehem Steel Corp. Roller-quenched and tempered carbon steel. Used in structural, heavy-machinery applications.

[d] Cold-drawn.

[e] Low-carbon. Common machining steels.

[f] Bar stock, medium carbon; higher-strength machining steel.

[g] Quenched and tempered.

Material	BH	Process description	S'_y, ksi/MPa	n'	K', ksi/MPa	σ'_f, ksi/MPa	b	ε'_f	c
				Steel (*cont.*)					
9262	260	Annealed	76/524	0.15	200/1379	151/1041	−0.07	0.16	−0.47
9262	280	Q&T	94/648	0.12	197/1358	177/1220	−0.07	0.41	−0.60
9262	410	Q&T	152/1048	0.09	292/2013	269/1855	−0.06	0.38	−0.65
A131 grade C[h]		HR plate	35/241	0.23	146/1007	219/1510	−0.16	0.67	−0.54
A36[i]		HR plate	41/282	0.18	128/883	287/1979	−0.17	0.89	−0.56
ABS grade DH36[j]		HR plate	51/352	0.15	130/896	165/1138	−0.11	1.23	−0.64
A736 class 3[k]		HR plate	79/544	0.07	119/820	167/1151	−0.08	1.70	−0.77
				Aluminum					
2024-T351		Solution treated, strain-hardened	62/427	0.07	95/655	160/1103	−0.12	0.22	−0.59

[h] Thick plate. Mostly for ship applications.

[i] Weldable hot-rolled carbon steel. Used in offshore structures, bridges, buildings, heavy machinery.

[j] ABS = American Bureau of Ships. HSLA weldable steel. Mainly for ships.

[k] Developed for LPNG tankers. Cu precipitation hardening steel. Very high toughness.

Sources: SAE J1099 [67] and Bethlehem Steel Corp [68], reprinted with permission.

TYPICAL MATERIAL PROPERTIES AND COSTS

These tables have been prepared for practice use with the problems of the text. The listings are only a sampling of hundreds of engineering materials available. Specific data should be obtained (especially prices) for real problems.

	Material	S_y, ksi/MPa	S_{fB}, ksi/MPa†	E, Mpsi/GPa‡	Density ρ, (lbm/in³)/(kg/m³)	Cost C_{mat}, ($/lbm)/($/kg)	C_{mach}, ($/in³)/($/m³)
			Nonferrous				
Wrought aluminum	1100 Hard (H 14) (99% min. Al)	22/152 (0.2%)	9/62 (5E + 08 cycles)	10.3/71	0.088/2.44E + 03	1.60/3.52	0.4/2.4E + 04
	2011 Temper T8 (5% Cu)	45/310 (0.2%)	18/124 (5E + 08 cycles)	10.3/71	0.102/2.83E + 03	3.40/7.48	0.5/3.1E + 04
	5050 (1.5% Mg) Hard (H 33)	29/200	14/97	10.3/71	0.098/2.72E + 03	1.60/3.52	0.4/2.4E + 04
Brass	Cartridge brass (70% Cu, 30% Zn) hard rod	63/434 (0.5%)	21/145 (5E + 08 cycles)	16/110	0.308/8.54E + 03	2.85/6.27	0.5/3.1E + 04
	Muntz metal (60% Cu, 40% Zn) hard rod	55/379 (0.5%)	20/138 (5E + 08 cycles)	15/110	0.308/8.54E + 03	2.85/6.27	0.5/3.1E + 04
			Ferrous				
Steel	1010 Cold-rolled (0.1% C)	44/303	53/365	30/207	0.280/7.77E + 03	0.65/1.43	0.9/5.5E + 04
	1040 Cold-rolled (0.4% C)	71/490	85/586	30/207	0.280/7.77E + 03	1.15/2.53	1.0/6E + 04
	2330 Drawn @ 400°F	195/1340	221/1524	30/207	0.280/7.77E + 03	1.65/3.64	1.1/6.7E + 04
	4130 Drawn @ 400°F	135/931	150/1034	30/207	0.283/7.85E + 03	1.45/3.20	1.3/7.9E + 04
	304 SS (austenitic wrought annealed)	35/241	85/586	28/193	0.290/8.03E + 03	2.25/4.95	1.8/1.1E + 05
	431 SS (martensitic wrought)	100/689	130/90	29/200	0.280/7.77E + 03	2.40/5.28	1.9/1.2E + 05
Cast iron	Gray (type 4; "40" indicates minimum strength in ksi)	$S_{fRB} = 18/12.4$	$S_{ut} = 45/310$ $S_{uc} = 140/97$	17/117	0.280/7.77E + 03	0.60/1.32	1.0/6E + 04

† For steels estimate S_{fRB} from S_{ut}. See Section 8.2.

‡ The shear modulus of elasticity ($\tau = G\gamma$) may be computed from $G = E/2(1 + \mu)$, which derives from the theory of elasticity. Here μ is Poisson's ratio, about 0.3 for steel, aluminum, and brass.

APPENDIX

G

BEAM DIAGRAMS AND FORMULAS*

Various other loadings may be obtained by superposition. (A few are indicated.)

Simply Supported Beams:

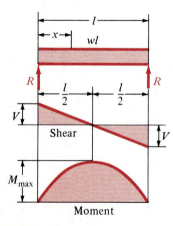

$$R = V = \frac{wl}{2}$$

$$V_x = w\left(\frac{l}{2} - x\right)$$

$$M_x = \frac{wx}{2}(l - x)$$

$$M_{max} \ (x = l/2) = \frac{wl^2}{8}$$

$$\Delta_x = \frac{wx}{24EI}(l^3 - 2lx^2 + x^3)$$

$$\Delta_{max} \ (x = l/2) = \frac{5wl^4}{384EI}$$

FIGURE G.1
Uniformly distributed load.

---

*Source: Manual of Steel Construction, American Institute of Steel Construction.

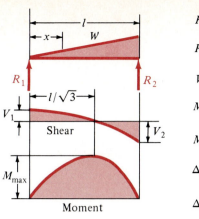

$$R_1 = V_1 = \frac{W}{3}$$

$$R_2 = V_2 = \frac{2W}{3}$$

$$V_x = \frac{W}{3} - \frac{Wx^2}{l^2}$$

$$M_x = \frac{Wx}{3l^2}(l^2 - x^2)$$

$$M_{max}\left(\text{at } x = \frac{l}{\sqrt{3}}\right) = \frac{2Wl}{9\sqrt{3}}$$

$$\Delta_x = \frac{Wx}{180EIl^2}(3x^4 - 10l^2x^2 + 7l^4)$$

$$\Delta_{max}\left(\text{at } x = l\sqrt{1 - \sqrt{\frac{8}{15}}}\right) = 0.01304\frac{Wl^3}{EI}$$

FIGURE G.2
Uniformly increasing load.

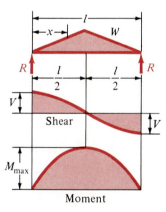

$$R = V = \frac{W}{2}$$

$$V_x\left(x < \frac{l}{2}\right) = \frac{W}{2l^2}(l^2 - 4x^2)$$

$$M_x\left(x < \frac{l}{2}\right) = Wx\left(\frac{1}{2} - \frac{2x^2}{3l^2}\right)$$

$$M_{max}(x = l/2) = \frac{Wl}{6}$$

$$\Delta_x\left(x < \frac{l}{2}\right) = \frac{Wx}{480EIl^2}(5l^2 - 4x^2)^2$$

$$\Delta_{max}(x = l/2) = \frac{Wl^3}{60EI}$$

FIGURE G.3
Uniformly increasing load to center.

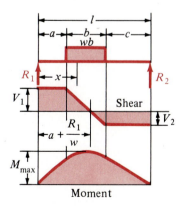

$$R_1 = V_1(\text{max, if } a < c) = \frac{wb}{2l}(2c + b)$$

$$R_2 = V_2(\text{max, if } a > c) = \frac{wb}{2l}(2a + b)$$

$$V_x \; (a < x < a + b) = R_1 - w(x - a)$$

$$M_x \; (x < a) = R_1 x$$

$$M_x \; (a < x < a + b) = R_1 x - \frac{w}{2}(x - a)^2$$

$$M_x \; (x > a + b) = R_2(l - x)$$

$$M_{max}\left(x = a + \frac{R_1}{w}\right) = R_1\left(a + \frac{R_1}{2w}\right)$$

For load at end, let $a = 0$. For more than one distributed load, use superposition.

FIGURE G.4
Partial uniform load.

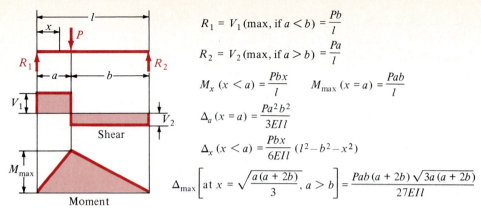

$$R_1 = V_1 \text{ (max, if } a < b) = \frac{Pb}{l}$$

$$R_2 = V_2 \text{ (max, if } a > b) = \frac{Pa}{l}$$

$$M_x \ (x < a) = \frac{Pbx}{l} \qquad M_{\max} \ (x = a) = \frac{Pab}{l}$$

$$\Delta_a \ (x = a) = \frac{Pa^2 b^2}{3EIl}$$

$$\Delta_x \ (x < a) = \frac{Pbx}{6EIl} \ (l^2 - b^2 - x^2)$$

$$\Delta_{\max} \left[\text{at } x = \sqrt{\frac{a(a + 2b)}{3}}, \, a > b \right] = \frac{Pab(a + 2b)\sqrt{3a(a + 2b)}}{27EIl}$$

For load at center, let $a = b = l/2$. For two or more loads, use superposition.

FIGURE G.5
Single load.

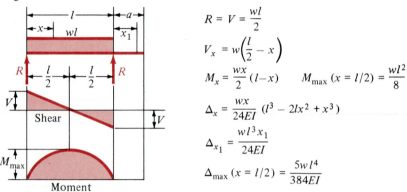

$$R = V = \frac{wl}{2}$$

$$V_x = w\left(\frac{l}{2} - x\right)$$

$$M_x = \frac{wx}{2}(l - x) \qquad M_{\max} \ (x = l/2) = \frac{wl^2}{8}$$

$$\Delta_x = \frac{wx}{24EI} \ (l^3 - 2lx^2 + x^3)$$

$$\Delta_{x_1} = \frac{wl^3 x_1}{24EI}$$

$$\Delta_{\max} \ (x = l/2) = \frac{5wl^4}{384EI}$$

FIGURE G.6
Overhanging beam, uniform load between supports.

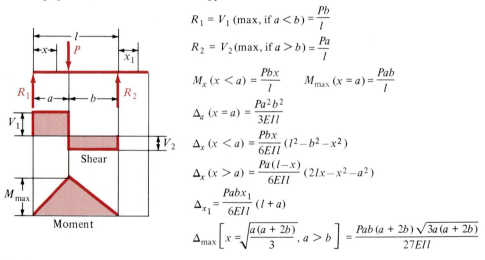

$$R_1 = V_1 \text{ (max, if } a < b) = \frac{Pb}{l}$$

$$R_2 = V_2 \text{ (max, if } a > b) = \frac{Pa}{l}$$

$$M_x \ (x < a) = \frac{Pbx}{l} \qquad M_{\max} \ (x = a) = \frac{Pab}{l}$$

$$\Delta_a \ (x = a) = \frac{Pa^2 b^2}{3EIl}$$

$$\Delta_x \ (x < a) = \frac{Pbx}{6EIl} \ (l^2 - b^2 - x^2)$$

$$\Delta_x \ (x > a) = \frac{Pa(l - x)}{6EIl} \ (2lx - x^2 - a^2)$$

$$\Delta_{x_1} = \frac{Pabx_1}{6EIl} \ (l + a)$$

$$\Delta_{\max} \left[x = \sqrt{\frac{a(a + 2b)}{3}}, \, a > b \right] = \frac{Pab(a + 2b)\sqrt{3a(a + 2b)}}{27EIl}$$

FIGURE G.7
Overhanging beam, single load between supports.

670

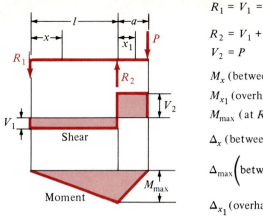

$$R_1 = V_1 = \frac{Pa}{l}$$

$$R_2 = V_1 + V_2 = \frac{P}{l}(l+a)$$

$$V_2 = P$$

$$M_x \text{ (between supports)} = \frac{Pax}{l}$$

$$M_{x_1} \text{ (overhang)} = P(a-x_1)$$

$$M_{max} \text{ (at } R_2) = Pa$$

$$\Delta_x \text{ (between supports)} = \frac{Pax}{6EIl}(l^2-x^2)$$

$$\Delta_{max}\left(\text{between supports, } x = \frac{l}{\sqrt{3}}\right) = \frac{Pal^2}{9\sqrt{3}EI}$$

$$\Delta_{x_1} \text{ (overhang)} = \frac{Px_1}{6EI}(2al + 3ax_1 - x_1^2)$$

$$\Delta_{max} \text{ (overhang at } x_1 = a) = \frac{Pa^2}{3EI}(l+a)$$

FIGURE G.8
Overhanging beam, load at overhang. One end built in.

One End Built In:

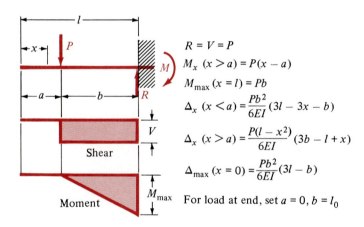

$$R = V = P$$

$$M_x \ (x > a) = P(x - a)$$

$$M_{max} \ (x = l) = Pb$$

$$\Delta_x \ (x < a) = \frac{Pb^2}{6EI}(3l - 3x - b)$$

$$\Delta_x \ (x > a) = \frac{P(l - x^2)}{6EI}(3b - l + x)$$

$$\Delta_{max} \ (x = 0) = \frac{Pb^2}{6EI}(3l - b)$$

For load at end, set $a = 0$, $b = l_0$

FIGURE G.9
Cantilever, single load.

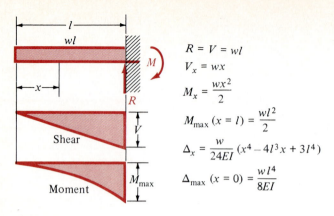

$$R = V = wl$$

$$V_x = wx$$

$$M_x = \frac{wx^2}{2}$$

$$M_{max} \, (x = l) = \frac{wl^2}{2}$$

$$\Delta_x = \frac{w}{24EI} \, (x^4 - 4l^3 x + 3l^4)$$

$$\Delta_{max} \, (x = 0) = \frac{wl^4}{8EI}$$

FIGURE G.10
Cantilever, uniform load.

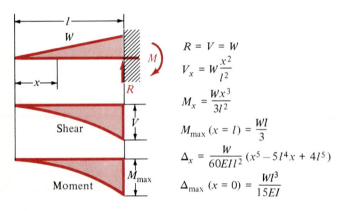

$$R = V = W$$

$$V_x = W \frac{x^2}{l^2}$$

$$M_x = \frac{Wx^3}{3l^2}$$

$$M_{max} \, (x = l) = \frac{Wl}{3}$$

$$\Delta_x = \frac{W}{60EIl^2} \, (x^5 - 5l^4 x + 4l^5)$$

$$\Delta_{max} \, (x = 0) = \frac{Wl^3}{15EI}$$

FIGURE G.11
Cantilever, uniformly increasing load.

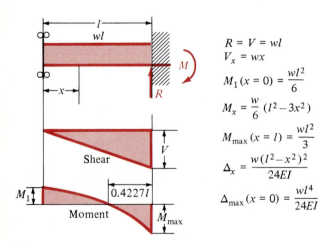

$$R = V = wl$$

$$V_x = wx$$

$$M_1 \, (x = 0) = \frac{wl^2}{6}$$

$$M_x = \frac{w}{6} \, (l^2 - 3x^2)$$

$$M_{max} \, (x = l) = \frac{wl^2}{3}$$

$$\Delta_x = \frac{w(l^2 - x^2)^2}{24EI}$$

$$\Delta_{max} \, (x = 0) = \frac{wl^4}{24EI}$$

FIGURE G.12
Uniform load, simple support at end.

672

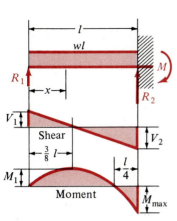

$$R_1 = V_1 = \frac{3wl}{8}$$

$$R_2 = V_2 \text{ (max)} = \frac{5wl}{8}$$

$$V_x = R_1 - wx$$

$$M_1\left(x = \tfrac{3}{8}\, l\right) = \frac{9}{128}\, wl^2$$

$$M_x = R_1 x - \frac{wx^2}{2}$$

$$M_{max} = \frac{wl^2}{8}$$

$$\Delta_x = \frac{wx}{48EI}\,(l^3 - 3lx^2 + 2x^3)$$

$$\Delta_{max}\left[x = \frac{l}{16}\,(1 + \sqrt{33}\,)\right] = \frac{wl^4}{185EI}$$

FIGURE G.13
Uniform load, simple support at end.

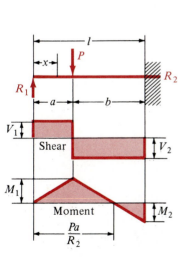

$$R_1 = V_1 = \frac{Pb^2}{2l^3}\,(a + 2l)$$

$$R_2 = V_2 = \frac{Pa}{2l^3}\,(3l^2 - a^2)$$

$$M_1\,(x = a) = R_1 a$$

$$M_2\,(x = l) = \frac{Pab}{2l^2}\,(a + l)$$

$$M_x\,(x < a) = R_1 x$$

$$M_x\,(x > a) = R_1 x - P(x - a)$$

$$\Delta_a\,(x = a) = \frac{Pa^2 b^3}{12EIl^3}\,(3l + a)$$

$$\Delta_x\,(x < a) = \frac{Pb^2 x}{12EIl^3}\,(3al^2 - 2lx^2 - ax^2)$$

$$\Delta_x\,(x > a) = \frac{Pa}{12EIl^3}\,(l-x)^2\,(3l^2 x - a^2 x - 2a^2 l)$$

$$\Delta_{max}\left(a < 0.414l \text{ at } x = l\,\frac{l^2 + a^2}{3l^2 - a^2}\right) = \frac{Pa}{3EI}\,\frac{(l^2 - a^2)^3}{(3l^2 - a^2)^2}$$

$$\Delta_{max}\left(a > 0.414l \text{ at } x = l\,\sqrt{\frac{a}{2l + a}}\right) = \frac{Pab^2}{6EI}\,\sqrt{\frac{a}{2l + a}}$$

FIGURE G.14
Single load, simple support at end. Both ends built in.

Both Ends Built In:

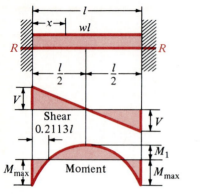

$$R = V = \frac{wl}{2}$$

$$V_x = w\left(\frac{l}{2} - x\right)$$

$$M_1\,(x = l/2) = \frac{wl^2}{24}$$

$$M_x = \frac{w}{12}(6lx - l^2 - 6x^2)$$

$$M_{max}\,(x = 0, l) = \frac{wl^2}{12}$$

$$\Delta_x = \frac{wx^2}{24EI}(l - x)^2$$

$$\Delta_{max}\,(x = l/2) = \frac{wl^4}{384EI}$$

FIGURE G.15
Uniform load.

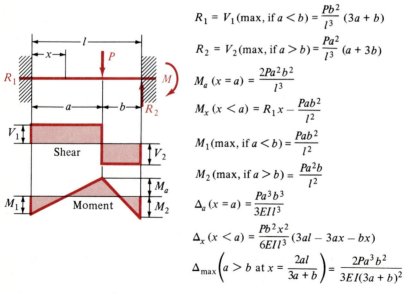

$$R_1 = V_1(\text{max, if } a < b) = \frac{Pb^2}{l^3}(3a + b)$$

$$R_2 = V_2(\text{max, if } a > b) = \frac{Pa^2}{l^3}(a + 3b)$$

$$M_a\,(x = a) = \frac{2Pa^2b^2}{l^3}$$

$$M_x\,(x < a) = R_1 x - \frac{Pab^2}{l^2}$$

$$M_1(\text{max, if } a < b) = \frac{Pab^2}{l^2}$$

$$M_2\,(\text{max, if } a > b) = \frac{Pa^2b}{l^2}$$

$$\Delta_a\,(x = a) = \frac{Pa^3b^3}{3EIl^3}$$

$$\Delta_x\,(x < a) = \frac{Pb^2x^2}{6EIl^3}(3al - 3ax - bx)$$

$$\Delta_{max}\left(a > b \text{ at } x = \frac{2al}{3a + b}\right) = \frac{2Pa^3b^2}{3EI(3a + b)^2}$$

For load at center, let $a = b = l/2$.

FIGURE G.16
Single load. Three simple supports, equal spans.

Three Simple Supports, Equal Spans:

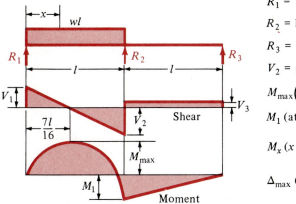

$$R_1 = V_1 = \tfrac{7}{16}\, wl$$

$$R_2 = V_2 + V_3 = \tfrac{5}{8}\, wl$$

$$R_3 = V_3 = -\tfrac{1}{16}\, wl$$

$$V_2 = \tfrac{9}{16}\, wl$$

$$M_{max}\left(x = \tfrac{7}{16}\, l\right) = \tfrac{49}{512}\, wl^2$$

$$M_1\, (\text{at } R_2) = \tfrac{1}{16} wl^2$$

$$M_x\, (x < l) = \frac{wx}{16}\, (7l - 8x)$$

$$\Delta_{max}\, (x = \tfrac{7}{16}\, l) = \frac{0.0092\, wl^4}{EI}$$

FIGURE G.17
Uniform load in one span.

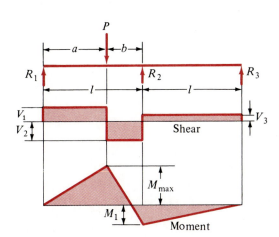

$$R_1 = V_1 = \frac{Pb}{4l^3}\left[4l^2 - a(l + a)\right]$$

$$R_2 = V_2 + V_3 = \frac{Pa}{2l^3}\left[2l^2 + b(l + a)\right]$$

$$R_3 = V_3 = -\frac{Pab}{4l^3}\, (l + a)$$

$$V_2 = \frac{Pa}{4l^3}\left[4l^2 + b(l + a)\right]$$

$$M_{max}\, (\text{at } P) = \frac{Pab}{4l^3}\left[4l^2 - a(l + a)\right]$$

$$M_1\, (\text{at } R_2) = \frac{Pab}{4l^2}\, (l + a)$$

FIGURE G.18
Single load.

REFERENCES

1. R. C. Juvinall, *Fundamentals of Machine Component Design*, Wiley, New York, 1967.
2. J. E. Shigley and L. D. Mitchell, *Mechanical Engineering Design*, 4th ed., McGraw-Hill, New York, 1983.
3. R. C. Johnson, *Optimum Design of Mechanical Elements*, 2d ed., Wiley, New York, 1980.
4. R. M. Phelan, *Fundamentals of Mechanical Design*, 3d ed., McGraw-Hill, New York, 1970.
5. J. P. Vidosic, "Design Stress Factors," *Proc. ASME*, vol. 55, May 1948, pp. 653–658.
6. R. T. Kent (ed.), *Mechanical Engineers' Handbook* [Part 1: Design and Production], 12th ed., Wiley, New York, 1950.
7. C. Lipson and R. C. Juvinall, *Handbook of Stress and Strength*, Macmillan, New York, 1963.
8. R. E. Peterson, *Stress Concentration Factors*, Wiley, New York, 1974.
9. G. Sines and J. L. Waisman (eds.), *Metal Fatigue*, McGraw-Hill, New York, 1958.
10. A. Hall, A. Holowenko, and H. Laughlin, *Theory and Problems of Machine Design*, Schaum's Outline Series, McGraw-Hill, New York, 1961.
11. S. P. Timoshenko and J. M. Gere, *Theory of Elastic Stability*, 2d ed., McGraw-Hill, New York.
12. E. J. Haug and J. S. Arora, *Applied Optimal Design*, Wiley-Interscience, New York, 1979.
13. R. C. Juvinall, *Engineering Considerations of Stress, Strain and Strength*, McGraw-Hill, New York, 1967.
14. H. O. Fuchs and R. I. Stephens, *Metal Fatigue in Engineering*, Wiley, New York, 1980.
15. A. Palmgren, "Die Lebensdauer von Kugellagern" ("Life of Ball Bearings"), *ZVDI*, vol. 68, 1924.
16. M. A. Miner, "Cumulative Damage in Fatigue," *Trans. ASME* (*J. Appl. Mech.*), vol. 67, September 1945, pp. A159–A164.
17. H. Neuber, "Theory of Stress Concentration for Shear-Strained Prismatical Bodies with Arbitrary Non-linear Stress-Strain Law," *J. Appl. Mech.*, vol. 28, December 1961, pp. 544–550.
18. G. S. Haviland, "Designing with Threaded Fasteners," *Mech. Eng.*, vol. 105, no. 10, October 1983, pp. 16–31.
19. G. H. Junker, "Principle of the Calculation of High-Duty Bolted Joints." Interpretation of Directive VDI 2230, English version published by SPS Technologies, Jenkintown, PA.
20. J. H. Bickford, *An Introduction to the Design and Behavior of Bolted Joints*, Marcel Dekker, New York, 1981.
21. K. N. Smith, P. Watson, and T. H. Topper, "A Stress-Strain Function for the Fatigue of Metals," *J. Mat.*, vol. 5, no. 4, December 1970, pp. 767–778.
22. J. Fisher and J. Struik, *Guide to Design Criteria for Bolted and Riveted Joints*, Wiley, New York, 1974.
23. J. Collins, *Failure of Materials in Mechanical Design*, Wiley, New York, 1981.

24. W. F. Brown, Jr., and J. E. Srawley, *Plane Strain Crack Toughness Testing of High Strength Metallic Materials, ASTM Special Technical Publication*, no. 410, Philadelphia, PA, 1966.

25. W. T. Thompson, *Theory of Vibration with Applications*, 2d ed., Prentice-Hall, Englewood Cliffs, NJ, 1981.

26. R. E. Walpole and R. H. Meyers, *Probability and Statistics for Engineers and Scientists*, 3d ed., Macmillan, New York, 1985.

27. L. Meirovitch, *Elements of Vibration Analysis*, McGraw-Hill, New York, 1975.

28. F. S. Tse, I. E. Morse, and R. T. Hinkle, *Mechanical Vibrations*, 2d ed., Allyn and Bacon, Needham Heights, MA, 1978.

29. J. M. Gere and S. P. Timoshenko, *Mechanics of Materials*, 2d ed., Wadsworth, Belmont, CA, 1984.

30. S. P. Timoshenko and J. M. Gere, *Theory of Elasticity*, 2d ed., McGraw-Hill, New York, 1961.

31. L. S. Marks, *Mechanical Engineers' Handbook*, 4th ed., McGraw-Hill, New York, 1941.

32. E. P. Popov, *Introduction to Mechanics of Solids*, Prentice-Hall, Englewood Cliffs, NJ, 1968

33. *Mach. Design* (1984 Materials Reference Issue), vol. 56, no. 8, April 19, 1984.

34. W. Weibull, "A Statistical Distribution Function of Wide Applicability," *Trans. AMSE (J. Appl. Mech.)*, vol. 73, September 1951, pp. 293–297.

35. R. M. Jones, *Mechanics of Composite Materials*, McGraw-Hill, New York, 1975.

36. J. R. Vinson and T. W. Chou, *Composite Materials and Their Use in Structures*, Halstead, New York, 1974.

37. A. T. Jones, "A Comparative Study of Viscous Damping Effects of Vibrating Beams," MSci thesis, M. E. Dept., Univ. of Nevada, Reno, 1965.

38. Z. D. Jastrzebski, *The Nature and Properties of Engineering Materials*, 2d ed., Wiley, New York, 1976.

39. J. Marin, *Mechanical Behavior of Engineering Materials*, Prentice-Hall, Englewood Cliffs, NJ, 1962.

40. A. M. Wahl, "Stresses in Heavy Closely Coiled Helical Springs," *Trans. ASME (J. Appl. Mech.)* vol. 51, no. 17, 1929, pp. 185–200. See also his *Mechanical Springs*, 2d ed., McGraw-Hill, New York, 1963.

41. C. T. Wang, *Applied Elasticity*, McGraw-Hill, New York, 1953.

42. R. C. Johnson, *Mechanical Design Synthesis: Substitute Creation Design and Organization*, 2d ed., Krieger, Melbourne, FL, 1978.

43. A. A. Griffith, "Phenomena of Flow and Fracture," *Phil. Trans. of the Royal Soc.*, vol. A, no. 221, 1921, pp. 163–198.

44. J. Marin, "Design for Fatigue Loading—Part 3," *Mach. Design*, vol. 29, no. 4, Feb. 21, 1957, p. 124–134.

45. J. C. Leslie, "Properties and Performance Requirements" in James Margolis (ed.), *Advanced Thermoset Composites*, Van Nostrand, New York, 1986, pp. 74–109.

46. J. Datsko, *Materials in Design and Manufacturing*, Malloy, Ann Arbor, MI, 1971.

47. R. G. Lambert, *Analysis of Fatigue under Random Vibration*, Bulletin 46, The Shock and Vibration Information Center, Naval Research Laboratory, Washington, DC, p. 55.

48. W. Gerber, "Bestimmung der zulossigen Spannungen in Eisenconstruction," *Z. Bayer Arch. Ing. Ver.*, vol. 6, 1874.

49. C. Lipson and N. J. Sheth, *Statistical Design and Analysis of Engineering Experiments*, McGraw-Hill, New York, 1972.

50. H. Tada, P. C. Paris, and G. R. Irwin, *The Stress Analysis of Cracks Handbook*, Del Research, Hellertown, PA, 1973.

51. G. C. M. Sih, *Handbook of Stress Intensity Factors*, Lehigh University, Bethlehem, PA, 1973.

52. F. A. McClintock and G. R. Irwin, "Plasticity Aspects of Fracture Mechanics," in *Fracturing Toughness Testing and Its Applications, ASTM Special Technical Publication*, no. 381, Philadelphia, PA, 1965, pp. 84–113.

53. R. Kugel, "A Relation between Theoretical Stress Concentration Factor and Fatigue Notch Factor Deduced from the Concept of Highly Stressed Volume," *Proc. ASTM*, vol. 61, 1961, pp. 732–748.

54. U. Hindhede et al., *Machine Design Fundamentals*, Wiley, New York, 1983.

55. M. F. Spotts, *Design of Machine Elements*, Prentice-Hall, Englewood Cliffs, NJ, 1978.

56. P. G. Forrest, *Fatigue of Metals*, Pergamon, Elmsford, NY, 1969.

57. G. M. Sinclair and T. J. Dolan, "Effect of Stress Amplitude on Statistical Variability in Fatigue Life of 75S–T6 Aluminum Alloy," *Trans. ASME, (J. Appl. Mech.)*, vol. 75, 1953, pp. 867–872.

58. R. W. Landgraf, "The Resistance of Metals to Cyclic Deformation," in *Achievement of High Fatigue Resistance in Metal and Alloys, ASTM Special Technical Publication*, no. 467, Philadelphia, PA, 1970, pp. 3–36.

59. J. A. Graham (ed.), *Fatigue Design Handbook*, Society of Automotive Engineers, Warrendale, PA, 1968.
60. H. J. Grover, "Fatigue of Aircraft Structures," NAVAIR, 01-1A-13, 1966.
61. J. A. Graham (ed.), *Fatigue Design Handbook*, Society of Automotive Engineers, Warrendale, PA, 1968.
62. H. S. Reemsnyder, *Simplified Stress-Life Fatigue Model*, Bethlehem Steel Corp., Bethlehem, PA, January 1985.
63. S. J. Stadnick and J. Morrow, "Techniques for Smooth Specimen Simulation of the Fatigue Behavior of Notched Members," in *Testing for Prediction of Material Performance in Structures and Components*, *ASTM Special Technical Publication*, no. 515, Philadelphia, PA, 1972, pp. 229–252.
64. R. W. Landgraf, "Cumulative Fatigue Damage under Complex Strain Histories," in *Cyclic Stress-Strain Behavior: Analysis, Experimentation, and Failure Prediction, ASTM Special Technical Publication*, no. 519, Philadelphia, PA, 1973, pp. 213–228.
65. H. S. Reemsnyder, "Evaluating the Effect of Residual Stresses on Notched Fatigue Resistance," Research Department, Bethlehem Steel Corp., Bethlehem, PA. From *Materials, Experimentation, and Design in Fatigue—Proceedings of Fatigue 1981*, Westbury Press, Guilford, England, 1981, pp. 273–295.
66. H. S. Reemsnyder, "Observations, Predictions and Prevention of Fatigue Cracking in Offshore Structures," in *Case Histories Involving Fatigue and Fracture Mechanics, ASTM Special Technical Publication*, no. 918, Philadelphia, PA, 1986.
67. "Technical Report on Fatigue Properties," SAE J1099, 1975.
68. G. A. Miller, T. R. Sharron, and H. S. Reemsnyder, *A Compilation of Strain-Cycle Fatigue Data for Bethlehem Steel Sheet and Plate Grades*, Bethlehem Steel Corp., Bethlehem, PA, 1986.
69. "Design of Transmission Shafting," an American National Standard, ASME Codes and Standards, *ANSI/ASME*, B106.1M, 1985.
70. D. B. Kececioglu and V. R. Lalli, "Reliability Approach to Rotating Component Design," *NASA TM*, D-7846, 1975.
71. V. C. Davies, H. J. Gough, and H. V. Pollard, "Discussion to 'The Strength of Metals under Combined Alternating Stresses'," *Proc. Inst. Mech. Eng.*, vol. 131, no. 3, 1935, pp. 3–103.
72. S. H. Loewenthal, "Proposed Design Procedure for Transmission Shafting under Fatigue Loading," *NASA TM*, 78927, 1978.
73. W. F. Brown, Jr., and J. E. Srawley, *Plane Strain Crack Toughness Testing of High Strength Metallic Materials, ASTM Special Technical Publication*, no. 410, Philadelphia, PA, 1966.
74. L. G. Johnson, *The Statistical Treatment of Fatigue Experiments*, American Elsevier, New York, 1964.
75. C. Lipson, G. C. Noll, and L. S. Clark, *Stress and Strength of Manufactured Parts*, McGraw-Hill, New York, 1950.
76. M. M. Frocht and H. N. Hill, "Stress-Concentration Factors around a Central Circular Hole in a Plate Loaded through Pin in the Hole," *Trans. ASME (J. Appl. Mech.)*, vol. 62, 1940, pp. A5–A9.
77. "Load and Resistance Factor Design," *LRFD*, American Institute of Steel Construction, 1986.
78. "Metallic Materials and Elements for Aerospace Vehicle Structures," *MIL-HANDBOOK*, U.S. Government Printing Office, Washington, DC.
79. S. B. Vemuri, D. Sutharshana, and S. R. Mettu, "Confronting Fatigue: A New Capability for CAD," *Mech. Eng.*, vol. 9, 1987, p. 80.
80. R. T. Hinkle and I. E. Morse, Jr., "Design of Helical Springs for Minimum Weight, Volume and Length," *Jour. Eng. Industry*, vol. 2, 1959, pp. 37–42.
81. J. E. Shigley and C. R. Misckhe, *Standard Handbook of Machine Design*, McGraw-Hill, New York, 1986.
82. *Design Handbook, Engineering Guide to Spring Design*, 1987 edition, Associated Spring, Barnes Group, Bristol, CT.
83. G. F. Leon and J. R. Payne, "An Overview of the U.S. PVRC Research Program on Bolted Flanged Connections," International Conference on Pressure Vessel Technology, Beijing, China, September 1988.
84. O. W. Blodgett, *Design of Weldments*, James F. Lincoln Arc Welding Foundation, Cleveland, OH, 1963.
85. R. B. McKee, "Fracture Strength of Fiber Reinforced Composites," Ph.D. thesis, College of Engineering, Univ. of California, Los Angeles, 1967.

INDEX

INDEX